Soil Microbiology

Soil Microbiology

Second Edition

Robert L. Tate III
Rutgers University

John Wiley & Sons, Inc.

New York | Chichester | Weinheim | Brisbane | Singapore | Toronto

The publication is designed to provide accurate and authoritative information in regard to the subject matter covered. It is sold with the understanding that the publisher is not engaged in rendering professional services. If professional advice or other expert assistance is required, the services of a competent professional person should be sought.

Library of Congress Cataloging-in-Publication Data:

Tate, Robert L., 1944-
    Soil microbiology / Robert L. Tate. —2nd ed.
        p. cm.
    Includes bibliographical references.
    ISBN 0-471-31791-8 (cloth : alk. paper)
    1. Soil microbiology.   I. Title.
QR111.T28 2000
579'.1757 — dc21   99-21922

Printed in the United States of America.

10 9 8 7 6 5 4 3 2 1

To
Ann, Robert, and Geoffrey

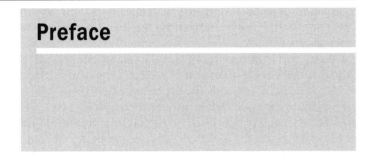

Preface

Slightly over four years ago, I was putting the finishing touches on the preface to the first edition of *Soil Microbiology*. As the final stages of this second edition of the treatise were completed and I gathered the thoughts for writing a second preface, questions of just why the rather long endeavor of writing another edition was necessary. Certainly, the basic tenets of the discipline are evolving, but not at a rapid rate. The objective of the first edition — "to provide the student with a strong basic knowledge of the biological, physical, and chemical properties of the soil and the microbes contained therein" — is unchanged. What has expanded significantly is our experience in using the concepts of soil microbiology to solve our expanding list of environmental problems. Thus, the most obvious differences between the two editions of this text is the inclusion of chapters on soil microbial diversity and bioremediation of soil systems. Although the "central dogma" underlying these topics is still evolving, opportunities now exist for including in a basic textbook an elucidation of the status of the topics and for highlighting areas of deficiency in our knowledge base.

When I first began contemplating the task of writing this edition, I asked many of my close associates what they felt was the greatest need of the text. Bioremediation and microbial diversity were frequently mentioned as possible topics. But one response stands out in my mind. An individual whose opinion I highly respect strongly stated that no "new words" should be added. Students were already faced with a "mountain of information" to master. I must admit that I too must agree, at least in part, with this observation. Students in my own classes commonly are overwhelmed by the vast amount of information before them. It is with this thought in mind that I went ahead and added more "words" to each of the chapters. My goal was to clarify the relationship of soil microbiology concepts to the general environmental matters with which students are experienced and to highlight new endeavors in the area of primary research. Since much of the foundation of soil microbiology derives from agriculturally based studies and the fact that our discipline remains a primary support for efficient production of food and fiber, the understanding of the agricultural environment is still a major part of our soil microbiology foundation, but this foundation is now extended greatly into more general soil concerns, such as urban soil management and soil stewardship and reclamation practices.

Last, this edition attempts to maintain the flavor of the first edition. Students need more than just a presentation of the "facts." Entries into the world of primary research literature are essential to provide a foundation for careers in our science. Thus, although complete review of the literature is not possible, an attempt was made to highlight a significant mass of current as well as historical research references. My apologies are offered to associates whose publications were overlooked, but such omissions are unavoidable under the circumstances of page limitations for a basic text and to meet the objective of not overwhelming students with masses of new information. Students are challenged to explore not only the summary of the principles provided in *Soil Microbiology* but also to delve into the rich variety of primary research that has been amassed supporting the principles of soil microbiology and pointing the way to future endeavors. It is only through such perhaps difficult but rewarding treks that students can truly master the challenges of the study of soil microbiology.

I take this opportunity to thank my colleagues and students who have provided the inspiration and conducted the research that has made this book possible. In particular, I also thank the current members of my soil microbiology research group, who have had to endure my deficiencies in providing true focus to their research endeavors as I put these words to paper. Their patience is gratefully acknowledged and highly appreciated. It is again my hope that this treatise will provide a basis for growth of the science of soil microbiology and an inspiration to the careers of young scientists.

Robert L. Tate III

New Brunswick, New Jersey
October 1999

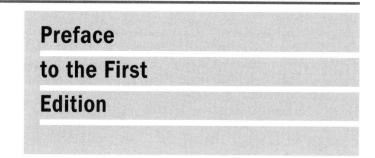

Preface to the First Edition

In the past, soil microbiology has been identified as a second tier science produced by the merger of soil science and microbiology — two disciplines of somewhat recent birth compared to the longer-lived basic sciences of chemistry, physics, mathematics, and biology. Now, soil microbiology has reached sufficient maturity to provide the underpinning of such newly emerging disciplines as environmental science and related studies. The necessities arising from managing the long undervalued soil ecosystem have produced a myriad of complex problems that can be solved only with an application of the primary principles of soil microbiology. Complex questions emerging from the necessities of reclamation of polluted or contaminated soil as well as the requirement to properly manage less affected, perhaps pristine terrestrial ecosystems require an appreciation of the delicate balance between the soil microbial community and its environment. This habitat of the soil microbe not only includes the minerals and organic matter of the soil matrix but also encompasses levels of life occurring therein or indeed thereupon.

It is with this concept of the growing necessity to understand the diversity of life contained within a gram of soil as well as the intricacies of the linkages of these living entities with terrestrial life in general that has provided the impetus for production of this treatise on soil microbiology. It is hoped that the words contained herein will provide a clear understanding of the principles upon which the discipline of soil microbiology is built, as well as illuminate their applicability to current soil-related environmental problems.

A concerted effort was made in assembling the material for this text to avoid the myopic impression that soil microbiology and related disciplines have developed strictly to meet the needs of agricultural scientists. Indeed, the roots of soil microbiology have long been fed by the requirements to provide food and fiber to a growing society, but the principles of the science are essential for guiding decisions of soil-related questions in all realms of our society. Hence, a conscious attempt has been made to illustrate the applicability of studies of the soil microbiology to the grander array of environmental science problems facing our society.

One further foundational assumption in writing *Soil Microbiology* is that the science of soil microbiology has grown to the point that it now is self-sustaining. That is, students of the science should be sufficiently versed in the basic sciences that their repetition within these pages would be redundant. Thus, inclusion of introductory chapters describing the basic tenets of soil science and microbiology would be as inappropriate in a soil microbiology text as would introductions to biology, chemistry, and physics. Students are expected to have a fundamental understanding of these foundational course materials before entering the realm of the soil microbe. Discussions of elementary chemical and biological principles in this text are presented only in the context of their applicability to the soil situation.

A book of this type cannot be prepared without the input of many friends and associates. I especially thank those unnamed colleagues and students who in their discussions with me have provided the seeds of inspiration that have matured into the concepts of our study. Last, the inputs of the many soil microbiologists who have conducted the experiments upon which the basic tenets of soil microbiology have been constructed must be gratefully acknowledged. It is the author's hope that assembly of this information in this treatise will provide the basis for extension of the introductory information contained herein far beyond our current images of our science.

Robert L. Tate III

New Brunswick, New Jersey
January 1994

Contents

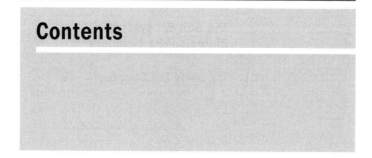

1   **The Soil Ecosystem: Physical and Chemical Boundaries**   **1**

11

**The Nitrogen Cycle: Soil-Based
Processes**   **314**

14 Denitrification 404

15 Sulfur, Phosphorus, and Mineral Cycles 433

Introduction

Members of the soil microbial community have been appreciated as interesting life forms. They have been exploited as producers of substances of great societal importance, such as a variety of antibiotics. This knowledge alone has been sufficient to produce the books occupying a multitude of library shelves. But such information alone is inadequate to provide solutions to the variety of microbially associated environmental problems facing society today. Environmental scientists and regulators as well as the lay public are faced with decisions regarding the proper stewardship of our soil resources as well as the management or reclamation of damaged systems. Solutions to these at times lifestyle-altering problems may be provided by considering soil to be a closed system with only external manifestations of the processes occurring therein deemed to be worthy of consideration. The most limited external implication could be the production of an aesthetically pleasing ecosystem. Such a viewpoint provides short-sighted, temporary solutions to long-term problems. Indeed, not only is an appreciation of the biological processes occurring in soil essential to achieve societal goals of caring for terrestrial systems, but true wisdom in decision-making is predicated on an understanding of how the individual soil biologically based processes combine to produce a vital, sustainable whole. The parts assembled into a viable soil system clearly produce a whole much more dynamic, much greater than can be predicted by their simple summation.

This thesis is underscored by the magnitude and multitude of anthropogenically affected soil sites currently demanding some form of reclamation management. For example, refuse from energy fuel or mineral recovery forms unsightly slag piles extending across the countryside. From within these piles, a yellow leachate with a pH of nearly 1 frequently flows. Nearly sterile soil and waters result from encounters with this deadly by-product of resource recovery. The questions become, "How best to prevent the production of leaching of the substance commonly known as acid mine drainage? How to restore the affected soils and waters to functional, aesthetic ecosystems? How can the remains of the mining industry be managed to prevent further environmental degradation?''

Perhaps less dramatic but of similar concern is the problem of evaluating potential difficulties of genetically engineered microorganisms (gems). The answers seem simple. An organic chemical has reached a soil system at toxic concentrations, but no microorganisms capable of decomposing the toxicant exist therein. Yet such microbes can be created in the laboratory through commonly available genetic manipulation procedures. Concerns involve unanticipated ecosystem degradation from the introduction of laboratory-created, alien organisms into established soil communities. "Will the introduced microbe survive sufficiently long to achieve the objective of its utilization? Will the unique gene carried by this gem be transferred to indigenous organisms, thereby creating an individual with capacity to wreak havoc on an otherwise stable soil system?'' These questions represent concerns to which soil microbiologists must respond. Resolution of the conflict requires a clear understanding of the behavior of alien microbes within a functioning soil ecosystem, as well as of the dynamics of gene transfer within soil populations.

These two initial examples relate to environmental problems whose impact involves the reclamation or management of a limited region of soil. Solutions to environmental problems affecting the totality our terrestrial system also rely on expansion of our knowledge and databases relating to soil microbial processes. For example, soil is a natural source and sink of greenhouse gases, such as methane and carbon dioxide. Basically, soil organic matter (humus) resources are the source and sink for these carbon compounds. Furthermore, the quantity of humus retained within a particular soil is the product of the physical, chemical, and biological properties of the system as well as any associated anthropogenic intervention. Therefore, all soil systems are characterized by occurrence of an equilibrium level of soil organic matter. Anthropogenic intervention into the ecosystem can result in a shift in the quantity of carbon sequestered in the soil. This situation is nowhere more obvious than in soils developed for intensive agricultural production. Historically, the yield of carbon dioxide to the atmosphere due to reduction in the quantity of soil humus in these soils has been significant. Thus, a simplistic means of managing greenhouse gases that could be proposed is to alter the management of soil systems so that they become a sink rather than a source of atmospheric carbon dioxide. An appreciation of the potential benefits of this process can be derived from consideration of variation of soil humus levels resulting from the conversion of intensive cultivation practices into reduced till or no-till agricultural soil management. Unfortunately, the questions associated with assessing the role of soil in managing greenhouse gasses are more complicated. Decisions related to greenhouse gas management and associated terrestrial effects are affected by, among other factors, the fact that soil temperatures are anticipated to increase due to global climate changes associated with the greenhouse phenomenon. Now, an interaction of human soil management decisions, changes in the chemistry of plant inputs due alteration of plant biomass composition by the elevated atmospheric temperatures, and alteration of soil physical properties (e.g., temperature and moisture) acting together create soil microbial community dynamics that are not as easily forecasted as was possible with alteration of agricultural soil management. An expanded comprehension of soil microbial dynamics and the effect of total-ecosystem processes on soil biological processes is needed.

It is with these types of concerns in mind that this treatise is presented. The overall object of the study is to provide the student with the strong basic knowledge of the biological, physical, and chemical properties of soil and the microbes contained therein necessary to provide the basis for sound environmental management and stewardship decisions.

Chapter **1**

The Soil Ecosystem: Physical and Chemical Boundaries

Soil Microbiology Defined: The world of the soil microbe presents a formidable challenge to those seeking to understand and/or manage the processes occurring therein. The soil ecosystem is the product of intricate interactions between a physical and chemical matrix of highly variable composition and living communities composed of essentially all life forms. Soil biological communities are sustained across the entire range of chemical and physical conditions within which life can exist and function.

Due in large part to the complexity of the soil community, soil microbiological experiments historically have involved study of individual microbial species purified and cultivated under controlled conditions in laboratory media. This simplification has allowed for a reduction in the complexity of the system of study and of data interpretation. Recently, use of consortia of bacteria or fungi for study of more complex processes, such as decomposition of xenobiotic compounds, has become reasonably common. These latter cultures exhibit some of the more complex interactions reflective of the soil ecosystem. Use of defined cultures or even simple mixtures of microbes is generally justified by the fact that it is necessary to attain maximum control of experimental variables in order to elucidate clearly the processes of interest.

As an appreciation of the nature of soils is developed, the inadequacy of experiments with axenic cultures (purified cultures of individual microbial species or strains) or defined mixed cultures (consortia) to define the world of the soil microbe becomes obvious. Generally, the basic properties of the biological processes measured in laboratory culture defy extrapolation to the more complex soil situation. Not only are stresses of the native soil system unmatched with defined culture conditions, but the microbes themselves change phenotypically and even genotypically in response to the laboratory growth conditions. Although the microbes studied may have been isolated from soil, a microbial variant with appropriate properties for optimal growth in test tubes or industrial reaction vessels is usually selected either spontaneously (generally, inadvertently as a result of laboratory culture methods) or through genetic manipulation of the microbial isolates. The former selection results from the rich genetic variations within the genome of each individual bacterial species. It is easy conceptually to consider each bacterial species to consist of a group of individuals with essentially identical genomes, but this viewpoint is far from reality. Bacterial species have similar traits and genetic composition, but their genome contains many genetic variations (mutations). Therefore, when members of a previously soil-resident bacterial species are selected for culture in the laboratory, mutant strains that previously existed in soil as a miniscule portion of the soil population of the species may be better able to grow under the laboratory conditions of the test tube than is the majority strain occurring in the soil community. Thus, a genetic variant becomes the strain studied in the laboratory. Further selection of spontaneously occurring variants may — or more likely does — occur in the laboratory culture. As a result, data collected from experiments with isolated microbial species or strains frequently only explain in part or mimic the processes occurring in soil. It is reasonable to conclude that soil microbiologists must expand the purview of experimental design to include the complexities — controllable and otherwise — of the soil ecosystem. The realm of soil microbiology must be defined to include both an understanding of the properties of the microbes themselves and an appreciation of the impact of variability of the soil environment on these traits. Soil microbial ecology must involve an evaluation of the behavior of organisms in their native habitats.

The Necessity of Soil Microbiological Studies: To the purist, elucidation of the principles of soil microbiology is immensely interesting and fully justified strictly by the information generated relating to the basic ecological interactions of soil systems. Yet, due to societal management of terrestrial ecosystems, our concern for soil biological processes reaches far beyond the realm of basic science. For example, our myopic exploitation of natural ecosystems has resulted in situations where the functioning of resident microbial populations is precluded or severely limited by mismanagement. Spills of toxic organic chemicals result in scarred landscapes. Similarly, products of metal processing have resulted in sites resembling moonscapes (Fig. 1.1a). Reclamation of such sites is attempted (e.g., Fig. 1.1b), but for attainment of such objectives, management plans must be developed which allow for establishment of essential soil biological processes.

In comparison with these obviously detrimental situations, development of soils for agricultural production may also reduce overall soil quality. Tillage initiates processes that may eventually reduce productivity of the soils. Loss of desirable soil structure and reduction of soil organic matter reserves are generally associated with implementation of intensive agricultural practices (see Tate, 1987).

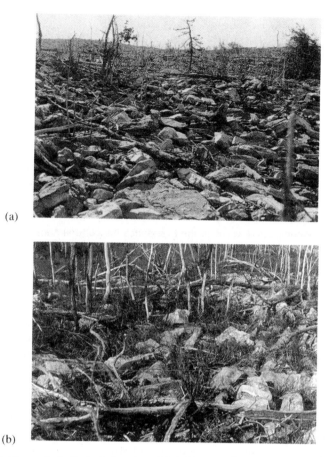

(a)

(b)

Fig. 1.1 Metal-impacted site in Palmerton, PA (USA). (a) Site without reclamation management. (b) Site following amendment of soil surface with a mixture of sludge, fly ash, lime, and grass seed.

A dramatic example of augmented soil organic matter decomposition due to cultivation of a previously undisturbed soil system is provided by observation of changes in soil properties resulting from cropping of peats. Conversion of the generally anoxic swampy soil to one where molecular oxygen is no longer limiting results in oxidation of the peat organic matter. This microbially catalyzed oxidation of the plant debris accumulated under the swampy conditions prior to site drainage to carbon dioxide and water results in loss of soil mass that is manifested by subsidence of the soil surface (Fig. 1.2) (see Tate, 1980).

Maintenance or restoration of basic soil quality must be a primary societal objective. Soil is a renewable resource, but the time frame of restoration of a badly damaged soil system may extend over several decades or even centuries.

Fig. 1.2   **Subsidence pole located in the Everglades Agricultural Area of South Florida (USA) showing loss of soil elevation between 1924 and 1979.**

Thus, in the short run, soil must be managed with the philosophy that it is an exhaustible resource. Proper soil stewardship cannot be achieved without a full appreciation of soil microbiological processes, for the totality of existence of a soil ecosystem relies upon sustenance of a functional soil microbial community. To develop appropriate management plans for restoration of damaged systems or for maintenance of ecosystem productivity, a clear understanding of *in situ* biological processes is necessary. This cannot be fully accomplished by using data generated from the observation of bacterial populations growing at an unnatural rate under the optimized conditions of the laboratory culture flask. Realistic estimates of metabolic activity and the kinetics of the process in the microbe's native environment must be developed. Procedures necessary for maintenance of the soil chemical and physical factors controlling expression of microbial activity in a balanced state are needed in order to ensure the long-term survival of critical populations for ecosystem remediation and sustenance. Such factors as temperature, pH, moisture, surface area, osmotic strength, and presence of predators all interact to select for the development and survival of a stable microbial community necessary for long-term ecosystem function.

The Challenge: Once it is conceived that reality for the microbe exists in its natural habitat, the challenge of defining that habitat emerges. For the soil microbiologist, this is not an easy task. Soil properties are highly variable. For example, microbes flourish in temperature-limited, native systems, such as hot-spring-impacted soils or in the cold desert soils of Antarctica. Highly productive soils may become biologically impoverished due to the impact of anthropogenically produced situations, such as those resulting from acid mine drainage or spills of xenobiotic organic compounds. Thus, it becomes abundantly clear that the soil ecosystem can be spoken of in general terms, but a true encounter with soil ecology must include an assessment of the length and breadth of these highly variable systems.

This assessment must start with differentiation of the entity to be studied. Soil microbiologists must realize that they are studying specific soil types, not simply "soil." The too commonly heard statement among soil microbiologists, "Soil was used in our experiments," would sound as strange to a soil scientist as would, "An animal was studied" would to a biologist. Currently, the U.S. soil classification system divides world soils into 12 orders (based on soil properties that reflect soil development). Examples of soil orders are aridisols (dry soils), andisols (volcanic soils), gelisols (permafrost-impacted soils), histosols (organic soils), inceptisols (embrionic soils), spodosols (typical of forest systems), and vertisols (characterized by soils with high-swelling clay contents). These orders are subdivided into suborders, great groups, subgroups, families, and series. Currently, over 15,000 soil series have been identified. (See Brady and Weil, 1999, or Miller and Gardiner, 1998, for a further discussion of soil taxonomy.) This situation of defining the specific soil unit of study is further complicated by the fact that a given field site may contain one or more soil series. Furthermore, biological activity within the same soil series varies with such ecosystem properties as plant type and distribution, seasonal climatic variation, and management practices—current and historic. Yet, the properties of the soil type underlying an ecosystem as defined and impacted by the aboveground communities must be assessed and understood.

To meet this need, the objectives of this chapter are (1) to define the properties delimiting the basic properties of soils and soil ecosystems, and (2) to elucidate the impact of the variability of chemical and physical properties of this system on microbial growth and development. Specific introduction to the biological aspects of soil ecosystems will be presented in Chapter 2.

1.1 The Soil Ecosystem

The properties of a soil ecosystem are not only a product of the nature of its physical, biological, and chemical components, but they are also modified by the interactions of these entities. The overall manifestation of a soil ecosystem can truly be said to be greater than the sum of its parts. Soil physical constituents are not immutable, but rather are modified by chemical and biological products of their associated life forms. Even in soil ecosystems where the needs of the microbial community are only minimally met, colonies of living cells develop slowly to the limit allowed by the physical and chemical parameters stressing the organisms. As will be elucidated in Chapter 2, the soil biological system modifies its environment to conditions more conducive for growth and community development. Thus, a soil ecosystem could be said to be evolving toward a steady-state condition dictated by the interactions of each of the mutable components.

Hence, any analysis of the structure of soil systems and the principles describing their function must commence with an evaluation of (1) the soil physical components, (2) their assembly into the physical whole, and (3) the implications of this structure on soil biological processes. The study must then proceed to a determination of the feedback mechanisms involved with living populations interacting with their physical environment. This analysis is necessary to develop fully the principles or maxims describing microbial interactions in soil and their importance to that portion of the system most commonly regarded aesthetically — the aboveground plant and animal community.

1.1.1   Soil Defined

A study of soil must be based on an understanding of the basic definition of soil combined with a conceptualization of the properties of soils that make each soil type unique. Simply stated, **soil is an unconsolidated mineral or organic substance on the earth's surface that provides a natural medium for growth of land plants.** This general picture of material covering terrestrial land masses suitable for plant growth must now be enlarged into a picture of the world in which soil microorganisms live and function.

Definition of Soil Types: Most soils are mineral soils — that is, soils that are composed primarily of mineral matter and whose physical properties are controlled by the proportions of the mineral particulates contained therein. These soils generally contain from less than 1 to approximately 4 percent organic matter, but they may consist of as much as 20 percent colloidal soil organic matter. Additionally, an organic layer up to 30 cm in depth may be found on the soil surface (e.g., a forest litter layer). These soils are to be contrasted with the much less commonly encountered organic soils (histosols). Organic soils contain more than 20 percent organic matter and are exemplified by peats and mucks.

The most obvious components of mineral soils are sand, silt, or clay, which are defined by their size distribution (Table 1.1). Note that for this separation of soil particles based on size distribution, the clay fraction is strictly defined by the particle size. Therefore, this fraction will contain clay minerals plus other soil components, which are not necessarily clay minerals but are of comparable size. A soil is classified into a specific textural category by the quantities of the mineral separates that it contains (Fig. 1.3). For example, a silt loam is a soil containing 50 percent or more silt and 12 to 27 percent clay, or 50 to 80 percent silt and less than 12 percent clay.

This disposition of soil by textural classification is useful for characterizing a particular soil of interest (either for field or laboratory experiments), but it is in reality the soil structure — that is, how these individual components are assembled — that best defines the habitat of soil microorganisms. Two soils may have nearly identical textural compositions yet exhibit extremely contrasting microbial activities. For example microbial properties of a silt loam existent

Table 1.1 Size Distribution of Soil Particle Classes

Particle Class	Subclass	Mean Diameter (mm)
Sand	Very coarse sand	2.00–1.0
	Coarse sand	1.0–0.5
	Medium sand	0.5–0.25
	Fine sand	0.25–0.1
	Very fine sand	0.1–0.05
Silt		0.05–0.002
Clay		<0.002

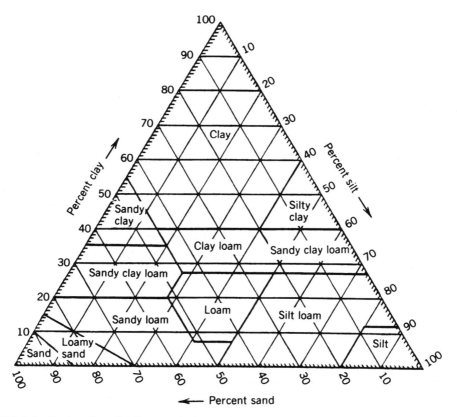

Fig. 1.3 Designation of soil textural classes by principal mineral component analysis (USDA Guidelines).

in an area flooded with acid mine drainage waters would reflect the prevailing stresses of oxygen deprivation and extremes of acid, whereas a silt loam from a well-maintained agricultural site would be free of such limitations. Furthermore, a soil with a well-developed aggregate structure in contrast to one where the structure has been lost due to intensive cultivation would be more conducive to microbial community development. Clearly, a soil classification system is required that is descriptive of *in situ* ecological influences on biological properties. This information is provided by the U.S. soil classification system discussed above. With this taxonomic system, soils are grouped by field and laboratory properties, among which are soil texture, moisture, pH, temperature regimes, and horizon development.

Soil Profile Development: An appreciation of soil horizon development and its effect on soil microbial activity is necessary to develop a complete picture of the impact of soil biological processes on the properties of an ecosystem. **A soil horizon is a layer of soil approximately parallel to the land surface and differing from adjacent layers physically, chemically, and biologically, or in characteristics such as color, structure, texture, consistency, biotic populations, and pH.** Soil horizonal development is clearly a product of current ecosystem properties as well as of site history. Each soil subgroup description includes an analysis of horizonal structure. Microbiological properties of a specific horizon may be studied in isolation from other portions of the soil profile. In such studies, the designated portion of the soil profile may be defined as an ecosystem in itself.

Most commonly studied soil horizons in microbiological research are the O, A, and B horizons, although as a result of complications of groundwater pollution, considerable research effort is being expended to understand biological processes occurring in deeper-lying aquifer materials. **O horizons are dominated by organic material** and are exemplified by forest litter layers. The **A horizons are mineral layers formed on the soil surface or below the O horizon. These regions are characterized by accumulations of organic materials intimately associated with soil mineral matter.** That is, A horizons are usually spoken of as surface soils. The colloidal organic matter concentrations tend to be maximized in this portion of the mineral soil profile. **B horizons are usually formed below an A or O horizon and are dominated by (1) carbonates, gypsum, or silica, alone or in combination; (2) evidence removal of carbonates; (3) concentrations of sesquioxides; (4) alterations that form silicate clay; (5) formation of granular, blocky, or prismatic structure; or (6) combinations of these.** In soils without organic surface accumulations, microbial activity tends to be maximized in the A horizon and declines precipitously in the B horizon.

1.1.2 Designation of Soil Ecosystems

Soil microbiologists frequently report results of study of a limited number of soil types and provide minimal descriptions of the sites from which the soil samples were collected. It is not unusual to see statements such as "a garden soil was used" in the literature. From the foregoing discussion, it is apparent not only that the physical and chemical properties of the soil are important for interpretation of the data, but also that properties must be recorded of the site from which the soil was collected as well as its history. Furthermore, statements simply indicating the ecosystem type from which a particular soil sample was collected are inadequate to allow proper interpretation of the data, extrapolation of the principles revealed by the research, and, frequently, repetition of the study. Although it may seem adequate to report that a grassland soil was evaluated in a particular experiment, application of proper soil classification procedures may reveal that a variety of soil types occur within the region classified by an ecologist as a grassland. That is, to conduct a meaningful examination of a soil ecosystem, the area to be sampled must be appropriately described.

Surface Community Impacts on Subsurface Microbial Activities: A complication to description of soil systems results from the fact that soils are continually changing. To the novice, it may seem that soil types and their properties represent a constant in their conceptualization of an ecosystem, but in reality, the soil life forms, climate, and mineral matter are continually altering those properties of the soil matrix that are used to define the soil type. Soil pH, salinity, aggregate structure, and cation exchange capacity are but a few examples of soil properties constantly being altered by the actions of the biological community (plant, animal, and microbial). For example, aboveground differences in plant community type and density can cause variations in belowground microbial and enzymological activities that may range over several orders of magnitude. This effect of plant biomass productivity on soil enzymes was shown through analysis of enzyme activity in a reasonably uniform organic soil (i.e., a single soil type) in the Everglades Agricultural Area (South Florida, U.S.A.). It was shown that acid phosphatase, invertase, xylanase, cellulase, and amylase activities varied by as much as 50-fold between noncultivated and soils cultivated to sugarcane, St. Augustinegrass [*Stenatophrum secundatum* (Walt) Ktze.], or paragrass [*Brachiaria mutica* (Forsk.) Stapf] (Duxbury and Tate, 1981). A limited understanding of the complexities of these organic soil-based systems would result if each soil sample (uncultivated or from grass or sugarcane fields) were simply classified as agricultural soils. Inputs from the aboveground plant community necessitate dividing the study area into at least three ecosystem types. Note that in the study cited, all aboveground biomass could have been grouped into a category of grasslands (i.e., both sugarcane and the St. Augustinegrass are by definition grasses). Similarly, as described in Section 1.2.2, the interaction of plants and soil microbes with soil particulates is essential for formation of soil aggregates.

Although plant communities are instrumental in soil development, it must be stressed that soil properties can be extremely different in seemingly related ecosystem types. This situation is best exemplified by comparing two large forest groupings, temperate vs. tropical forest soils. Temperate forest soils may be moderately acidic (pH 3.5–4.5) and possess a well-defined horizonal development (Fig. 1.4). They may also have a well-developed surface litter layer plus a subsurface horizon (spodic horizon) with an accumulation of organic matter. Clearing of forests and cultivation of such soils in the temperate regions has resulted in moderately productive agricultural systems. In contrast, tropical rain-forest soils generally lack a surface organic horizon; are usually highly acidic, nutrient-poor soils; and are poor candidates for agricultural development.

Thus, it may be concluded that our operational definition of a soil must include the concept of their being composed of unconsolidated mineral or organic matter as stated above, but it must also encompass analysis of the ecosystem properties affecting soil development. The microbiologist must realize that soil is a continuously developing entity with properties reflective of the site in which it has evolved and is still evolving.

State Factor Theory of Soil Development: Interactions of total ecosystem properties on soil development were described succinctly by Jenny (1941) in his State Factory Theory. Factors that have been shown to moderate soil development include regional climate, biotic factors (properties of the organisms occurring in the aboveground ecosystem and the soil), topography, parent

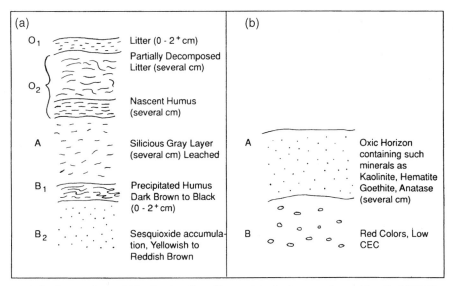

Fig. 1.4 Comparison of idealized soil profile models for temperate and tropical forest soils. (a) Spodosol surface horizons with accumulated organic matter, and leached A horizon. Spodosols are exemplified by acidic forest soils. (b) Latosilization as would be seen in an oxidol (tropical forest ecosystem). These soils are generally highly leached, acidic soils with no significant humus layer. See Fanning and Fanning (1989) for further discussion of these soil types.

material, and time. The theory was recently modified to include anthropogenic influences (Amundson and Jenny, 1991). Therefore, it must be appreciated when examining a specific ecosystem that the properties of the soil that control the biological activity and in reality make the soil unique are a product of a variety of variables interacting over a time frame of centuries. Thus, the current ecosystem type as well as those preceding it have combined to produce the soil of interest.

Optimization of Study Site Size: Application of this basic understanding of the variability of a soil ecosystem to field studies is solely limited by the viewpoint of the definer. For example, a site for the study of microbial processes could be designated to encompasses only a microscopic portion of the soil profile (e.g., the mineral and organic components surrounding a few microbial colonies—i.e., a microsite). Since such colonies frequently consist of a few cells at most, this ecosystem would have boundaries of several microns. Whereas this microsite represents the entire world to the growing microbe—and an appreciation of the situation around the microbe is important for understanding total ecosystem function—practicality dictates that scientists designate their ecosystem of interest on a larger basis.

For the soil microbiologist, soils ecosystems are frequently defined by aboveground system properties: for example, climate (desert vs. swamp), topography (aspect along a hill), plant community type (forest vs. grassland), or even proximity to the growing plant (within-row vs. between-row sampling of agricultural systems). In each of these situations, the soil microbiologist is attempting to define a functional unit of a larger ecosystem (in the extreme, total land mass) that is small enough to allow representative soil sampling and study.

The depth within a soil profile from which samples are collected also affects the type of result gained from study of the soil. Soil scientists are likely to study the A or O horizon of the soil profile since microbial activity is generally maximal therein, but it must be remembered that distinct properties and microbial communities have also developed within all soil horizons. Therefore, for a scientific analysis, a subsurface portion of the soil could be experimentally isolated for study and defined to represent a specific ecosystem. For example, in order to gain a complete picture of the bioremediation potential of a site contaminated with a biodegradable pollutant, it would be desirable to sample microbial activities within the vadose zone (that portion of the soil profile below the rooting zone, but above the water-saturated zone) as well as surface soils if the chemical contamination has proceeded below the root zone, especially if the objective is to prevent groundwater contamination.

1.1.3 Implications of Definition of the Soil Ecosystem

Generally, the reason for study of a soil system is to develop principles that allow extrapolation of the results to large soil systems or soils in general. This capability depends on the extent of the soil system selected for study. Frequently, the greater the variability of physical or chemical properties in the soils and the better the principles underlying their impact on biological properties are elucidated, the greater the applicability of the discoveries to soil systems in general. Unfortunately, simply because of the logistics involved in soil sampling and laboratory analyses, arbitrary limitations of the extent of the ecosystem area studied are inevitable. The underlying requirement is that the soil samples collected should reflect the heterogeneity of the soil site to the maximum degree possible.

The better the soil samples reflect site variation, the higher the potential for elucidating the full range of the microbial processes occurring therein and the properties of the soil controlling these activities. The smaller the portion of the overall system studied, the less the potential for valid extrapolation of the data to provide a description of the total ecosystem. Thus, the soil microbiologist faces the dilemma of selecting a soil site small enough to be sampled in a representative manner without being overwhelmed by the number of soil samples to be studied, but sufficiently expansive that the data collected will be at least to some degree representative of the total ecosystem. (See Tate, 1987, for a more detailed discussion of soil-sampling procedure problems associated with study of larger ecosystems.)

The more limited the dimensions used to define an ecosystem of interest, the more important become considerations of the impact of external processes on the integrity of the system. For example, shallow aquifer sediments may be examined, but it must be considered that energy derived from oxidation of carbonaceous compounds is dependent upon the rate that the fixed carbon is leached through the soil profile. Drawing boundaries that define the portion of the larger ecosystem to be studied will clearly enable some aspects of the ecosystem to be self-contained whereas others must be supplied from external sources. The extreme example of this situation is the micro-system introduced above: that is, microbial colonies existent upon a soil granule. Assuming that this colony is formed by the most commonly occurring bacterial grouping within soil — heterotrophs — the carbon oxidized for energy must originate from beyond the world of the bacterial microcolony. Substrates for carbon-based energy metabolism are ultimately produced photosynthetically by above-ground plant communities. From a short-term view, both in time and space, an endogenous supply of carbonaceous substrate could be considered to exist in the form of native soil organic matter, but this soil organic carbon resource originated as plant biomass and once exhausted, it can be replenished only by movement of plant photosynthate into the soil microsite.

1.2 The Micro-Ecosystem

Existence of aboveground communities literally rests on the stability and function of soil processes occurring at the microscopic level. Because microorganisms range in size from less than a micron to a few microns, individual soil particles have a major impact on their function. The challenging aspect of this observation is that the biological processes occurring at the micron level have profound implications on total ecosystem development and longevity. For example, a series of bacterial colonies consisting of as few as a half-dozen cells each may determine the fate of a potentially ecosystem-destructive xenobiotic compound. Hence, a clear understanding of the interactions of the soil microbial community and elemental soil components (sand, silt, clay, and colloidal organic matter) is imperative to comprehend total ecosystem function.

1.2.1 Interaction of Individual Soil Components with the Biotic System

The basic components of soil are sand, silt, clay, and humic substances. Although all soil particulates provide surfaces for microbial colonial development, clay and colloidal organic matter have the smallest diameters and therefore present the largest surface area for interaction with soil microbes and their products. The relationship between estimated particle diameter and surface area is demonstrated in Fig. 1.5. The interactions of clay and humic

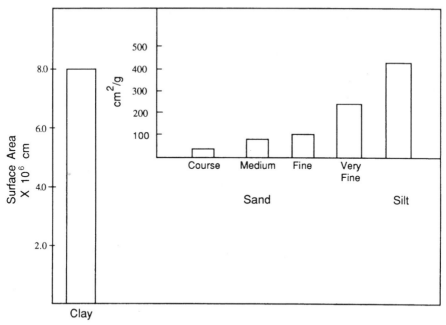

Fig. 1.5 Variation of surface area of soil particles. The particles are assumed to have spherical shapes. Data are selected from Foth and Turk, 1972.

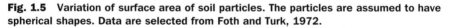

substances with microbes, enzymes, and nutrients are further accentuated by their ionic associations. The relative contributions of humic substances and clays to microbial processes are detailed below.

1.2.1.1 Clay and Ecosystem Function

Structural Properties of Clay Minerals: Clay minerals are major controllers of soil biological and chemical interactions because of their small size, regular crystalline structure, and surface charge. Thus, it is of interest to explore these aspects of clay minerals. A more complete description of clays can be found in basic soil science texts such as Brady and Weil (1999) and Miller and Gardiner (1998).

The magnitude of importance of any property of clay minerals on soil processes is highly accentuated by their surface area. The surface area of one gram of clay particles is at least a thousandfold that of one gram of sand. Thus, even if the clay particles constitute as little as one-tenth of the soil particles, this 1000-fold difference in surface area highly accentuates or imbalances the effect of clay on surface processes in the soil. Soil water relationships and their impact on soil microbial activity, and consideration of the distribution of negative charges on soil minerals are good examples of the disproportionate importance of soil clay minerals on biological processes. Soil microorganisms are commonly associated with the surfaces of soil particles. Water films, whose existence is critical for function of soil organisms, are distributed around the soil particles. Due to the physical interactions between the bipolar water molecule and the particle surfaces, water molecules closest to the particle surface are held more rigidly to the particle and thus are less available to the microbes. Therefore, in soils with higher surface areas, more water is needed to coat the particles before a sufficiently thick layer of water is formed that some water is available to the microbes—and growing plants. (See the above-cited soil science texts for an introduction to water dynamics in soil systems.) Furthermore, the greater the surface area of soil particles, the greater the number of negative charges that are available for chemical interactions, organic and inorganic, to alter the concentrations of the chemicals in solutions, that is, sorption, desorption, precipitation, and dissolution phenomena.

Clay minerals are secondary minerals; that is, except for mica, clay minerals are formed in soil from the weathering of primary minerals or from the products of their weathering. Most clays are crystalline and are composed of layers of oxygen atoms with silicon and aluminum atoms with oxygen atoms associated through ionic bonding. For crystalline clays, these silicate and aluminum layers are stacked (like a deck of cards) into clay particles, termed *micelles*. As will be noted below, less regularly oriented amorphous clays are also found in some soils. Common groups of clays are layer silicate clays, hydrous oxides of iron and aluminum, plus allophane and other amorphous clays.

Layer silicate clays are primarily crystalline clays with specific structures and predictable properties. These clays have a platelike structure formed from flat sheets of silicon, aluminum, magnesium, and iron linked by oxygen and hydroxyl groups. The major silicate clays are commonly referred to as phylosilicates due to their leaflike structure. The crystalline structure can be summarized as being composed of two distinct crystalline sheets as follows:

- A silica-dominated sheet composed of silicon atoms surrounded by four oxygen atoms in a tetrahedryl arrangement. The interlocking array of silicon and shared oxygen atoms is a tetrahedron sheet.

- An aluminum- or magnesium-dominated sheet composed of aluminum and magnesium atoms surrounded by six oxygen atoms in an octahedryl arrangement. Thus, the layer is referred to as an octahedryl sheet. If the layer is dominated by aluminum, it is a dioctahedryl sheet, and if by magnesium, a trioctahedryl sheet.

The tetrahedryl and octahedryl sheets are bound together in regular crystals by shared oxygen in the different layers. Major groups of these layer silicate clays are

- 1:1 — one tetrahedryl (Si) sheet per octahedryl (Al) sheet. Kaolinite is the most prominent member of this group.

- 2:1 — one octahedryl sheet sandwiched between two tetrahedryl sheets as exemplified by smectite (montmorillonite)

- 2:1:1 — typically 2:1 crystals as above alternating with Mg-dominated trioctahedryl layers as exemplified by chlorites

Both a variable (depending on soil pH) and a fixed negative charge are associated with layer silicate clays. The pH-dependent charge results from ionization of the hydroxyl groups on the surface and edges of the crystal. The fixed charge is the product of isomorphous substitution for the aluminum or silicon atoms within the crystalline structure. For example, the aluminum atom is only slightly larger than the silicon atom. Therefore, aluminum can fit into the center of the tetrahedron in place of the silicon without altering the basic physical structure of the crystal. The aluminum atom carries a $+3$ charge compared to the $+4$ charge of the silicon. Thus, one of the negative charges of the shared oxygen atoms is not satisfied, resulting in a net negative charge of the crystal.

The negative charges of clay minerals contribute to the soil cation exchange capacity and are instrumental in interactions with soil organic components and microbes. The quantity of negative charge associated with clay minerals is a property of the specific clay type. Thus, identification of the type of clay in a soil site is essential for predicting the chemical and biological dynamics occurring therein.

Gibbsite ($Al_2O_3.3H_2O$) and Goethite ($Fe_2O_3.H_2O$) are examples of hydrous oxides of iron and aluminum. These clays are found primarily in the highly weathered soils of the topics and subtropics, but also occur in temperate regions. They are less studied than are the layer silicate clays. Hydrous oxides of iron and aluminum may be crystalline or amorphous. They are not sticky clays and their charge varies with pH. At high pH they have a slightly negative charge, whereas at the highly acidic pH values common to the weathered tropical and subtropical soils where they generally occur, they are positively charged. This contrasts to the layer silicate clays, which are normally negatively charged.

Allophane and other amorphous material clays are much less studied than the two other clay mineral groups. The most significant members of this class of clays are allophane and imogolite. They are aluminum silicates and are prevalent in andisols (volcanic ash derived soils). Thus, examples of locations where they could be encountered are the Northwestern United States, Hawaii, and Japan.

Adsorption of Organic Substances to Clay Particles: Clay minerals provide a surface for sorption of a variety of organic substances. Not only do whole microbial cells and viruses sorb to soil clay particles, but also a diversity of organic compounds have been demonstrated to interact with clay surfaces. Examples of the types of organic compounds reacting ionically with clay particles includes carbohydrate phosphates (Goring and Bartholomew, 1951), nucleotides (Goring and Bartholomew, 1952), humic and fulvic acids (Greenland, 1971; Kodema and Schnitzer, 1974), and aromatic compounds as exemplified by *p*-cresol (Boyd and King, 1984). Most biologically produced compounds interact with clay particles. The environmental quality concerns relating to clay-organic matter interactions involve the bioavailability of the sorbed substances and for biodecomposable substances, the impact of sorption on degradation kinetics. Tightly sorbed organic substances, especially those existing within soil micropores, may be essentially unavailable to the biological community — i.e., they may be sequestered (see Chapter 16 for a more complete discussion of this topic). Biodegradation susceptibility of sorbed organic compounds may be reduced or enhanced (see below). Thus, the adsorption potential of any organic compound entering soil should be evaluated in relation to estimation of the effect of this process on its longevity *in situ*.

It must be remembered that both substances entering the soil system from external sources (e.g., pesticides and other soil amendments) as well as those synthesized *in situ* may be sorbed onto clay surfaces. Sorensen (1972) found that amino acid metabolites synthesized in montmorillonite-amended soil following cellulose, hemicellulose, or glucose amendment were bound by soil clays. Clay-amended soil retained two to three times more carbohydrate-derived amino acid carbon than did nonamended soil over a three-year incubation period. In a related study, ^{14}C-labeled cellulose was added to seven soils with silt plus clay contents ranging from 8 to 75 percent (Sorensen, 1975). As in the previously cited study, the quantities of amino acid carbon synthesized from the cellulose remaining in the soil after 30 days' incubation ranged from 6 percent in the lowest silt plus clay-containing soil to 18 percent in the highest silt plus clay-containing soil, implying a stabilizing or protective effect due to sorption to clay minerals.

Association of organic compounds with clay particles is not a totally random process. There is an enrichment of specific carbon moieties within clay particles. This is exemplified by evaluation of the properties of humic acids associated with clay particles. In native soil samples, aliphatic structures are more common in humic acids derived from the clay fraction than other soil fractions (Arshad and Lowe, 1966; Schnitzer et al., 1988; Shulton and Schnitzer, 1990). Using chemical, infrared, and ^{13}C nuclear magnetic resonance, and gas chromatographic mass spectrometric analysis, Schnitzer et al. (1988) found that about half of the clay-associated organic matter consisted of humic materials, the remainder being mainly long-chain aliphatics.

Clay-Organic Matter Adsorption Mechanisms: An immediate conceptual problem raised by this list of materials sorbed to clay particles is the observation that both the most commonly occurring clays in temperate region soils and many of the sorbed organic compounds are negatively charged at the predominant soil pH values. Thus, mechanisms must exist to reduce the repulsive forces between the like charged entities. Examples of such mechanisms are reduction of the quantity of negative charges (pH modification or methylation of the carboxyl groups on the organic matter) and cationic bridging of negatively charged functional groups of the humic substances with the clays. The latter process is depicted by a model proposed by Greenland (1971) (Fig. 1.6). The negative charges on the organic and mineral materials are linked in soils by divalent cations.

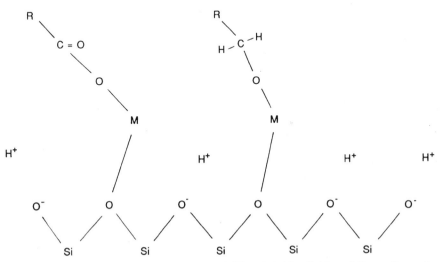

Fig. 1.6 Schematic drawing of divalent metal (M) salt bridge linking of the surface of a mica-type clay with a negatively charged organic compound and the distribution of a hydrogen ion cloud around the clay surface.

Organic compounds retained on clay particles through cationic bridges can clearly desorb from the surface with changes in ionic strength and cation content of the soil solution. A more permanent association of organic compounds to clay particles can occur. This association between clay particles and organic materials results from interlayer clay-organic matter complex formation. This was demonstrated to occur in two New Zealand soils by Theng et al. (1986). Humic substances with a polymethylene chain structure were found to be regularly interstratified with mica-smectite in highly acidic soils containing low microbial activity.

Any process that alters the negative charges of either the clay or organic matter reduces the repulsive forces. This can be accomplished by altering soil pH or through methylation of the carboxyl groups. Nayak et al. (1990) demonstrated that reduction of the acidic groups of humic substances through methylation increased fixation with montmorillonite. The increased association with the clay particles resulted not only from a reduction of interparticle repulsive forces but also from a reduction in hydration energies.

Biodegradation and Clay Interactions: Association of clay and soil organic matter affects soil micro-structure (see Section 1.2.2.1 for elucidation of soil structural interactions) as well as the decomposition rate of sorbed organic compounds. Studies of the impact of clay on decomposition of a wide variety of organic compounds have shown that purified clay minerals when added to growing microbial cultures can stimulate, decrease, or have no effect of decomposition of organic materials (see Kunc and Stotzky, 1974; Huang and Tate, 1997). Kunc and Stotzky (1974) found no clear relationship between the type of organic substance and the type of clay relating to the effect on biodegradation except with aldehydes, whose decomposition was accelerated by clay minerals. Lynch and Cotnoir (1956) noted that bentonite inhibited decomposition of a variety of substrates but that illite and kaolinite had little effect on decomposition. Similarly, Olnes and Clapp (1972) found that the kinetics of the decomposition of the montmorillonite-dextran complex contained periods that were affected to a greater degree by the montmorillonite content, whereas other portions of the degradation curve were affected more by the dextran concentration.

Clearly, it is difficult to make general statements regarding the impact of clay sorption on biological decomposition kinetics. The basis for these seemingly contradictory observations can be revealed by examining the effect of clay on bioavailability of sorbed organic compounds. Two ramifications of clay-organic matter interaction can be pictured depending upon the relationship of the quantity of substrate present in the system and the ability of the organism to interact with it. Carbon and energy sources for microbes may exist in their ecosystem at concentrations too low for significant interaction between them and the microbes to occur. These materials may become sufficiently concentrated on the clay surface that the quantities desorbed from the clay and thus occurring in the water surrounding the clay particle are sufficient to induce microbial catabolism. In contrast, for substrates present in high concentration in soils with high clay contents, their concentration in solution could be reduced by adsorption to the clay particle to a sufficient extent to preclude or greatly reduce the rate of decomposition. In this latter situation, strong binding of the carbonaceous substance to the clay particle reduces the probability of its collision with and uptake by the degrader population. The rate of desorption from the clay thus becomes the limiting factor determining the rate of decomposition of the carbonaceous energy source.

Thus, the impact of clay sorption in soil ecosystems is more a function of residual concentrations of organic compound in interstitial waters and in the microsite containing the microbial colony than is the apparent concentration when the bulk soil sample is analyzed (i.e., microsite concentrations of the compound of interest may far exceed macrosite levels). Thus, decomposition kinetics of the substance reflect the microsite concentrations rather than those that would be predicted from total ecosystem content of the carbonaceous growth substance.

Not only do clay minerals alter decomposition kinetics of organic substances in soil through sorptive-desorption reactions, but they may also indirectly affect microbial metabolism by altering their chemical environment. For example, due to the formation of an hydrogen ion cloud around the clay particle, the pH of the microsite is reduced (see Fig. 1.6). Thus, enzyme-catalyzed reactions in clay soils may appear to have pH optima lower than that detected in the absence of clay particles (Fig. 1.7).

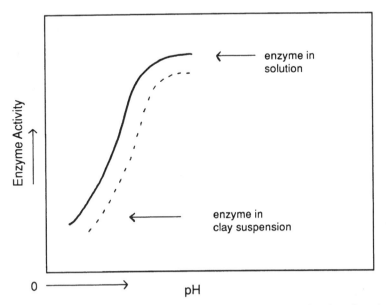

Fig. 1.7 Hypothetical data reflecting alteration of apparent pH optima by microsite variation in pH due to clay micelles.

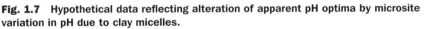

1.2.1.2 Humic Substances and Ecosystem Function

Soil organic matter is commonly considered to be equivalent to soil humus (Stevenson, 1994). The soil humus fraction consists of all organic compounds normally found in soil (with the exception of recognizable plant or animal components), their partial decomposition products, and microbial biomass. Studies of the nature of soil organic matter are complicated by the diversity of materials present and the fact that the definition does not coincide with what can be reasonably separated and analyzed; for example, clear separation of plant and animal biochemicals from the bulk soil organic matter is not easily accomplished. The mixture of chemical substituents of soil humus ranges from simple monomeric components of living cells, such as organic acids, amino acids, and saccharides, to the random polymers classified as humic acids. Essentially any organic compound contained in or synthesized by living cells occurs in soil, although some may be present in minuscule amounts due to their being relatively easily decomposed by soil microbes. Microbial biomass function in soils will be presented in Chapter 2, whereas the role of these materials in nutrient cycling will be analyzed in Chapter 9.

Soil Humic Substances: Of more interest for the present discussion is that portion of the soil organic matter classified as humic substances. **Humic substances are a series of high molecular-weight, brown to black substances formed by secondary synthesis reactions in soil.** This organic matter fraction is not designated or isolated from soil based on a function or role in the soil ecosystem. It is a portion of soil humus that is defined strictly by its solubility in acidic or alkaline solutions. Humic substances are separated from soil in alkaline solutions. Acidification of the alkaline extract to a pH between 1 and 2 results in precipitation of the humic acids. Fulvic acids are soluble in both acid and alkali. Humin is that portion of the colloidal soil organic matter fraction that is not solubolized with alkaline solutions. (See Stevenson, 1994, for a description of extraction methods and properties of these soil organic components.)

To understand the impact of humic substances, especially humic acids, on microbial function in the soil ecosystem, an appreciation of the structural components interacting with soil particles and the microbial community is necessary. (For a more complete discussion of theories regarding structure and origin of these compounds, see Stevenson, 1994.) Humic and fulvic acids are heterogeneous, random polymers of a variety of aromatic and aliphatic components. No single chemical structure can be proposed to represent these molecules. Indeed, because of the fact that a wide diversity of aromatic and aliphatic compounds are linked via a variety of covalent linkages forming humic acids with molecular weights in the thousands of daltons, it can reasonably be concluded that few if any of the molecules are identical. In contrast, fulvic acids are more oxidized than humic acids, that is, they have a higher oxygen content and a lower carbon content and are of lower molecular weight.

Structural Implications of Humic Acids: The nature of the constituents of humic substances determines their importance to total ecosystem function. The heterogeneity and complexity of humic and fulvic acids has limited the level of detail of understanding of the specific chemical reactions between humic and nonhumic soil organic components. An appreciation for the types of reactions occurring between microbial cells or individual biochemicals and humic acids has developed from our basic concept of the molecular content of humic substances and the use of related model compounds.

Our concept of humic acids structure and *in situ* interactions in soil has recently expanded through the application of spectroscopic analytical techniques, such as ^{13}C nuclear magnetic resonance (NMR) spectroscopy. (See Preston, 1996, for a recent review of the utility and limitations of NMR techniques in elucidating the elements of humic acid structure.) Data derived from NMR-based analytical procedures suggest that aromatic rings comprise a smaller portion of the humic acid molecule than was previously believed (e.g., see Hatcher et al., 1981, and Malcolm, 1990). Hatcher et al. (1981) found that aromaticities of humic acids from a variety of soils collected from Prince Edward Island ranged from 35 to 92 percent. Significant aliphatic structures were generally noted. These were not detected by chemical oxidation procedures. These authors also suggested that phenolic carbons may be minor components of humic acids. They were previously considered to be major substituents. This conclusion was based on the observation that NMR spectra did not contain intense signals corresponding to the presence of these molecules. Schnitzer and Preston (1986) suggest that since the presence of phenolic hydroxyls is confirmed by several independent methods, failure to detect them by NMR analysis merely indicates a shift in the NMR signal resulting from these moieties. These studies exemplify the evolving nature of our understanding of humic acids and their structure, and the limitations imposed by the inability to develop detailed structural formulae for these molecules. (For a more detailed analysis of humic and fulvic acids structure and origin in soil, see Stevenson, 1994.)

The primary conclusion regarding the structure of humic substances and their importance to the soil microbiologist is derived from the complexity of their composition. They contain varying ratios of phenolic, methoxyl, aromatic, hydrocarbon, amino acid, and nitrogen moieties plus covalently linked polysaccharides and proteins. The ratio of these substituents present is dependent on the source of the humic acid preparation. Great variation in composition occurs between humic acids prepared from soil, stream and marine samples (Malcolm, 1990). Significant differences in humic acids have also been noted for humic acids prepared from a variety of soil types—alfisols and mollisols (Novak and Smeck, 1991)—and soil management histories (Stearman et al., 1989).

Humic Substances, Organic Compounds, and Biodecomposition: Humic substances, although a variety of microorganisms have been suggested to be capable of metabolizing them, are unlikely contributors to the energy resources of soil microbes. This conclusion results primarily from observations of the complexity of the structure of the molecules — they contain a large variety of substituents that are essentially randomly assembled. (See Chapter 4 for further elucidation of microbial energy metabolism in soil.) Humic substances affect soil biological processes to the greatest extent by interacting with soil minerals in soil aggregate formation (see Section 1.2.2) and by their role in removal of biodegradable substances from the pool of easily metabolized substrates (i.e., humification). Enzymatic interaction with humified substrates is inhibited physically by the imposing humic acid physical structure.

As was discussed above with clays, humic substances sorb organic compounds through temporary associations. For these organic matter interactions, sorption results from π-π interactions, van der Waals forces, and hydrogen bonding as well as ionic associations. More permanent linkage of organic compounds to humic substances also occurs due to formation of covalent bonds. Examples of organic compounds that may be linked covalently to humic acids include amino acids, peptides, proteins, aromatic compounds, and polysaccharides.

Covalent linkage of phenolic compounds to humic acids may be catalyzed by phenoloxidase (Bollag et al. 1980; Dec and Bollag, 1997; Liu et al. 1985). Carbon-carbon and carbon-nitrogen linkages are catalyzed through free radical formation by these laccase-type enzymes (Fig. 1.8). The linkage of aromatic ring-containing compounds occurs in *ortho-ortho* or *para-para*, or *ortho-para* type associations by this biologically catalyzed process (Sjoblad and Bollag, 1981). Aside from being a mechanism for the formation of humic substances *in situ*, these oxidative processes provide a means for the stabilization of a variety of xenobiotic compounds, such as pesticides, into soil organic matter.

The impact of the covalent association of organic compounds with humic substances on their biodecomposition kinetics is more predictable than was described above for clay-organic matter interactions. Humification of organic compounds results in their stabilization in soil. For example, in a study of the impact on microbial degradation by nucleophilic addition of amino groups to the aromatic nuclei or quinones, Bondietti et al. (1972) found that bonding of amino sugar units to model humic acids resulted in major reductions in the decomposition rate. Over a 12-week incubation period, more than 70 percent of free glucosamine was oxidized to carbon dioxide, while only 15 to 23 percent of the humic acid-model polymer-bound amino sugar was oxidized. The importance of covalent linkage of the humic acids and sugar was shown by the fact that simply mixing the amino sugar with the model humic acid molecule did not reduce the oxidation rate.

It is important to note that substances covalently bonded to humic acid can still be mineralized but that the rate of oxidation is altered. Basically, what is under consideration here is the fate of the carbon of the humified organic compound after it has been incorporated into the humic acid molecule. Since it is part of the larger substance — that is, humic acid — its decomposition kinetics tend to resemble the slow mineralization rate of humic acid molecule.

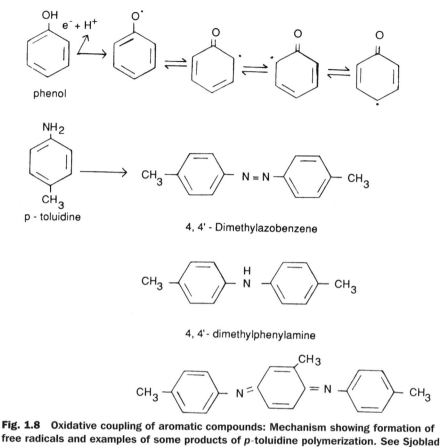

Fig. 1.8 Oxidative coupling of aromatic compounds: Mechanism showing formation of free radicals and examples of some products of *p*-toluidine polymerization. See Sjoblad and Bollag (1981) for further discussion and examples.

This observation has environmental significance. Covalent bonding to humic acids could be a viable mechanism for renovation of xenobiotic compound-contaminated soils if the substances are permanently bound or if they are released in an inactivated (i.e., partially or totally catabolized) form. Unfortunately, this condition is not always met. Release of humic acid-bound pesticides intact can occur (Bartha, 1980; Lichtenstein, 1980; Still et al., 1980). Thus, from an agricultural viewpoint, difficulties may arise from release of a pesticide that had been previously applied to an appropriate crop and stabilized in soil humic substances several years after application. Particular concern would be precipitated should this pesticide reappear in a crop where use would be inappropriate either from a crop-sensitivity or regulatory view.

It must be stressed that in most cases, the more probable fate of a biodecomposable substance added to soil is mineralization, either before humification or even subsequent to humification. For example, Haider and

Martin (1988) found that *Phanerochaete chrysosporium* could mineralize xenobiotic compounds at a reasonable rate (13 to 56 percent of xenobiotic carbon as carbon dioxide over an 18-day period). Similarly, Saxena and Bartha (1983a) found that humus-bound 3,4-dichloroaniline (a biodegradation intermediate) was mineralized at a rate approximately the same as that of the humic acid molecule itself. Adsorbed 3,4-dichloroaniline was mineralized faster than humic acid molecules. Of note is the observation that fertilization of agricultural soil with anhydrous ammonia could mobilize humus-bound 3,4-dichloroaniline (Saxena and Bartha, 1983b).

The size and complexity of the humic acid molecule creates the potential for a stable, noncovalent association between it and simple organic molecules. Using the classical formula for the humic acid molecule, Schulten and Schnitzer (1997) have described a three-dimensional structure of the humic acid molecule that can accommodate noncovalent associations of simple organic molecules within voids in the molecule. These associations are sufficiently stable energetically that the longevity of the associations between the simple and complex organic substances resembles that of covalently associated molecules (Dec and Bollag, 1997). Hydrophobic compounds have been shown to associate with hydrophobic domains, micelles, within the humic acid molecule (von Wandruska, 1998). Such associations may participate in reduction of bioavailability of toxicants through sequestration (see Chapter 16). Environmental concerns relating to these noncovalent associations relate to the need to assess the extent of their occurrence in native soil ecosystems, to evaluate the potential for release of the toxicant from the association, and to quantify the environmental risk resulting from the freeing of the toxicant from the humic acid mycelle.

1.2.2 Soil Aggregate Structure and Biological Systems

Interest in the microstructure of soil is derived from an ecological viewpoint as well as an operational laboratory angle. It is useful in evaluating field processes to understand the physical relationship of biotic and abiotic soil components *in situ* and how this association affects overall ecosystem development. Furthermore, due to the fact that it is difficult to analyze soil microbial processes in the field, techniques are required to fractionate soil into functionally related components in the laboratory so that the nature of the biological processes occurring therein and the factors controlling their rate can be determined.

1.2.2.1 Native Soil Aggregate Structure and Its Impact on Ecosystem Function

The world of the soil microbe does not consist of individual sand, silt, or clay particles or even of organic matter *in toto*;, rather, it is a mixture of these substances assembled in a seemingly random order. Yet, microorganisms may appear to be associated with one specific ecosystem component in this complex environment. For example, Fisk et al. (1999), using a combination of specific stain and micromorphological techniques, found that soil-amended bacteria tended to become associated with variably charged mineral oxides and organic matter within the intergrain microaggregates of a Freehold sandy loam and occurred along coated mineral surfaces. Also, rhizosphere organisms may be growing essentially on the root surface. Microbes may be physically attached to partially decomposed plant debris. But even in these situations, the metabolism of soil microbes is affected by the association of the soil particle to which they are attached within larger groupings called aggregates. Indirect impacts of soil aggregation on the metabolic rates of individual microbial colonies result from the fact that the degree of soil aggregate formation controls biological process by limiting such soil properties as water infiltration and availability, oxygen tension, and nutrient movement.

Soil structure is defined at the level of the microbial cell by the association of soil particulates and colloidal organic matter into soil aggregates. **Soil aggregates are random combinations of soil organic and mineral components assembled into micro- ($<50\,\mu m$ mean diameter) and macroaggregates ($>50\,\mu m$ mean diameter particles).** By their nature, soil aggregates are a product of interactions of the soil microbial community, the soil parent material, the aboveground plant community, and ecosystem history.

Soil Aggregate Formation: Assembly of soil particulates into aggregates is an excellent example of abiotic and abiontic organic matter plus living cells and mineral components interacting to affect soil physical properties. A conceptual model of the biological contributions to soil aggregate formation can be described as a coupling of feedback-controlled processes (Fig. 1.9). Biological participants are higher plants as well as soil microbes. The physical variable altered by the synergistic interactions of the biological entities is degree of association of soil particulates into at least transiently stable structures. Microorganisms metabolize photosynthetically fixed carbon to produce the compounds responsible for binding of soil mineral matter together, polysaccharides and humic substances. Improvement in the size and number of aggregates formed enhances plant community growth, which in turn stimulates even further the microbial development, which may result in even greater improvements in soil structure. In contrast, poor soil structure retards air diffusion, water movement, and root growth, thereby resulting in a less active soil microbial community than would be attained under more optimal growth conditions.

A conceptual model describing the interaction of soil biotic and mineral components to form water stable aggregates, which explains the impact of soil system management and the duration of structural relationships, was developed by Tisdall and Oades (1982). Organic binding agents were divided into three categories: transient (mainly polysaccharides), temporary (roots and fungal hyphae), and persistent (humic substances associated with polyvalent metal

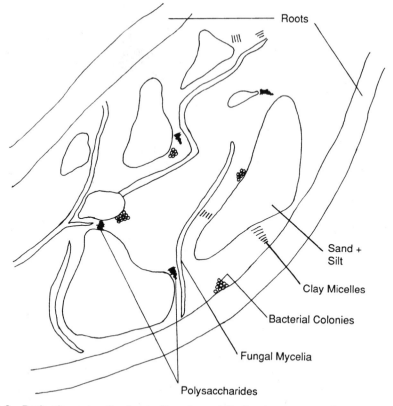

Fig. 1.9 **Basic elements of soil substituents contributing to aggregate formation (not drawn to scale).**

cations). Soil aggregates are formed in a hierarchical arrangement by stabiliz-ation of smaller aggregates into larger structures by different binding agents (Waters and Oades, 1991). Microaggregates are stabilized by persistent bind-ing agents, whereas polysaccharides and biomass (microbial and root) result in the less persistent association of the microaggregates into macroaggregates (Beare et al., 1997; Elliott, 1986; Hu et al., 1995). Microaggregates consist primarily of clay and humified organic materials linked by polyvalent metals (Edwards and Bremner, 1967). Due to the biodegradation resistance of the persistent binding agents, the microaggregate structure is long term and appears to be more a characteristic of the soil than a product of soil management procedures. In contrast, the assemblage of microaggregates into macroaggregates is less stable and affected more by soil management due to its impact on soil fungal and root development, including mycorrhizal associations (Hu et al., 1995). This model emphasizes the importance of polysaccharides and humic substances in soil aggregate formation, but it must be appreciated that any organic compound that stimulates or decreases microbial growth in soil indirectly affects aggregate formation by affecting the growth dynamics of the fungal community.

Role of Carbohydrates in Soil Aggregate Formation: The importance of polysaccharides as binding agents for adhesion of soil mineral components into aggregates has been demonstrated by assessing the correlation of polysaccharide levels with quantities of water-stable aggregates, quantifying the effect of periodate oxidation of polysaccharides on aggregation, and analyzing the influence of polysaccharide adsorption on the mechanical strength of clay minerals. For example, Chaney and Swift (1984) found that total organic matter, total carbohydrate, and humic material had highly or very highly significant correlations with aggregate stability in 120 soils. To rule out other soil physical properties as effectors of aggregate formation, 26 agricultural soils were evaluated. Only organic matter and associated properties correlated significantly with aggregate stability. Other soil parameters examined were sand, silt, clay, and iron contents, and cation exchange capacity.

Direct evidence for the role of polysaccharides in soil has been gained by assessing the effect of their selective destruction on soil aggregation. Periodate treatment of soil results in oxidation of polysaccharides contained therein. Not all polysaccharides are destroyed by such treatment. The proportion of polysaccharides oxidized depends on the duration of the treatment and the degree of contact of the sugar with the oxidizing agent. The polysaccharide may be occluded by close association within soil particulate structures. Even with these limitations, a major role of polysaccharides in aggregate formation is revealed by this method. Periodate treatment of soil from a long-term grass field destroyed microaggregates greater than 45 μm in diameter along with the carbohydrates (Cheshire et al., 1983). Residual carbohydrate was enriched in sugars commonly found in plant materials. This suggested preferential destruction of microbial polysaccharides by the periodate. The relationship between soil polysaccharide and aggregate stability was verified in 15 soils and 7 soil series (Cheshire et al., 1984).

Chenu and Guérif (1991) assessed the strength of the association between polysaccharides by assessing the mechanical strength of mixtures of kaolinite or montmorillonite plus or minus a polysaccharide, scleroglucan. The polysaccharide increased the mechanical strength in mixtures of both clays through the formation of polymer bridges between the clay particles and from the progressive coating of the clay surface by the polysaccharide. The primary effect of the polysaccharide on soil aggregation was therefore the result of increased strength of interparticle bonds.

Humic Acids and Aggregate Stability: Microaggregate formation results from interactions between humic substances and clay particles (e.g., Capriel et al., 1990; Dinel et al., 1991; Schnitzer and Kodema, 1992). Aliphatic side chains are particularly instrumental in this process. A high correlation between soil aliphatic fractions (Capriel et al., 1990) and aliphatic enriched humic fractions (Dinel et al. 1991) with aggregate stability underscores the importance of these fractions in microaggregate formation. As was noted above, humic-associated aliphatics can form interlayer associations with clays (Theng et al., 1986), and these materials are enriched in soil clay fractions (Arshad and Lowe, 1966; Schnitzer et al., 1988). The fact that these associations are more a property of the soil itself and not due to management is supported by the observation that although the quantities of aliphatic fraction extracted with supercritical-hexane varied with management, the chemical properties revealed by ^{13}C nuclear magnetic resonance and Fourier-transform infrared differences were independent of management (Schnitzer et al., 1988).

Soil Management and Aggregate Stability: The stability of microaggregates and macroaggregates reflects the biodegradation susceptibility of the primary organic polymers or biological biomass instrumental in their formation. Microaggregates stabilized through association of humic substances with clay particles are reasonably resistant to management procedures and are more reflective of soil physical properties.

Soil management typically results in modification of factors controlling soil microbial activity. For example, cultivation of soil can cause catabolism of organic matter previously occluded by soil minerals, increases in soil aeration, and modification of soil pH through liming. Portions of the soil organic fractions could be depleted under appropriate conditions in a matter of days or weeks. Therefore, soil management procedures that result in a decrease in soil organic components result in a reduction of quantities of the water-stable macroaggregates produced by their interaction with soil particulates. The best examples of such declines in soil structural properties are found in studies of the effect of long-term cultivation on soil properties. For example, Tiessen and Stewart (1983) evaluated the effect of cultivation for between 4 and 90 years on soil particle size distribution of three north American grassland soils (Cryoborolls). They noted the continuous depletion of organic matter and the loss of particles $> 50 \, \mu m$ diameter with time of cultivation, indicating that the more degradable metabolizable soil organic matter components supporting the structure of soil aggregates were being depleted with time.

Conversely, soil management procedures chosen to increase microbial activity, resulting in an increase in biomass and polysaccharide synthesis, cause improvements in soil aggregate structure. This was reflected in the study of Tisdall and Oades (1982), where the duration and magnitude of the improvement in soil microstructure was shown to be dependent on the nature of the metabolized carbon substrate amended to the soil samples. Maximum aggregation occurred in soil amended with the rapidly metabolized substrate, glucose, whereas the more lasting but less dramatic improvement in soil structure was observed with more slowly metabolized cellulose. Although cellulose is much more biodegradation-resistant than is glucose (days compared to months for depletion from soil), each of the materials could be considered to be rapidly catabolized when compared with humic substances (degradation times measured in decades or more). Reduction of tillage intervention with crop production is another example of a means of manipulating soil communities and organic matter levels in order to improve aggregate structure. Changing from an intensive cultivation management strategy to no- or minimum-till operation results in increases of soil humus levels and augmented microbial — including fungal — biomass, which coincides with improvements in the soil aggregate structure.

The impact of carbon inputs from crop growth on aggregate formation also can vary with the species grown. For example, Angers (1992) in a comparison of the effect of alfalfa (*Medicago sativa* L.) or corn (*Zea mays* L.) growth to the aggregate structure of soils from bare fields found maximum aggregation under alfalfa with minimal changes in this soil property during the study in the fallow field and that cropped to corn. Most likely, the effect on soil aggregation resulted from a combination of variation in quantities and quality of organic matter available to the soil microbial community plus the resultant alteration of the activity of these populations.

Plant biomass may be incorporated into soils through natural cycling processes, as is observed in grasslands where a distinctive thatch layer can develop (see Tate, 1987) or in agricultural soils when crop residues are returned to the soil as well as when exogenously produced organic materials, such as composts and sludges, are amended to soil. The effect of such amendment on soil aggregate formation can be substantial. For example, Glauser et al. (1988) found an increase of water-stable aggregates to 85 percent on soils receiving sludge amendments, compared to 45 percent in nonamended soils.

1.2.2.2 Separation of Soil Particulates by Density Fractionation

A primary objective of many soil microbiological endeavors is to assess the potential for a metabolic process to occur (e.g., nitrogen mineralization) by analyzing specific soil components or reactions occurring in isolated soil fractions. Achieving this objective is most often precluded by the fact that soil fractionation procedures tend to be based on physical or chemical properties, not biological functional relationships. A procedure that appears to have potential to overcome this limitation is density fractionation. This procedure is still a physical differentiation method — soil is separated by density in sodium iodide solutions (Spycher et al., 1983; Strickland and Sollins, 1987) — but a functional partitioning of soil organic matter is achieved. For example, suspension of whole soil in a high-density sodium iodide solution yields a light and a heavy soil fraction. The light fraction is a reservoir of labile plant carbon and nutrients since it is enriched in partially decomposed plant material, including dead root fragments and fungal hyphae. Organic matter contained in the heavy fraction is enriched in humic acids. Combination of water dispersion and density fractionation demonstrated enrichment of organo-mineral particles in the light clay-size fraction with organic carbon characterized by wide carbon:nitrogen ratios and low amounts of alkali-extractable carbon (Young and Spycher, 1979). Similarly Sollins et al. (1984) noted enrichment of biodecomposable plant remains and microbial remnants in the light fraction compared with more biodegradation-resistant materials in the heavy fractions. Dalal and Henry (1988) found in an examination of cultivation effects of carbohydrate distribution in soils that the light fraction was enriched with carbohydrate carbon (27 to 43 percent of the carbohydrate carbon was found in this fraction). Cultivation resulted in a rapid decline in the light fraction and a substantial loss of carbohydrate, especially in clayey soils. These observations are significant ecologically since the primary energy source driving soil microbial processes is labile plant carbon. Thus, assessment of reactions occurring in the light soil density fraction and contrasting them to rates found in the heavy fraction may provide a good indication of metabolic potential in soil ecosystems.

1.3   The Macro-Ecosystem

An appreciation of variation in soil properties on a macro-scale is equally important to understanding the microsite characteristics described above. Both the vertical and horizontal variation in soil properties limiting microbial activity must be quantified.

Although a considerable interest is developing in estimating microbial function in deep aquifer materials, most soil microbiology research is directed at developing an understanding of processes occurring at the point of contact between the aboveground and belowground communities. Thus, soils from the O, A, and to some extent the upper portions of the B horizon are most frequently examined. Interactions in the B horizon are especially important in sites where the O horizon is nonexistent and the A horizon is thin or has been lost through erosion.

The importance of these major surface soil horizons is derived from the fact that the primary energy source for the soil microbial community is derived from photosynthate entering soil primarily as plant biomass or root exudates. Although some water-soluble plant carbon may leach through the soil profile, especially in soils with measurable organic horizons, the primary microbial encounter with organic carbon-containing materials is in the top few centimeters of the soil profile. Thus, a major concern in assessing the microbial community is the distribution of carbon inputs on the soil surface — that is, plant population growth and dispersion of fixed carbon within the soil profile due to past biomass productivity and its transformation and redistribution within the soil matrix. Secondary considerations relate to variation of physical and chemical growth affecting parameters across the soil surface and throughout the profile. Again, many of these result from activity of the aboveground community. Other factors that must be considered are temperature (which varies diurnally as well as seasonally and is affected by plant density and degree of canopy closure) and moisture (which may also vary seasonally and is affected by topography and plant canopy development). Distribution and amount of microbial activity in soil horizons can also be related to soil parent material and the soil developmental history as well as its pH and salinity. Distribution and range of variation of each of these factors must be considered when extrapolating microbiological data to the ecosystem level.

Variation in soil microbial activity in surface soils is generally controlled by inputs of energy in the form of photosynthetically fixed carbon contained in or exuding from plant biomass. The impact of this energy supply distribution is exemplified by examining the sources of fixed carbon in a temperate forest soil. Within the forest soil, microbial energy is derived from oxidation of fixed carbon contained in litter and from decomposition of root exudates, sloughed cells, and fine roots. The contribution of fine roots to the soil nutrient pool is frequently overlooked, but it can constitute a major source of nutrients and energy for soil microbes (See McClaugherty et al., 1982; McClaugherty et al., 1984; Vogt et al., 1983 for examples). In two temperate forest stands, fine root production was roughly equivalent to leaf biomass production (McClaugherty et al., 1982). If an O horizon has developed, microbial activity in the A horizon is also influenced by fixed carbon carried in leachates and by that portion of the plant biomass mixed into the soil surface by earthworm and other animal activity. Thus, maximal soil microbial activity would be anticipated to occur in the vicinity of the tree or shrub (and grass if it occurs on the forest floor) and under areas of litter accumulation and decomposition.

A similar distribution pattern of plant biomass is observed in a grassland soil. Grass biomass enters soils as leafy material in the thatch and as root exudates and decaying root biomass in a dense root structure. Highest quantities of water-stable aggregates are generally found in soils collected from grassland ecosystems compared to soils from other major ecosystem types, if all other limiting factors are equivalent. This results from the high input of biomass, the consistent production of this fixed carbon throughout the growing season, and the proportion of readily decomposable components constituting the plant debris as compared to alternate ecosystem types such as forest or agricultural soils. For example, Jenkinson and Ayanaba (1977) noted that 70 percent of ryegrass (*Lolium multiflora*) decomposed with a half life of 0.25 years in agricultural soils of Rothamsted, England. The remainder had a half life of eight years.

With agricultural soils, the organic matter inputs are commonly reasonably equal to those of the grassland in quality, but their distribution and quantities are more limited than is observed with the grassland. Thus, soil structural development is reduced. In intensively cultivated sites, reduction of colloidal organic matter levels in the soil results in a loss of soil microstructure (see Tate, 1987 for further analysis of limitations and soil structure in this ecosystem type). Analysis of microbial activity in these soils must be conducted with an understanding of crop planting patterns, cropping history, and tillage practices.

In widely spaced grass crops, such as sugarcane (*Saccharum* spp. L.), biomass inputs to the microbial community between rows is minimal.

Organic matter accumulates in swamp ecosystems. Organic matter inputs from the aboveground plant community are limited by the high soil water table, as is the decomposition rate of the decaying plant biomass. The accretion of organic matter indicates that biomass production occurs more rapidly than decomposition. Other organic inputs to the swamp system originate from erosion of soils from adjacent upland areas. Since the A horizon soils contain most of the colloidal soil organic matter, washing of these soils down slope results in an importing of this organic matter into the swamp environment.

1.4   Concluding Comments

The act of allowing a soil sample to flow between one's fingers does not leave the impression that a living, respiring community is functioning therein. It could rather easily be concluded that the utmost importance in the various soil types must be attributed to their physical components — that is, their sand, silt, clay, and, perhaps, rock and pebble contents. Although such a lack of apparent importance of biological entities is at first disappointing, such an observation is critical for the development of our concepts of soil microbiology. It is too easy for soil microbiologists to concentrate on the extremely diverse community of microbes existing in a soil sample and their vast array of genotypic and phenotypic properties, and to forget that these organisms are of no benefit to the total ecosystem if they are incapable of functioning within the soil physical matrix. The physical and chemical properties and their variability (macro- and microsite) must always be appreciated if a true understanding of the limitations and capabilities of the soil microbial community is to be fully developed.

The variability of the soil environment and its impact on life processes is both intriguing and frustrating. As microbiologists are keenly aware of the fact that there exists no typical microbe for even the simplest physiological or ecological study, they must also appreciate the fact that there is no standard soil. Soil components and their interactions are at least as variable as are the characteristics of their associated microbial populations. Thus, a complete understanding of soil microbial behavior and its contributions to total ecosystem function must include an appreciation of its physical environment — both the mix of individual soil components as well as the physical impediments presented to the soil microbes by the assembly of these discrete particles into a defined soil structure, including horizonal development as well as aggregate formation. Assessment of the soil type and its management history is as important to the conducting of "good science" as is provision of a genus or species name to the microbes studied. That is, a complete soil microbiological study must result not only in an appreciation of the nuances of the biological aspects of the soil ecosystems, but also in a conceptualization of the impact of variations in the physical and chemical environment on the expression of the living capacities.

References

Amundson, R., and H. Jenny. 1991. The place of humans in the state factor theory of ecosystems and their soils. Soil Sci. 151:99–109.

Angers, D. A. 1992. Changes in soil aggregation and organic carbon under corn and alfalfa. Soil Sci. Soc. Am. J. 56:1244–1249.

Arshad, M. A., and L. E. Lowe. 1966. Fractionation and characterization of naturally occurring organo-clay complexes. Soil Sci. Soc. Am. Proc. 30:731–735.

Bartha, R. 1980. Pesticide residues in humus. ASM News 46:356–360.

Beare, M. H., S. Hus, D. C. Coleman, and P. F. Hendrix. 1997. Influences of mycelial fungi on soil aggregation and organic matter storage in conventional and no-tillage soils. Appl. Soil Ecol. 5:211–219.

Bollag, J.-M., S. Y. Liu, and R. D. Minard. 1980. Cross-coupling of phenolic humus constituents and 2,4-dichlorophenol. Soil Sci. Soc. Am. J. 44:52–56.

Bondietti, E., J. P. Martin, and K. Haider. 1972. Stabilization of amino sugar units in humic-type polymers. Soil Sci. Soc. Am. Proc. 36:597–602.

Boyd, S. A., and R. King. 1984. Adsorption of labile organic compounds by soil. Soil Sci. 137:115–119.

Brady, N. C., and R. R. Weil. 1999. The Nature and Property of Soils. Prentice Hall. NJ. 881 pp.

Capriel, P., T. Beck, H. Borchert, and P. Härter. 1990. Relationship between soil aliphatic fraction extracted with supercritical hexane, soil microbial biomass, and soil aggregate stability. Soil Sci. Soc. Am. J. 54:415–420.

Chaney, K., and R. S. Swift. 1984. The influence of organic matter on aggregate stability in some British soils. J. Soil Sci. 35:223–230.

Chenu, C., and J. Guérif. 1991. Mechanical strength of clay minerals as influenced by an adsorbed polysaccharide. Soil Sci. Soc. Am. J. 55:1076–1080.

Cheshire, M. V., G. P. Sparling, and C. M. Mundie. 1983. Effect of periodate treatment of soil on carbohydrate constituents and soil aggregation. J. Soil Sci. 34:105–112.

Cheshire, M. V., G. P. Sparling, and C. M. Mundie. 1984. Influence of soil type, crop and air drying on residual carbohydrate content and aggregate stability after treatment with periodate and tetraborate. Plant Soil 76:339–347.

Dalal, R. C., and R. J. Henry. 1988. Cultivation effects on carbohydrate contents of soil and sol fractions. Soil Sci. Soc. Am. J. 52:1361–1365.

Dec, J., and J.-M. Bollag. 1997. Determination of covalent and noncovalent binding interactions between xenobiotic chemicals and soil. Soil Sci. 162:858–874.

Dinel, H., G. R. Mehuys, and M. Lévesque. 1991. Influence of humic and fibric materials on the aggregation and aggregate stability of a lacustrine silty clay. Soil Sci. 151:146–158.

Duxbury, J. M., and R. L. Tate III. 1981. The effect of soil depth and crop cover on enzymatic activities in Pahokee muck. Soil Sci. Soc. Am. J. 45:322–328.

Edwards, A. P., and J. M. Bremner. 1967. Microaggregates in soil. J. Soil Sci. 18:64–73.

Elliott, E. T. 1986. Aggregate structure and carbon, nitrogen, and phosphorus in native and cultivated soils. Soil Sci. Soc. Am. J. 50:627–633.

Fanning, D. S., and M. C. B. Fanning. 1989. Soil Morphology, Genesis, and Classification. John Wiley & Sons, NY. 395 pp.

Fisk, A. C., S. L. Murphy, and R. L. Tate III. 1999. Microscopic observations of bacterial sorption in soil cores. Biol. Fert. Soil 28:111–116.

Froth, H. D., and L. M. Turk. 1972. Fundamentals of Soil Science. John Wiley & Sons. NY.

Glauser, R., H. E. Doner, and E. A. Paul. 1988. Soil aggregate stability as a function of particle size in sludge-treated soils. Soil Sci. 146:37–43.

Goring, C. A. I., and W. V. Bartholomew. 1951. Microbial products and soil organic matter: III. Adsorption of carbohydrate phosphates by clays. Soil Sci. Soc. Am. Proc. 15:189–194.

Goring, C. A. I., and W. V. Bartholomew. 1952. Adsorption of mononucleotides, nucleic acids, and nucleoproteins by clays. Soil Sci. 74:149–164.

Greenland, D. J. 1971. Interactions between humic and fulvic acids and clays. Soil Sci. 111:34–41.

Haider, K., and J. P. Martin. 1988. Mineralization of 14-C labelled humic acids and of humic-acid bound ^{14}C-xenobiotics by *Phanerochaete chrysosporium*. Soil Biol. Biochem. 20:425–429.

Hatcher, P. G., M. Schnitzer, L. W. Dennis, and G. E. Marciel. 1981. Aromaticity of humic substances in soils. Soil Sci. Soc. Am. J. 45:1089–1094.

Hu, S., D. C. Coleman, M. H. Beare, and P. F. Hendrix. 1995. Soil carbohydrates in aggrading and degrading agroecosystems — Influences of fungi and aggregates. Agric. Ecosyst. Environ. 54:77–88.

Hwang, S., and R. L. Tate III. 1997. Interactions of clay minerals with *Arthrobacter crystallopoietes*: Starvation, survival and 2-hydroxypyridine catabolism. Biol. Fert. Soils 24:335–340.

Jenkinson, D. S., and A. Ayanaba. 1977. Decomposition of carbon-14 labeled plant material under tropical conditions. Soil Sci. Soc. Am. J. 41:912–915.

Jenny, H. 1941. Factors of soil formation; a system of quantitative pedology. McGraw-Hill Book Co., NY.

Kodema, H., and M. Schnitzer. 1974. Adsorption of fulvic acid by non-expanding clay minerals. Trans. Int. Congr. Soil Sci. 10th. 2:51–56.

Kunc, F., And G. Stotzky. 1974. Effect of clay minerals on heterotrophic microbial activity in soil. Soil Sci. 118:186–195.

Lichtenstein, E. P. 1980. "Bound" residues in soils and transfer of soil residues in crops. Residue Rev. 76:147–153.

Liu, S.-Y., A. J. Greyer, R. D. Minard, and J.-M. Bollag. 1985. Enzyme-catalyzed complex-formation of amino acid esters and phenolic humus constituents. Soil Sci. Soc. Am. J. 49:337–342.

Lynch, D. L., and L. J. Cotnoir, Jr. 1956. The influence of clay minerals on the breakdown of certain organic substrates. Soil Sci. Soc. Am. J. Proc. 20:367–370.

Malcolm, R. L. 1990. The uniqueness of humic substances in each of soil, stream, and marine environments. Anal. Chem. Acta 232:19–30.

McClaugherty, C. A., J. D. Aber, and J. M. Melillo. 1982. The role of fine roots in the organic matter and nitrogen budgets of two forested ecosystems. Ecology. 63:1481–1490.

McClaugherty, C. A., J. D. Aber, and J. M. Melillo. 1984. Decomposition dynamics of fine roots in forested ecosystem. Oikos 42:378–386.

Miller, R. W., and D. T. Gardiner. 1998. Soils in Our Environment. Prentice Hall, NJ. 736 pp.

Nayak, D. C., C. Varadachari, and K. Ghosh. 1990. Influence of organic acidic functional groups of humic substances in complexation with clay minerals. Soil Sci. 149:268–272.

Novak, J. M., and N. E. Smeck. 1991. Comparisons of humic substances extracted from contiguous Alfisols and Mollisols in Southwestern Ohio. Soil Sci. Soc. Am. J. 55:96–102.

Olnes, A., and C. E. Clapp. 1972. Microbial degradation of a montmorillonite-dextran complex. Soil Sci. Soc. Am. J. 36:179–181.

Preston, C. M. 1996. Applications of NMR to soil organic matter analysis: History and prospects. Soil Sci. 161:144–166.

Saxena, A., and R. Bartha. 1983a. Microbial mineralization of humic acid-3,4-dichloro-aniline complexes. Soil Biol. Biochem. 15:59–62.

Saxena, A., and R. Bartha. 1983b. Binding of 3,4-dichloroaniline by humic acid and soil: Mechanism and exchangeability. Soil Sci. 136:111–116.

Schnitzer, M., and H. Kodema. 1992. Interactions between organic and inorganic components in particle-size fractions separated from four soils. Soil Sci. Soc. Am. J. 56:1099–1105.

Schnitzer, M., and C. M. Preston. 1986. Analysis of humic acids by solution and solid-state carbon-13 nuclear magnetic resonance. Soil Sci. Soc. Am. J. 50:326–331.

Schnitzer, M., J. A. Ripmeester, and H. Kodama. 1988. Characterization of the organic matter associated with a soil clay. Soil Sci. 145:448–454.

Shulten, H.-R., and M. Schnitzer. 1990. Aliphatics in soil organic matter in fine-clay fractions. Soil Sci. Soc. Am. J. 54:98–105.

Schulten, H.-R., and M. Schnitzer. 1997. Chemical model structures for soil organic matter and soils. Soil Sci. 162:115–130.

Sjoblad, R. D., and J.-M. Bollag. 1981. Oxidative coupling of aromatic compounds by enzymes from soil microorganisms. *In* E. A. Paul and J. N. Ladd (eds.), Soil Biochemistry 5:113–152. Marcel Dekker, NY.

Sollins, P., G. Spycher, and C. A. Glassman. 1984. Net nitrogen mineralization from light- and heavy-fraction soil organic matter. Soil Biol. Biochem. 16:31–37.

Sorensen, L. H. 1972. Stabilization of newly formed amino acid metabolites in soil by clay minerals. Soil Sci. 114:5–11.

Sorensen, L. H. 1975. The influence of clay on the rate of decay of amino acid metabolites synthesized in soils during decomposition of cellulose. Soil Biol. Biochem. 7:171–177.

Spycher, G., P. Sollins, and S. Rose. 1983. Carbon and nitrogen in the light fraction of a forest soil: Vertical distribution and seasonal patterns. Soil Sci. 135:79–87.

Stearman, G. K., R. J. Lewis, L. J. Tortorelli, and D. D. Tyler. 1989. Characterization of humic acid from no-tilled and tilled soils using carbon-13 nuclear magnetic resonance. Soil Sci. Soc. Am. J. 53:744–749.

Stevenson, F. J. 1994. Humus Chemistry. Genesis, Composition. Reactions. John Wiley & Sons, NY.

Still, C. C., T.-S. Hsu, and R. Bartha. 1980. Soil-bound 3,4–dichloroaniline: Source of contamination in rice grain. Bull. Environ. Contam. Toxicol. 24:550–554.

Strickland, T. C., and P. Sollins. 1987. Improved method for separating light- and heavy-fraction organic material from soil. Soil Sci. Soc. Am. J. 51:1390–1393.

Tate, R. L. III. 1980. Microbial oxidation of organic matter of histosols. 4:169–201. *In* M. Alexander (ed.), Advances in Microbial Ecology. Plenum Press, NY.

Tate, R. L. III. 1987. Soil Organic Matter. Biological and Ecological Effects. John Wiley & Sons, NY.

Theng, B. K. G., G. J. Churchman, and R. H. Newman. 1986. The occurrence of interlayer clay-organic complexes in two New Zealand soils. Soil Sci. 142:262–266.

Tiessen, H., and J. W. B. Stewart. 1983. Particle-size fractions and their use in studies of soil organic matter: II. Cultivation effects on organic matter composition in size fractions. Soil Sci. Soc. Am. J. 47:509–514.

Tisdall, J. M., and J. M. Oades. 1982. Organic matter and water-stable aggregates in soils. J. Soil Sci. 33:141–163.

Vogt, K. A., C. C. Grier, C. E. Meier, and M. R. Keyes. 1983. Organic matter and nutrient dynamics in forest floors of young and mature *Abies amabilis* stands in Western Washington, as affected by fine-root inputs. Ecol. Monogr. 52:139–157.

Von Wandruszka, R. 1998. The micellar model of humic acid: Evidence from pyrene fluorescence measurements. Soil Sci. 163: 921–930.

Waters, A. G., and J. M. Oades. 1991. Organic matter in water-stable aggregates. Pp. 163–174. *In* W. S. Wilson (ed.), Advances in Soil Organic Matter Research: The Impact on Agriculture and the Environment. The Royal Society of Chemistry, Thomas Graham House. Cambridge, England.

Young, J. L., and G. Spycher. 1979. Water-dispersible soil organic mineral particles: I. Carbon and nitrogen distribution. Soil Sci. Soc. Am. J. 43:324–328.

The Soil Ecosystem: Biological Participants

Except for an occasional insect or earthworm, once visible traces of plant biomass are removed, any mineral soil appears to be a lifeless mass. Yet, even desert soils are teaming with microscopic lifeforms. The seemingly lifeless mass of clay, sand, and silt, which easily slips between our fingers, is home for a complex microbial community, including bacteria, fungi, protozoa, and viruses. This intricate association of microbial cells and soil particulates is critical for total ecosystem sustenance. The integrity of the total ecosystem, above- and belowground, rests on the stability, resilience, and function of the soil microbial community. Destruction of the soil microbiota through mismanagement or environmental pollution results in decline or even death of the aboveground plant and animal populations. Thus, development of an understanding of soil microbes, their properties, and the nature of the interactions with and within their environment is essential.

Clearly, study of biological processes in soil presents a unique challenge. The properties of the mineral components, the heterogeneous nature of their assembly into soil aggregates, and the minute size of the living organisms minimizes the potential for measurement of microbial processes directly in the field. Thus, to evaluate native soil processes, representative soil samples must be collected and handled in a manner that preserves the integrity of the biological community and maintains its function at a level representative of *in situ* rates (i.e., levels of activities previously existing in the field). Hence, this chapter is presented with the objective of elucidating the nature of the resident living components of soil and of determining their properties in relationship to total ecosystem function. This evaluation of soil life forms will include a presentation of some commonly used analytical procedures available for quantifying soil microbial populations and the consequences of their use.

2.1   The Living Soil Component

Microorganisms constitute less than 0.5 percent (w/w) of the soil mass, yet, they have a major impact on soil properties and processes. Aboveground manifestations of an ecosystem could not develop without extensive outside intervention were the soil microbial biomass to be destroyed. For example, all macro- and micro-nutrients that are essential for plant growth would have to be supplied from exogenous sources. Appreciation of this fact has historically stimulated interest in soil microbes from three viewpoints: (1) identification of the organisms present and elucidation of their relationship with other members of the soil community — that is, basic ecological studies; (2) assessment of the activity of microbes participating in processes of agricultural importance, such as nitrogen fixation, nitrogen mineralization, or pesticide decomposition; and (3) selection of organisms capable of producing anthropogenically important products, for example, antibiotics. More recently, with the observation that human activities have caused a decline in the quality of soil resources, considerable effort has been expended in the study of soil microbial processes associated with maintaining ecosystem stability or renovation of damaged sites.

Since soil bacterial and fungal communities are prime participants in each of these areas, the most commonly studied soil living fraction has been these organisms. Thus, much of what is described in this chapter relates to these primary decomposers; yet, it must be remembered that interactions with protozoa, nematodes and other higher life forms helps to maintain the vigor and productivity of the bacterial and fungal communities. (See Chapter 6 for a more detailed discussions of these interactions.)

Were the transformations of mineral and organic components occurring in soil strictly the product of chemical reactions, soil system processes would be both predictable and limited in extent. Knowledge of the conditions existent at any specific point in time, the quantities of the reactants and products currently present, and the properties of the reactants generally allows prediction of the rate and extent of a chemically driven reaction. This situation is contrasted with systems driven by biological processes, wherein expressed and uninduced metabolic capability contained within the genome of living and resting cells and the capacity for rapid cellular replication provide for potential development of biological processes far greater than those predicted by analysis of existent chemical and biochemical reactants and their products.

2.1.1   Biological and Genetic Implications of Occurrence of Living Cells in Soil

To gain a full appreciation of the consequences of biological activities in soil, several unique properties of the living component of soil must be examined in greater detail. Properties of the microbial community that confer special properties to soil processes are that (1) living organisms possess a diverse metabolic capacity encoded in the microbial genome that is not necessarily expressed phenotypically, (2) they replicate, (3) they have a definite cellular structure, (4) they form resting structures, (5) they catalyze reactions that change the soil, and (6) they are a link between the aboveground and belowground processes.

2.1.1.1 Gene Pool Potential

A unique property of the living component of soil is that the biochemical potential contained therein is significantly greater than that expressed at any particular point in time. Many of the genes encoded in the genome of both active and inactive soil microbes are not expressed under the existing environmental conditions. Furthermore, many microorganisms are quiescent; that is, they are present in a resting state. Thus, essentially none of the products of the genes contained in these inactive microbes are produced. Therefore, it is concluded that two major pools of potential metabolic activity found in soil are (1) active microbial cells with noninduced enzymatic capability, and (2) cells present in an inactive or resting state or present with insufficient population densities to produce measurable effects on the expressed or measurable levels of metabolic activity. In either situation, genotypic traits exist within the soil microbial biomass that: when they are induced, could result in a biological community with entirely different properties than currently exist.

Induction of Metabolic Activities in Soils: Synthesis of requisite enzymes for previously unexpressed metabolic activities is induced primarily by changes in the physical or chemical properties of the soil. Physical modification of the soil properties can be exemplified by the development of anoxic conditions. Imposition of an oxygen-free status commonly occurs as the result of flooding of a soil where the excess soil moisture is retained under reasonably static conditions. The lack of water movement within and overlying the soil limits the influx of molecular oxygen due to mass movement. The only source of molecular oxygen becomes diffusion from the atmosphere through the water into soil pores—an extremely slow process. This situation coupled with an actively respiring aerobic microbial community results in depletion of soil-resident free oxygen. Prior to exhaustion of the free oxygen supplies, aerobic metabolism predominates. Without a continued oxygen supply, the aerobes must enter into a resting stage or expire, when the anaerobes become active. Hence, enzymes associated with anaerobic respiration are induced. Similarly, chemical stimulation of enzyme synthesis may be exemplified by the reaction of the cellular enzyme synthesis mechanisms to soil amendment—naturally through invasion by plant or animal tissues and their products or to anthropogenic intervention as exemplified by sludge or compost amendment or entry of xenobiotic compounds, such as pesticides, into the soil site.

A metabolic activity stimulated by the development of anoxic conditions is denitrification. An examination of the kinetics of induction of this activity can serve as a model of changes in other biologically catalyzed processes by this type of extreme modification of life-controlling soil properties. Nitrogen oxides are converted to nitrous oxide and dinitrogen (i.e., denitrified) at a low level in most soils. Following a soil-pore–saturating rainfall, should all other requirements for denitrification be available, nitrogen oxides are reduced to dinitrogen and/or nitrous oxide at rates several orders of magnitude greater than occurred prior to flooding of the soil. This stimulation of denitrification enzyme activity results from induction of denitrification enzymes under the oxygen-limited conditions imposed by water saturation of a soil. Smith and Tiedje (1979) found that upon oxygen depletion, two distinct denitrification rates were observed: phase 1, where nitrogen oxide reduction proceeded at rates supported by the quantities of previously existing denitrification enzymes, and phase 2, attained after 4 to 8 hours of anoxic conditions and attributed to full derepression of denitrifying enzyme activity in the absence of significant microbial growth. Should conditions conducive for denitrification persist, eventually the soil denitrification capacity would be enhanced further by increases in population densities of the responsible bacteria. A variety of studies have shown field responses of denitrification activity reflective of these laboratory observations (Christensen et al, 1991a; Flühler et al., 1976; Focht, 1974; Parsons et al., 1991; Sexton et al., 1985).

Examples of chemical modification — that is, provision of metabolic inducers — can be associated with pesticide use (as indicated above), amendment of soil with sulfur, and any other situation where energy-yielding substrates are introduced into the ecosystem. As with denitrification, a lag in enzyme activity of a few hours following soil amendment is observed before a rapid increase in the enzyme activity is detected.

In summary, enzyme induction provides a mechanism for a rapid, short-term response of the microbial community to ecosystem perturbation. New enzymatic activity can be expressed in existing populations in as short a timeframe as 30 minutes to a few hours. After removal of the stimulus, enzymatic activities could return to preinduced levels as quickly as they were induced. The longevity of any soil enzymatic activity depends upon the stability of the proteins involved, their synthesis rate, and the rate they are decomposed by cellular or extra-cellular proteolytic activity. Therefore, enzyme induction could be viewed as a rapid biochemical response for maintenance of ecosystem homeostasis.

Cellular Replication and Soil Properties: The conditions predicating the importance of cellular proliferation in the dynamics of metabolic function in any soil follow:

1. Microbial biomass in soil is a mixture of actively growing and resting cells.

2. A wide range of population densities exist in soil. Populations of individual species constituting the microbial community vary over several orders of magnitude — from as few as 100 or fewer individuals per gram of soil to perhaps a million or more organisms per gram.

3. The relative sizes of various population densities are determined by their ability to compete with companion organisms.

4. At least the minimal chemical, physical, and biological requirements of the individual microbial species must be met before growth and replication can occur.

5. The chemical, physical, and biological properties of any soil site are in a state of flux. As these properties change or are altered, the activities and nature of the microbial community are also modified.

In contrast to the relatively rapid response associated with induction of degradative pathways discussed above, cellular proliferation is a slow adaptation to ecosystem modification, but its impact on the ecosystem may be much more lasting. Enzymes or enzyme systems may be synthesized at rates that result in detectable levels of activity in a matter of hours. In contrast, changes in a metabolic rate occurring as a result of cellular proliferation may require several days or months to become detectable. The time limitations involved with cellular replication are shown when a bacterially catalyzed process is considered. These cells reproduce by binary fission, with growth rates in soil of from less than one to about two or three cell divisions per day. Thus, since production of populations of several thousand cells may be necessary to yield a detectable change in metabolic activity, several days may be required for development of a sufficient cell mass to have a measurable impact on the soil community. Indeed, it may take several weeks or months of incubation in the presence of the growth-inducing substance before population densities of a few hundred individuals develop in situations where only a few cells with the requisite metabolic capacity existed per g of soil prior to soil amendment. Once the microbial populations have developed, they may survive as resting cells or spores for decades or longer, even though the conditions causing their development no longer exist. Thus, a large population of inactive or resting cells may persist in a soil in which they are no longer capable of growth. Long-term survival is more commonly the exception (i.e., most nonfunctional, non-spore–forming biomass will be decomposed in periods of a few months to a year), but even this level of survival of a phenotype in soil exceeds that resulting from enzyme induction and subsequent repression of the activity.

Relationship of Enzyme Induction and Cellular Proliferation on the Concept of Steady State Conditions: The potential for replication of microbial cells and induction of nonexpressed activities requires a consideration of the concept of steady-state conditions in soil. For most scientists, an understanding of steady-state conditions is based on experience with chemical reactions. In the hypothetical reaction:

$$A + B \rightleftharpoons C + D$$

an equilibrium state, or steady state situation, is reached, which is dependent upon the reaction constants for the forward and reverse reaction and the concentrations of the reactants and the products; that is, the forward and back reactions are in equilibrium.

Within the reasonably defined situation of the test tube, predictable levels of all chemical species can be reached. Such is not the situation in a soil ecosystem. The concentrations of reactants, products, and levels of the enzymes involved are highly variable. Parameters affecting reaction conditions change with time, temperature, and certainly season. Therefore, a biological concept of steady state applicable to soil biological populations must be developed.

One option would be to define steady state by the growth status of the microbial populations (i.e., population densities are unchanging). In axenic culture, this situation could be represented by stationary growth phase (see Chapter 4), but this condition is never really achieved in soil by actively respiring microbes. Growth of microbial cells is encouraged by consumption of microbial biomass by predator populations, by augmentation of availability of carbon and energy supplies induced by diffusion of fixed carbon substances to the microbial cells, by mass transfer of nutrient-laden water through the soil pores, or through root growth and death associated with plant community growth and decline, as well as by variation in soil physical properties, such as temperature and moisture. Furthermore, cell longevity may be reduced by the stress of coping with the harsh conditions of the environment. Thus, the microbial biomass could be envisioned as being in a constant state of flux.

Since both the microbial populations and their environment are steadily, albeit at times very slowly, changing, a chronological dimension to the definition of steady state must be applied. The fact that a variety of soil properties that control microbial population densities change regularly on a daily and seasonal level suggests that steady-state conditions may be determined by consideration of changes in mean population densities with time. For example, temperature cycles daily as well as seasonally. Nutrient levels available to microbial populations frequently relate to plant growing season since photosynthetically fixed carbon is the primary carbon and energy source for most soil microbes. To accommodate this regular cycling or fluctuation of soil properties controlling microbial activity, a steady state could be said to exist in a soil ecosystem when the cycles of microbial activity or proliferation are replicated on a regular (time) basis (e.g., seasonal, plant growth status, or annually).

Based on even this periodic variation of soil microbial properties and activity, a question could be raised by the purist regarding whether steady-state conditions really ever occur in soils. In reality, a true steady-state condition is rarely, or perhaps never, achieved in the highly variable soil environment, but many of our concepts and models of soil systems are based on an assumption of steady state. Perhaps a more scientifically defensible expression of the status of the soil microbial community would be to modify our concept of steady state by designation of the occurrence of a *quasi* steady state rather than an absolute steady-state condition, for truly, a constant level of biological activity is never reached in any soil system, even on an annual basis. It would be rare to observe a field situation where exactly the same physical and chemical conditions impinging on microbial activity occur repeatedly year after year.

Similar considerations must be taken when evaluating the microbial activity at the field level at any single point in time. Again, as indicated in Chapter 1, the physical properties are extremely variable across a field surface, dependent upon such ecosystem-related properties as the distribution and nature of the plant community. Variation in microbial activity becomes even more acute as the level of the microsite is approached. This heterogeneity in levels of activity of a microbial process in soil has been clearly demonstrated (Christensen et al., 1991b; Parkin, 1987; Parkin et al., 1987). Denitrification requires anoxic conditions plus metabolizable carbon sources and nitrate to occur. These necessities vary spatially and chronologically in soil. Parkin (1987) found "hot spots" of high specific denitrification in soil, which were linked to microsites with elevated particulate organic carbon. In a soil core where a high denitrification rate was measured, 85 percent of the denitrification capacity of a 98-g sample was found in an 0.08-g subsample of the core. Thus, it is easy to postulate that depending upon the distribution of the microbial colonies and the substrates necessary for their function, activity could vary from essentially zero (no microbial cells present) to extremely high levels (large colonies in the vicinity of active plant roots). Again, the mean value of the activity in question over the field of study could be defined as the steady-state level of activity for the duration of the study. The "hot" or "cold" spots of activity may vary in intensity and location with time, but the overall mean of metabolic activity in the field could remain reasonably constant.

In conclusion, a clear picture of the nature of steady state in a soil ecosystem could be likened to the daily conditions of any large city. When taken as a whole, the activity of the population appears to differ little on a daily basis (i.e., quasi steady state); cars and busses are streaming into the city and the sidewalks are clogged with people rushing to and from their offices. On any given weekday, the numbers of individuals associated with any given location appears to be reasonably constant. Yet, at the vantage point of a street corner, the situation appears to be nearly chaotic. Such is the activity of field soil as exemplified by the reactions occurring along a transect of any field site. Biological activities may be extremely variable along the transect (varying over several orders of magnitude), but an average value representative of the overall soil ecosystem along the transect can be calculated. Furthermore, the activity may be highly variable during the growing season, but again, a value representative of the process rates for the season may be estimated. In our model of the city, this variability could be likened to that involved in comparing weekdays with weekends. Operationally the highly variable soil systems could be defined as being at a quasi steady-state level. An apparently chaotic system can be reduced to some semblance of reproducible order when integrated along a long transect or over the time frame of a growing or annual season. It is this potential for predictability and representation that makes possible the description of an ecosystem whose whole existence is based on processes occurring in highly variable microsites.

2.1.1.2 Cell Structure and Biochemical Stability in Soil

Amendment of soil with essentially any biodecomposable substrate, assuming conditions are appropriate for microbial metabolism to occur, results in a reasonably rapid depletion of the added material. In apparent contradiction to that observation is that fact that a wide variety of decomposable organic compounds contained in living cells are readily isolated from soil. That is, an apparent stability of labile substances is observed. These contrasting observations result from the definite cellular structure of microbial biomass. Readily decomposable organic compounds are protected within the cellular structure. Hence, longevity of the biochemicals is not determined totally by the capacity of the microbial community to synthesize the enzymes necessary for their catabolism, but rather, decomposition kinetics are determined, at least in part, by the capacity of the microbial biomass to breach the protective cellular barriers. In situations where the carbonaceous substrates are encased within protective cell walls, microbial populations may be starving in a soil containing large reserves of carbon and energy supplies that can support microbial cellular growth but are locked within protective cell walls.

2.1.1.3 Resting Structures and Soil Respiration

Another unique property of the soil biological system is that the cells may enter a metabolic state that allows them to survive under adverse conditions. Two levels of nonreplicating state of viable cells exist in soil. In the absence of appropriate energy supplies, microbes may enter a quiescent or reduced metabolic state or they may actually form resistant structures (e.g., spores, endospores, or sclerotia).

Whereas the impact of true resting structures, such as spores, on soil is minimal until they outgrow, forming active vegetative cells, those cells in a resting (low-activity) state have a continuous but low-level effect on the conditions of the microsite. These latter microbes must continue to consume energy to maintain their cellular structure. Thus, by oxidizing internal energy sources, they continue to consume free oxygen (respire) and produce carbon dioxide. At least a minimal localized effect of the carbon dioxide generated from respiration associated with cellular maintenance would be expected to occur. Generation of this gas alters the atmosphere of the microsite where the resting microbial cell resides and increases the acidity of the interstitial soil water. For example, nitrifiers in the absence of their energy source, ammonium or nitrite, or any heterotroph existing in the absence of fixed carbon substrates may enter this situation where the only metabolic processes occurring in the cell are those associated with maintaining the cell structure. This form of existence could be pictured as a type of starvation where ultimate survival depends upon an eventual influx of an energy-supplying nutrient. Interestingly, microbes may exist for extended periods of time (decades to centuries for

Arthrobacter sp.) under these conditions. For these microbes, cellular metabolic rates are reduced to those levels necessary to maintain a minimal cell structure (i.e., the level of enzyme-synthesis machinery and metabolic tools necessary to synthesize the enzymes and cellular products necessary to take advantage of proper growth conditions when they develop). Considering the chronological as well as physical heterogeneity of organic carbon distribution within soil, it is logical to conclude that the majority of soil microbes are found either in this reduced metabolic state or in resting structures.

2.1.1.4 Soil Mineral Transformation by Microbial Cells

Microbes are also unique in soil in that they can alter the solubility of soil mineral components, reduce organic compounds to essentially undetectable levels, modify soil structure, oxidize inorganic compounds, and use a variety of soil components as electron acceptors; that is, merely through growth and metabolism, soil microbes alter their environment. For example, soil structure is enhanced by production of polysaccharides, which may link soil particles into macroaggregates (Chapter 1). In contrast, soil structure may be lost through oxidation of the colloidal organic matter supporting soil aggregate structure by the same microbes. A less obvious but nonetheless important aspect of the impact of microbial metabolism on soil particulates is the dissolution of soil minerals resulting from organic or inorganic acid production. A variety of organic acids, including acetic acid, citric acid, and lactic acid, carbonic acid (carbon dioxide), and mineral acids (sulfuric acid and nitric acid), are produced by the soil microbial biomass. These acids are direct contributors to the weathering of soil minerals.

Soil organic and inorganic components may also be oxidized or reduced as energy sources or electron acceptors for the microbial community. All nonphotosynthetic microorganisms must oxidize growth substrates for energy. Common reactions are the oxidation of ammonium to nitrite or nitrate, organic carbon oxidation to carbon dioxide, and elemental sulfur conversion to sulfuric acid (see Chapter 4).

The impact of these oxidative processes on soil is magnified by the nature of the final acceptor of the electrons produced by oxidation of the growth-supporting substances. A basic principle of biochemistry is that all oxidative reactions must be balanced by reductive processes. Thus, some environmental substances must be reduced as a result of microbial growth. Denitrifiers oxidize organic carbon while reducing nitrate to nitrous oxide and dinitrogen. Aerobic catabolism of organic carbon most commonly involves reduction of oxygen to water. Furthermore, carbon oxidation may result also in the reduction of sulfur oxides to sulfide or elemental sulfur.

Practical implications of this basic biological principle become apparent when it is considered that both oxidative and reductive processes have major implications in reclamation of contaminated soils. Whereas aerobic, heterotrophic metabolism is used to purify soils contaminated by organic carbon-based compounds, reductive processes may be exploited for renovation of metal-contaminated and acid-impacted sites. For example, sulfate reduction to hydrogen sulfide may be facilitated in renovation of acid mine drainage-impacted sites (Mills, 1985). Similarly, hydrogen sulfide production may be used to encourage removal of heavy metals from waters or immobilize these metals in soil systems through precipitation as metal sulfides (see Wildung and Garland, 1985). Frequently, a wetland-type system is developed to accomplish these tasks. Passage of metal-bearing water via overland flow through swampy ecosystems where hydrogen sulfide is generated results in precipitation of the metal sulfides.

2.1.1.5 Microbial Link to Aboveground Communities

It is relatively easy to forget that the processes occurring in soil cannot occur at maximum rates in the absence of fixed carbon inputs from the aboveground plant community. Some microbial activity is supported by catabolism of colloidal soil organic matter and more slowly decomposed biological components (complex polysaccharides, lignin, and lignin products), but maximal biomass productivity is derived from catabolism of more easily oxidized substrates, such as simple polysaccharides and proteins. Since the primary, natural source of such compounds in soil is plant biomass, a nearly essential dependency develops between the autotrophic aboveground community and soil microscopic life.

Similarly, the aboveground community is dependent upon the decomposer population in soil to mineralize macro- and micronutrients contained within dead plant biomass. For example, the organic nitrogen in the plant biomass must be converted to ammonium and nitrate before it can again be incorporated into new plant biomass. The facilitators of this cyclic process are the soil microbial community. (See Chapter 12 for further discussion of these processes.)

2.1.2 Implications on Microbial Properties of Handling of Soil Samples

An appreciation of the unique properties of living organisms in soil is mandatory for study and characterization of the versatile and rather fragile material called soil. Soil samples to be used to characterize native processes must not be treated in a manner destructive to the living systems contained therein. Of first consideration when collecting soil samples is the necessity of remembering that the prime mover in the system is biological. Thus, the sample should be maintained so that biotic components are preserved and induction of major changes in their composition is avoided.

A primary example of a commonly used procedure in soil sample collection that necessarily results in alteration of soil biological activity is provided by an evaluation of the problems encountered by air-drying soil samples. Historically, soil scientists have preserved soil samples by drying them. With this procedure, moist soil samples are incubated in the laboratory under conditions that allow soil moisture to approach ambient atmospheric levels. The final level of moisture retained in the soil may be considerably below that normally occurring under field conditions, in that periodic influxes of water from rainfall or dew formation are precluded. This procedure is appropriate for preservation of many chemical properties of soil, but it is disastrous for evaluation of most soil microbial processes.

Air-drying of soil creates a new ecosystem that may be quite different from that existing in the field from which the soil was collected. For soil microbes to function there must be a coating of water on the soil particles (see Chapter 5). When soils are dried, sensitive microbes may die or be induced to enter into a resting stage. The fact that major changes in soil properties occur during soil desiccation is suggested by the observation that a burst of microbial respiration occurs when air-dried soils are remoistened. This increased respiration results from the oxidation of organic carbon liberated from soil aggregates disrupted by soil drying and from catabolism of the dead microbial biomass. This results in changes in the levels and nature of soil enzymatic activities, modification of soil organic components, and changes in the composition of the microbial community. It must be noted that it was stated above that the ecosystem "*may* be different from that existing in the field." This situation may occur naturally in dry seasons or periods of drought.

Thus, modification of the physical condition of soil can have a dramatic effect on the nature of the active microbial populations. This situation is particularly acute if these are the microbial communities of interest. For example, anaerobic organisms are killed by contact with molecular oxygen. A gradient of sensitivity of such populations occurs (see chapter 3), but in studies where quantification of such populations or their activities is desired (especially with anoxic soil samples such as swampy material), oxygenation of the soil must be avoided. Modification of the aeration status of soils from that existing *in situ* under the oxygen-free or oxygen-limited conditions may also cause increased population densities of aerobic microbes. A pool of partially decomposed fixed carbon substrates accumulates under oxygen-limited conditions. These substances provide carbon and energy for the aerobic microbes once free oxygen is introduced into the system, thereby resulting in an augmentation of their population densities.

2.2 Measurement of Soil Microbial Biomass

Soil microbial biomass is a primary catalyst of biogeochemical processes as well as an energy and nutrient reservoir. For example, its significance is exemplified not only by the infinite array of biochemical transformations catalyzed in soil but also by the quantities of fixed nitrogen it contains. Anderson and Domsch (1980) found that in 26 agricultural soils, total soil nitrogen contained in the microbial biomass ranged from 0.5 to 15.3 percent, with an average of approximately 5 percent. This nitrogen becomes available to the aboveground community upon death and decay of the microbial cells. (See Chapter 10 for further discussion of this cycling of microbial nitrogen.) In soils not receiving exogenously supplied fixed nitrogen (for example, nonagriculture soils), this nitrogen pool is sufficiently large that its stability is a prime factor in controlling the flux of fixed nitrogen through the ecosystem. In disturbed soils, such as when grassland soils are initially cultivated, a decline in soil microbial biomass results in a release of large quantities of fixed nitrogen. The fixed nitrogen not incorporated into newly synthesized microbial cells or aboveground biomass is lost from the ecosystem through leaching to groundwater or through runoff, thereby creating a potential for nitrogen pollution of surface and groundwater. Hence, it is frequently useful to estimate the size of the soil microbial biomass and its stability.

Microbial biomass is readily assessed in simple ecosystems, such as the growth flask, but its measurement in soil is complicated by complexity of the system and the fact that these cells constitute a small portion of the total soil mass. One could propose to estimate soil microbial biomass by counting the number of individual living units, for example, cells for bacteria or mass of mycelium for fungi, in a soil sample. Unfortunately, it is essentially impossible to isolate all of the microbial cells from soil. Thus, a surrogate indicator of the quantity of living cells present in a soil sample is needed. This substance should be a reasonably easily quantified cellular component that can be extracted from soil. An ideal indicator of microbial biomass should be

- Present in all cells
- Occur in all microbial cells at the same concentration, regardless of species designation
- Be present in the cells at the same concentration, independent of growth status
- Be rapidly decomposed upon death of the cell
- Be quantitatively extractable from soil
- Be easily assayed

The latter trait is necessitated by the large number of samples that are generally processed and by the fact that the biomass may change during storage of the samples. A rapid, easily conducted procedure can allow avoidance of long storage times between sample collection and analysis. A variety of candidate substances for estimating soil microbial biomass have been used, but none meets all of the above-listed criteria as an ideal indicator.

Soil microbial biomass has been commonly estimated through

- Direct counting of the microbes (Sonderstrom, 1977; Paul and Johnson, 1977; Rosser, 1980; and Lundgren, 1981)
- Analysis of specific cellular components, such as adenosine triphosphate (ATP), phospholipids, or muramic acid (Ausmus, 1971, Findlay et al., 1989; King and White, 1977; Paul and Johnson, 1977; Fazio et al., 1979; Jenkinson et al., 1979; Verstraete et al., 1983)

- Measurement of specific microbial processes, such as nitrogen mineralization (Alef et al., 1988), or reduction of dimethylsulfoxide to dimethylsulfide (Alef and Kleiner, 1989)
- Measurement of respiration rates (Anderson and Domsch, 1973)
- Direct analysis of cellular components solubilized by chloroform fumigation (Brookes et al., 1982; Vance et al., 1987a; Tate et al., 1988; Sparling and West, 1988) or the carbon dioxide produced by respiration of these products (chloroform fumigation-incubation method) (Jenkinson, 1976; Jenkinson and Powlson, 1980)

Of these, the most commonly encountered procedures are direct counts, ATP analysis, respiration methods, and variations on the chloroform fumigation procedure. As indicated below, these procedures are commonly used singly, but examples of studies are published where comparable results are noted using the methods in parallel on the same samples (e.g., Fritze et al., 1996). It must be noted that comparable results should not always be anticipated since the various procedures measure different parameters of the microbial biomass and therefore should not be anticipated to vary in concert. For example, the aerobic respiration methods commonly assess glucose metabolism, whereas the chloroform fumigation relates to the quantities of cell carbon released by the chloroform and subsequently mineralized by the surviving microbial community. Thus, these will be examined in greater detail to show the strengths and problems associated with soil microbial biomass assessment. (A detailed presentation of the methods can be found in Parkinson and Paul, 1982.)

ATP Measure of Soil Microbial Biomass: For these estimates, soil is homogenized in a buffer solution to extract the ATP (Fig. 2.1). Generally, an acidic buffer is used to minimize solubilization of soil humic substances, which interfere with measurement of the ATP. The extractant is removed from the soil suspension by centrifugation or filtration, and the ATP is quantified by measuring light production with the firefly luciferin-luciferase system. Soil microbial biomass carbon is calculated by multiplying the measured ATP concentrations by a constant that is proportional to the amount of ATP contained per unit of microbial biomass as determined using a standard bacterial cell culture cultivated in laboratory media. The procedure is dependent upon the extraction efficiency of ATP from the soil sample and selection of an appropriate constant for final calculations.

An assumption underlying this procedure is that there is negligible variation in ATP contents of microbial cells. The validity of this assumption is questionable since cellular ATP contents are highly dependent upon the metabolic status of the cell. Actively metabolizing microbial cells contain considerably more ATP than is found in resting cells. Recall that a major portion of the bacterial cells in surface soils exist in a resting, or inactive stage. Furthermore the proportion of the cells that are active is unknown, but it is reasonable to assume that this value varies between soils collected from

SOIL EXTRACTION

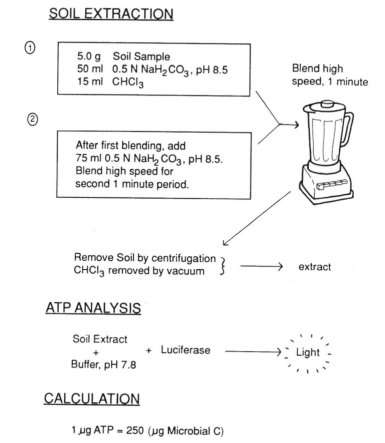

ⓘ

5.0 g   Soil Sample
50 ml   0.5 N NaH$_2$CO$_3$, pH 8.5
15 ml   CHCl$_3$

Blend high
speed, 1 minute

②

After first blending, add
75 ml 0.5 N NaH$_2$CO$_3$, pH 8.5.
Blend high speed for
second 1 minute period.

Remove Soil by centrifugation
CHCl$_3$ removed by vacuum } ⟶ extract

ATP ANALYSIS

Soil Extract
     +   + Luciferase ⟶ Light
Buffer, pH 7.8

CALCULATION

1 μg ATP = 250 (μg Microbial C)

Fig. 2.1   Outline of a typical ATP analysis procedure for estimating soil microbial biomass (see Parkinson and Paul, 1982 for further details).

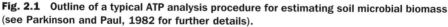

different ecosystem types and within soils from the same ecosystem, depending on such soil properties as energy inputs and moisture levels. (For further information on the use of ATP for microbial biomass measurements, see Ausmus, 1971; Jenkinson et al., 1979; Ahmed and Oades, 1984; Paul and Johnson, 1977; and Verstraete et al., 1983; Verstraeten et al., 1983.)

Soil Aerobic Respiration Measurements: The rate at which fixed carbon substrates are oxidized to carbon dioxide in a soil sample is proportional to the quantities of organisms mediating the reaction (Anderson and Domsch, 1978) (Fig. 2.2). With this procedure, soil samples must be collected and any plant roots or macroscopic biomass removed to negate carbon dioxide contributions from these sources. With the soil respiration procedure, an easily metabolizable carbon source, such as glucose, is added to the soil sample and carbon dioxide

1. Determination of Optimal Glucose Concentration

Select lowest glucose concentration producing optimal CO_2 flux

2. Assessment of Respiration Rates of Test Samples

3. Calculation

Microbial Biomass = 40.04 (CO_2 flux) + 0.37

Fig. 2.2 **Outline of aerobic respiration method for estimating soil microbial biomass (see Parkinson and Paul, 1982 for further details).**

evolution measured with infrared analysis, through gas chromatography, or, should sufficiently large quantities of carbon dioxide be produced, by titration of the gas collected in alkaline solutions. Alternatively, ^{14}C-labeled substrates, (e.g., ^{14}C-glucose) may be added to the soil sample and the ^{14}C-labeled carbon dioxide produced collected and quantified.

This procedure may be used to differentiate fungal and bacterial respiration in soil by amending the samples with population-specific antibiotics (Anderson and Domsch, 1973; Anderson and Domsch, 1975). Streptomycin is generally used to inhibit bacterial respiration and cycloheximide for eucaryotic respiration. Each antibiotic should be used singly and in combination since a proportion of the microbial population is resistant to both antibiotics. Furthermore, several concentrations of antibiotic must be tested with each soil studied to assure use of optimal antibiotic levels. Inhibition of microbial activity is proportional to the quantity of antibiotic dissolved in the soil interstitial water. Antibiotics sorbed onto soil organic matter or clays is inactivated. Thus, the effective concentration of the antibiotic in soil may be significantly lower than the total quantity added. Higher antibiotic concentrations are required for inhibition of microbial populations in soils with high clay or colloidal organic matter contents than would be necessitated in a sandy soil, for example. Similarly, too much antibiotic must not be added to the soil since excessive antibiotic concentrations may inhibit the activity of nontarget microbes.

A major consideration in use of any technique involving measurement of carbon dioxide from soil is the soil pH (e.g., see Martens, 1987). Microbial biomass may be underestimated by failure to recover all of the carbon dioxide produced during the incubation period. In neutral or alkaline soils, carbon dioxide is retained in soil as bicarbonate or carbonate due to the following equilibrium:

$$CO_2 + H_2O \rightleftharpoons HCO_3^- + H^+ \rightleftharpoons CO_3^{2-} + 2H^+$$

Chloroform-Fumigation (Extraction or Incubation) Technique: Because of the general simplicity of this technique and the fact that it is relatively inexpensive, chloroform-fumigation procedures are the most commonly employed methods for estimation of soil microbial biomass. Thus, the principles, applicability, and problems associated with the technique must be examined in detail.

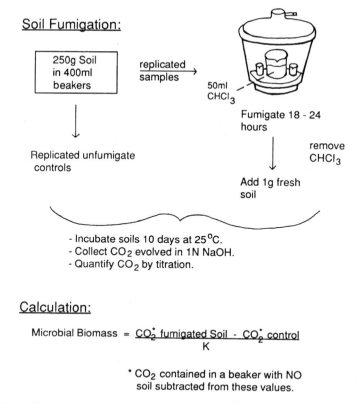

Fig. 2.3 Outline of the chloroform fumigation-incubation procedure for estimation of soil microbial biomass (see Parkinson and Paul, 1982 for further details).

Chloroform-fumigation procedures (Fig. 2.3) are based on (1) the disruption of cellular membranes by chloroform and (2) quantification of the carbon contained in the biomass of the cells killed by the chloroform vapors. Generally, chloroform vapors are used. Following fumigation, soils may be extracted directly and soluble carbon (e.g., Deluca, 1998; Sparling and West, 1988; Tate et al., 1988; Vance et al., 1987a), nitrogen compounds (e.g., Azam et al., 1988; Dahlin and Witter, 1998; Gunapala and Scow, 1998), or phosphate (Brookes et al., 1982; Brookes et al., 1984; Hedley and Stewart, 1982) measured (estimates of microbial biomass carbon, nitrogen, or phosphorus, respectively). Alternatively, for determination of microbial biomass carbon, the soils may be incubated to allow biological conversion of the dead microbial biomass to carbon dioxide to occur. This procedure is based upon assumption that the biomass carbon of the cells killed by the chloroform vapors is used by the surviving soil microbial biomass as carbon and energy sources. (Note that aerobic incubation conditions are required. Anaerobic conditions result in incomplete oxidation of the cellular carbon of the dead cells.) To assure existence of adequate microbial populations during the incubation period, the fumigated and control soil samples are usually inoculated by amendment of the chloroform-treated soil with small quantities of non-fumigated soil (see Chapman, 1987). For assessment of carbon dioxide production, soils are sealed in respiration chambers. Carbon dioxide is collected either in alkaline solutions and titrated or assayed with gas chromatographic procedures.

For each of the procedures — incubation or direct extraction methods — assay of nonfumigated soil serves as a control. The total biomass present is calculated by subtracting quantities of carbon dioxide produced in non-fumigated soil samples (controls) from that yielded in the fumigated soil samples and dividing by a constant. The constant is a value representative of the killing efficiency of the fumigation procedure. Most frequently, for assessment of microbial biomass carbon, killing efficiency is considered to be about 40 percent (i.e., a constant of 0.4). Different values for the constant are used when quantifying microbial biomass nitrogen (see Azam et al., 1988) and subsurface soils (e.g., Dictor et al., 1998; Tessier et al., 1998). Since the killing efficacy of the chloroform varies with such soil properties as moisture and type, it is imperative that the formula used to convert the raw data to biomass carbon be stated so that the data may be recalculated should it be necessary to compare them with values from other studies where alternative calculation procedures were used.

Some variation in incubation period of the soil samples following chloroform fumigation has been used. The treatment and control samples may be incubated concurrently for 10 days and total carbon dioxide yielded assayed, or the control sample could be incubated for 20 days with carbon dioxide evolved between days 10 and 20 being used as the correction factor in the calculations. In this situation, for the calculations, the carbon dioxide produced in control samples during the last 10 days of a 20-day incubation period is subtracted from that yielded in the fumigated samples during the first 10 days of incubation. This procedure avoids inclusion of the commonly observed burst of microbial activity in disturbed soils samples (as would be anticipated to occur with control soils) in the final calculation. Failure to compensate for this soil disturbance-induced carbon dioxide production may result in an underestimation of total soil microbial biomass.

Aside from providing a reasonable estimation of the quantities of cellular components (carbon, nitrogen, and phosphorus) contained in microbial biomass, chloroform fumigation of soils has proven to be valuable tool for demonstration of assimilation of soil amendments by the soil microbial community. For example, Kassim et al. (1982) quantified the incorporation of glucose, acetate, pyruvate, uracil, uridine, amino acids, and a variety of polysaccharides into microbial biomass. To accomplish this, ^{14}C-labeled substrates were added to soil samples and then the amended soil samples were incubated for varying time periods to allow their incorporation into various soil organic matter fractions, including microbial biomass. Following incubation, the nonreacted ^{14}C-labeled amendment was washed from the soil. For analysis of the microbial biomass-incorporated materials, the soil samples were fumigated with chloroform. Assessment of the excess ^{14}C-labeled carbon dioxide produced during the incubation period in the fumigated samples provided an estimation of the quantity of the metabolized carbon incorporated into microbial cellular components.

The chloroform fumigation procedure is an effective means for estimating microbial biomass in most soils, but difficulties have been encountered with application of the technique to acidic soils (Coûteaux et al., 1989; Vance et al., 1987a; Vance et al., 1987b), to wet soils (Ross, 1987; Ross, 1988), and in plant biomass-amended soils (Martens, 1985). A common trait among these situations is that the capacity of the microbial community to recover following fumigation and subsequent catabolism of the dead microbial cell substance has been affected. Large quantities of available carbon preexisting in the soil samples may result in minimal differences between quantities of carbon dioxide produced by control and fumigated soils (i.e., the carbon dioxide production from the killed biomass oxidation is not significantly greater than the large quantities produced from catabolism of native soil organic matter.). In extremely acidic soils, data suggest that the nature of the population (fungal vs. bacterial contributions to carbon metabolism) may differ between fumigated and control soils (Tate, 1991). In either case, modification of the incubation procedure and/or method of calculation of the microbial biomass may be necessary (Vance et al., 1987c).

Direct Estimation of Microbial Biomass: Soil bacterial and fungal biomass may be viewed directly in preparations of soil samples. The refraction index of microbial cells is not sufficiently different from that of some soil particulate organic components to allow clear differentiation. Thus, staining of the preparation is necessary. A variety of methods have been developed to enhance microscopic resolution of microbial cells in soil. Nonspecific staining [e.g. acridine orange (e.g., Mills and Bell, 1986); fluorescent staining of respiring bacteria (e.g., Rodriguez et al., 1992), ethidium bromide (Rosser, 1980)] may be used for general measurements of soil bacteria or fungi (e.g., see Fig. 2.4). Fluorescent-labeled antibodies are useful for quantifying individual species populations.

An advantage of direct-observation procedures is the capacity to detect and identify organisms at the species level and to differentiate active from inactive microbial populations. Metabolically active bacteria may be detected by treating soil samples with tetrazolium salts (accumulation of water-insoluble formazan accumulates in the cells), labeling cells with radioactive substrates

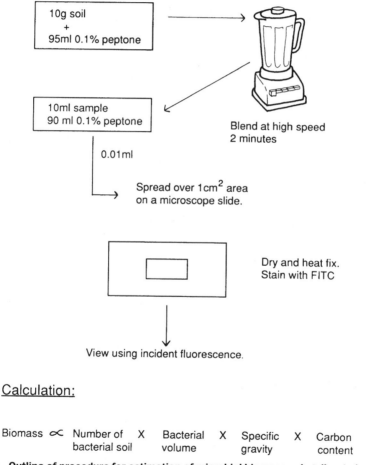

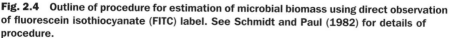

Calculation:

Biomass ∝ Number of   X   Bacterial   X   Specific   X   Carbon
bacterial soil   volume   gravity   content

Fig. 2.4 Outline of procedure for estimation of microbial biomass using direct observation of fluorescein isothiocyanate (FITC) label. See Schmidt and Paul (1982) for details of procedure.

(autoradiography) (e.g., see Roszak and Colwell, 1987) or by counting synthetically active bacteria. With the latter procedure, soil samples are incubated with naladixic acid, an inhibitor of DNA synthesis. The antibiotic amendment causes an elongation of metabolically active cells. The active cells can grow, but they cannot divide. These microbial cells may be detected in the treated soil preparations with acridine orange stain and epifluorescence microscopy.

Once the microbial cells have been quantified in several microscopic fields, the values are converted to biomass by multiplying the number of cells per unit of soil by their volume times the cell density. The calculation of the number of cells per unit of soil requires knowledge of the area of the microscope field and dilution of soil used in preparing the slide. Biovolume may be calculated from cellular dimensions estimated during their enumeration. Difficulties are encountered in converting the cell volume to cell mass (Bakken and Olsen, 1983; Bratbak, 1985; Bratbak and Dundas, 1984; Van Veen and Paul, 1979). Such conversion requires assumption of the impact of moisture on changes in cell dimension during sample preparation.

Once the counts of cells existent on the prepared slides are determined, along with the difficulty of conversion of biovolume to biomass, problems arise regarding the validity of extrapolation of data collected from the limited area of the microscope slide to field-scale dimensions. For these studies, small quantities of soil (perhaps less than a gram of soil) are suspended and a portion of the suspension is spread in a thin layer on a microscope slide, or alternatively, the liquid used to suspend the soil is extracted and the microbes contained therein collected on an appropriate filter medium. The slide or filter is dried, the microbial cells stained, and a portion of the preparation examined under the microscope. Since each of the procedures for estimation of microbial biomass requires the use of soil samples of limited size, some concern about extrapolation of the data to represent field or even ecosystem values is appropriate. Greatest care must be exercised with extrapolation of the direct count data since only very small soil samples can be studied practically.

Limitations of Microbial Biomass Measurements: A variety of inaccuracies are intrinsically associated with estimation of soil microbial biomass. (Note that in every case the word *estimate* not *measure* is used in conjunction with the value produced from the microbial biomass assessments.) In selecting procedures, determination of constants for calculation of results, and in interpreting the data, basic properties of the soil must be considered. These include the soil moisture level, pH, and organic matter contents referred to above as well as soil texture. Extraction efficiency of charged materials from soils with high clay contents may be reduced. Furthermore, high levels of colloidal soil organic matter may sorb products of interest or interfere with analysis (ATP-luciferin-luciferase assay).

It should also be noted that the calculation procedure for each of the procedures described above requires use of a constant. Selection of the control value assumes a killing efficiency for chloroform fumigation, concentration of ATP in cells of soil microbes, or a specific effect of soil drying on biovolume for direct counting procedures. All such assumptions have a leveling effect on the data, for they may be true for some samples or cells and not for others.

The above difficulties must be considered by the investigator in data analysis but are to a certain degree beyond experimental control. Other procedures that are within design control involve laboratory and field procedural variables. Nearly all procedures require some degree of sample storage prior to analysis. Once soil samples are collected, creation of a new ecosystem commences. Soil properties limiting microbial activity in the field site are removed (e.g., perhaps temperature, soil structural occlusion of organic matter, moisture variation) and new delimiters of microbial population densities and activities are created. Such changes must be minimized. The most reasonable means of reducing artifacts due to sample preparation procedures is to assay the microbial biomass as quickly after collection as possible (preferably within minutes to a few hours). Should this not be possible, storage at 4°C and avoidance of direct sunlight are mandatory. Some studies suggest that soil samples can be stored at $-20°C$ for extended periods without major changes in some measures of microbial biomass and activity (e.g., Stenberg et al., 1998). It is imperative that control studies be conducted to validate storage procedures since utility of a particular storage method may vary with soil or ecosystem source as well as with assay method. Good laboratory procedure requires documentation of all sample preparation procedures so that any compromising of the results can be determined.

Finally, the experimenter must ask the question, "What is being measured by the procedure used?" Some methods provide an estimation of the total microbial populations present (e.g., chloroform-fumigation procedures or ATP analysis), whereas other methods quantify only the microbes capable of a particular reaction. For example, utilization of glucose in soil aerobic respiration measurements provides a good estimate of the biomass of organisms capable of oxidizing glucose as an energy source. In the former situation (total biomass quantification), both active and inactive microbial populations are quantified whereas in the latter (soil aerobic respiration assessment), only those capable of catabolizing a particular biochemical process are enumerated.

2.3 The Nature of the Soil Inhabitants

All six major groups of microorganisms plus higher animals interact in soil to catalyze the biogeochemical processes occurring therein. Soil bacteria, fungi, actinomycetes, protozoa, algae, viruses, nematodes, and mite populations contribute to development of the total community. The arrangement of these classes of organisms may be evaluated in a hierarchical manner. This association provides the basis for fundamental viewpoints in examining native ecosystems. The bottom of this arrangement is the **individual** organism. The summation of the individuals is a **population**. The totality of all populations of different organisms in soil constitutes the **community**, whereas the interaction of all biological components with the abiotic portions of the environment constitutes the **ecosystem**.

Soil ecological research can be divided by the portion of this hierarchy stressed in the study. Such research may have an autecological or synecological basis. **An examination of the behavior of a single species in an ecosystem, including a study of the effect of chemical, biological, and physical aspects of the ecosystem on the population, is an autecological study.** In contrast, soil may be examined as a whole. In this situation, soil is being treated as if it were a single "tissue." Thus, **the relationships between the environment and community of organisms (synecology)** is being stressed. For synecological research, processes are of more concern than the identity of the individual microbes involved. For example, denitrification kinetics and its variation between soil types may be studied (synecology) or an experimenter may concentrate on the identity of the individual denitrifying species existent in a soil site (autecology).

2.4 Autecological Soil Microbiology

Quantification of microbial biomass is useful for expressing ecosystem potential (e.g., nutrient or energy cycling, or even general biological capacity), but most evaluations of soil community activity require elucidation of specific biological or biochemical processes and their kinetics. The processes themselves may be the center of the experiments (synecology) (see Section 2.5), or elucidation of the specific microbes and the system properties controlling their activity may be the pivotal point of the research (autecology). The latter aspect of ecosystem evaluation is the topic of this section.

Initial analysis of the basis for autecological research is a determination of what constitutes a bacterial species. An autecological study by strict definition involves assessment of the presence or activity of individual species. This is reasonably readily accomplished with higher organisms where species designations are more easily assayed, but the fluidity of bacterial species complicates such research for the soil microbiologist. This difficulty arises from the reasonably easy transfer of genetic material between bacteria (e.g., transduction, conjugation, transformation). The definition of a bacterial species resides in the genetic material contained therein and its expression as it affects cellular morphology, colonial structure, and metabolic capacity. Should the genetic material resulting in distinctive properties that define one of the bacterial species be readily transferred into another species, the distinction between the two groups of bacteria becomes questionable at best. Such problems with speciation of common soil bacteria are reasonably common.

Questionable or difficult separation of bacterial isolates into species has led to other more easily derived groupings of bacteria. Thus, autecological research may involve the study of bacterial groupings separated by phenotypic traits (functional or even colonial groupings; i.e., guilds) rather than by species designation (Mills and Bell, 1986). Alternatively, an autecological study, in principle, may involve analysis of rDNA similarity with an assumed percent similarity as the basis for species designation (see Chapter 3). Thus, the discussion of autecological research that follows is based on the expanded definition of autecology that allows for the study of guilds (or arbitrarily defined DNA similarities) as well as the more traditional species designations.

Limitations to Autecological Research: It must be noted that the compromise involved in delineation of bacterial species incorporates a limitation into the value of the data derived from such studies. A primary objective of any ecological study is not only to understand the particular system of study, but to develop principles applicable to a variety of situations. Taxonomic designation of the living organisms studied has provided a common link between research projects. The phrase "a lion is a lion no matter where it is encountered" can be readily accepted. To some degree, such a statement could be made for many of the historically studied bacterial species, but a generalization of this type regarding bacteria is questionable. Thus, use of the less restrictively derived guilds limits intersystem and interlaboratory comparisons.

A second difficulty associated with autecological research results from the minute dimensions of soil microbes. Autecological research provides meaningful data for understanding biological processes, but extrapolation of such research on an ecosystem basis is limited. Data are collected clearly describing the interactions of microbial colonies and their environment, but the area affected by an individual microbial colony in soil encompasses a few cubic microns at best. This inherently small size of the microsite affected by the individual microbe necessitates collection of large quantities of data before a clear picture of the effect of the totality of soil microbial colonies emerges (Tate, 1986). Development of automated microbiological procedures for species identification, expanded use of fluorescent antibodies (including monoclonal antibodies), and application of DNA homology analysis to detection of individual microbial species and quantifying their density in soils will result in an expansion of autecological soil microbiological research.

2.4.1 Viable Counts/Enrichment Cultures

Prior to the recent development of procedures for detecting nucleic acid sequences specific for selected microbes and their application to physically and chemically complex systems like soil, the initial step in any autecological study of soil involved growth of the microbes of interest in axenic culture. Indeed, this remains the initiation point — and in many cases the primary limitation — for most such research today. Exceptions to this observation are exemplified by studies in which the experimenter's interest resides simply with detecting selected microbes and estimating their population density. In that situation, indicators such as species-specific activities or DNA-DNA hybridization techniques eliminate the necessity of some cultural studies.

Microbes may be isolated from soil using nonselective techniques, such as growth on soil extract agar (perhaps this could better be termed *less selective* since growth in any laboratory medium allows growth of only a portion of the microbial population), elective growth, and enrichment cultures. With the latter two procedures, microbes with certain desired physiological properties are directly sought. With enrichment cultures, increases in microbes existing in low population densities are favored prior to isolation of the axenic cultures, whereas with the former, an extract of soil may be cultured directly on selective media.

The first concern in enumerating specific soil microbes, especially bacteria, results from their population density. Typically bacteria exist in soil at densities of 10^8 to 10^9 colony-forming units per g dry soil. These are associated with 10^7 to 10^9 actinomycete propagules per g dry soil and about 10^6 fungal and protozoan propagules each per g dry soil. Direct plating on a nonselective agar medium would clearly be unfruitful. Soil samples must be diluted to reduce colonial growth on agar plates to 20 to 300 colonies per plate (Fig. 2.5). Consideration of the dilution factor plus the quantity of diluent added to the growth medium allows calculation of viable microbes in the original soil sample. A result of this necessity to dilute soil samples is that the data from such studies are highly variable. A slight variation in dilution precision — especially at the lower dilutions — causes large differences in estimated propagule densities. Variations as large as plus or minus 100 percent are not uncommon.

With viable counts of microbes from any environmental sample, a qualification must be made regarding the meaning of *viable*. Those bacteria classified as viable for any specific experiment in reality only include the propagules capable of growth on the media used and only under the particular incubation conditions employed. Comparison of data from direct microscopic analysis of soil samples with results from viable plate counts reveals that in most cases only about 0.1 to 10 percent of the observed bacteria grow in culture.

A further complication associated with microbiological analyses of individual species in soil samples results from the variation in microbial growth rates. Typically, those organisms that form visible colonies during the most commonly used incubation times are those with the shortest generation times. Thus, many organisms that are important *in situ* are overgrown. This difficulty may be overcome by using selective media (for example, containing antibiotics to which the desired population is resistant but most soil bacteria or fungi are sensitive, or by using growth-selective substrates) or enrichment cultures, again based on an unique property of the microbe(s) of interest.

Viable plate counts are also used for enumeration of soil fungal populations. Estimates of fungal population densities in soil are confounded by their prolific spore production. Immense quantities of spores may exist in soil with few or none of the organisms existent as hyphae. Generally, fungal population densities are best estimated by direct microscopic observation.

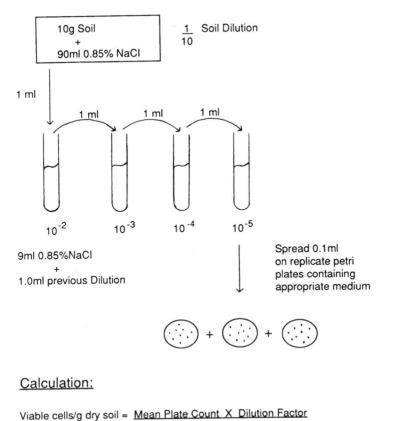

Calculation:

Viable cells/g dry soil = $\dfrac{\text{Mean Plate Count X Dilution Factor}}{\text{Dry weight of soil}}$

Fig. 2.5 Outline of viable plate count method for estimating numbers of organisms per gram of soil.

Most-Probable-Number Procedures: A variation of the viable count procedure is most-probable-number estimations of microbial populations. Many microbial populations (e.g., a variety of protozoa, nitrifiers and sulfur reducing bacteria) are difficult to grow in defined media. Also, when all potential contributors to a given metabolic function, perhaps nitrification, are to be quantified, it must be realized that some individual species contributing to the activity may have been grown in axenic culture, but the probability frequently exists that other nonculturable strains occur in soil. Most-probable-number procedures may be used to provide a measure of the density of such populations (Fig. 2.6).

Most-probable-number procedures are based on the determination of the dilution of the soil sample beyond which no propagule of the population to be quantified can be detected—that is, the extinction point. Therefore, with this

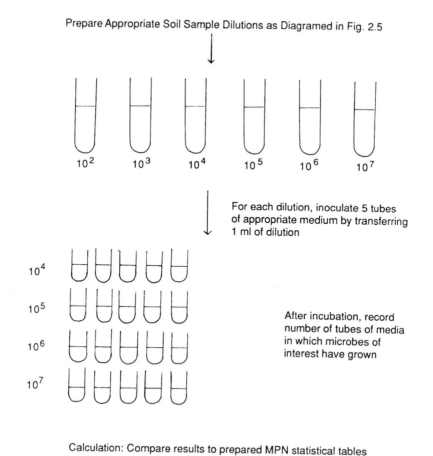

Prepare Appropriate Soil Sample Dilutions as Diagramed in Fig. 2.5

10^2 10^3 10^4 10^5 10^6 10^7

For each dilution, inoculate 5 tubes of appropriate medium by transferring 1 ml of dilution

10^4
10^5
10^6
10^7

After incubation, record number of tubes of media in which microbes of interest have grown

Calculation: Compare results to prepared MPN statistical tables

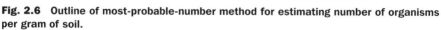

Fig. 2.6 Outline of most-probable-number method for estimating number of organisms per gram of soil.

method, soil must be diluted until no further propagules of the organisms of interest are present in the highest dilution prepared and multiple tubes of culture media (e.g., 5 per dilution) are inoculated from each dilution. After an appropriate incubation period, a trait common to all of the organisms to be quantified is measured (visual observation of protozoan cells, nitrate or nitrite production for nitrifiers, or black color produced by precipitation of iron sulfide, for the examples listed above). Population densities are determined using prepared tables relating to the statistical probability of the presence of the organisms of interest to the number of positive samples for each dilution (e.g., Alexander, 1982). This procedure retains the problems of selectivity of media and incubation conditions. Furthermore, due to the nature of the calculation procedure and the range of dilutions used, confidence limits are frequently as great as plus and minus 300 percent.

2.4.2 Intrinsic Limitations of Viable Count Procedures

Each step in any viable count procedure (soil sample collection and storage, preparation of dilutions, colony development) contains major impediments for production of an accurate population census. The most fundamental of these is the necessity of collecting site-representative soil samples. Once attained, the soil must be stored in a manner that precludes major changes in the population density. Fortunately, considering the intrinsic variation of the dilution procedure itself, some change in the microbial populations can occur before statistically significant effects are detected. That is, with reasonable sample storage procedures, changes in microbial population densities are usually less than the inherent variation of plate count or most-probable-number data. For most studies, storage of the soil samples for limited times at 4°C results in minimal changes in the data.

Two sources of data variation incurred during preparation of soil dilutions may be controlled by careful attention to proper laboratory procedures and experimental protocol (these are variability due to dilution procedure precision and the potential for microbial growth during dilution preparation). A third source of error is basic to the procedure—association of microbial cells with soil particle—and likely must be considered during data interpretation, though little can be done to prevent the problem. Association of microbial propagules with soil particulates may result in an underestimation of their number. Typically, microbial colonies in soil consist of a few cells adsorbed to a soil particle or linked through production of a polysaccharide slime. Incomplete mixing of the soil in the diluent could result in data that are an estimate of soil particle-bound colonies rather than separated microbial propagules. This data compromise can be minimized by thoroughly suspending and mixing the soil sample in the diluent, but realistically it cannot be eliminated.

A problem frequently encountered even when extreme caution is used in preparing the dilutions is microbial growth in the dilution media. Microorganisms generally have generation times of several hours in native soil samples. Thus, it might be considered that time lapse from preparation of the initial dilution to plating of the sample has little effect on the colonies developing on the plate. Unfortunately, this assumption is not true. Mixing of the soil sample in the dilution medium liberates metabolizable organic material from the soil. The solubolized organic carbon may precipitate a few rapid divisions of the soil heterotrophic microbes under the less restrictive conditions of the dilution tube. This problem is particularly acute when the diluent contains mineralizable organic carbon. For example, it is not unusual to prepare dilutions in 0.1 percent (w/v) amino acids or 0.1 strength of a growth medium such as trypticase soy broth to minimize the impact of dilution on fragile microbial cells. Hence, it is desirable to minimize the time lapse between preparation of the initial dilution and inoculation of the growth medium.

Interpretation of viable count data must be based on the knowledge that (1) not all soil microbes are capable of growth on the medium and (2) those organism detected are a mixture of organisms that are active and those that were inactive *in situ*. As indicated above, the number of colonies developing on the growth media represents only the microbes capable of growth on the selected medium under the prevailing incubation conditions. This is particularly true when selective media are used. These media are necessarily more stringent than those selected for estimation of total microbial populations (e.g., bacterial or fungal population densities). For example, it is not unusual to attempt to isolate organisms capable of using a particular pesticide as a carbon and energy source. A medium is prepared containing nitrogen, phosphorus, sulfur, and trace minerals necessary for growth and buffered at an appropriate pH for population development. A concentration of pesticide is added to the salts medium that is sufficient for colonial development, but not toxic. This medium appears to be ideal for the study, yet no microbes grow or only a few species are isolated from a soil sample in which it is known that a diverse population capable of catabolizing the pesticide exists. A limiting factor associated with the medium is that all microbes that require vitamins or amino acids are precluded from growth.

A further caveat regarding interpretation of viable count data relates to the nature of the state of the microbial populations in soil. Production of a colony on the test medium indicates that the particular organism was viable in the soil of interest, but nothing is indicated regarding the activity of the microbe in the native soil sample. Active growing cells, nongrowing cells, and resting cells will develop colonies under appropriate conditions. Thus, isolation of a particular organism in axenic culture solely means that the organism was present in the soil sample, not that it contributed to the metabolic activity expressed in the soil at the time of sampling.

2.4.3 DNA Hybridization/PCR Procedures

DNA hybridization is a valuable procedure for detection of specific genetic sequences. This technique is useful in screening microbial isolates from soil for specific genotypic strains as well as for detection of specific gene sequences in DNA samples extracted directly from soil (e.g., Sayler and Layton, 1990). Combination of the method with the polymerase chain reaction (pcr) provides a sensitive means of detecting genes present in a few microbes in a soil sample (e.g., Steffan and Atlas, 1991). Evaluation of DNA sequences unique to individual microbial species, strains, or even guilds defined by a trait such as nitrogen fixation allows estimation of their population densities.

The polymerase chain reaction provides for amplification of individual nucleic acid sequences without the necessity of cloning them (Fig. 2.7). Chemically synthesized primers are hybridized to specific homologous DNA sequences adjacent to the target DNA regions. Thermally stable DNA polymerase is used to duplicate the target sequences. The thermal stability of the polymerase preserves the enzyme activity during the repetitive sequence of

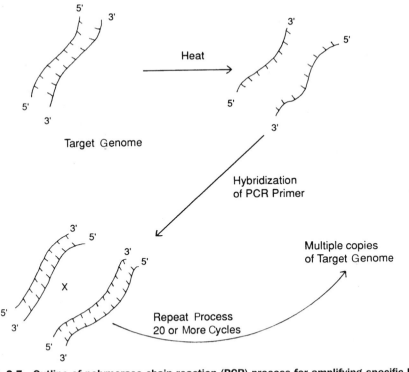

Fig. 2.7 Outline of polymerase chain reaction (PCR) process for amplifying specific DNA sequences.

melting of the DNA necessary for yielding the exponential increase in the target DNA.

For application of this procedure to soil systems, the bacteria whose populations are to be monitored may be separated from the soil prior to analysis, or the cells may be lysed *in situ* and the liberated DNA extracted. Humic acids must be removed from the sample since they inhibit the PCR reaction (e.g., Tsai and Olson, 1992a). For detection of microbes in low numbers, the extracted DNA may be concentrated by filtration (Bej et al., 1991a). Jacobson and Rasmussen (1992) separated bacteria from soil for PCR analysis using cation exchange resin.

The detection limits of hybridization procedures are low. For example, DNA-DNA hybridization limits of 4.3×10^4 cells per g dry soil (approximately 0.2 pg of hydrolyzable DNA in a $1 \mu g$ DNA sample) were measured for *Bradyrhizobium japonicum* populations in soil (Holben et al., 1988). PCR increases the assay sensitivity so that as few as one colony-forming unit can be detected per g of soil (Josephson et al., 1991; Pillai et al., 1991; Steffan and Atlas, 1988). Steffan and Atlas (1988) reported a 1000-fold increase in sensitivity with PCR compared to analyzing nonamplified DNA. In their study of *Pseudomonas cepacia* AC1100 as few as 100 colony-forming units per 100 g sediment were detected in a background population of 10^{11} cells of diverse nontarget organisms. Tsai and Olson (1992b) were able to detect low copy numbers of 16S ribosomal gene fragments of *Escherichia coli* seeded in soils and sediments using PCR amplification. As few as 3 cells per g soil were detected. Picard et al. (1992) detected 0.2×10^5 genomes of *Frankia* sp. per g soil using PCR techniques. (For further discussion of these procedures, see reviews by Arnheim and Erlich, 1992; Bej et al., 1991b; Gibbs, 1990; Sayler and Layton, 1990; Steffan and Atlas, 1991.)

2.4.4 Expression of Population Density per Unit of Soil

The ultimate goal for most ecological studies is to compare data from divergent soil ecosystems and to evaluate the impact of differences between the sites in physical, chemical, or biological properties on the activities. Traditionally, soil population densities and enzymatic activities as well as chemical concentrations have been expressed on a per-gram-of-dry-soil basis. This method of data presentation may be reasonable with some systems, but with the diversity of soils encountered in field studies, consideration of the bulk density and the volumes of soil represented is important. This conclusion is based on the observation that, in reality, soil processes occur in a three-dimensional system — that is, a volume of soil. The quantity of soil contained within that volume is highly variable. The bulk density of most mineral soils ranges from about 0.9 to 1.2. Thus, activity per g dry soil is approximately the activity per cubic cm of soil. In contrast, bulk densities of soils containing high levels of organic matter or organic soils may be quite low. For example, organic soils frequently have bulk densities of 0.1 g soil per cm^3 or less. Assuming a bulk density of 0.1, any activity expressed per g of dry soil would represent 10 cubic cm of soil within the native environment. Thus, were the results with the low-bulk–density organic soil compared on mass of soil basis with those similarly expressed for an average mineral soil, a tenfold error in conceptualization of the field situation would result from the different field volumes represented. Hence, it is important when evaluating published data, or preparing one's own data for publication, to include a study of the site soil bulk density so that activities or populations in comparable quantities of soil can be evaluated.

2.4.5 Products of Soil Autecological Research

To the individual interested in studying unusual microbes or even in evaluating the diversity of microbial life forms, soil is a virtual treasure chest. Each gram of soil contains a mixture of both anaerobes and aerobes, heterotrophs and autotrophs, protozoa, nematodes, and viruses. Seemingly independent of current soil situation, we find organisms with temperature optima from cold-loving psychrotrophic bacteria to those best adapted to high temperatures, thermophiles. Similar adaptability of populations to soil pH and salinity are found within the population of a single soil.

Provision of a list of most frequently encountered bacterial or fungal species, historically the most commonly studied soil populations, can only be criticized by the omissions rather than by its inclusiveness. For not only is there a vast array of species naturally present and functioning within any soil ecosystem, but propagules gain entry by transport through air, water, or sediment transport. Further, a variety of anthropogenic activities (such as those associated with waste disposal) augment the range of microbial species present. Commonly encountered soil bacterial species include representatives from the genera *Acinetobacter, Agrobacterium, Alcaligenes, Arthrobacter, Bacillus, Brevibacterium, Caulobacter, Cellulomonas, Clostridium, Corynebacterium, Flavobacterium, Hyphomycrobium, Metallogenium, Micrococcus, Mycobacterium, Pseudomonas, Sarcina, Streptococcus,* and *Xanthomonas.* These heterotrophic bacterial genera are augmented in soil by autotrophic and mixotrophic representatives, including a variety of nitrifiers, *Thiobacillus* species, and iron bacteria. Actinomycete populations include the commonly encountered *Nocardia* sp. and *Streptomyces* species. Fungal populations comprise a variety of slow- and fast-growing species. Perhaps the most commonly encountered are *Penicillium* and *Aspergillus* species, plus representatives of the *Zygomycetes* and the mycorrhizae-associated ascomycetes and basidiomycetes.

The objectives of experiments requiring identification of soil populations range from isolation of individuals with traits of interest to elucidation of the diversity of soil microbes by measurement of tens to hundreds of different microbial species. The latter studies include evaluation of the interactions between the diverse populations present as well as a determination of the effect of external stimuli, including cultural impact, such as of clear-cutting of forests or of industrial pollution, on microbial diversity.

Diversity studies of soil microbes have been limited in the past due to the difficulty of determining function *in situ* and the vast quantities of data necessary for identification of the species present or to group the organisms into guilds based on structural or physiological similarity. Development of automated techniques for cultural characterization and data analysis has made such studies more practical (see Russek-Cohen and Colwell, 1986: Mills and Bell, 1986; Holder-Franklin, 1986; Holder-Franklin and Tate, 1986). These experiments have provided an understanding of the genetic potential within a given ecosystem and an indication of the resilience of the system to external perturbation—that is, the capacity for homeostatic stability. Some systems with low species diversity, such as soils receiving thermal hot spring outflows, appear to be reasonably stable, but the classic maxim has been that the greater the species diversity, the higher the probability that the community has the capacity to overcome intrusions such as influx of contaminants or pH or temperature alterations. Much of this stability is derived from latent genotypic capability.

2.5 Principles and Products of Synecological Research

Although the sciences of microbiology and ecology have traditionally involved association of a genus and species name with the living creatures studied, as indicated above, such activities have more limited value in soil ecosystem characterization. This is due in part to taxonomic problems of defining a bacterial species and the capacity of soil organisms to exist in active, inactive, or resting states. Thus, the presence of an organism does not necessarily prove its current participation in ecosystem-driving or -defining activities. Thus, synecological studies are of growing importance in defining soil activity. This process microbiology research involves assessment of metabolic activities and enzymatic activities, as well as the properties of the soil ecosystem resulting from and controlling these activities.

As with the evaluation of specific microbial populations discussed above, a prime concern in examining soil as a "tissue" is collection of a representative soil sample. Since soil enzyme analysis or respiration measurements generally require larger samples than are necessary for detection of individual microbial species (grams of soil for a synecological study vs. perhaps milligrams or less for an autecological analysis), sample heterogeneity difficulties are reduced. Composited soil samples collected in a manner to average system variability become more practical; for example, a composited soil sample from a corn field may contain equal quantities of soil from between rows and within rows. Considerations with soil collection may reside more with concern for above-ground plant distribution affect or soil horizon influences than is possible with autecological studies.

2.6 Interphase Between Study of the Individual and Community Microbiology

Neither the study of individual species present in soil and their metabolic capability nor consideration of soil metabolic activities solely provides the conceptual basis upon which to construct the science of soil microbiology. Axenic cultures of soil microorganisms provides a detailed understanding of the nuances of the metabolic intricacy of the "players" in the soil ecosystem. Evaluation of enzymatic and respiratory activities of isolated soil samples demonstrates site-specific and perhaps even chronologically limited system traits, but without an understanding of the organisms present, little is revealed regarding the resilience of the community. That is, both the axenic culture and the soil sample are requisites for the soil microbiologist.

This conclusion is demonstrated quite clearly when the impact of soil heterogeneity on growth factor limitations and microbial interactions are considered. Microorganisms require growth factors (i.e., vitamins and amino acids) for growth. As will be discussed in Chapter 4, a greater proportion of soil bacteria require these factors than are capable of synthesizing their metabolic needs *de novo*. The effect of this observation can be seen by examining a hypothetical reaction catalyzed by a microorganism requiring a vitamin. Assume that in the test tube the following reaction sequence is catalyzed by a single microbe:

$$A \rightarrow B \rightarrow C \rightarrow D$$

Let us hypothesize that for this reaction to go to completion, a growth factor is required for conversion of product B to product C. This may not be considered in laboratory experimentation because the growth media typically used are supplemented with yeast extract or comparable vitamin and amino acid sources to provide all nondefined growth needs. Thus, experimental results are published with the conclusion that the bacterial isolate, perhaps one commonly occurring in soil, is perfectly capable of mineralizing the compound of concern, yielding carbon dioxide and water. That is, the compound of interest is concluded to be biodegradable. Unfortunately, the requisite growth factor may be missing from the soil microsite. Therefore, instead of observing complete mineralization as the laboratory data would have predicted, product B accumulates. Thus, a substance that laboratory data indicated is biodegradable yields intermediates that accumulate in the ecosystem.

Laboratory data could also predict environmental stability of a product when complete decomposition is observed in native soils. This situation occurs in ecosystems where no single microbial species posses the enzymatic capacity to convert the substance to carbon dioxide and water, but a consortium exists that can catalyze the conversion. Again, consider the above generic reaction sequence:

$$A \rightarrow B \rightarrow C \rightarrow D$$

In this situation, substance A may be converted to product B by a one-microbial species. A second or even a third microorganism could be postulated to be required for transformation of products B to C and C to D. Thus, data from axenic culture indicates environmental stability of substance A, whereas amendment of soil with substance A results in complete conversion of substance A to product D.

2.7 Concluding Comments

This analysis of the soil biological community provides justification for reaching a seemingly heretical conclusion that the total potential of a soil ecosystem exceeds the sum of its parts. Conceptually, soil is readily separated into a mass of mineral and chemical components as discussed in Chapter 1, plus a variety of living organisms. It is these latter participants in the system that transform the soil mineral matrix into the ever-changing, life-supporting substance essential for our terrestrial existence. Although a vast array of biological processes occur in a soil sample, a much greater potential is contained within the unexpressed microbial genome. Thus, development of an understanding of the intricacies of the interaction of the biological dimension of soil with its physical and chemical aspects is essential for proper stewardship of our ecosystem. This conceptual understanding is based on the results of a variety of autecological and synecological studies of soil organisms and processes. As introduced in the brief overview of some methods available for analysis of soil microbes, validity and interpretation of the data collected in these ecological studies must be predicated on a firm appreciation of the methodological problems associated with soil microbiological research. Even with their inherent limitations, such studies provide the foundation for further incursions into evaluation of the ever-changing world of the soil microbe.

References

Ahmed, M., and J. M. Oades. 1984. Distribution of organic matter and adenosine-triphosphate after fractionation of soils by physical procedures. Soil Biol. Biochem. 16:465–470.

Alef, K. T. Beck, L. Zelles, and D. Kleiner. 1988. A comparison of methods to estimate microbial biomass and N-mineralization in agricultural and grassland soils. Soil Biol. Biochem. 20:561–565.

Alef, K., and D. Kleiner. 1989. Rapid and sensitive determination of microbial activity in soils and in soil aggregates by dimethylsulfoxide reduction. Biol. Fert. Soils 8:349–355.

Alexander, M. 1982. Most Probable Number Method for Microbial Populations. *In* A. L. Page (ed.), Methods of Soil Analysis, Part 2. pp. 815–820. Agronomy Society of America, Madison, WI.

Anderson, J. P. E., and K. H. Domsch. 1973. Quantification of bacterial and fungal contributions to soil respiration. Arch. Mikrobiol. 93:113–127.

Anderson, J. P. E., and K. H. Domsch. 1975. Measurement of bacterial and fungal contributions to respiration of selected agricultural and forest soils. Can J. Microbiol. 21:314–322.

Anderson, J. P. E., and K. H. Domsch. 1978. A physiological method for the quantitative measurement of microbial biomass in soils. Soil Biol. Biochem. 10:215–221

Anderson, J. P. E., and K. H. Domsch. 1980. Quantities of plant nutrients in the microbial biomass of selected soils. Soil Sci. 130:211–216.

Arnheim, N., and H. Erlich. 1992. Polymerase chain reaction strategy. Annu. Rev. Biochem. 61:131–156.

Ausmus, B. S. 1971. Adenosine triphosphate — a measure of active microbial biomass. Biologie du Sol 14:8–9.

Azam, F., R. L. Mulvaney, and F. J. Stevenson. 1988. Determination of *in situ* K_N by the chloroform fumigation-incubation method and mineralization of biomass N under anaerobic conditions. Plant Soil 111:87–93.

Bakken, L. R. and R. A. Olsen. 1983. Buoyant densities and dry-matter contents of microorganisms: Conversion of a measured biovolume into biomass. Appl. Environ. Microbiol. 45:1188–1195.

Bej, A. K., M. H. Mahbubani, J. L. Dicersare, and R. M. Atlas. 1991a. Polymerase chain reaction-gene probe detection of microorganisms by using filter-concentrated samples. Appl. Environ. Microbiol. 57:3529–3534.

Bej, A. K., M. H. Mahbubani, and R. M. Atlas. 1991b. Amplification of nucleic acids by polymerase chain reaction (PCR) and other methods and their applications. CRC Crit. Rev. Biochem. Mol. Biol. 26:301–334.

Bratbak, G. 1985. Bacterial biovolume and biomass estimations. Appl. Environ. Microbiol. 49:1488–1493.

Bratbak, G., and I. Dundas. 1984. Bacterial dry matter content and biomass estimations. Appl. Environ. Microbiol. 48:755–757.

Brookes, P. C., D. S. Powlson, and D. S. Jenkinson. 1982. Measurement of microbial biomass phosphorus in soil. Soil Biol. Biochem. 14:319–329.

Brookes, P. C., D. S. Powlson, and D. S. Jenkinson. 1984. Phosphorus in the soil microbial biomass. Soil Biol. Biochem. 16:169–175.

Chapman, S. J. 1987. Inoculum in the fumigation method for soil biomass determination. Soil Biol. Biochem. 19:83–87.

Christensen, S., S. Simkins, and J. M. Tiedje. 1991a. Temporal patterns of soil denitrification: Their stability and causes. Soil Sci. Soc. Am. J. 54:1614–1618.

Christensen, S., S. Simkins, and J. M. Tiedje. 1991b. Spatial variation in denitrification: Dependency of active centers on the soil environment. Soil Sci. Soc. Am. J. 1608–1613.

Coûteaux, M. M., R. Henkinet, P. Pitta, P. Bottner, G. Billès, L. Palka, and G. Vannier. 1989. Native carbon mineralization of an acid organic soil after use of the chloroform-fumigation method to estimate microbial biomass. Biol. Fert. Soils 8:172–177.

Dahlin, S., and E. Witter. 1998. Can the low microbial biomass C-to-organic-C ration in an acid and a metal contaminated soil be explained by differences in the substrate utilization efficiency and maintenance requirements. Soil Biol. Biochem. 30:633–641.

Deluca, T.H. 1998. Relationship of 0.5 M K_2SO_4 extractable anthrone-reactive carbon to indices of microbial activity in forest soils. Soil Biol. Biochem. 10:1293–1299.

Dictor, M. C., L. Tessier, and G. Soulas. 1998. Reassessment of the K-ec coefficient of the fumigation-extraction method in a soil profile. Soil Biol. Biochem. 30:119–127.

Fazio, S. D., W. R. Mayberry, and D. C. White. 1979. Muramic acid assay in sediments. Appl. Environ. Microbiol. 38:349–350.

Findlay, R. H., G. M. King, and L. Watling. 1989. Efficacy of phospholipid analysis in determining microbial biomass in sediments. Appl. Environ. Microbiol. 55:2888–2893.

Flühler, H., M. S. Ardakani, T. E. Szuszkiewicz, and L. H. Stolzy. 1976. Field-measured nitrous oxide concentrations, redox potentials, oxygen diffusion rates, and oxygen partial pressures in relation to denitrification. Soil Sci. 122:107–114.

Focht. D. D. 1974. The effect of temperature, pH, and aeration on the production of nitrous oxide and gaseous nitrogen—a zero-order kinetic model. Soil Sci. 118:173–179.

Fritze, H., P. Vanhala, J. Pietkäinen, and E. Mälkönen. 1996. Vitality fertilization of Scots pine stands growing along a gradient of heavy metal pollution: Short-term effects on microbial biomass and respiration rate of the humus layer. Fresenius J. Anal. Chem. 354:750–755.

Gibbs, R. A. 1990. DNA amplification by the polymerase chain reaction. Anal. Chem. 62:1202–1214.

Gunapala, N., and K. M. Scow. 1998. Dynamics of soil microbial biomass and activity in conventional and organic farming systems. Soil Biol. Biochem. 30:805–816.

Hedley, M. J., and J. W. B. Stewart. 1982. Method to measure microbial biomass phosphorus in soils. Soil Biol. Biochem. 14:337–385.

Holben, W. E., J. K. Jansson, B. K. Chelm, and J. M. Tiedje. 1988. DNA probe method for the detection of specific microorganisms in the soil bacterial community. Appl. Environ. Microbiol. 54:703–711.

Holder-Franklin, M. A. 1986. Ecological relationships of microbiota in water and soil as revealed by diversity measurements. pp. 93–132. *In* R. L. Tate III (eds.), Microbial Autecology: A Method for Environmental Studies. John Wiley & Sons, NY.

Holder-Franklin, M. A., and R. L. Tate III. 1986. Introduction of the computer into autecological studies. pp. 76–91. In R. L. Tate III (ed.), Microbial Autecology: A Method for Environmental Studies. John Wiley & Sons, NY.

Jacobsen, C. S., and O. F. Rasmussen. 1992. Development and application of a new method to extract bacterial DNA from soil based on separation of bacteria from soil with cation-exchange resin. Appl. Environ. Microbiol. 58:2458–2462.

Jenkinson, D. S. 1976. The effects of biocidal treatments on metabolism in soil. IV. The decomposition of fumigated organisms in soil. Soil Biol. Biochem. 8:203–208.

Jenkinson, D. S., S. A. Davidson, and D. S. Powlson. 1979. Adenosine triphosphate and microbial biomass in soil. Soil Biol. Biochem. 11:521–527.

Jenkinson, D. S., and D. S. Powlson. 1980. Measurement of microbial biomass in intact soil cores and sieved soil. Soil Biol. Biochem. 12:579–581.

Josephson, K. L., S. D. Pillai, J. Way, C. P. Gerba, and I. L. Pepper. 1991. Fecal coliforms in soil detected by polymerase chain reaction and DNA-DNA hybridizations. Soil Sci. Soc. Am. J. 55:1326–1332.

Kassim, G., J. P. Martin, and K. Haider. 1982. Incorporation of a wide variety of organic substrate carbons into soil biomass as estimated by the fumigation procedure. Soil Sci. Soc. Am. J. 45:1106–1112.

King, J. D., and D. C. White. 1977. Muramic acid as a measure of microbial biomass in estuarine and marine samples. Appl. Environ. Microbiol. 33:777–783.

Lundgren, B. 1981. Fluorescein diacetate as a stain of metabolically active bacteria in soil. Oikos 36:17–22.

Martens, R. 1985. Limitations in the application of the fumigation technique for biomass estimations in amended soils. Soil Biol. Biochem. 17:57–63.

Martens, R. 1987. Estimation of microbial biomass in soil by the respiration method: Importance of soil pH and flushing methods for the measurement of respired CO_2. Soil Biol. Biochem. 19:77–81.

Mills, A. L. 1985. Acid mine drainage: Microbial impact on the recovery of soil and water ecosystems. pp. 35–81. *In* R. L. Tate III and D. A. Klein (eds.), Soil Reclamation Processes: Microbiological Analyses and Applications. Marcel Dekker, NY.

Mills, A. L., and P. E. Bell. 1986. Determination of individual organisms and their activities in situ. pp. 27–60. *In* R. L. Tate III (ed.) Microbial Autecology: A Method for Environmental Studies. John Wiley & Sons, NY.

Parkin, T. B. 1987. Soil microsites as a source of denitrification variability. Soil Sci. Soc. Am. J. 51:1194–1199.

Parkin, T. B., J. L. Starr, and J. J. Meisinger. 1987. Influence of sample size on measurement of soil denitrification. Soil Sci. Soc. Am. J. 51:1492–1507.

Parkinson, D., and E. A. Paul. 1982. Microbial biomass. Pp. 821–830. *In* A. L. Page (Ed.), Methods of Soil Analysis. Part 2. Chemical and Microbiological Properties. American Society of Agronomy. Madison, WI.

Parsons, L. L, R. E. Murray, and M. S. Smith. 1991. Soil denitrification dynamics: Spatial and temporal variations of enzyme activity, populations, and nitrogen gas loss. Soil Sci. Soc. Am. J. 55:90–95.

Paul, E. A., and R. L. Johnson. 1977. Microscopic counting and adenosine 5′-triphosphate measurement in determining microbial growth in soils. Appl. Environ. Microbiol. 34:263–269.

Picard, C., C. Ponsonnet, E. Paget. X. Nesme, and P. Simonet. 1992. Detection and enumeration of bacteria in soil by direct DNA extraction and polymerase chain reaction. Appl. Environ. Microbiol. 58:2717–2722.

Pillai, S. D., K. L. Josephson, R. L. Bailey, C. P. Gerba, and I. L. Pepper. 1991. Rapid method for processing soil samples for polymerase chain reaction amplification of specific gene sequences. Appl. Environ. Microbiol. 57:2183–2186.

Rodriguez, G. G., D. Phipps, K. Ishiguro, and H. F. Ridgway. 1992. Use of a fluorescent redox probe for direct visualization of actively respiring bacteria. Appl. Environ. Microbiol. 58:1801–1808.

Ross, R. J. 1987. Soil microbial biomass estimated by the fumigation-incubation procedure: Seasonal fluctuations and influences of soil moisture content. Soil Biol. Biochem. 19:397–404.

Ross, D. J. 1988. Modifications to the fumigation procedure to measure microbial biomass C in wet soils under pasture: Influence on estimates on seasonal fluctuations in the soil biomass. Soil Biol. Biochem. 20:377–383.

Rosser, D. J. 1980. Ethidium bromide a general purpose fluorescent stain for nucleic acid in bacteria and eukaryotes and its use in microbial ecology studies. Soil Biol. Biochem. 12:329–336.

Roszak, D. B., and R. R. Colwell. 1987. Metabolic activity of bacterial cells enumerated by direct viable count. Appl. Environ. Microbiol. 53:2889–2983.

Russek-Cohen, R., and R. R. Colwell. 1986. Application of numerical taxonomy procedures in microbial ecology. pp. 133–146. *In* R. L. Tate, III (ed.), Microbial Autecology: A Method for Environmental Studies. John Wiley & Sons, NY.

Sayler, G. S., and A. C. Layton. 1990. Environmental application of nucleic acid hybridization. Annu. Rev. Microbiol. 44:625–648.

Schmidt, E. L., and E. A. Paul. 1982. Microscopic methods for soil microorganisms. pp. 803–814. *In* A. L. Page (ed.), Methods of Soil Analysis, Part 2. Agronomy Society of America, Madison, WI.

Sexton, A. J., T. B. Parkin, and J. M. Tiedje. 1985. Temporal response of soil denitrification rates to rainfall and irrigation. Soil Sci. Soc. Am. J. 49:99–103.

Smith, M. S., and J. M. Tiedje. 1979. Phases of denitrification following oxygen depletion in soil. Soil Biol. Biochem. 11:261–267.

Sonderstrom, B. E. 1977. Vital Staining of fungi in pure cultures and in soil with fluorescein diacetate. Soil Biol. Biochem. 9:59–63.

Sparling, G. P., and A. W. West. 1988. A direct extraction method to estimate soil microbial C: Calibration *in situ* using microbial respiration and ^{14}C labelled cells. Soil Biol. Biochem. 20:337–343.

Steffan, R. J., and R. M. Atlas. 1988. DNA amplification to enhance detection of genetically engineered bacteria in environmental samples. Appl. Environ. Microbiol. 54:2185–2191.

Steffan, R. J., and R. M. Atlas. 1991. Polymerase chain reaction: Applications in environmental microbiology. Annu. Rev. Microbiol. 45:137–161.

Stenberg, B., M. Johansson, M. Pell, K. Sjodahlsvensson, J. Stenstrom, and L. Torstensson. 1998. Microbial biomass and activities in soil as affected by frozen and cold storage. Soil Biol. Biochem. 30:393–402.

Tate, K. R., D. J. Ross, and C. W. Feltham. 1988. A direct extraction method to estimate soil microbial C: Effects of experimental variables and some different calibration procedures. Soil Biol. Biochem. 20:329–335.

Tate, R. L., III. 1996. Importance of autecology in microbial ecology. pp. 1–26. *In* R. L. Tate III (ed.), Microbial Autecology: A method for Environment Studies. John Wiley & Sons, NY.

Tate, R. L., III. 1991. Microbial biomass measurement in acidic soils: Effect of fungal:bacterial activity ratios and soil amendment. Soil Sci. 152:220–225.

Tessier, L., E. G. Gregorich, and E. Topp. 1998. Spatial variability of soil microbial biomass measured by the fumigation extraction method, and K-ec as affected by depth and manure application. Soil Biol. Biochem. 30:1369–1377.

Tsai, Y.-L., and B. H. Olson. 1992a. Rapid method for separation of bacterial DNA from humic substances in sediments for polymerase chain reaction. Appl. Environ. Microbiol. 58:2292–2295.

Tsai, Y.-L., and B. H. Olson. 1992b. Detection of low numbers of bacterial cells in soils and sediments by polymerase chain reaction. Appl. Environ. Microbiol. 58:754–757.

Vance, E. D., P. C. Brookes, and D. S. Jenkinson. 1987a. An extraction method for measuring soil microbial biomass C. Soil Biol. Biochem. 19:703–707.

Vance, E. D., P. C. Brookes, and D. S. Jenkinson. 1987b. Microbial biomass measurements in forest soils: The use of the chloroform fumigation-incubation method in strongly acid soils. Soil Biol. Biochem. 19:697–702.

Vance, E. D., P. C. Brookes, and D. S. Jenkinson. 1987c. Microbial biomass measurements in forest soils: Determination of K_C values and tests of hypotheses to explain the failure of the chloroform fumigation-incubation method in acid soils. Soil Biol. Biochem. 19:689–696.

Vanhala, P., H. Fritze, and S. Neuvonen. 1996. Prolonged simulated acid rain treatment in the subarctic: Effect on the soil respiration rate and microbial biomass. Biol. Fert. Soils. 23:7–14.

Van Veen, J. A., and E. A. Paul. 1979. Conversion of biovolume measurements of soil microorganisms, grown under various moisture tensions, to biomass and their nutrient content. Appl. Environ. Microbiol. 37:686–692.

Verstraete, W., H. Van De Werf, K. Kucnirowicz, and M. Ilaiwi. 1983. Specific measurement of soil microbial ATP. Soil Biol. Biochem. 15:391–396.

Verstraeten, L. M. J., K. De Conick, K. Vlassak, W. Verstraete, H. Van De Werf, and M. Ilaiwi. 1983. ATP content of soils estimated by two contrasting extraction methods. Soil Biol. Biochem. 15:397–402.

Wildung, R. E., T. R. Garland. 1985. Microbial development on oil shale waste: Influence on geochemistry. pp. 107–139. *In* R. L. Tate III and D. A. Klein (eds.), Soil Reclamation Processes: Microbiological Analyses and Applications. Marcel Dekker, NY.

Microbial Diversity
of Soil
Ecosystems

For a complex ecosystem such as soil, the question of definition of the portion of the community to be studied as an indicator of the biological diversity is not trivial. Each trophic level of the biological community has a clear impact on the nature of the ecosystem and the dynamics of the biological processes occurring therein. Animals may contribute to decomposition of complex organic matter by breaking the larger plant components into smaller pieces, thereby increasing their surface area, or even by decomposing the plant biomass directly. Their mere presence in the soil alters the physical structure of the soil sufficiently to alter the rates of biological processes occurring therein. These contributions to overall decomposition rates are exemplified by the impact of the community of earthworms and their resultant contribution to soil pore structure (see Section 7.3.2). Each biological entity within the soil system must and does affect the general structure of the biological community (i.e., total community diversity) and its activity. Unfortunately, the complexity of the total soil biological community precludes its study *in toto*. Thus, to date, soil biological diversity studies have been directed primarily at evaluating bacterial and perhaps fungal community diversity. A portion of the justification for the limited viewpoint of such analyses resides in the fact that these portions of the total biological community are the most easily studied. But perhaps a better justification of using bacterial species as indicators of soil diversity resides in the vast number of different species existent in a gram of soil and their primary role in biological decomposition of organic matter. The number of bacterial species in a soil sample is far greater than the few that can be cultivated in the laboratory. Using reassociation analysis of DNA, it has been shown that the diversity of the total soil bacterial community is as much as 170-fold greater than that revealed by collection of bacterial isolates from soil. Furthermore, pristine soil samples can contain in the order of 10,000 bacterial types (Torsvic et al., 1996). These types of studies have led to estimates of 4,000 to 7,000 different genome equivalents g^{-1} soil (Torsvik et al., 1990).

Further support for using microbial diversity to assess soil function is derived from the fact that decomposer populations in soil serve as an integrator of the impact of all physical, chemical, and biological properties of the soil on soil biological processes. The amount and extent of their activities and the specific species present are the result of the totality of all soil properties affecting their capability of functioning. In each microsite, the species that is functioning is the one best able to adjust to the chemical, physical, and chemical properties of the microsite. Thus, it could be concluded that the totality of species (species diversity *per se*) reflects a community of organisms selected by, and adapted to the specific soil system. Bacterial and fungal community structure can therefore be considered to be a basic descriptor of a specific soil system. Thus, this chapter is presented with the objective of evaluating the methods available for assessing soil microbial diversity, of determining the limitations of the various methods, and of elucidating the types of information that are and are not provided by assessing soil microbial diversity.

3.1 Classical Studies of Soil Microbial Diversity: Numerical Taxonomy

Early studies of soil bacterial diversity have been based on the same procedural philosophies as were the diversity analysis of higher animal or plant populations. The basic steps of the procedure involved isolating and identifying the numerically dominant bacterial species occurring in the soil of interest. The data were then analyzed by assessment of degree of similarity of the species distribution between soil samples with methods that provided a cluster analysis and by calculation of indicators of species diversity such as the Shannon Weaver index. (See Russek-Cohen and Colwell, 1986, for a review of this topic). These types of studies are time-consuming and labor-intensive in that large numbers of bacterial strains must be purified from soil and characterized physiologically.

Difficulties with the assessment of variations in soil microbial community structure through use of numerical taxonomic methods beyond the concerns of time and labor that must be devoted to their accomplishment involve inherent limitations of cultivation of active members of the soil biological community. The procedure relies on being able to isolate representative samples of soil bacterial populations and the capability of determining bacterial species designations for the isolates.

Limitations Due to Inability to Grow All Soil Bacteria in Laboratory Culture: The cultivation procedures necessary to select the axenic cultures characterized in the diversity analysis cannot be expected to yield bacterial strains in proportion to their existence in soil samples. A primary maxim in evaluating soil bacterial populations is the fact that only a small proportion of the microbes occurring in soil can be grown on laboratory media. Thus, the organisms originally isolated for the diversity characterization are a subset of the organisms existing *in situ*. The cultivated organisms are those that can utilize the energy source provided in the medium under the physical and chemical limitations of the growth medium. Furthermore, the organisms generally selected for study are those that grow rapidly on the isolation medium. Thus, slow-growing microbes or those that produce small colonies are commonly overlooked. These observations lead to the conclusion that an underlying assumption of the diversity analysis procedure using cultivated microbes is false. That is, it could be assumed that the microbes growing in culture are representative of the microbial population in general or at least of the dominant populations in the soil sample of interest. In reality, all that can be said about the strains selected is that propagules existed in the soil system, they may or may not have been active *in situ* (in surface soils, most of the microbes are inactive or are in a resting state), and they may have been among the numerically dominant populations. Other species or strains may have been dominant but incapable of growth on the media used to select the bacterial isolates. Thus, the only conclusion that can be made is that the organisms isolated from the soil samples are the numerically dominant strains *that could be cultivated*.

Problems of Bacterial Species Definition: An inherent difficulty associated with all studies relating to species identification of soil bacteria is the definition of a bacterial species. Bacterial species can be loosely defined as a collection of closely related bacterial strains. Classically, the strains forming a particular bacterial species have been grouped based upon possession of a series of common physiological capabilities, structural properties, and miscellaneous traits — such as motility and growth characteristics (for example, pH or temperature tolerance ranges). A small percent of the potential "species" in the bacterial world have been classified. Of greater concern to these definitions is that the entity that might be classed as a particular bacterial species is mutable. That is, its basic properties change. Mutations or transfer of genetic material from one species to another can readily blur the distinctions that were originally used to define the individual species. Thus, bacterial taxonomic procedures are limited by the small proportion of species described and by the fact that species definitions in bacteria are rather blurred and to a large degree artificial.

Arguably, it could be stated that specific species designations are of greater significance when considering public health problems then when evaluating soil populations. This statement is based on the observation that it is overall metabolic capability of the community and degree of similarity of the individuals that define the functional soil system rather than the existence of nonchanging individual bacterial species — because they simply do not exist in the complex soil community. Therefore, although the above-cited methods and examples of studies have provided an indication of the basic variation in the microbial populations in soil, other techniques are needed to truly characterize the diversity of the members of the community.

Alternatives to Bacterial Strain Isolation: Ideally, a method for analysis of soil microbial diversity should involve evaluation of the total population — those that grow on laboratory media plus those that don't, those that have been described as known species and those that possibly never will be, and those that are active in the community as it currently exists plus those that become active when the community is stressed. Since the simple task of separating the total microbial community from soil is insurmountable, a reasonable compromise to achieve this ideal objective would be to use a surrogate to assess microbial diversity in soil. This surrogate indicator of microbial diversity would have to occur only in the living cells, vary in a meaningful way in relationship to the overall microbial diversity in the system, and be sufficiently variable that groups could be constructed based on similarity. Essentially, with these alternative assessments, the soil sample is treated as if it were a single tissue type and the extracted substances as indicators of the diverse capabilities contained within that tissue. Candidates for use in estimating soil microbial diversity include their metabolic capabilities and cell chemical components.

3.2 Biochemical Measures of Soil Microbial Diversity

Fundamentally, to be aligned with classical considerations of biological diversity, assessments of soil microbial diversity should be based on a delineation that varies in parallel with actual species diversity. Furthermore, the population sampling of the system of interest should either be all-inclusive or at least fully representative of the diversity of the community. Few systems, above- or belowground, exist wherein enumeration of the totality of species present is possible, so a means of assuring that those species detected are a true subsample of the community is necessary. In a true species diversity analysis of any ecosystem, the truth of the above tenets is assumed. But the foregoing discussion relating to bacterial species identity and procedures associated with isolation of bacterial strains from soil samples reveals the limited validity of the assumptions. Thus, an estimate of the biological diversity of a soil community must reside upon selection of a surrogate assessor that meets the above specifications of an ideal measure as closely as possible.

Further requirements for this ideal measure of biological diversity are of a more practical nature and are derived primarily from the complexity and variability of the medium itself, soil. The procedure should be quick and easy to conduct so that a large number of samples can be processed. This requirement is likely the largest impediment to general application of classical ecological and species analysis procedures by methods such as those entailing numerical analysis of populations. An extension of this specification for our analytical procedure is that the method should be reasonably economic to conduct and require only moderate specialized equipment and skill so that it can be generally applied to the study of the world's soil ecosystems. These requirements are only partially fulfilled by methods currently available. The procedures most commonly used to estimate soil biological diversity are metabolic diversity (BIOLOG), phospholipid fatty acid analysis, and DNA analytical procedures. For each of these techniques, the nature of the soil biological variable assessed, the limitations of the data, and the potential for provision of useful information about community structural variation in soil systems will be analyzed below.

3.3 Metabolic Diversity of Soil Systems

To avoid (at least in part) the limitations inherent in considering the metabolic capabilities of axenic-grown bacterial strains isolated from soil samples, the metabolic diversity assessment procedures described herein are predicated on the consideration that soil can be treated as if it were a tissue. Then the general metabolic diversity in extracts of soil samples can be assessed. To use any metabolic property to estimate changes in microbial community structure of soil samples, there must be sufficient variation in the physical, chemical, and biological properties of soils that the biochemical processes occurring will be characteristic of that particular system and sufficiently different from those of other soil ecosystems that the procedure can serve some diagnostic role in comparing soil ecosystems; that is, the use of metabolic capability of soil samples as an indication of microbial diversity changes is predicated on the assumption that significant differences in this parameter occurs between soil samples. The truth of this assumption is not obvious when specific metabolic processes are considered. First, the prime energy source for soil ecosystems is photosynthetically fixed carbon. Second, although there are differences in the biochemical makeup of different plant species or of the same plant species growing under varying conditions, the general array of biochemicals encountered by soil microbes decomposing them is not greatly different from ecosystem to ecosystem. Therefore, soil microbes are facing a reasonably common group of biochemicals of varying concentrations in almost all ecosystems. It could thus be anticipated that all soil microbial communities would possess a reasonably similar arsenal of metabolic capacities to recover energy and nutrients from these substances. Therefore variations in metabolic diversity between soils could be predicted to be minimal. Fortunately, metabolic diversity analysis does provide information that varies with ecosystem type and the conditions of particular ecosystems.

An important caveat must be expressed, though. It must be noted that current methods for quantifying soil metabolic diversity (e.g., the BIOLOG procedure) are not necessarily based on assessment of metabolic activities playing an active role in community function in the soil sample when it was collected. Rather the existence of an organism or an enzyme in soil extracts that is actively synthesized or whose synthesis can be induced during incubation of the assay mixture is examined. Thus, as will be expanded upon below, although the data are reproducible, the relationship between the metabolic profiles obtained by the assay procedure and actual soil expressed activities has yet to be demonstrated.

Assessment of metabolic diversity and the practicality of the procedure resides upon the capability of providing a "representative" extract of soil microbes and enzymes and assessing the metabolic abilities contained in the extract in a rapid and efficient manner. The core of the technique is the metabolic assay procedure. Commonly used methods are based on a redox system that allows concurrent characterization of the capability of microbes and enzymes communities to oxidize a large number of organic substances. A commonly used procedure for assessing metabolic activity of soil extracts or of purified bacterial isolates is the BIOLOG method. This technique has been found to provide reproducible profiles of bacterial communities (e.g., Haack et al., 1995). With this technique, the capability of the microbes in a soil extract to oxidize a variety of organic substrates is assessed by measuring reduction of a tetrazolium dye. The organic test substrates (95 for the most commonly used procedure) plus the tetrazolium indicator are distributed individually in the wells of microtiter plates. The oxidation of each substrate is determined by measuring the color development due to reduction of the tetrazolium dye. Commonly, the color change is recorded at intervals so that a rate of metabolism of each substrate is quantified. As discussed in greater detail below, the combination of substrates oxidized for each soil sample can be analyzed with procedures such as principle component analysis to determine the degree of variation in metabolic diversity between the soil communities. Reproducible metabolic diversity patterns have been provided for soil ecosystems under a variety of cropping conditions and different plant types (Bossio and Scow, 1995; Garland and Mills, 1991; Garland and Mills, 1994; Lupwayi et al., 1998; Windig, 1994; Zak et al., 1994) as well as sites contaminated with heavy metals from smelter operations (Kelly and Tate 1998b; Knight et al., 1997). By comparing variations in agricultural management of soils related to rice cultivation (flooded or drained with rice straw incorporated or burned), Bossio and Skow (1995) demonstrated the importance of carbon source on determination of microbial community structure using metabolic diversity analysis (BIOLOG). Similarly, Zak et al. (1994) found that metabolic diversity analysis detected shifts in microbial community structure along a moisture gradient in soils at the Jornada Long-Term Ecological Research Site in the Northern Chihuahuan Desert.

As is indicated below, two general variations of evaluation of metabolic capabilities of the bacterial community of soils using the BIOLOG procedure have been used: characterization of bacterial isolates grown axenically in the laboratory, and evaluation of mixed populations contained in soil extracts. Both methods have limitations that must be considered during data analysis and interpretations.

Since the BIOLOG procedure was developed for bacterial species identification, one of its primary strengths is in detecting the presence of specific species or strains in environmental samples [e.g., *Vibrio anguillarum* (Kuhn et al., 1996) and halophiles (Garabito et al., 1998)]. Furthermore, the procedure can be applied to numerical taxonomic studies as described above (e.g., Behrendt et al., 1997; Timonen et al., 1998). These analyses provide a means of differentiating large numbers of bacterial isolates rapidly, but they suffer from the limitations discussed above in regards to the use of any technique that requires isolation of bacterial strains from soil. Due to the limitations of isolating representative microbes from soil described above, it is difficult, perhaps even impossible, to relate the results of the study to elucidation of the specific members of the active, functioning soil microbial community.

A reasonable compromise to overcome this limitation is to inoculate soil extracts directly into the wells of the BIOLOG plates, without any purification of the microbial species. With this technique, an underlying assumption is that enzymes and microbes active (currently) in the field will be detected by the assay procedure. That is, the data will to some degree reflect actual metabolic activities expressed in the field. As discussed above, reproducible patterns reflecting metabolic diversity ('fingerprints') that are characteristic of communities characteristics of specific soil systems and that vary predictably with specific soil properties, such as soil moisture, are produced. BIOLOG analyses provide sensitive indicators of community changes to carbon substrate inputs—be they due to plant species variation effects on rhizosphere exudates (e.g., Ellis et al., 1995; Garland, 1996a; Grayston et al., 1998) or chemical pollutants (e.g., Fuller et al., 1997; Thompson et al., 1999). In the Thompson et al. (1999) study, amendment of soil with 1,2-dichlorobenzene reduced total bacterial counts, viable fungal hyphal length, and the number of carbon compounds metabolized on the BIOLOG plate.

"Metabolic Diversity" vs. "Functional Diversity" Nomenclature: The results from a metabolic profile type test, such as BIOLOG, could be described by a number of terms. Herein the results have been called *metabolic diversity*. They could also be referred to as *functional diversity*. Use of the term *function diversity* has been criticized (Kennedy and Gewin, 1997; Konopka et al., 1998). As succinctly stated by Konopka et al. (1998), "As a measure of the functional diversity of microbial communities, this approach suffers because the tested substrates do not accurately represent the types of substrates present in the ecosystems, and the metabolic redundancy of species implies that changes in the response may only crudely represent the actual microbial population dynamics." Thus, care must be used in applying the term *function* to these data because of additional implications of the word *function*. This word not only relates to a metabolic capability but it also carries implications relating to the expression of the activity *in situ*. That is, use of the term *metabolic diversity* only implies a potential capability, whereas *functional diversity* actually suggests an active expression of the capability. Further complexities associated with the term *functional diversity* relate to the trophic level of the soil system under consideration. For example, functional diversity considered at the microsite level can relate to expressed and unexpressed metabolic activities whereas a consideration of all trophic levels within the soil ecosystem could expand these functions from the biochemical level to the point of including such processes as biological interactions — predators, prey, or parasitic interactions — or even physical transformations of the soil, such as the impact of earthworm activity on soil structure and the resultant effect on water infiltration. (See Kennedy and Gewin, 1997, for further discussion of this concept.)

Concerns with Interpretation of Metabolic Diversity Data: A primary question relating to such data is not "Do the data relate to the nature of the ecosystem being sampled?" — they do. The response to this initial question was adequately documented by the references cited above. The primary concern is "Can the patterns of metabolism of the individual substrates in the BIOLOG plate be used to specifically describe how the various metabolic processes are occurring in the field site?" — generally they cannot. This conclusion is based on the following observations. The substrates used in the BIOLOG plates were initially selected because they were useful for taxonomic identification of bacterial species — not because they appear in any particular environment. Thus, these substrates may or may not be present in a soil ecosystem. Furthermore, if present in soil, the particular organic substrates used in the metabolic test may or may not exist in the field soils at concentrations that would affect the composition of the microbial community. Thus, unless a definite role in the field can be established for metabolism of any particular substrate in the BIOLOG plate, the data should be considered to represent only a "fingerprint" indicative of the community in question. At the initial level, occurrence of a positive test for metabolism of a particular substance on the BIOLOG plate does not necessarily translate to indicating that it is metabolized in the field.

Further research is needed to determine the relationship of the metabolism of each individual substrate on the plate to field soil properties. Since reaction rates for the various metabolic processes measured by the BIOLOG procedure are recorded, it is tempting to relate the relationship of the various reaction rates to field potentials. Can the ability to catabolize at a particular rate or not to use a particular organic compound be extrapolated back to the field situation? Again, it must be stressed that the data for positive wells can at best be considered to provide only an indication of potential activity. The organism or enzyme (or the potential to synthesize the enzyme) causing the activity detected in the specific microtiter well can reasonably be assumed to have existed in the soil sample (if the soil samples and extracts were handled in a manner to preclude contamination in the laboratory), but no inference can be gained regarding its function since the environmental conditions may not have been appropriate for it to function in the soil sample. The organism could have been existing in a resting or an active state. Therefore, although a positive reaction has occurred at a measured rate in the test, there is no assurance that the rate has any relationship to "the real-world" situation. Even less can be concluded from the negative data. The state of no activity could have been detected because of a lack of organisms capable of catalyzing the process in the original soil samples. Alternatively, the organisms may have been present in the soil extract but not in the particular aliquot dispensed into the well containing the substance of interest. Furthermore, the organism could have been present in the soil and the test well, but may not have been capable of functioning under the particular incubation conditions or with the medium used in the test.

A further difficulty with this procedure resides with the unstated and unfulfilled assumption that the aliquots of soil extract dispensed into each microtiter well are identical. Also, it is assumed that any growth or changes in the microbial populations after they are added to the wells will be consistent between the wells. Both assumptions are unfulfilled with the procedure. Smalla et al. (1998), using denaturing gradient gel electrophoresis and temperature gradient electrophoresis assessment of the diversity of ribotypes in inocula and individual wells of BIOLOG plates, showed that the patterns do not necessarily reflect the functional potential of the numerically dominant community members. Analysis of microbial populations of the various test wells reveals that significant differences in the microbial community occur between the wells of the same plate inoculated with the same extract. Thus, at least a portion of the variation in occurrence of oxidation of specific organic substrates between the soil extracts results from the probability of an organism or enzyme requisite for catalyzing the process reaching the well containing the substrate to be oxidized. In conclusion, differences in the data relate to procedural differences that are the result of random variation in the distribution of inoculant and are necessarily not due to fundamental differences in the soil sample. The practical extension of this observation is the need to replicate all analysis.

A fundamental conceptual difficulty further complicates interpretation of metabolic diversity data associated with evaluation of environmental samples. This complication involves the amount of time that the inoculated microtiter plates are incubated before the results are recorded. As indicated above, the procedure was developed to aid in taxonomic identification of pure cultures of bacteria. To accomplish this task, the bacteria are grown in a defined medium, washed, and suspended at a *specific population density* in a specified buffer. Then, each well of the BIOLOG plate is inoculated with a few microliters of this cell suspension. Therefore, the wells of the plate contain approximately the same population density of organisms at about the same developmental stage. The inoculated plates are incubated under prescribed conditions for a set time to allow for appropriate reactions to occur. That is, a set incubation time can be used because of the uniformity of the inoculum and the reasonable uniformity of survival and/or replication of the cells in each well. The point to be stressed is that the biological content of each well of the BIOLOG plate is reasonably consistent. In contrast, because of the heterogeneity of the microbial populations in soil extracts, as noted above it is unlikely that exactly the same array of species will be added to each well of the BIOLOG plate. More importantly, because of the variety of metabolic states of the microbes extracted from soil, even if the same species occurs in each well, the metabolic state may vary; that is, some may be active respiring whereas others may be in a resting state. These species variation and population development differences in the BIOLOG wells inoculated with direct extracts from soil samples have been documented (Garland, 1996b; Haack et al., 1995; Lindstrom et al., 1998). The practical consequences of these variations is that it cannot be assumed that the reaction rate between the various wells on a plate or between comparable plates will be the same. Thus, color development of the plates cannot be analyzed after a prescribed inoculation time. Yet, to compare the results for different soil samples, comparable development of the plates to be compared is required. This is not possible using a set incubation time for the individual, inoculated plates. To overcome this limitation, Garland (1996b) developed the average well color development method for analyzing the data. With this method, the results on the BIOLOG plates are compared based on the extent of the reaction, thereby eliminating or reducing the vagaries of using a constant incubation time. Although not without its criticisms (e.g., see Howard, 1997), the method does allow comparison of reasonably comparable "metabolic fingerprints" of various environmental samples. It must be noted that even if equivalent population densities are added to the wells of the BIOLOG plate, the development time for the reactions to occur may vary. An example of this is found in a study of Zn contaminated soils. Kelly and Tate (1998a) found that color development times in BIOLOG plates inoculated with extracts from Zn contaminated soils to be considerably longer than those from control sites.

Considering all of these difficulties with the procedure, what useful information is provided by the metabolic diversity data? At this point, it can be reasonably concluded that the method provides a reproducible "fingerprint" that can be used to assess the condition of a soil system at a particular time and place, and to evaluate the impact of change or management of the system on the soil biological community. Insufficient data have been recorded thus far to assess long-term stability of the "fingerprint" and the amount of change that can occur within a specific system thoughout the annual cycle without altering, significantly, the fingerprint. To what degree the fingerprint is a characteristic or identifier of the system, and to what degree the fingerprint changes with modification of the controllers of the biological activity of the system have not been determined. Again, will the purported fingerprint be stable over a growing season, several growing seasons, or until the site conditions have been altered sufficiently that the soil is supporting a new type of ecosystem? As with any newly emerging procedure, the range of utility and degree of valid extrapolation of the results must be established. Metabolic diversity provides an useful assessment of the biological status of a particular soil ecosystem and the impact of change on that system.

3.4 Phospholipid Fatty Acid Analysis

A second indicator of variation in biological diversity in soils involves determination of the diversity of phospholipid fatty acids contained in soil samples. To utilize a surrogate, such as phospholipid fatty acids, as an indicator of biological diversity, the group of chemicals must meet five requirements:

- It must be common to all groups of organisms of interest.
- The substances must vary in a predictable manner between groups of microbes.
- These microbial groups must vary between soil systems in a reproducible and predictable manner.
- The substance must be unstable outside of the cell so that quantities of the chemicals extracted from the soil represent living organisms only.
- It must be possible to isolate representative samples of the cell component from the soil.

Phospholipid fatty acids in soil and microorganisms meet all of these criteria. Furthermore, since the microbes do not have to be removed from soil prior to extraction of the fatty acids, the extract likely represents the totality of the fatty acids contained in the soil microbial population. Note that phospholipid fatty acid analysis has been used to characterize axenically grown bacterial isolates from soil for assessment of microbial community diversity (e.g., Germida et al., 1998), but as with the BIOLOG procedure, a meaningful characterization of soil community dynamics can be attained by direct extraction of the phospholipid fatty acids from the soil. It is assumed that all phospholipid fatty acids in soil are equally accessible to the extracting agent and that the phospholipid fatty acid profile of the extract is representative of the total phospholipid fatty acid component of the soil. That is, the diversity of the phospholipid fatty acids extracted from soil is concluded to be representative of that occurring *in situ* and the procedure is considered to be reproducible.

The phospholipid fatty acid profile of extracts of soil reveals changes in the microbial community structure of the entire soil sample. For this assay, the soil microbial phospholipid fatty acids are extracted from soil with solvent mixtures such as chloroform, methanol, and phosphate buffer. (See White et al., 1979, for a detailed description of the extraction procedure.) Following extraction, the chloroform fraction is concentrated and the fatty acids are separated on a silicic acid column. (See method of King et al., 1977.) The polar lipid fraction is collected, dried under nitrogen, saponified, and methylated. The fatty acid methyl esters are then identified [e.g., with the MIDI protocol (MIDI, 1995)]. The MIDI system identifies and quantifies fatty acids of 9 to 20 carbon chain lengths. The nomenclature for the fatty acids commonly used with the procedure consists of a serices of numbers and Greek letters (e.g. 16:1 ω7t), referring to the total number of carbon atoms in the chain:the number of double bonds, the position of the double bond from the methyl end of the chain, and the conformation. As with any procedure, some fatty acids are not separated by the chromatographic methods used and are simply referred to as mixtures in the data pool. Different subgroups of microorganisms (e.g., fungi, mycorrhizal fungi, gram positive bacteria, actinomycetes, gram negative bacteria) are represented in the profile by the quantities of specific fatty acids. For example 16:1 ω5c is suggested to be an indicator for arbuscular mycorrhizal fungi (Haack et al., 1995), 16:0 10 of nocardioform actinomycetes, and TBSA 18:0 10 Me and 17:0 10 Me of actinomycetes (Frostegård et al., 1993b) and 18:2 ω6c of fungi (Guckert et al., 1985). Thus, the resultant phospholipid fatty acid profile provides an indication of not only changes in diversity of the biological community (by changes in the overall composition of the profile) but also of the changes in these groups of microbes. This procedure has been shown to be a sensitive indicator of alteration of general soil biological community structure (Frostegård et al., 1993a) and metal-amended soils or in soil impacted by metal smelters (Frostegård et al., 1996; Kelly and Tate, 1998b; Pennanen et al., 1996). The method is sufficiently sensitive that impact of root herbivory on rhizosphere communities (Denton et al., 1999) as well as shifts in community structure due to soil management (Bardgett et al., 1997; Bossio et al, 1998; Frostegård et al. 1993c; Klamer and Bååth, 1998) can be detected. A useful adjunct to the analysis of phospholipid fatty acid diversity is that total phospholipid fatty acid content of the extract has been used to estimate soil microbial biomass (e.g., Denton et al., 1999; Klamer and Bååth, 1998).

Neither the metabolic diversity or phospholipid fatty acid diversity analysis provides any indication of the variation in actual species diversity of the soil samples. As was indicated above, the final product of each merely represents the variation in either activities or chemicals that are hoped to parallel the degree of variation in true species diversity. The phospholipid fatty acid profile is likely more reflective of true variation of the populations since a representative sample of the indicator fatty acids is extracted directly from the soil samples. Recall that with the metabolism-based procedure, the activity quantified may or may not have been expressed in soil, and since an extraction of the microbial populations is involved, a differential separation of the microbial cells and enzymes is likely. Although neither technique fulfills the requisite of simplicity and economy stated above, probably the best procedure in using metabolism and phospholipid fatty acid indictors of soil biological diversity is to use both methods concurrently. Use of two independent assessors of diversity provides a clearer picture of the variation of diversity within the soil site. Furthermore, the phospholipid fatty acid-based study provides some additional data describing the distribution of large groups of soil microbes.

3.5 Nucleic Acid-Based Analysis of Soil Microbial Diversity

Several properties of the DNA molecule contribute to its utility for assessment of soil biological diversity. It meets the fundamental requirements stated above for a surrogate marker for such assessments in that it is present in all living cells and its survival outside the living cell is limited. That is, we can reasonably assume that the DNA isolated form soil represents that present within the soil community. From the view of the DNA molecule itself, two structural properties that are of great utility are the fact that it is composed of two complementary strands and that phosphorus makes up a major portion of the molecule. The value of the existence of two complementary strands is their utility in amplifying the quantity of specific DNA present in the sample and the fact that the rate of reassociation (reannealing) of the DNA can serve as an indicator of its diversity. The phosphorus atom adds to the utility of DNA assessment because the radioactive isotope of phosphorus (^{32}P) can be used to visualize and quantify specific genetic markers in the DNA preparations. DNA sequences do change. That is, genes do mutate. Fortunately, the mutation rate is not uniform throughout the genome. A specific property of the genome of particular value in assessing species variation in soil samples is the fact that particular regions of the DNA molecule are highly conserved. Thus, changes in these less variable regions of the genome, such as the DNA that encodes 16S ribosomal RNA, can be used as markers for specific organisms and indicators of the diversity of the DNA extracted from soil. (See Chapter 2 for examples of DNA analysis-based methods used to detect specific microbial species and strains in soil.)

A variety of techniques have been developed for characterizing DNA extracted from the soil microbial community. They all have some limitations in common. A primary consideration is that DNA analyses provide an indication of the variation of the genome of the microbial community, but unless markers specific to an individual species are used in the analysis, no information is provided regarding individual species identities. Diversities can be estimated by assuming that similarities in DNA sequence greater than 70 percent represent individual species.

Once an estimate of the number of species present in a sample plus perhaps the identity of some of the individuals has been determined, the question of their contribution to ecosystem function arises. Due to the great redundancy existing in soil microbial communities, the occurrence of resting or inactive cells, and the fact that all DNA is extracted from the soil sample, it is clear that these analyses of DNA sequences and reannealing rates reveal little information about the identity of the active microbes in the soils. Conclusions requiring activity estimates of microbes through use of nucleic acid analysis necessitates quantification of moieties synthesized only in active cells. Such a nucleic acid fraction is the messenger RNA (mRNA). mRNA is synthesized only when the particular enzyme encoded by the DNA is required by the cell. Furthermore, it has a short residence time in the cell. Thus, mRNA exists in the cell for only those short periods of time for which the enzyme activity is required. Due to difficulties of preparing a clean preparation of mRNA from soil samples, these methods are still in their infancy.

A variation on these procedures that is used to indicate when specific enzymatic activities are being expressed in a soil sample is to use reporter genes.

In this case, genes (reporter genes) whose enzymatic product is easily detected are inserted into genome to form an association with the specific genes whose expression in soil is of interest. Thus, when the gene of interest is expressed, so is the more easily quantified reporter gene. A commonly used reporter gene is that resulting in the production of β-galactosidase (the *lac z* gene). Therefore, if a specific genetically engineered organism containing a reporter gene is added to soil, expression of its activity *in situ* can be specifically detected. This method only indicates that the enzymes encoded by the portion of the genome in question were synthesized in the soil — not necessarily that replication of the soil-amended microbial strain occurred. Until better methods have been developed to isolate and to analyze the variability in the mRNA fraction, DNA and RNA analytical procedures are limited to revealing the general diversity of the soil biological genome and not of the active portion thereof.

As with each of the procedures used in soil community diversity analysis, axenically cultivated microbial isolates may be evaluated or DNA may be extracted directly from soil samples. The former procedure, use of microbial isolates, has all of the limitations described above for the metabolic or phospholipid fatty acid diversity analyses. Analysis of soil DNA extracts is complicated by the heterogeneity of the chemical components of soil and the potential for interference with the extraction and detection of the DNA. A primary concern is the potential for interference in DNA analytical procedures by any humic acids contained in the extracts. Thus, the extraction methods selected must either minimize coextraction of humic acids or involve a purification step to remove the humic acid fraction. Several good methods of DNA extraction from soil that minimize difficulties associated with humic acids (e.g., see Tebbe and Vahjen, 1993; Tsai and Olson, 1992; Moyer et al., 1996; Zhou et al., 1996) have been developed. Examples of current studies in which these procedures have been used to evaluate unique properties of soil communities or the impact of various management procedures include, for Hawaiian soil bacterial communities, Nüsslein and Tiedje (1998); for impacts of agricultural management, Borneman et al. (1996) and Øvreås and Torsvik (1998); for Amazonian soils and impact of deforestation, Borneman and Triplett (1997); for grassland soils, Felske and Akkermans (1998); and for effects of elevated atmospheric carbon dioxide levels on rhizosphere community structure, Griffiths et al. (1998).

3.6 Conclusions: Utility and Limitations of Diversity Analysis Procedures

A primary conclusion from study of each of the major types of methods used to characterize soil microbial community diversity is that no single technique provides an all-inclusive, definitive description of soil microbial community structure and the impact of variation in community defining soil properties on the composition of this community. Each technique yields an indication of the variation of a particular parameter associated with microbial communities — ability to metabolize specific arrays of organic substances, nucleic acid or phospholipid fatty acid profiles. These properties of the community vary independently of each other. Therefore, some correspondence in the "answers" provided by the techniques may occur, but cannot be expected. Many studies are thus conducted using combinations of the procedures to provide a more clear understanding of community dynamics. For example, Fritze et al. (1997) found no effect of fertilization of metal-polluted forest soil microbial community structure with the study of metabolic diversity (BI-OLOG), yet the treatments were clearly distinguished with phospholipid fatty acid analysis. Examples of other studies combining metabolic and phospholipid fatty acid analysis diversities include bioremediation effects on microbial communities (Siciliano et al., 1998) and agricultural management (Bossio and Skow, 1998). Similarly, examples of metabolic diversity (BIOLOG) analyses and DNA characterization combined studies include agricultural soil character-ization (Øvreås and Torsvik, 1998) and pesticide application on soils (Engelen et al., 1998). Phospholipid fatty acid analyses and DNA studies have been combined to study such soil management concerns as soil amendments (e.g., Griffiths et al., 1999).

None of these diversity analyses techniques is specifically designed to detect microbes definitely active at the time of soil sampling. Changes due to management detected by any of the procedures suggest that a shift in microbial population composition or capabilities must have occurred, but the microbes may have been temporarily activated and then induced to enter a resting state.

Even beyond these considerations of when components of the microbial community are active, a more fundamental conceptual need for soil microbial diversity analyses is the development of a fundamental concept of just what a change in community diversity really means. Assuming that problems asso-ciated with defining a bacterial species can be overcome, the fact that soil communities are highly redundant — and that much of the redundancy results from inactive or resting cells — complicates the development of an absolute understanding of the importance of changes in soil microbial diversity. True advances in investigations of these types of studies await assemblage of sufficient data to allow emergence of a concept defining the true importance in increases or decreases in soil microbial diversity, specifically that portion of the population actually functioning in the ecosystem.

References

Bardgett, R. D., D. K. Leemans, R. Cook, and P. J. Hobbs. 1997. Seasonality of the soil biota of grazed and ungrazed hill grasslands. Soil Biol. Biochem. 29:1285–1294.

Behrendt, U., T. Muller, and W. Seyfarth. 1997. The influence of extensification in grassland management on the populations of micro-organisms in the phyllosphere of grasses. Microbiol. Res. 152:75–85.

Borneman, J., P. W. Skroch, K. M. O'Sullivan, J. A. Palus, N. G. Rumjanek, J. L.

Jansen, J. Nienhuis, and E. W. Triplett. 1996. Molecular microbial diversity of an agricultural soil in Wisconsin. Appl. Environ. Microbiol. 62:1935–1943.

Borneman, J., and E. W. Triplett. 1997. Molecular microbial diversity in soils from eastern Amazonia: Evidence for unusual microorganisms and microbial population shifts associated with deforestation. Appl. Environ. Microbiol. 63:2647–2653.

Bossio, D. A., and K. M. Scow. 1995. Impact of carbon and flooding on the metabolic diversity of microbial communities in soils. Appl. Environ. Microbiol. 61:4043–4050.

Bossio, D. A., and K. M. Skow. 1998. Impacts of carbon and flooding on soil microbial communities–phospholipid fatty acid profiles and substrate utilization patterns. Microbial Ecol. 35:265–278.

Bossio, D. A., K. M. Scow, N. Gunapala, and K. J. Graham. 1998. Determinants of soil microbial communities — effects of agricultural management, season, and soil type of phospholipid fatty acid profiles. Microbial Ecol. 36:1–12.

Denton, C. S., R. D. Bardgett, R. Cook, and P. J. Hobbs. 1999. Low amounts of root herbivory positively influence the rhizosphere microbial community in a temperate grassland soil. Soil Biol. Biochem. 31:155–165.

Ellis, R. J., I. P. Thompson, and M. J. Bailey. 1995. Metabolic profiling as a means of characterizing plant-associated microbial communities. FEMS Microbiol. Ecol. 16:9–17.

Engelen, B., K. Meinken, F. von Wintzingerode, H. Heuer, H.-P. Malkomes and H. Backhaus. 1998. Monitoring impact of a pesticide treatment on bacterial soil communities by metabolic and genetic fingerprinting in addition to conventional testing procedures. Appl. Environ. Microbiol. 64:2814–2821.

Felske, A., A. D. L. Akkermans. 1998. Spatial homogeneity of abundant bacterial 26S rRNA molecules in grassland soils. Microbial Ecol. 36:31–36.

Fritze, H., T. Pennanen, and P. Vanhala. 1997. Impact of fertilizers on the humus layer microbial community of scots pine stands growing along a gradient of heavy metal pollution. Pp. 68–83 *In* H. Insam and A. Rangger (eds.), Microbial Communities. Functional Versus Structural Approaches. Springer-Verlag, NY.

Frostegård, A., E. Bååth, and A. Tunlid. 1993a. Shifts in structure of soil microbial communities in limed forests as revealed by phospholipid fatty acid analysis. Soil Biol. Biochem. 25:723–730.

Frostegård, A., A. Tunleid, and E. Bååth 1993b. Phospholipid fatty acid composition, biomass, and activity of microbial communities from two different soil types experimentally exposed to different heavy metals. Appl. Environ. Microbiol. 59:3605–3617.

Frostegård, A., A. Tunleid, and E. Bååth 1993c. Shifts in the structure of soil microbial communities in limed forests as revealed by phospholipid fatty acid analysis. Soil Biol. Biochem. 25:723–730.

Frostegård, A., A. Tunleid, and E. Bååth 1996. Changes in microbial community structure during long term incubation in two soils experimentally contaminated with metals. Soil Biol. Biochem. 28:55–63.

Fuller, M. E., K. M. Scow., S. Lau, and H. Ferris. 1997. Trichloroethylene (TCE) and toluene effects on the structure and function of the soil community. Soil Biol. Biochem. 29:75–89.

Garabito, M. J., M. C. Marquez, and A. Ventosa. 1998. Halotolerant *Bacillus* diversity in hypersaline environments. Can. J. Microbiol. 44:95–102.

Garland, J. L. 1996a. Patterns of potential C source utilization by rhizosphere communities. Soil Biol. Biochem. 28:223–230.

Garland, J. L. 1996b. Analytical approaches to the characterization of samples of microbial communities using patterns of potential C source utilization. Soil Biol. Biochem. 28:213–221.

Garland, J. L., and A. L. Mills. 1991. Classification and characterization of heterotrophic microbial communities on the basis of patterns of community-level sole-carbon-source-utilization. Appl. Environ. Microbiol. 57:2351–2359.

Garland, J. L., and A. L. Mills. 1994. A community-level physiological approach for studying microbial communities.Pp. 77–83 *In* K. Ritz, J. Dighton, and K. E. Giller (eds.), Beyond the Biomass. Wiley-Sayce, London.

Germida, J. J., S. D. Siciliano, J. R. Defreitas, and A. M. Seib. 1998. Diversity of root-associated bacteria associated with field grown canola (*Brassica napus* L.) and wheat (*Triticum aestivum* L.). FEMS Microbiol. Ecol. 26:43–50.

Grayston, S. J., S. Q. Wang, C. D. Campbell, and A. C. Edwards. 1998. Selective influence of plant species on microbial diversity in the rhizosphere. Soil Biol. Biochem. 30:369–378.

Griffiths, B. S., K. Ritz, N. Ebblewhite, and G. Dobson. 1999. Soil microbial community structure: Effects of substrate loading rates. Soil Biol. Biochem. 31:145–153.

Griffiths, B. S., K. Ritz, N. Ebblewhite, E. Paterson, and K. Killham. 1998. Ryegrass rhizosphere microbial community structure under elevated carbon dioxide concentrations, with observations on wheat rhizosphere. Soil Biol. Biochem. 30:315–321.

Guckert, J., C. Antworth, P. Nichols, and D. White. 1985. Phospholipid, ester-linked fatty acid profiles as reproducible assays for changes in prokaryotic community structure of estuarine sediments. FEMS Microbiol. Ecol. 31:147–158.

Haack, S. K., H. Garchow, M. J. Klug, and L. J. Forney. 1995. Analysis of factors affecting the accuracy, reproducibility, and interpretation of microbial community carbon source utilization patterns. Appl. Environ. Microbiol. 61:1458–1468.

Howard, P. J. A. 1997. Analysis of data from BIOLOG plates: Comments on the method of Garland and Mills (Letter to the Editor). Soil Biol. Biochem. 29:1755–1757.

Kelly, J. J., and R. L. Tate III. 1998a. Use of BIOLOG for the analysis of microbial communities from zinc-contaminated soils. J. Environ. Qual. 27:600–608.

Kelly, J. J., and R. L. Tate III. 1998b. Effects of heavy metal contamination and remediation on soil microbial communities in the vicinity of a zinc smelter. J. Environ. Qual. 27:609–617.

Kennedy. A. C., and V. L. Gewin. 1997. Soil microbial diversity: Present and future considerations. Soil Sci. 162:607–617.

King, J. D., D. C. White, and C. W. Taylor. 1977. Use of lipid composition and metabolism to examine structure and activity of estuarine detrital microflora. Appl. Environ. Microbiol. 33:1177–1183.

Klamer, M., and E. Bååth 1998. Microbial community dynamics during composting of straw material studied using phospholipid fatty acid analysis. FEMS Microbiol. Ecol. 27:9–20.

Knight, B. P., S. P. McGrath, and A. M. Chaudri. 1997. Biomass carbon measurements and substrate utilization patterns of microbial populations from soils amended with cadmium, copper, or zinc. Appl. Environ. Microbiol. 63:39–43.

Konopka, A., L. Oliver, and R. F. Turco. 1998. The use of carbon substrate utilization patterns in environmental and ecological microbiology. Microbial Ecol. 35:103–115.

Kuhn, I., B. Austin, D. A. Austin, A. R. Blanch, P. A. D. Grimont, J. Jovre, S. Koblavi, J. L. Larsen, R. Mollby, and K. Pedersen. 1996. Diversity of *Vibrio anguillarum* isolates from different geographical and biological habitats determined by the sue of a combination of eight different typing methods. Syst. Appl. Microbiol. 19:442–450.

Lindstrom, J. E., R. P. Barry, and J. F. Braddock. 1998. Microbial community analysis—A kinetic approach to constructing potential C source utilization patterns. Soil Biol. Biochem. 30:231–239.

Lupwayi, N. Z., W. A. Rice, and G. W. Clayton. 1998. Soil microbial diversity and community structure under wheat as influenced by tillage and crop rotation. Soil Biol. Biochem. 30:1733–1741.

MIDI. 1995. Sherlock Microbial Identification System Operating Manual: Version 5. MIDI, Newark, DE.

Moyer, C. L., J. M. Tiedje, F. C. Dobbs, and D. M. Karl. 1996. A computer-simulated restriction fragment length polymorphism analysis of bacterial small-subunit rRNA genes: Efficacy of selected tetrameric restriction enzymes for studies of microbial diversity in nature. Appl. Environ. Microbiol. 62:2501–2507.

Nüsslein, K., and J. M. Tiedje. 1998. Characterization of the dominant and rare members of a young Hawaiian soil bacterial community with small-subunit ribosomal DNA amplified from DNA fractionated on the basis of its guanine and cytosine composition. Appl. Environ. Microbiol. 64:1283–1289.

Øvreås, L., J. S. Jensen, F. L. Daae, and V. Torsvik. 1998. Microbial community changes in a perturbed agricultural soil investigated by molecular and physiological approaches. Appl. Environ. Microbiol. 64:2739–2742.

Øvreås, L., and V. Torsvik. 1998. Microbial diversity and community structure in two different agricultural soil communities. Microbial Ecol. 36:303–315.

Pennanen, T., A. Frostegård, H. Fritze, and E. Bååth 1996. Phospholipid fatty acid composition and heavy metal tolerance of soil microbial communities along two heavy metal polluted gradients in coniferous forests. Appl. Environ. Microbiol. 62:420–428.

Russek-Cohen, E., and R. R. Colwell. 1986. Applications of numerical taxonomy procedures in microbial ecology. Pp. 133–146. *In* R. L. Tate III (ed.), Microbial Autecology. A Method for Environmental Studies. Wiley-Interscience, NY.

Siciliano, S. D., J. J. Germida, and S. D. Siciano. 1998. BIOLOG analysis and fatty acid methy ester profiles indicate that pseudomonad inoculants that promote phytoremediation alter the root-associated microbial community of *Bromus biebersteinii.* Soil Biol. Biochem. 30:1717–1723.

Smalla, K., U. Wachtendorf, H. Heuer, W.-T. Liu, and L. Forney. 1998. Analysis of BIOLOG GN substrate utilization patterns by microbial communities. Appl. Environ. Microbiol. 64:1220–1225.

Tebbe, C. C., and W. Vahjen. 1993. Interference of humic acids and DNA extracted directly from soil in detection and transformation of recombinant DNA from bacteria and yeast. Appl. Environ. Microbiol. 59:2657–2665.

Thompson, I. P., M. J. Bailey, R. J. Ellis, N. Maguire, and A. A. Meharg. 1999. Response of soil microbial communities to single and multiple doses of an organic pollutant. Soil Biol. Biochem. 31:95–105.

Timonen, S., K. S. Jorgensen, K. Haahtela, and R. Sen. 1998. Bacterial community structure at defined locations of *Pinus sylvestris-Suillus bovinus,* and *Pinus sylvestris-Paxillus involutus* mycorrhizospheres in dry pine forest humus and nursery peat. Can. J. Microbiol. 44:499–513.

Torsvik, V., Goksøyr, and F. L. Daae. 1990. High diversity in DNA of soil bacteria. Appl. Environ. Microbiol. 56:782–787.

Torsvik, V., R. Sorheim, and J. Goksoyr. 1996. Total bacterial diversity in soil and sediment communities. J. Indust. Microbiol. Biotehcnol. 17:170–178.

Tsai, Y.-L. and B. H. Olson. 1992. Rapid method for separation of bacterial DNA from humic substances in sediments for polymerase chain reaction. Appl. Environ. Microbiol. 58:2292–2295.

White, D. C., W. M. Davis, J.S. Nickels, J. D. King, and R. J. Bobbie. 1979. Determination of the sedimentary microbial biomass by extractable lipid phosphate. Oecologia 40:51–62.

Windig, A., 1994. Fingerprinting bacterial soil communities using Biolog microtitre plates. Pp. 85–94. *In* K. Ritz, J. Dighton, and K. E. Giller (eds.), Beyond the Biomass. Wiley-Sayce, London.

Zak, J. C., M. R. Willig, D. L. Moorhead, and H. G. Wildman. 1994. Functional diversity of microbial communities—a quantitative approach. 26:1101–1108.

Zhou, J., M. A. Bruns, and J. M. Tiedje. 1996. DNA recovery from soils of diverse composition. Appl. Environ. Microbiol. 62:316–322.

Chapter 4

Energy Transformations and Metabolic Activities of Soil Microbes

A microbe's habitat must not only have physical and chemical properties that are conducive to sustenance of cellular viability (Chapter 5), but must also contain the requisite energy source(s) to drive cellular metabolic processes as well as essential macro- and micronutrients for biomass synthesis. Assumed in this statement is that the microbes themselves possess the genotypic traits necessary to function in their habitat and that prevailing conditions are sufficient to allow the expression of these traits. Prerequisites for population development must also include a means to cope with a variety of biological stresses, such as competition and predation (see Chapter 7).

An appreciation of this multitude of primary requirements for microbial existence in soil provides a portion of the foundation necessary for evaluation of microbial metabolism and energy transformations in soil. Therefore, the objective of this chapter is to elucidate the basic growth and energy recovery mechanisms of the soil microbial community. This will be achieved by (1) evaluating the applicability and implications of laboratory-derived principles of microbial growth kinetics to soil ecosystem function, and (2) elucidating the diversity of energy-yielding processes existent in soil microbial communities. Examples of extension and application of the principles revealed in this chapter are found in the chapters involved with soil-based biogeochemical cycle process and that describing soil bioremediation processes (Chapter 16). It is assumed in this quest that all participants are familiar with basic microbiological and biochemical principles. A clear understanding of the fundamentals of intermediary metabolism, energy-yielding processes in cellular metabolism, regulation of gene expression, and microbial growth parameters is required. Elementary microbiology and biochemistry textbooks should be consulted should deficiencies be noted. The study that follows is prepared with an emphasis on the elucidation of the nuances of the repercussion of soil physical, chemical, and biological realities on microbial metabolic function.

4.1 Microbial Growth Kinetics in Soil

Since the primary benefits to microbes in soil are growth and/or maintenance of population densities, consideration of general microbial growth kinetics and specific examples of these replication and survival patterns in native soil ecosystems is a necessary starting point for analysis of soil physiological processes. To achieve this goal, the applicability of laboratory-derived relationships to field situations will be examined.

Open vs. Closed Biological Systems: In the laboratory, microbial growth kinetics are assessed as closed systems (e.g., the culture tube or batch soil incubation chambers) or as open systems (e.g., a chemostat or in some primary aspects, soil columns) (Fig. 4.1). A closed ecosystem is self-contained in that external contributions to microbial function are excluded, whereas open

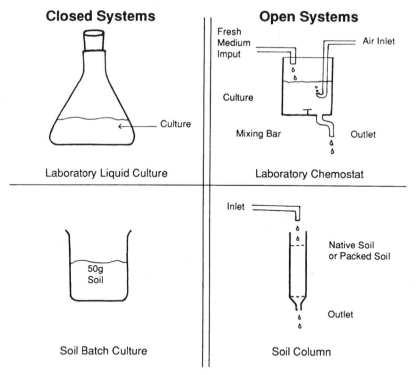

Fig. 4.1 Examples of laboratory-incubated open and closed systems for study of microbial processes.

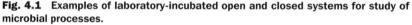

systems receive inputs from processes occurring external to the defined boundaries of the site. As a consequence, in a closed system, nutrient supplies can be limited and metabolic waste materials or by-products can accumulate. In open systems, microbial cells can be constantly bathed in growth nutrients if supplied by the external source, and wastes do not necessarily accumulate in the vicinity of the colony or individual cell.

Although it is essentially impossible to exclude external influences from any portion of a soil site, conditions may exist locally where microbial population dynamics mimic those of a culture vessel incubated in the laboratory. This situation can be exemplified by microbial community development in soil surrounding a portion of decaying insect biomass or plant debris. Growth of the microbial population is initially controlled locally by the quantities of readily metabolized substrates exuded from the animal or plant tissues. As the microbes grow and replicate, these easily metabolized substituents are depleted or are reduced to growth-limiting concentrations. Then, cellular function is controlled by the rate of solubilization or release of nutrients from within the chitinous cell walls of the insect or the cellulosic walls of the plant material — that is, solubilization of simple sugars by chitinase or release of nutrients protected by cell structures through breaching of protective barriers, respectively. Upon depletion of all energy and nutrient supply pools in the vicinity of the microbial colony (i.e., total catabolism of the fixed carbon contained in the plant or animal debris), microbial growth ceases and the population becomes quiescent. This state could be achieved prematurely, should growth-limiting cellular waste products accumulate to inhibitory levels prior to nutrient depletion, or if another essential nutrient supply is exhausted (e.g., free oxygen for aerobic microbes). At this point, continued activity of the microbial population depends on an influx of nutrients from outside its boundaries, as described above.

The artificiality of this comparison of the native soil condition with closed cultural systems emerges when the totality of soil properties influencing microbial growth are considered. In the site described above, the primary delimiters of microbial replication were energy and nutrients, but imposition of moisture limitations, deficiencies in fixed nitrogen resources, or even difficulties with temperature modulation can easily be envisioned. Even with this small list of soil properties affecting microbial function, it is clear that both internal and external forces are involved in sustenance of the soil biological community. In a soil site contiguous with a variety of different soil habitats, stresses and the means of their relief for microbial populations may easily originate in regions beyond the artificial boundaries of an arbitrarily defined closed system. Hence, in soil, closed ecosystem properties may be invoked to simplify data interpretation, but the experimenter must always bear in mind that the microbes of interest are functioning in a system of interconnected habitats.

Not only is a closed system an imperfect descriptor of the rates of biologically catalyzed processes in soils, but the soil ecosystem is also poorly described by the kinetics of an open system. (As will be shown below, chemostat systems are generally used to develop mathematical models of an open biological system.) True, many delimiters of biological activity in soil are imposed from outside sources (e.g., nutrient and energy sources, moisture). This external control of microbial growth could result from a variety of conditions, such as from fixed carbon contributions of root exudates in rhizosphere soils, leaching of soluble fixed carbon sources from soil surface organic matter accumulations (e.g., forest litter layers, thatch accumulated in grassland soils), or anthropogenically created situations such as where industrial or domestic waste systems enter into a soil ecosystem (e.g., sewage or factory effluents).

These similarities to nutrient control of microbial growth in a chemostat are matched by an equally significant array of dissimilarities. A good example of these differences is the fate of microbial biomass. The influx of nutrients into the system can control biomass production in a manner reminiscent of those existent in a chemostat until space limitations are imposed. In a chemostat, microbial biomass remains constant although the cells are in an active growth state. New cells are washed from the system as rapidly as they are produced. This condition is not replicated in soil. Microbial cells tend to be trapped within the soil particulate structure. Thus, nutrients may flow to the cells and waste products be removed by the dynamic fluid system, but microbial growth eventually approaches the space limitations imposed by the microsite structure.

4.1.1 Microbial Growth Phases: Laboratory-Observed Microbial Growth Compared to Soil Population Dynamics

The intricacies of microbial growth and the impact of soil properties on this process are best exemplified by contrasting microbial growth principles derived from laboratory study to those occurring in the soil ecosystem. The generalized microbial growth curve (Fig. 4.2) is usually divided into four phases; lag, exponential growth, stationary, and decline phase. Each of these portions of the growth curve can be observed in soil ecosystems. Yet it must be noted that large changes in soil microbial populations are unusual. A major change in population density generally occurs only with an input of easily metabolized organic matter. This pattern of population development is most commonly associated with soil sites where the native microbial growth rates are stimulated by alteration of soil properties, such as introduction of plant debris into a carbon- or energy-limited community, through oxygenation of an anaerobic system, or through increased availability of native soil organic matter through disruption of the soil structure that precludes its catabolism. Once the previously existing impediments to microbial growth are removed, the cells replicate until a new limitation to community development is encountered.

Lag Phase: A lag phase occurs when conditions are appropriate for microbial growth, yet there is a delay before a measurable change in population

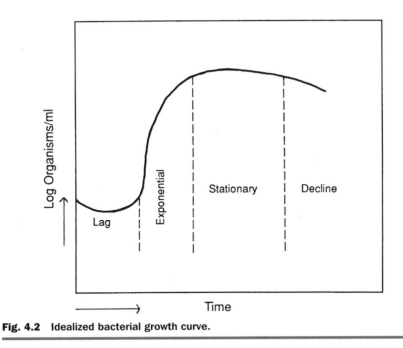

Fig. 4.2 Idealized bacterial growth curve.

density is detectable. This cellular division could be due to site amendment with metabolizable substances that were previously not present (e.g., pesticides, xenobiotic compounds) or alteration of the physical conditions (e.g., imposition of anaerobic conditions, which would stimulate growth of obligate anaerobes, or reduction of soil pH sufficiently to favor acidophilic populations).

In axenic cultures, a sensitive measure of cellular replication is usually employed. Hence, an accurate assessment of the lag time is gained. This assay sensitivity is difficult to achieve in soil. Most methods for quantifying soil microbial population density do not detect small changes in microbial population densities. Thus, several cell-division cycles must occur before a change in the microbial populations is recorded. Therefore it could be concluded that at least a portion of the delays in response of soil populations due to system modification is the product of difficulties in quantifying microbial population densities in soil.

A variety of methods are available for assessing increases in microbial biomass and biomass of individual microbial species in soil. (see Chapter 2 for an analysis of the methods for detecting microbial populations), including techniques for assessing the synthesis rates for various biomass specific compounds, such as lipids [^{14}C-acetate incorporation into microbial lipids (Barnhart and Vestal, 1983; Federle et al., 1983; Tate, 1985)] and nucleic acids [^{3}H-thymidine incorporation (Christensen et al., 1989)]. Accuracy of the data yielded from these procedures is controlled by such factors as

- The variability of biomass within soil samples
- The extraction efficiency for recovery of the population or cellular component assayed from the soil sample
- The minimal populations that can be detected with each procedure

A variety of implications of the occurrence of a lag phase for microbial growth in soils on interpretation of data describing soil process rates can be conceived. A positive aspect of the occurrence of a lag phase in soil microbial growth kinetics can be exemplified by considerations associated with use of biodegradable soil amendment, for example, pesticides. Sufficient longevity of the pesticide in soil for it to interact with the target species is requisite. Greatest time is allowed if the population density of the microbes catalyzing destruction of the soil amendment is minimal. With populations of microbes capable of mineralizing the soil amendment of a few hundred or even thousands of organisms per g soil, several cell divisions may be required before detectable declines in the pesticide concentration occur. Thus, lag periods of days to weeks could be anticipated with pristine systems. In such systems, the chemical amendment would have an adequate longevity to accomplish the desired ends before its equally desirable biodegradation occurs. Efficacy of repeated uses of the soil amendment in the system would be dependent on the longevity of the microbial populations involved in its decomposition.

A negative environmental implication of extended delays in biological decomposition of soil amendments is the potential for subsoil or groundwater contamination. Again, biodegradable compounds may be added to soil purposely or accidentally through mismanagement or through spills. To minimize the environmental impact in these situations, rapid biodegradation of the contaminant is necessary. The longer the lag in development of requisite biochemical activity, the greater the potential for leaching of the substance to a site where its presence is inappropriate and its decomposition is less likely to occur (for example, the A horizon to subsoil regions where decomposition is limited or groundwater contamination can occur).

This lag in development of new biological activity may be exploited during laboratory estimation of *in situ* biochemical activities. Proper collection and storage of soil samples is essential for assessment of soil processes. Unfortunately, disturbance of the soil matrix induces growth of many microbial populations. Fortunately, the delay in growth (lag phase) and use of methods such as storage at 4°C to reduce the microbial growth rate may afford the experimenter several hours to days to complete the laboratory assessments before statistically significant changes in population densities occur. The length of storage allowed before data are compromised by microbial growth depends upon the properties of the species of interest and the sensitivity of the assay method. The more sensitive the assay (i.e., the capacity to detect changes in population density of a few cells per gram soil) or the more prolific the microbes of interest (i.e., growth rates of a few hours under the storage conditions), the greater will be the problems associated with estimation of *in situ* soil microbial population densities. Less storage stability is anticipated for samples to be assayed using polymerase chain reaction procedures or fluorescent antibody detection of microbial cells, since these methods are sufficiently sensitive to allow detection of increases of population densities of a few cells per g of soil. In contrast, considerable increases in populations quantified by most-probable-number procedures is possible before significant changes in data are demonstrable. Confidence limits of ± 300 percent are not unusual with these types of procedures. In either case, a control study must be conducted to determine the acceptability of the chosen storage procedure for the soils and populations in question.

Exponential Growth Phase: Conceptually, occurrence of a period of exponential growth, a blooming per se, of specific microbial populations in soil is easily envisioned. Amendment of soil with essentially any biodegradable substance at a nontoxic, growth-supporting concentration creates a new niche for the resident microbes. The genetic diversity of the indigenous community as well as the commonly occurring immigration of propagules from adjacent as well as distant ecosystems (e.g., via wind and water erosion) virtually assure inoculation of the incorporated materials with organisms capable of their mineralization.

Other less appreciated, but perhaps more commonly occurring processes inducing microbes to enter an exponential growth phase are associated with indigenous ecosystem processes. Intrusion of predators into a microsite where bacterial growth is controlled by space limitations may reduce the population densities sufficiently to allow several rounds of cell division. Indeed, balanced feeding of the predator could maintain the prey population in an active growth phase for extended time periods. When this occurs, the microbial population could be dividing at a relatively rapid rate, but ingestion of the newly produced cells by the predators would prevent an observable increase in prey population density.

Similarly, a burst of cell division occurs following freeze-thaw cycles in soil. Freezing and thawing of soil results in death of susceptible microbial populations. For example, Biederbeck and Campbell (1971) found that one freeze-thaw cycle in recently cropped soil resulted in death of 92 percent of the bacterial population, 55 percent of the fungi, and 33 percent of the actinomycetes. The organic substances constituting the cell mass of the deceased organisms provided carbon and energy as well as a variety of cofactors for replication of the surviving microbial populations. This nutrient reservoir plus the reduction in the limitations of the space available for microbial growth results in a rapid increase in microbial biomass.

Soil disturbance, natural through activity of soil animals or the result of anthropogenic intervention such as cultivation, may also induce at least a temporary increases in microbial biomass. The disturbance may release occluded organic matter (i.e., substances trapped in soil aggregates) and/or relieve space or free oxygen limitations to growth. For example, plowing of a meadow soil resulted in a decrease in native organic matter and an increase in microbial biomass carbon in the top 6 cm of soil by 40 to 50 percent (Angers et al., 1992). Although initially increases in microbial populations result from cultivation of virgin soils, ultimately, due to depletion of easily metabolized organic matter resources, the microbial biomass level in the soil declines. (see Tate, 1987, for a review of this topic.)

Stationary Phase: For many microbial species in soil, stationary phase is their primary mode of existence. Growth is limited to the degree that new cell production is dependent upon accompanying cell death. For example, although most of the surface area in soil is not occupied by microbial colonies, colonial development is controlled by the space available for expansion in the region of available nutrient and energy resources. Organisms lacking the capacity to migrate to new habitats soon become constrained by the accessible volume of the soil pore in which they are developing. Hence, division of the parent cells depends upon production of space by death and lysis of companion cells. Similar situations may result from the immobilization of a growth-limiting, essential nutrient into microbial biomass. Again, replication is controlled by the release of the limiting nutrient from biomass via cellular death and decomposition.

Decline (Death Phase): This portion of the bacterial growth curve is essential for the maintenance of soil community homeostasis. Under relatively constant conditions, system survival depends upon maintenance of the *status quo*. This includes a balance in microbial population size. Population densities may increase in response to ecosystem perturbation, but to maintain community stability, the augmented population must return to its preexisting density. This conclusion is predicated on the assumption that a permanent change or relatively long-lived modification of system dynamics has not occurred. In that situation, all microbial population densities are modulated to the new levels optimal for the newly established system parameters. In community with inputs and outputs in equilibrium, population increases are dampened by such mechanisms as death and decay of the newly synthesized cell mass, predation by protozoa or other predators, and subsequent decay of their populations.

Diauxic Growth: In laboratory culture, microbes are generally grown in a medium containing a single carbon and energy source plus a variety of cofactors. The growth curve depicted in Fig. 4.2 is characteristics of such a situation. Since soil microbes rarely, if ever, encounter chemically pure growth substrates, this growth model must be modified to reflect the more complex carbon and energy supplies existent in soil and the microbial response to them. If the microorganism lacks the genetic capacity to metabolize more than one of the fixed carbon substrates present in its habitat, the presence of multiple organic compounds is of little consequence. This situation is rare. Of more common occurrence are microbes with the means of catabolizing a variety of carbon sources, many of which are concurrently present in the growth environment.

In a uniform system with no other limitations, a principle of microbial physiology that applies to microbial growth kinetics in the presence of multiple carbon and energy sources is catabolite repression. Catabolite repression, also called the glucose effect, occurs when the presence of one growth substrate represses synthesis of enzymes necessary for metabolism of alternate compounds. Thus, utilization of a second substrate for growth cannot occur until the initial substance is exhausted or reduced in concentration below growth supportive levels. This results in diauxic growth kinetics (Fig. 4.3). The growth pattern reflects successive exhaustion of the available substrates. Once the first nutrient and energy resource is depleted, a lag in growth is observed before the microbes commence to use the second substrate for energy and carbon. During the lag period, the microbes are synthesizing the enzymes needed for mineralization of the second growth-supporting substrate. This transition between energy and nutrient sources can be repeated several times should the microbes be existing in a complex mixture of metabolizable organic compounds.

Demonstration of diauxic growth in the heterogeneous world of soil microbes is difficult. A preferred carbon and energy source for a particular soil population may become exhausted in one microsite while an adjacent site retains a reasonable supply of the substance. Thus, in a cross-section of a soil

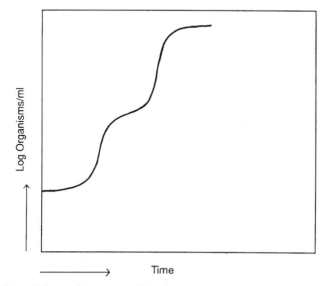

Fig. 4.3   **Bacterial growth curve resulting from sequential catabolism of two energy substrates (diauxic growth).**

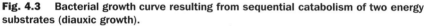

community, a variety of microsites can be envisioned to occur wherein many different organic substances are being concurrently metabolized. Studies in pure culture may predict occurrence of diauxic growth kinetics, but the true field data may not reflect its occurrence because of the nonhomogeneity of the field soil. A more specialized growth curve as exemplified by diauxic growth kinetics may be more significant in a soil contaminated with a mixture of organic substance where sequential mineralization of the substituents could occur.

4.1.2 Mathematical Representation of Soil Microbial Growth

A variety of complex mathematical models have been developed to describe microbial metabolism in soil ecosystems [e.g., see Tate (1987) for a review of this topic]. These relationships are commonly based on combinations of simple microbial growth relationships derived from assessment of population behavior in laboratory culture. One set of such equations that may be used to elucidate the nuances of microbial growth in soil are the classical Monod equations (Fig. 4.4).

Growth Rate Relationships to Microbial Biomass: Equation 1 ($dM/dt = \mu M$) depicts the fundamental association of growth rate with total population biomass present. The specific growth rate (μ) is not only intrinsic to the microbe itself (i.e., there is a maximal growth rate characteristic of the

Classical Monod Equations for Microbial Growth

1. Equation for Logarithmic Growth:

$$dM/dt = \mu M$$

2. Growth Substrate Impact on Specific Growth Rate

$$\mu = \frac{\mu_{max} [S]}{K_s + [S]}$$

3. Growth Yield

$$y = - \frac{d[S]}{dM}$$

Where μ = specific growth rate, M = microbial biomass, t = time, μ_{max} = maximum growth rate, K_s = saturation coefficient, y = growth yield

Fig. 4.4 Classical Monod equations describing microbial growth, cell mass yield, and growth substrate interactions.

organism functioning under optimal conditions) but is also controlled by the physical and chemical conditions of the environment in which the organism exists. In soil, the chemical and physical limitations of the habitat result in microbial population replication at a rate far below its maximum capabilities.

Since bacteria divide by binary fission, the growth rate is related to the generation time as follows:

$$T_d = \text{generation time} = \ln 2/\mu$$

Under the harsh conditions of the soil environment, generations times of several hours to days are common. These extended generation times compare to division cycles of as short a duration as 20 minutes in laboratory culture conditions.

While individual microbial populations in a soil community may be actively replicating, changes in cell mass as well as total microbial biomass may remain unchanged. Individual populations frequently increase or decline, but it is not uncommon to detect a relatively constant level of microbial biomass in an undisturbed soil site. This phenomenon is explained by the fact that (1) most microbes in the surface soil are not replicating—they are in some form of resting state, (2) some of the microbial populations are replicating, and (3) others are declining due to cell death, predation, and so forth. Thus, a small portion of the microbial community could be actively dividing, but their increase in number results in an insignificant alteration of the total soil biomass due to the quasi buffering effect of the inactive and declining portions of the microbial community. Similarly, microbial growth can be shown to be occurring, yet population densities appear to be unchanging. Microbes that decompose fixed carbon in soil are fed upon by a variety of predators and parasites. Thus, the microbes could be consumed by predators at a rate approximating their growth rate. Thus, quantification of growth rate by a method such as ^{3}H incorporation into DNA would indicate an actively multiplying microbial community whereas cell numbers would remain constant.

Growth Substrate Effects on Microbial Growth Rate: Although it is reasonable to assume that the microbial growth rate in soil is proportional to the concentration of growth-limiting carbon and energy substrates as indicated by Monod equation 2 (Fig. 4.4), the relationship is not applicable to soil communities. It is essentially impossible to determine the concentration of growth substrate occurring in the soil microsites where microbial growth occurs. Assumptions underlying the use of concentrations of total substrate in a soil sample as a predictor of microbial growth rate are (1) the substrates and microbes are evenly distributed throughout the soil sample and (2) all of the carbon and energy sources present in the culture (or vicinity) of the microbial cell are equally available for catabolism. These assumptions are clearly inapplicable in a heterogeneous system such as soil.

Along with the variability in microbial oxidation of their energy sources resulting from the admixture of living cells, soil particulate mineral matter, and colloidal organic matter, heterogeneity of fixed carbon distribution within the soil matrix also results from the exhaustion of nutrients in microsites due to the activity of resident microbial populations. Once the local nutrient supply is depleted, microbial growth rates are controlled by the nutrient input rate. Growth kinetics then are described by diffusion equations (static systems) or flow dynamics (where energy resources are contained in water flowing through soil pores—dynamic systems).

Similarly, adequate energy or carbon sources may be available to the microbes in soil microsites but may occur in a form that precludes microbial attack. That is, they may be water insoluble or retained within water-insoluble structures. Decomposition of these materials and hence the growth rate of microbial populations are controlled by the surface area of the particulate material, the rate of breaching of the protective barriers, or even the dissolution rate of the particulate substance.

Surface-area–controlled microbial growth is exemplified by the oxidation rate of elemental sulfur particles. Sulfur is minimally soluble in water, but its oxidation to sulfate provides an excellent energy source to a variety of sulfur-oxidizing bacteria. Physical attachment of the bacteria to the sulfur particle is not only required for oxidation of this substance but is frequently the rate-controlling factor. Early studies by Vogler and Umbreit (1941) revealed the necessity for this physical interaction. Recently, in an examination of the kinetics of oxidation of sulfur by *Thiobacillus ferrooxidans*, Espejo and Romero (1987) noted that only the bacteria attached to the sulfur particles were capable of growth. Therefore, since the elemental sulfur particles are not soluble in water and direct contact of the microbial cells is necessary for its oxidation, it can be concluded that the surface area of the particles controls not only substrate oxidation rate but also subsequent microbial biomass production. (See Chapter 15 for a more detailed examination of the role of surface area of sulfur particles in determination of their oxidation rates.) Thus, in this situation, microbial growth rate is proportional to substrate surface area, not its concentration.

In contrast, Stucki and Alexander (1987) found that dissolution rate, not surface area of the particles, controlled the decomposition of biphenyl by *Moraxella* sp. *Pseudomonas* sp., and *Flavobacterium* sp. Once the water-soluble substrates were depleted, dissolution rate of the hydrocarbon apparently controlled the microbial growth rates.

Many biodecomposable xenobiotic compounds are detected in soils at micro- and nanogram concentrations per gram soil. This is especially true of a variety of herbicides that are effective at low concentrations. Due in part to the common practice of evaluating biodegradation potential in culture media containing at least a thousand fold these concentrations [greater than 0.05 percent (wt/wt)], it has been assumed that microorganisms could not mineralize the minute quantities of organic substrates. Several laboratory studies have shown that mineralization of nanogram concentrations of a variety of organic substrates in culture (Simkins and Alexander, 1984), in sewage (Simkins and Alexander, 1984; Simkins and Alexander, 1985), and in soil (Scow et al.,

1986). In the latter study, soil was amended with phenol or aniline and incubated in biometer flasks at 70 percent or -0.3 bar soil moisture. The soil moisture level resembled one that would be commonly encountered in a well-drained field soil. The phenol and aniline were mineralized, but not at rates resembling Monod kinetics. A two-compartment model provided the best fit of the data. These data suggest that biological oxidation can occur in soil at concentrations that would not be anticipated to support synthesis of cell mass.

Growth Yield as a Measure of Biomass Production in Soil: In culture, the quantity of biomass yielded is proportional to the rate of substrate oxidation (Monod equation 3, Fig. 4.4). This relationship is applicable to microbial growth under well-defined cultural conditions where biomass production is totally determined by oxidation of a known energy source, such as occurs in a chemostat. Because of the simplicity of the relationship and the general ability to quantify microbial biomass and growth substances, application of this relationship to soil ecosystems is tempting. Unfortunately, the idealized conditions of the chemostat are rarely, if ever, replicated in soil. Use of this relationship with soil systems is precluded (1) by the variable contributions of energy resources to biomass production in soil, (2) by difficulties in attribution of biomass yields with oxidation of a specific growth substance, and (3) by complexities in linking biomass synthesis and growth substrate oxidation chronologically in complex soil ecosystems. In soil, a variable proportion of the energy yielded from oxidation of the growth substrate by the slowly replicating cells is utilized for cellular maintenance (maintenance energy). Thus, a clear relationship between energy production and biomass yields is not always evident in soil ecosystems. Furthermore, with the number of fixed carbon and other energy sources available in the soil microbe's habitat, at least a portion of the growth assessed in an experiment could result from oxidation of energy sources not quantified or even considered in the experimental design. Over the duration of an experiment, microbes may catalyze a number of different energy sources.

A further complication associated with complex soil systems relates to the capability to attribute changes in concentration of energy substrates directly with microbial growth yields. The concurrent observation of products of microbial metabolism in soil is not sufficient evidence to assume their linkage in a cause-and-effect relationship. For example, nitrate is commonly detected in soil samples. This compound is the product of energy production by nitrifiers that oxidize nitrite to nitrate. It could be easily assumed that the nitrite oxidizers in a soil sample arose from the production of nitrate associated with them in their habitat, but this nitrate could as easily have entered the soil from external sources (carried with percolating water) or have been produced by previous generations of nitrite oxidizers. Direct linkage of the two variables, nitrifier biomass production and nitrite oxidation, is appropriate only if their augmentation occurs concurrently. Since this prerequisite cannot be generally assumed to be met, application of the third Monod equation to analysis of soil systems should be avoided.

4.1.3 Uncoupling Energy Production from Microbial Biomass Synthesis

Cometabolism: Organic carbon compounds may also be oxidized in soils in a manner in which energy production, substrate oxidation, and microbial biomass yields are uncoupled. That is, a microbe produces enzymes that catalyze the oxidation of a compound, but no metabolic benefit is provided to the microbe. The energy required to maintain the microbial cell results from the oxidation of a second substrate, which may or may not be chemically related to the cometabolized material. A variety of aromatic ring-containing compounds are catabolized in soil through this mechanism. For example, catabolism of a variety of polychlorinated biphenyls (PCBs) constituting the mixture Arochlor was stimulated by the addition of biphenyl to soil samples (Focht and Brunner, 1985). In the presence of the biphenyl, between 48 and 49 percent of the PCBs were converted to carbon dioxide, compared to less than 2 percent in the absence of the biphenyl. Similar cometabolic responses have been noted for aniline, phenol, and their monochlorinated derivatives (Janke and Ihn, 1989); catabolism of trichloroethylene in the presence of methane (Henry and Grbic-Galic, 1991); and chloroparaffinic hydrocarbons (Beam and Perry, 1973; Beam and Perry, 1974). Cometabolism is particularly important in the metabolism of halogenated organic compounds (e.g., Boyle, 1989; Wackett, 1995).

A fortuitous observation from the view of environmental reclamation is the fact that the energy source and the cometabolized substance need not be present in equivalent amounts. Growth-supporting concentrations of an energy-producing substrate can facilitate the catabolism of nanogram concentrations of the cometabolized substrate (e.g., see Schmidt and Alexander, 1985; Schmidt et al., 1987; Wang et al., 1984; Wiggins and Alexander, 1988). Hence, amendment of soil with metabolizable substrates may encourage decomposition of contaminating organic compounds present in trace quantities.

Maintenance Energy: A proportion of the energy derived from oxidation of growth substrates is used for cell maintenance. This is termed *maintenance energy* (the portion of the energy yielded by substrate metabolism that is not used for biosynthetic purposes). Maintenance energy is used to provide for such basic cellular functions as motility and maintenance of internal osmotic pressure. Under energy-limiting conditions, the growth yield of microbial cells is reduced by the amount of energy expended to meet these needs. The slower the generation time of a microbe, the larger the proportion of energy required to maintain cellular integrity. Because of the long generation times of soil microbial populations and the observation that these organisms are generally energy limited, significant reductions in growth yield results from maintenance energy expenditures by these populations.

An example of the impact of maintenance energy expenditures on growth yield of microbes growing at reduced growth rates was provided by Traore et al. (1983). Reduction of the apparent growth rate in chemostat culture from

0.206 per hour to 0.0125 per hour reduced the growth yield of *Desulfovibrio vulgaris* from 6.65 to 2.06 g per mol. That is, at the lowest growth rate, 68 percent of the energy was used for cellular maintenance. This phenomenon combined with the low-free energy efficiencies explain how relatively high respiration rates may be detected in soil with a minimal increase in microbial biomass. (Free-energy efficiency equals the free energy captured by the microbe divided by the free energy available in the substrate oxidized for energy. See the section "Chemoautotrophic Existence in Soils" later in this chapter for a comparison of free-energy efficiencies of soil microbes.) The microbes are replicating, but they are doing so rather inefficiently.

4.2 Implications of Microbial Energy and Carbon Transformation Capacities on Soil Biological Processes

Although elucidation of biochemical pathways utilized by soil microbes and the physiological control of the expression of these capacities generally is studied in axenic culture, the real value of microbial metabolic processes resides in their expression by the organisms in their native environment.

The metabolic diversity of the soil microbial community has long been appreciated. In fact, the list of organic substances that can be mineralized by soil microbes was so extensive that it was considered conceptually to be essentially endless. Soil microbes were deemed to be infallible (the microbial infallibility principle) in their capability to rid our soils of unwanted organic compounds. Thus, it was commonly believed in the mid-twentieth century that microbes that are capable of decomposing essentially any organic compound could be expected to occur in soil. This opinion provided the justification for the optimistic consideration of soil as a vast repository for processing human organic wastes. With the advent of the chemical era and initiation of what could be termed a time of excessive exploitation of soil resources, this concept of microbial infallibility has frequently been transformed into a feeling of extreme microbial fallibility. Many difficulties have arisen because organic compounds, which contain chemical linkages rarely if ever encountered in biologically synthesized compounds, are produced industrially and added to soil systems on a routine basis. These same compounds may have a negative or positive effect on the function of the soil microbial community, but the primary environmental impact of consideration is their extended persistence. Accordingly, the concept of microbial infallibility has to be modified to include the concept of "biological" synthesis; that is, the soil microbial community is considered to be capable of decomposing all biologically synthesized organic compounds (or synthetic chemicals that mimic the covalent linkages produced biologically), assuming all other physical and chemical requirements for microbial function are met.

To develop soil management techniques that optimize native biological processes as well as to produce beneficial xenobiotic compounds that are biologically decomposed in all target ecosystems, an appreciation of the diversity of metabolic potential and its control in soil ecosystems must be evolved. Thus, this discussion is presented with the objective of elucidating the impact of microbial metabolic and energy transformation capacities on ecosystem function and resilience.

4.2.1 Energy Acquisition in Soil Ecosystems

Soil microbes, as do all living entities, derive the energy necessary for cellular function through oxidation reactions. A basic principle of biochemistry is that all oxidation must be balanced by reductions; that is, electrons generated in the oxidation processes must be transported to an electron acceptor. Common terminal electron acceptors include oxygen, nitrate, nitrite, sulfate, and a variety of organic compounds, including acetate and pyruvate.

The quantity of energy provided to the growing cell in the form of adenosine triphosphate (ATP) is dependent upon the intermediate steps between the initial oxidative reaction and the final electron acceptor. A direct transfer of the electron from the oxidized substrate to an organic recipient as occurs in glycolysis (Fig. 4.5) yields a single ATP molecule per electron transfer, whereas passage of the electron through a series of intermediate reduction and oxidation reactions (Fig. 4.6), as occurs in respiratory metabolism of glucose through the citric acid cycle, yields as many as 3 ATP molecules per electron transferred. Thus, the incomplete oxidation of glucose to organic intermediates by glycolysis produces a net gain of 2 ATP molecules, whereas a maximum of 38 ATP molecules is produced by the complete conversion of glucose to carbon dioxide (Fig. 4.7).

Aerobic versus Anoxic Processes: Environmental metabolic processes can conveniently be grouped into two rather broad categories: aerobic and anaerobic conversions. Maximal biological energy is made available for cellular function when free oxygen is the terminal electron acceptor. Inversely, incomplete oxidation of fixed carbon substrates in the absence of free oxygen results in reduced energy recovery and reduced biomass production. Based on the above discussion of substrate-level phosphorylation and the recovery of energy through an electron-transport chain, such conclusions are reasonable, and are applicable to the majority of metabolic processes.

Fortunately, the underlying assumption of this generalization — cytochrome-based respiration (the basic process when molecular oxygen serves as the terminal electron acceptor) only occurs under aerobic conditions — is *not* true. Denitrification (e.g., Koike and Hattori, 1975; Kristjansson et al., 1978) and sulfate reduction (e.g., see Kim and Akagi, 1985; Postgate, 1984), two conversions that require anoxic conditions, yield energy to the growing cell through cytochrome-based electron-transfer chains similar to those depicted in Fig. 4.5. Thus, for these two anoxic processes, the quantity of cell mass produced approaches that achieved when free oxygen is the terminal acceptor.

EMBDEN - Meyerhof Pathway

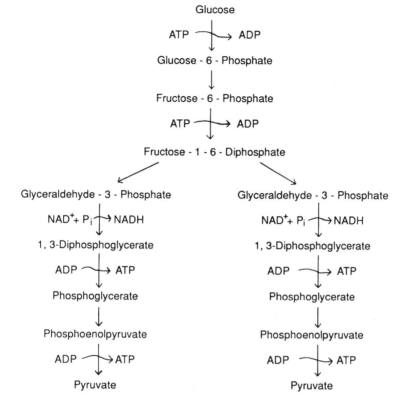

Fig. 4.5 Glycolytic conversion of glucose to pyruvate. Note the net production of 2 ATP and 2 NADH molecules. Under fermentative conditions, organic intermediates such as acetaldehyde or pyruvate serve as the terminal electron acceptors.

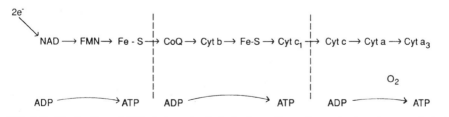

Fig. 4.6 Production of ATP via transport of electrons through an electron-transport chain with molecular oxygen as the terminal electron acceptor.

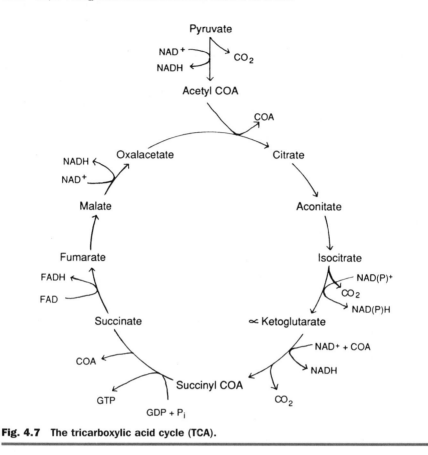

Fig. 4.7   The tricarboxylic acid cycle (TCA).

The potential for exploiting these anoxic respiratory processes for bioremediation of soils is discussed in Chapter 16.

A basic principle to be considered in these dissimilatory processes (oxidation of carbon substrates coupled to cytochrome oxidation under anoxic conditions) is that the utilization of the specific electron acceptors is mutually exclusive. That is, the dissimilatory reduction of nitrate or sulfate does not occur until the molecular oxygen supply in the microsite of the microbial cell is exhausted. This results in a reduction in the local reduction-oxidation potential sufficiently to allow utilization of the alternate electron acceptors. Each of these substances functions at specific reduction-oxidation potentials (Table 4.1). That is, the redox potential of the environment of the microbial cells is perched at the specific value associated with a given electron acceptor until it is exhausted, then the potential drops to a new level characteristic of the next electron acceptor. For example, in a system containing nitrogen oxides (but no oxygen), the reduction potential remains at the level dictated by the nitrogen oxides until all of them have been reduced through denitrification,

Table 4.1   Reduction Potentials of Some Common Electron Acceptors Functioning in Soil Biological Systems

Redox couple	E_0 (volts)
O_2/H_2O	0.82
Fe^{3+}/Fe^{2+}	0.77
NO_3^-/NO_2^-	0.42
Cytochrome c Oxidized/Reduced	0.25
Acetaldehyde/Ethanol	0.20
Cytochrome b Oxidized/Reduced	0.07
SO_4^{2-}/S^{2-}	-0.22
$NAD^+/NADH$	-0.32
CO_2/CH_4	-0.35
H^+/H_2	-0.42

then organic substances or other acceptors (e.g., sulfate), if present, become the primary terminal electron acceptors for the active biological community.

A complex example of this phenomenon involves the sequential oxidation and reduction of growth substrates and terminal electron acceptors in a flooded soil containing nitrate, ammonium, manganese (II), and ferrous iron (Patrick and Jugsujinda, 1992). No overlap in the oxidation or reduction of nitrate and manganese was detected, and little overlap in the transformation of manganese and ferrous ions occurred.

The environmental implications of the potential for dissimilatory oxidation of fixed carbon with nitrate or sulfate as the final electron acceptor include the fact that greater biomass is supported by the processes than occurs with classical fermentation reactions, and that the organic substances providing the energy are completely oxidized to carbon dioxide and water. The latter trait is particularly significant in bioremediation processes (Tiedje et al., 1984; Chapter 16). This is especially important in situations where the oxygen supply is exhausted due to the rapid catabolism of the energy-supplying substrate. Should nitrogen oxides or sulfate be present, the contaminant could still be reduced rapidly to carbon dioxide and water even under the anoxic conditions.

The potential for exploiting dissimilatory processes under anoxic conditions for decomposition of soil contaminants is underscored by the variety of substrates catabolized by denitrifiers and sulfate reducers. Denitrifiers have been shown to oxidize a variety of aromatic ring-containing compounds, including phthalic acid (Aftring et al., 1981; Nozawa and Maruyama, 1988), toluene (Altenschmidt and Fuchs, 1991; Evans et al., 1991a), xylene (Evans et al., 1991b), benzene and a variety of alkyl benzenes (Hutchins, 1991), resorcyclic acids and resorcinol (Kludge et al., 1990), plus the more complex polycyclic aromatic hydrocarbons (McNally et al., 1998; Mihelcic and Luthy, 1988a; Mihelcic and Luthy, 1988b). An added advantage of oxidation of these substances via denitrification is that a second soil pollutant, nitrate, is also removed from the system through its conversion to nitrous oxide and dinitrogen. Similarly, sulfate can serve as a terminal electron acceptor for oxidation of such compounds as chlorophenol (Häggblom and Young, 1990), aniline and dihydroxybenzenes (Schnell et al., 1989), benzoate (Tsaki et al., 1991), catechol (Szewzk and Pfennig, 1987), and *m*-cresol (Ramanand and Suflita, 1991).

4.2.2 Microbial Contributions to Soil Energy and Carbon Transformations

The diversity of habitats available to soil microbes is reflected in their versatility in carbon and energy metabolic capacities. Essentially all microbial groupings (Table 4.2) are found at least to a limited extent in soil ecosystems. Some are important contributors to total microbial biomass and are versatile in their capacity to oxidize fixed carbon substrates (e.g., the heterotrophs), whereas others, although limited contributors to total soil microbial biomass, are important because of their pivotal role in soil biogeochemical processes (e.g., nitrifiers). Others pique the interest of soil microbiologists not because of the imposition of their presence on soil system function, but more due to their unusual or precarious existence in the system (e.g., soil algae or photosynthetic bacteria).

To gain a more complete understanding of metabolic properties of soil microbes, the function and biological limitations of autotrophs and heterotrophs (aerobic and anaerobic) will be examined. Although these organisms are found in a variety of habitats, their function in soil acquires unique characteristics imposed by the soil physical and chemical environment.

Table 4.2 Some Groupings of Soil Microbes Based on Their Primary Energy Source

Microbial Group	Energy Source	Carbon Source	Terminal Electron Acceptors	Examples
Heterotrophs				
Aerobes	Fixed carbon	Fixed carbon	O_2, NO_x, SO_4^{2-}	*Arthrobacter* sp., *Pseudomonas* sp., *Bacillus* sp., dentrifiers
Anaerobes	Fixed carbon	Fixed carbon	Fixed carbon	*Clostridium* sp.
Fermentors	Fixed carbon	Fixed carbon or CO_2	Fixed carbon	Enterics, lactic acid bacteria
Autotrophs				
Chemoautotrophs	Inorganic chemicals (aerobes)	CO_2	O_2, NO_3^-	*Nitrifiers Thiobacillus*
	H_2 (anaerobes)	CO_2	CO_2	Methanogens
Photoautotrophs	Light	CO_2		Algae, photosynthetic bacteria

Chemoautotrophic Existence in Soils: Chemoautotrophic bacteria gain their energy through the oxidation of inorganic compounds. Primary electron sources are ammonium, nitrite, hydrogen, and a variety of sulfur-based anions. A characteristic of these microbes is their specificity for particular energy sources. For example, autotrophs oxidizing nitrogenous compounds use only ammonium or nitrite as electron sources, whereas sulfur autotrophs are restricted to oxidation of sulfur compounds. Furthermore, the specificity generally extends to individual compounds within each grouping—for example, nitrifiers are divided into ammonium and nitrite oxidizers. Chemoautotrophs may assimilate simple organic compounds, but their primary cellular structure is provided through the reduction of carbon dioxide. Many of the autotrophic bacteria are obligate autotrophs; that is, they can use only carbon dioxide as a carbon source. Facultative autotrophs are capable of growing either autotrophically or heterotrophically.

Chemoautotrophic organisms are of pivotal importance to soil biotic function. Although their contribution of fixed carbon to the ecosystem is minor compared to that incorporated by higher plants, oxidation of their inorganic substrates is frequently central to the completion of various soil biogeochemical cycles. For example, ammonium oxidation to nitrate is primarily catalyzed by autotrophic nitrifiers in soil.

Although neither heterotrophs or chemoautotrophs capture all of the energy released from oxidation of their substrate, chemoautotrophic bacteria are especially inefficient in energy recovery (free energy efficiency). The free energy efficiency is generally measured in culture by quantifying the heat of combustion of the cells produced by metabolism of a specific substrate and the heat of combustion of the quantity of substrate oxidized in this process. This is not measured directly in soil because of the difficulty in determining the specific substrate catabolized among the vast array available to the cell. The free energy efficiency can be as low as 5 to 10 percent for oxidation of nitrite to nitrate by *Nitrobacter* species (Alexander, 1977) to 50 percent for the oxidation of elemental sulfur to sulfate by *Thiobacillus thiooxidans.*

For the obligate autotrophs, all of the cellular substituents must be synthesized from the reduction of carbon dioxide. Considering that the conversion of ammonium to nitrate yields 66 kcal of energy per mole of substrate oxidized, and the oxidation of nitrate to nitrate produces only 20 kcal of energy per mole of substrate, and that the cellular substituents of both the ammonium and nitrite oxidizers are the same, approximately threefold more nitrite must be oxidized to produce the same biomass of nitrite oxidizers as for the ammonium oxidizers. Approximately 35 nitrogen atoms are oxidized per carbon fixed for the ammonium oxidizers as compared to about 100 nitrogens for the nitrite oxidizers. Thus, the microorganisms using these low-energy substrates must oxidize large quantities of their substrates to produce the same biomass as those growing by more efficient metabolic processes.

In summary, chemoautotrophs are of minor importance as a contributor to biomass in soil ecosystems. Their primary contribution to ecosystem development resides in their capacity to oxidize soil mineral components. They are the primary if not sole contributor to nitrate formation, catalyze some sulfur oxidations, participate in oxidation of ferrous iron to ferric ion, are instrumental in genesis of acid mine wastes, contribute to corrosion and weathering of rocks and statues, and prevent the accumulation of molecular hydrogen.

Anaerobic Heterotrophs and the Soil Ecosystem: Anaerobes are those microorganisms that grow and reproduce in the absence of free oxygen. Organisms using cytochrome-based electron transport systems, even when functioning under anoxic conditions, are by definition excluded from this class of organisms. Although there is in reality a gradient of sensitivity to free oxygen, oxygen is lethal to anaerobes.

Anaerobic microbes may be divided into two classes, strict or obligate anaerobes, and facultative anaerobes. Strict anaerobes do not grow or survive in the vegetative phase for extended periods in the presence of molecular oxygen. Examples of this group of organisms include the clostridia (which are ubiquitous in soil), methanogens, and some protozoa (which occur in the rumen and sediments). No anaerobic fungi have been reported to occur in soil although anaerobic fungi may be found in the rumen (Billon-Grand et al, 1991; Dore and Stahl, 1991; Gordon and Phillips, 1989; Lowe et al., 1987; Teunissen et al., 1991; Webb and Theodorou, 1991). Facultative anaerobes are capable of growth in the presence or absence of free oxygen. Examples of these organisms include the commonly encountered enteric bacteria.

In spite of their sensitivity to molecular oxygen and the abundance of this molecule in our terrestrial system, anaerobic microbes are nearly ubiquitous in soil systems. In fact, in a well-aggregated soil, anaerobic and aerobic metabolism occur concurrently. Aerobes frequently function on the surface of a soil aggregate at a rate sufficient to exhaust free oxygen supplies such that the internal portions of the granule are anaerobic. Since it is not uncommon for the nutrients requisite to support growth of anaerobic microbes to be contained within the aggregate structure, these populations not only survive but contribute significantly to overall soil metabolic activity.

Flooding of the soil, as can result from a heavy rainfall, may result in an expansion of the anaerobic microbial biomass. It must be stressed that a flooded soil is not necessarily an anaerobic soil. The oxygen tension of a flooded soil system depends on (1) the oxygen diffusion rate in water, (2) the rate of consumption of this molecule by the living community plus any depletion due to chemical processes, as well as (3) the potential for oxygenation of the sites through influxes of oxygen-bearing water. That is, a flooded system in which the water level is maintained by inputs of oxygenated waters may not become anoxic. Soil sites commonly associated with oxygen-depleted conditions are swamps or marshes, rice paddies, sediments, and any other sites receiving inputs of easily decomposed organic matter. The latter soils include a variety of environmentally stressed situations such as buried organic matter in landfills or soils affected by spills of biodecomposable societal products.

The products of complete mineralization of an organic substance containing carbon, hydrogen, oxygen, and sulfur in the absence of oxygen are carbon dioxide, methane, ammonium ion, hydrogen sulfide, and water. This list of products contrasts to the carbon dioxide, ammonium ion, water, and hydrogen sulfide yielded by aerobic decomposition of the same compound. Of more general importance to soil systems are the products of incomplete mineralization of organic substances by facultative and obligate anaerobic organisms. In these situations, alcohols and organic acids accumulate. These fermentations leave a major portion of the energy contained in the substrates in the fermentation products. Thus, the cell yield is minimal compared to what would be produced through complete, aerobic oxidation of the fixed carbon sources.

Methanogenesis: Methane generation provides an excellent example of complete mineralization of organic compounds under anaerobic conditions and the interaction of a variety of anaerobic bacteria to achieve these ends. Complex substrates are decomposed by anaerobic and fermentative bacteria to the simple substrates that can be converted to methane. Methane synthesis is continuous in marshes and bogs where the extremely reducing environment necessary for the process (-200 to -1000 mV) commonly exists.

Methane production is catalyzed by a highly specialized group of obligately anaerobic microbes, the methanogenic bacteria. This unique group of anaerobic bacteria includes the following species: *Methanobacterium arbophilicum, M. formicum, M. ruminatum, M. mobile, M. thermoautotrophicum, Methanococcus vanielli, Methanosarcina barkeri,* and *Methanospirillum hungatii.* These organisms reduce carbon dioxide to methane with hydrogen generally serving as the electron donor by the following reaction:

$$4H_2 + CO_2 \rightarrow CH_4 + 2H_2O$$

Therefore, hydrogen is serving as a sole source of reducing power for both methanogenesis and cell carbon synthesis. Other substrates that can be converted to methane include methanol, formate, acetate, and methylamines. With the latter substrates, the carbon substrate serves as both the electron donor and the final electron acceptor. For example, acetate is converted to methane and carbon dioxide as follows:

$$CH_3COO^- + H^+ \rightarrow CH_4 + CO_2$$

The acetate, carbon dioxide, and hydrogen used by these methanogenic bacteria is generated by the fermentation of complex polymers, including cellulose and other polysaccharides, by obligate anaerobes, and fermentative bacteria (Fig. 4.8).

Aerobic Heterotrophs and the Soil Ecosystem: The soil microbial community is primarily driven by the energy contained within photosynthetically fixed carbon. That is, the soil ecosystem is indirectly solar powered. When it is considered that most other carbon pools entering soil are products of decomposition of plant carbon (e.g., animal biomass, composts, and sludges,

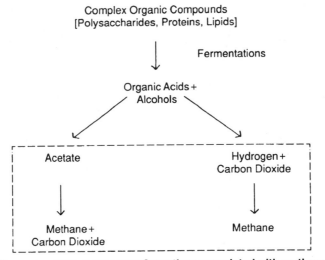

Fig. 4.8 **General organic matter transformations associated with methanogenesis.
Processes enclosed in the box are catalyzed by methanogens.**

as well as petroleum), the intense dependence of belowground life processes on plant productivity is highlighted.

Biomass carbon entering soil can be divided into two primary pools, biodegradation resistant, and readily or easily metabolized carbon. Clearly, the biodegradation-resistant categorization is a relative term in that these substances (e.g., lignified materials) are decomposed in soil, albeit slowly. The readily metabolized fractions are defined as those fixed carbon materials that provide a carbon and energy source for the soil microbial community with a minimal energy expenditure by the microbe. Included in these energy expenditures are the activities necessary to synthesize catabolic enzymes as well as those necessary to activate the substrate molecule to enhance its degradation susceptibility—for example, the phosphorylation of hexoses with adenosine triphosphate to form hexose-monophosphates (Fig. 4.5).

The biodegradation-resistant substances may contain significant chemical energy reserves to support microbial growth, but a diversity of enzymes is necessary to convert the compounds to forms whose energy can be readily recovered by the microbial community. Thus, the energy expenditures for synthesis of the enzymes that are necessary to convert these substances to forms from which the microbes may recover energy commonly exceed the potential energy that can be recovered. The biodegradation-resistant group of compounds include plant lignin as well as soil humic acids.

It is not unusual to know a priori that a certain metabolic activity must be present in a soil sample, but due to difficulties associated with analysis of a complex mixture of abiotic and biotic components, detection or quantification of the process is precluded. Fortunately, the properties of comparative biochemistry dictate that metabolic reactions occurring in bacteria or fungi that can be studied in laboratory culture are common with minor variations to those occurring in the soil microbial community. Thus, if a fixed carbon substance is amended to a soil sample and its decomposition is documented, it is reasonable to conclude that the decomposition is occurring, with minor variations, by commonly described biochemical pathways — for example, Embden Meyerhoff Parnas, citric acid cycle, and so on.

This observation allows construction of a model predictive of the fate of the primary classes of organic carbon compounds entering soil (Fig. 4.9). As in other ecosystems, in soil, most fixed carbon will be catabolized to simple organic acids that can be activated to acetyl-coA and fed into the tricarboxylic acid cycle. For example, proteins are hydrolyzed by proteases and deaminated to organic acids. Aromatic ring-containing substances are hydroxylated to facilitate ring cleavage (Fig. 4.10). In this process, molecular oxygen is incorporated directly into the aromatic ring.

Hence, it is logical to predict that within any soil sample containing viable microbial populations, all of the common enzymes to intermediary metabolism associated with these cycles plus those activities, such as proteases and hydrolases necessary to produce the precursors of these cycles, are present in that soil sample.

Humification: The complexity of the soil system, including the potential for a variety of spontaneous chemical reactions, provides a number of fates for reactive biochemical intermediates that may not be detected in laboratory culture. Essentially any reactive organic compound entering soil could be incorporated into soil humic acids; that is, it can be humified. Since many of the reactions associated with covalent linkage of organic compounds to humic acids are chemically catalyzed, the reaction rate is controlled in large part by the reactivity of the substrates and the probability of interaction of the reactants. In typical soils containing less than 2 percent organic matter, the probability of such reactions is limited, although significant. Since in most systems the controlling factor in humification is the probability of collision of the reactants, the survival time of the molecule in soil becomes a major predictor of whether the substance will be decomposed totally or at least partially, or whether it will be humified. The more long-lasting organic components, such as complex polyaromatics in petroleum or lignin, are more likely to be humified than the more ephemeral proteins and simple sugars. See Tate (1986) for further discussion of humification processes.

4.3 Concluding Comments

The soil microbial community is clearly distinguished by its diversity in metabolic capacity. Along with the common variety of fixed carbon substances catabolized by living organisms, soil microbes also oxidize a number of soil

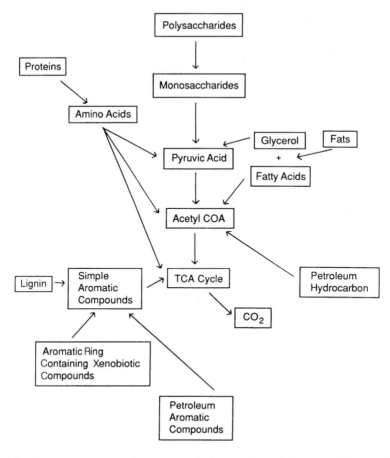

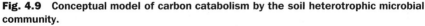

Fig. 4.9   **Conceptual model of carbon catabolism by the soil heterotrophic microbial community.**

minerals (e.g., sulfides), metals (e.g., ferrous ion), and several cations and anions containing atoms capable of existing in numerous oxidation states (e.g., ammonium and thiosulfate). Combined with the capacity to utilize several different terminal electron acceptors (e.g., molecular oxygen, nitrate, nitrite, sulfate, and numerous organic compounds), microbial life is capable of existing in soil under essentially any chemical and physical conditions that allow for sustenance of cellular integrity.

The societal legacy of this myriad of metabolic capabilities is the potential for the soil microbial community to survive severe chemical insults and return to previously existing conditions. Influxes of acidic solutions of sulfate (acid mine drainage) may be ameliorated by anaerobic reduction of the sulfate to sulfide (i.e., the sulfate is used as a terminal electron acceptor). Toxic organic

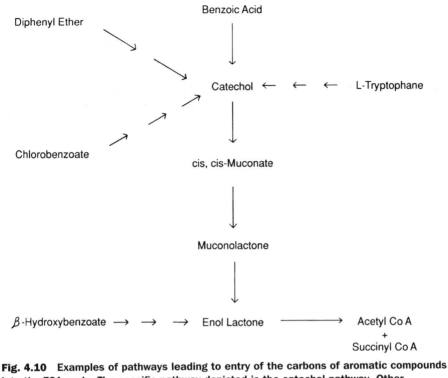

Fig. 4.10 Examples of pathways leading to entry of the carbons of aromatic compounds into the TCA cycle. The specific pathway depicted is the catechol pathway. Other aromatic ring-degrading processes include the gentisate and protocatechuate pathways.

compounds may be mineralized in the presence or absence of molecular oxygen. Heavy-metal contaminants can be removed from interstitial waters by precipitation with the sulfide produced by sulfate reduction. Survivability and sustenance of the soil microbial community is built to a large extent on its vast catalog of catabolic capabilities.

References

Aftring, R. P., B. E. Chalker, and B. F. Taylor. 1981. Degradation of phthalic acids by denitrifying, mixed cultures of bacteria. Appl. Environ. Microbiol. 41:11771183.

Alexander, M. 1977. Introduction to Soil Microbiology. John Wiley & Sons, NY. 467 pp.

Altenschmidt, U., and G. Fuchs. 1991. Anaerobic degradation of toluene in denitrifying *Pseudomonas* sp.: Indication for toluene methylhydroxylation and benzoyl-CoA as central aromatic intermediate. Arch. Microbiol. 156:152158.

Angers, D. A., A. Pesant, and J. Vigneux. 1992. Early cropping-induced changes in soil aggregation, organic matter, and microbial biomass. Soil Sci. Soc. Am. J. 56:115119.

Barnhart, C. L. H., and J. R. Vestal. 1983. Effects of environmental toxicants on metabolic activity of natural microbial communities. Appl. Environ. Microbiol. 46:970977.

Beam, H. W., and J. J. Perry. 1973. Co-metabolism as a factor in microbial degradation of cycloparaffinic hydrocarbons. Arch. Mikrobiol. 91:8790.

Beam, H. W., and J. J. Perry. 1974. Microbial degradation of cycloparaffinic hydrocarbons via co-metabolism and commensalism. J. Gen. Microbiol. 82:163169.

Biederbeck, V. O., and C. A. Campbell. 1971. Influence of simulated fall and spring conditions on the soil system. I. Effect on soil microflora. Soil Sci. Soc. Am. Proc. 35:474-479.

Billon-Grand, G., J. B. Fiol, A. Breton, A. Bruyre, and Z. Oulhaj. 1991. DNA of some anaerobic rumen fungi: G + C content determination. FEMS Microbiol. Lett. 82:267270.

Boyle, M. 1989. The environmental microbiology of chlorinated aromatic decomposition. J. Environ. Qual. 18:395402.

Christensen, H., D. Funck-Jensen, and A. Kjoller. 1989. Growth rate of rhizosphere bacteria measured directly by the tritiated thymidine incorporation technique. Soil Biol. Biochem. 21:113118.

Dore, J., and D. A. Stahl. 1991. Phylogeny of anaerobic rumen *Chytridiomycetes* inferred from small subunit ribosomal RNA sequence comparisons. Can. J. Bot. 69:1964-1971.

Espejo, R. T., and P. Romero, 1987. Growth of *Thiobacillus ferrooxidans* on elemental sulfur. Appl. Environ. Microbiol. 53:19071912.

Evans, P. J., D. T. Mang, K. S. Kim, and L. Y. Young. 1991a. Anaerobic degradation of toluene by a denitrifying bacterium. Appl. Environ. Microbiol. 57:11391145.

Evans, P. J., D. T. Mang, and L. Y. Young. 1991b. Degradation of toluene and *m*-xylene and transformation of *o*-xylene by denitrifying enrichment cultures. Appl. Environ. Microbiol. 57:450454.

Federle, T. W., M. A. Hullar, R. J. Livingston, D. A. Meeter, and D. C. White. 1983. Spatial distribution of biochemical parameters indicating biomass and community composition of microbial assemblies in estuarine mud flat sediments. Appl. Environ. Microbiol. 45:5863.

Focht, D. D., and W. Brunner. 1985. Kinetics of biphenyl and polychlorinated biphenyl metabolism in soil. Appl. Environ. Microbiol. 50:10581063.

Gordon, G. L. R., and M. W. Phillips. 1989. Degradation and utilization of cellulose and straw by three different anaerobic fungi from the ovine rumen. Appl. Environ. Microbiol. 55:17031710.

Häggblom, M. M., and L. Y. Young. 1990. Chlorophenol degradation coupled to sulfate reduction. Appl. Environ. Microbiol. 56:32553260.

Henry, S. M., and D. D. Grbic-Galic. 1991. Influence of endogenous and exogenous electron donors and trichloroethylene oxidation toxicity on trichloroethylene oxidation by methanotrophic cultures from a groundwater aquifer. Appl. Environ. Microbiol. 57:236244.

Hutchins, S. R. 1991. Biodegradation of monoaromatic hydrocarbons by aquifer microorganisms using oxygen, nitrate, or nitrous oxide as the terminal electron acceptor. Appl. Environ. Microbiol. 57:24032407.

Janke, D., and W. Ihn. 1989. Cometabolic turnover of aniline, phenol, and some of their monochlorinated derivatives by the *Rhodococcus* mutant strain AM 144. Arch. Microbiol. 152:347352.

Kim, J.-H., and J. M. Akagi. 1985. Characterization of a trithionate reductase system from *Desulfovibrio vulgaris*. J. Bacteriol. 163:472475.

Kludge, C., A. Tschech, and G. Fuchs. 1990. Anaerobic metabolism of resorcyclic acids (*m*-dihydroxybenzoic acids) and resorcinol (1,3benzenediol) in a fermenting and in a denitrifying bacterium. Arch. Microbiol. 155:6874.

Koike, I., and A. Hattori. 1975. Energy yield of denitrification: An estimate from growth yield in continuous cultures of *Pseudomonas denitrificans* under nitrate-, nitrite-, and nitrous oxide-limited conditions. J. Gen. Microbiol. 88:1119.

Kristjansson, J. K., B. Walter, and T. C. Hollocher. 1978. Respiration-dependent proton translocation and the transport of nitrate and nitrite in *Paracoccus denitrificans* and other denitrifying bacteria. Biochemistry 17:5014-5019.

Lowe, S. E., M. K. Theodorou, and A. P. J. Trinci. 1987. Growth and fermentation of an anaerobic rumen fungus on various carbon sources and effect of temperature on development. Appl. Environ. Microbiol. 53:12101215.

McNally, D. L., J. R. Mihelcic, and D. R. Lueking. 1998. Biodegradation of three- and four-ring polycyclic aromatic hydrocarbons under aerobic and denitrifying conditions. Environ. Sci. Technol. 32:26332639.

Mihelcic, J. R., and R. G. Luthy. 1988a. Degradation of polycyclic aromatic hydrocarbon compounds under various redox conditions in soil-water systems. Appl. Environ. Microbiol. 54:11821187.

Mihelcic, J. R., and R. G. Luthy. 1988b. Microbial degradation of acenaphthene and naphthalene under denitrification conditions in soil-water systems. Appl. Environ. Microbiol. 54:11881198.

Nozawa, T., and Y. Maruyama. 1988. Anaerobic metabolism of phthalate and other aromatic compounds by a denitrifying bacterium. J. Bacteriol. 170:57785784.

Patrick, W. H., Jr., and A. Jugsujinda. 1992. Sequential reduction and oxidation of inorganic nitrogen, manganese, and iron in flooded soil. Soil Sci. Soc. Am. J. 56:10711073.

Postgate, J. R. 1984. The Sulphate-Reducing Bacteria. Cambridge University Press, NY. 208 pp.

Ramanand, K., and J. M. Suflita. 1991. Anaerobic degradation of *m*-cresol in anoxic aquifer slurries: Carboxylation reactions in a sulfate-reducing bacterial enrichment. Appl. Environ. Microbiol. 57:16891695.

Schmidt, S. K., and M. Alexander. 1985. Effects of dissolved organic carbon and second substrates on the biodegradation of organic compounds at low concentrations. Appl. Environ. Microbiol. 49:822827.

Schmidt, S. K., K. M. Scow, and M. Alexander. 1987. Kinetics of *p*-nitrophenol mineralization by a *Pseudomonas* sp.: Effects of second substrates. Appl. Environ. Microbiol. 53:26172623.

Schnell, S., F. Bak, and N. Pfennig. 1989. Anaerobic degradation of aniline and dihydroxybenzenes by newly isolated sulfate-reducing bacteria and description of *Desulfobacterium anilini*. Arch. Microbiol. 152:556563.

Scow, K. M., S. Simkins, and M. Alexander. 1986. Kinetics of mineralization of organic compounds at low concentrations in soils. Appl. Environ. Microbiol. 51:10821085.

Simkins, S., and M. Alexander. 1984. Models for mineralization kinetics with the variables of substrate concentrations and population density. Appl. Environ. Microbiol. 47:12991306.

Simkins, G., and M. Alexander. 1985. Nonlinear estimation of the parameters of Monod kinetics that best describe mineralization of several substrate concentrations by dissimilar bacterial densities. Appl. Environ. Microbiol. 50:816824.

Stucki, G., and M. Alexander. 1987. Role of dissolution rate and solubility in biodegradation of aromatic compounds. Appl. Environ. Microbiol. 53:292297.

Szewzk, R., and N. Pfennig. 1987. Complete oxidation of catechol by the strictly anaerobic sulfate-reduction *Desulfobacterium catecholicum* new species. Arch. Microbiol. 147:163168.

Tate, R. L., III. 1985. Carbon mineralization acidic, xeric forest soils: Induction of new activities. Appl. Environ. Microbiol. 50:454-459.

Tate, R. L., III. 1987. Soil Organic Matter: Biological and Ecological Effects. John Wiley & Sons, NY. 291 pp.

Teunissen, M. J. H. J. M. Op den Camp, C. G. Orpin, J. H. J. Huis in 't veld, and G. D. Vogels. 1991. Comparison of growth characteristics of anaerobic fungi isolated from ruminant and non-ruminant herbivores during cultivation in a defined medium. J. Gen. Microbiol. 137:14011408.

Tiedje, J. M., A. J. Sexstone, T. B. Parkin, and N. P. Revsbech. 1984. Anaerobic processes in soil. Plant Soil 76:197212.

Traore, A. G., C. E. Hutchikian, J. LeGall, and J-P. Belaich. 1982. Microcalorimetric studies of the growth of sulfate-reducing bacteria: Comparison of the growth parameters of some *Desulfouibri. Speaes. J. Bacterial.* 149:606611.

Tsaki, M., Y. Kamagata, K. Nakamura, and E. Mikami. 1991. Isolation and characterization of a thermophilic benzoate-degrading, sulfate-reducing bacterium, *Desulfotomaculum thermobenzoicum* sp. nov. Arch. Microbiol. 155:348352.

Volger, K. G., and W. W. Umbreit. 1941. The necessity for direct contact in sulfur oxidation by *Thiobacillus thiooxidans.* Soil Sci. 51:331337.

Wackett, L. P. 1995. Bacterial co-metabolism of halogenated organic compounds. *In* L. Y. Young and C. F. Cerniglia (eds.), Microbial Transformation and Degradation of Toxic Organic Chemicals. Pp. 217241. Wiley-Liss, New York.

Wang, Y.-S., R. V. Subba-Rao, and M. Alexander. 1984. Effects of substrate concentration and organic and inorganic compounds on the occurrence and rate of mineralization and cometabolism. Appl. Environ. Microbiol. 47:11951200.

Webb, J., and M. K. Theodorou. 1991. *Neocallimastic hurleyensis* sp. nov., an anaerobic fungus from the ovine rumen. Can. J. Bot. 69:12201224.

Wiggins, B. A., and M. Alexander. 1988. Role of chemical concentration and second carbon sources in acclimation of microbial communities for biodegradation. Appl. Environ. Microbiol. 54:28032807.

Process
Control
in Soil

Accurate prediction of the response of soil biological communities to environmental perturbation is predicated upon the premise that an accurate or at least a reasonably representative understanding of ecosystem properties controlling soil processes has been developed. This foundational prerequisite presents a paradox for those associated with the study of soil biological activity within the confines of environmental and soil sciences. Microbiological principles and metabolic reactions are understood at the individual cellular and molecular levels (i.e., the microlevel), whereas soil properties delimiting the expression of microbial potential are assessed at the macrolevel (e.g., ranging from several grams of soil to a total ecosystem characterization). Furthermore, the environmental scientist requires integration of the microsite and macrosite realms on an ecosystem-wide or terrestrial level. These three spheres of concern (microsite, ecosystem-wide, worldwide) are not always easily merged, but such reconciliation of data is essential.

Macro- and Microsite Reality and Ecosystem Heterogeneity: The primary difficulty associated with the merging of microbial and soil chemical and physical analyses results from the high degree of variability of soil properties both in a vertical and horizontal plane within the soil profile (see Chapter 1). A false impression of the microsite traits may be gained from study of the relatively large soil samples. Values for soil properties, such as pH or cation exchange capacity, measured with as little as ten grams of soil may be quite different from the actual situation within which the microbes exist.

Comparable problems are associated with extrapolation from the level of the manageable soil sample to provision of an ecosystem-wide model. Unfortunately, as with evaluation of the dynamics of the microbial world, soil properties of large soil site are sufficiently variable that it is difficult, if not impossible, to collect a soil sample that would be a reliable representative sample of total site variability. In conclusion, it must be remembered that a primary conflict is associated with evaluating process control of microbially catalyzed processes in complex soil sites. That is, although microbiology is a science directed toward study of the minute, it must fit into the soil science framework of total ecosystem or even worldwide relationships.

This conflict in scientific purview is exemplified by studies of carbon dioxide evolution from soil. Respiratory carbon dioxide generation serves as a measure of biological activity at the micro- or macrosite level (i.e., μ^3 or cm^3), but such activity also has implications on quantities of greenhouse gases generated and their impact on terrestrial climates, which in turn control conditions at the soil microsite where the greenhouse gas originated. Soil microbial respiration is directly controlled by such factors as temperature, moisture, and organic matter inputs, which are affected, to a large degree, by the greenhouse-related phenomena. To quantify the contribution of soil microbial respiration to atmospheric levels of greenhouse gases, localized assessments must be conducted in a manner that allows their integration into the more comprehensive, world situation. Hence, any "real-world"–based soil microbiological study must be designed with an appreciation of the limitations and hazards of extrapolation of data relating to the realm of the microbe to that of higher plants and animals.

This telescoping nature of the application of soil microbiological research reveals a characteristic and, to some degree, an intellectual trap of our discipline. Data are necessarily a summation of a variety of reactions occurring at highly variable soil microsites. Thus, a composite representation of soil processes must result in the emphasis of dominant reactions, but other processes of major importance to ecosystem integrity and function may be overlooked or the importance of their contribution to system complexity minimized. For example, the bulk of terrestrial soils are aerobic. This condition is a product of soil pores being contiguous with the atmosphere. Molecular oxygen diffuses freely into the pores of nonwaterlogged soils as well as being transferred deep into the soil matrix through mass action. Mass transfer of atmospheric gases into soil can occur when groundwater is drained from the soil or as a result of the percolation of oxygen-bearing irrigation or rainwater through soil pores. Thus, it is easy to emphasize aerobic processes in conceptual models at the detriment of obligatorily anaerobic reactions. Yet, we know that anaerobic soil bacteria occur and are active in essentially all soils and that the products of anaerobic microbial activity are essential for ecosystem survival (e.g., denitrification). Thus, an important caveat in study of the rates of soil processes is that the potential for a biologically catalyzed reaction to occur must not be excluded simply from the results of an appraisal of overall system properties. Characterization must include a macro- and microsite evaluation.

The Black Box Viewpoint of Soil Systems: An additional conflict in development of soil microbiological concepts relates to contrasting images of ecosystem function derived from evaluation of the activity of individuals or from assessment of total ecosystem activities. The biological and structural heterogeneity of soils necessitates development of a dual understanding of the central object of soil biological studies. Evaluation of community activity can be likened to a box that conceals all microprocesses occurring therein, revealing only external effects of manipulation—that is, condition analogous to the engineering black box. In this instance, the identities and quantities of only the substances entering or exiting the box can be determined. Some degree of understanding of internal process rates can be developed for modeling of the ecosystem through study of this "black box." Alternatively, details of properties of the individual organisms (reactions and their associated rates within the black box) may be elucidated. These sorts of data reveal an intriguing picture of the adaptation of microbes to their environment. Unfortunately, individual enzymes and processes can be evaluated ad infinitum and still not produce a clear means of assemblage of the data or the concepts derived from their analysis into an image of the working whole. Thus, for soil microbiological research, total ecosystem studies must be developed with a clear understanding of the limitations imposed by ignorance of the "minutia." Similarly, any evaluation of microsite processes, must be conducted with a view of how the "micro"-processes are affected by the "macro" world in which they occur.

All soil processes are the summation of the activity of a variety of individuals, albeit a cell, multicellular complex, or even collection of isoenzymes. The individual in the soil must be appreciated and knowledge of the microsite conditions impinging on its existence are essential, yet appreciation of the individual is only of value for ecosystem evaluation in light of the revelation it provides in discerning overall site community dynamics and their impact on ecosystem function.

Hence, the topics broached in this chapter will be presented with an appreciation of the impact of the individual microbes and the properties of their habitat on total ecosystem dynamics. Initial studies will involve an evaluation of the concept of a limiting factor and experimental quantification of the impact of various environmental properties on biological activities. This will be followed by elucidation of the variation of specific soil chemical and physical traits. Soil properties interacting with biological entities to be evaluated herein include nutrient requirements, soil moisture, oxygen tension, redox potential, pH, and temperature.

5.1 Microbial Response to Abiotic Limitations: General Considerations

5.1.1 Definition of Limitations to Biological Activity

The classical definition of an environmental factor that determines the rate of a biological activity can be summarized by Liebig's law of the minimum: **"Under steady-state conditions, the essential material available in amounts** most closely approaching the critical minimum needed by a given organisms will tend to be the limiting one" (Dommergues et al., 1978). This principle was originally developed to describe the impact of variation of nutrients on microbial growth and activity, but it has been expanded in recent years to include a number of growth limitors, including temperature and a variety of inorganic and organic metabolic inhibitors.

It must be stressed that the application of Liebig's law to a particular ecosystem is appropriate only if the system is at steady state. This condition is necessitated by the fact that the organisms and their interactions cannot be adequately assessed in relationship to rate-controlling soil conditions except at steady state; otherwise, populations and/or their activities may be increasing to reach the augmented density or activity level allowed by the new prevailing environmental conditions, or they may be declining to accommodate newly imposed conditions.

Although it is conceptually attractive, Liebig's law is of limited applicability for describing the dynamics of soil ecosystems. This results from three conditions:

- Soil systems are so complex that it is rare for the microbes to have the luxury of dealing with a single stress factor.

- This combined stress produces different responses by the microbial community than would be anticipated to result from any soil property acting singly on the microbial cells.

- The structure of the soil ecosystem makes it difficult to determine the exact conditions at the microsite where the organisms are living.

Implications of Interaction of Limiting Factors on Microbial Activity: It is rare for one soil property to preclude life processes. Exceptions may be associated to some degree to portions of the dry deserts of Antarctica (Benoit and Hall, 1970; Cameron, 1972) or incidence of extreme anthropogenic intervention (e.g., acidification due to acid mine drainage, nuclear power plant cooling water outflows, or soils receiving massive influxes of heavy-metal–containing mine wastes). Similarly, it is difficult to predict the existence of a terrestrial habitat where no limitations to microbial growth occur—that is, a system where microbial growth and replication can occur at will. Essentially, all soil microbes are subject to compromises in the rates at which they function. More commonly, a combination of several soil physical and chemical conditions stress soil microbial populations sufficiently to cause them to operate at suboptimal levels. The soil microbes must adapt to an array of stresses rather than reacting to one imposing problem. For example, Wildung et al. (1975) documented the interactive nature of temperature and water content on soil respiration. Soil temperature was shown to alter respiration rates at moisture contents as low as -106 to -88 bar suction. Similarly,

Wardle and Parkinson (1990) found that lowest ecosystem stability was most associated with low temperature and low moisture conditions.

Population Responses to Stress Factor Interaction: The range of tolerance of a life process to alteration of environmental conditions may also vary due to interaction of two factors. It may be predicted from laboratory studies that a microbial population of interest would be unaffected by a particular soil property, yet in native soils, the growth of the organisms is precluded under seemingly acceptable conditions or its range of occurrence is truncated. For example, fungi grow in culture within the pH range of 3.0 to 8.0, but in soil the practical range is closer to 3.0 to 6.5. In this case, the impact of a chemical controller is altered by the microbe's capacity to compete with cohabiting biological populations. Thus, the activity of soil fungi is controlled by soil pH plus the organisms' ability to compete with soil bacteria. The latter populations are better competitors at the higher pH than are the fungi.

Similarly, some environmental properties may compensate for cellular inadequacies, thereby allowing microbial populations to develop outside the range predicted from laboratory studies for its maximum, minimum, or optimal activity. For example, it is not unusual to isolate soil microbes capable of growth at 30°C in culture but not at 37°C in a particular medium. If growth factors are added to the culture medium, the organisms grow quite well at the higher temperature. This results from the fact that an enzyme essential for the production of the growth factor is temperature sensitive — that is, it is inactive at the higher temperature. Availability of this obligatory growth factor in the environment allows the previously inactive organism to proliferate in the seemingly detrimental habitat.

Spatial Heterogeneity and Determination of Limitations to Microbial Growth: A further difficulty with application of Liebig's law of the minimum to conditions in native soil systems is more of a practical concern than a conceptual difficulty. Due to soil spatial and temporal variability, data may imply nonexistent restrictions to microbial growth. Basically, these observations result from the fact that a microorganism must encounter an inhibitory or stimulatory soil component before its growth or activity is affected. Microbes grow as microcolonies in soil microsites where their needs are met. Thus, microbial growth is limited by the distribution of carbon, nitrogen, and other nutrients; aeration; anoxic conditions; and so forth. Any troublesome soil property occurring beyond the perimeter of the colony is nonexistent from the viewpoint of the cells. Furthermore, environmental conditions vary on a daily, seasonal, and annual basis. This is easily envisioned by consideration of the rise in temperature in the surface centimeter of soil on an early Spring day. In the early morning hours, the temperature may be suboptimal for microbial activity; perhaps the soil is frozen. As the hours pass, soil temperature increases, resulting in a stimulation of biological processes until some optimal or maximal rate is reached. On the rare hot days, temperature combined in association with the resultant increased desiccation rate could present a situation that is inhibitory to the life processes occurring in the very top portion of the soil profile. Similar temporal effects could be described for other variables, such as nutrient concentration, pH, and moisture. Nutrients may be leached through soil during rainfall events that clearly cause dramatic changes in soil moisture levels. Similarly pH variation may occur at the microsite level due to acid production by the microbes themselves or result from a periodic influx of human products, such as acid rain or acid mine drainage waters. Thus, analysis of the soil site at a particular time of day or season of the year may indicate that the growth-promoting or inhibiting conditions do not exist, whereas if the measurements were taken a few hours earlier or later, contrasting results would be obtained.

These observations indicate the necessity of modification of our concept of a limiting factor to fit the complexities of soil. A more appropriate basis for consideration of control of soil microbial process is the following: **The growth of a soil microbiological population at steady state depends on a combination of several limiting factors acting collectively and interdependently at the soil microsite wherein the microbe is functioning.** For examples, denitrifiers in soil are limited by variation in easily metabolizable carbon levels, nitrate concentrations, as well as oxygen concentrations. Autotrophic nitrifiers are frequently controlled by soil pH, ammonium concentrations, and oxygen tension. It is likely reasonable to assume that none of the soil properties at the microsite where the microbes grow are optimal for the organism. Thus, it can be concluded that the specific level of activity of each individual is controlled by the combination effect of nearly all physical, chemical, and biological properties of the site. Exceptions would be those times or situations when one soil property occurs at such an extreme value that its effect overrides all others (e.g., soil moisture so low that minimal cellular respiration is possible, or temperatures approaching the maximum allowable level). In the bulk of the situations, though, it is the impact of the combination of limitations that determines the extent of microbial population development in a soil site.

5.1.2 Elucidation of Limiting Factors in Soil

Under most circumstances, the complexities of the soil environment as well as the microscopic and submicroscopic size of the soil microbial community, precludes clear determination of limiting factors in the field. In some situations, field observations are quite useful (e.g., Antarctic dry deserts or anthropogenically affected sites where extremes in physical or chemical conditions are easily discerned), but for the vast majority of the terrestrial ecosystem, controlled, laboratory-based study of the system is required to quantify the contribution of each individual soil property to biological reaction rates. Options for determining biological activities in soil include variation of single or combinations of properties in laboratory-incubated soil samples or assessment of field activities in several diverse systems.

Limitations of Laboratory-Generated Data: Scientific conditioning creates a desire of experimenters to study the effect of variation of each individual soil property on biological processes under controlled conditions in the laboratory. This offers the security of simplicity and reproducibility. In laboratory experiments, soil conditions may be varied singly or in concert, and the resultant impact on microbial activity assessed accurately. Easily manipulated factors include pH, temperature, moisture, and nutrient levels. These conceptually alluring studies may yield results that describe the capacity of existing microbial populations in the soil system to cope with alteration of their habitat in the short term. The utility of such data for prediction of the resiliency of the native soil population to long-term perturbation is frequently limited. To achieve the latter objective, soil samples should be incubated for a sufficient time period to allow the microbial community to adjust to the newly imposed conditions, particularly if the chemical or physical properties of the soil are being altered.

The limitations resulting from failure to allow community development in laboratory study is shown by evaluating the impact of varying soil pH beyond ranges normally detected in the soil ecosystem studied. It is frequently desirable to determine the pH range over which an enzymatically or biologically catalyzed process occurs. It could appear reasonable that the impact of variation in this parameter could readily be observed by varying the pH of one or a few representative soils. For this experiment, soil could be amended with different amounts of sulfuric acid, yielding soil samples with pH varying at set intervals over the pH range of interest. The biological activity would then be measured in each soil sample and its variation with pH assessed. The results of such an experiment are reasonably predictable. An ecological maxim is that the active microbial population in any soil sample is that which is best capable of coping with the prevailing soil pH. Thus, it can be concluded a priori that major changes in the pH of an individual soil sample may result in inhibition of the currently existing microbial strains. Such a study (i.e., adjustment of soil pH in the laboratory) must lead to the conclusion that the biological community in that particular soil sample has a defined pH range in which it is capable of functioning similar to the generic curve plotted in Figure 5.1. This type of experiment fails to allow consideration of the potential for development of microbial communities better adapted to function at the newly imposed soil pH. Since acidophilic or alkylinophilic microbes may exist in low numbers in the original soil sample, and generation times of soil microbes are measured in major portions of or multiples of days, extended incubation of the soil sample at the newly established pH is usually necessary for acclimation of the microbial community to occur.

Complications of data interpretation associated with adjustment of soil pH result not only from population selection, but also from the buffer capacity of the soil. Sufficient acid can be added to a soil sample to reach an acidic pH selected for a particular experimental design, but if the pH of the soil sample is measured a few hours after initial adjustment, it is commonly found that the

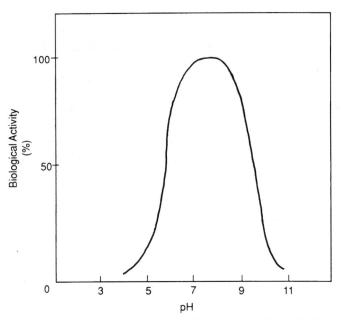

Fig. 5.1 Example of variation of a hypothetical biological activity with changes in soil pH. For this representation, an optimum of approximately pH 7 with rapid declines in activity above and below this value is depicted.

pH level may have returned to a value approaching the preamendment level. Both the microbial community and the soil buffer capacity serve to minimize the effect of soil acidification. Thus, not only is it necessary to incubate the soils for sufficient period to allow selection of acid-resistant microbial populations, but the natural tendency for the soil to return to native conditions must be considered. Changes of the soil pH during the incubation period may allow continued activity of the native population rather than development of acidophiles or acid-tolerant organisms. Thus, to prevent misinterpretation of the laboratory studies, soil pH of the amended soils must be monitored throughout the study. In conclusion, native soil samples may be adjusted to mimic variation in a particular soil property, but the data produced will reflect capabilities of the native soil community to cope with the change, not the capabilities of an ecologically adapted community. To achieve a measure of the latter situation, the pH-adjusted soil samples would have to be incubated for months or longer to allow the results of the selective processes to develop.

Evaluation of Soil Properties and Microbial Variability in Native Samples: An alternative available to overcome this potential for laboratory-generated misunderstandings of the complexity of soil community resiliency and adaptability is to study microbial population dynamics in several native soil samples that already exhibit variation of the property of concern. Microbial populations in native soils could be reasonably anticipated to have adapted to the soil site. Thus, if pH interactions with microbial activity were to be studied, it could be assumed that the microbial community existent in each soil is the one best adapted to function therein. Selection of appropriate soil samples for these studies is complicated. Except under fortuitous situations, it is rare to find a series of soil sites where only one or a few of soil properties vary in an otherwise constant environment. Thus, for study of native soil samples, not only is it necessary to assess the biological activity of interest, but all possible chemical and physical soil properties that could impinge upon that activity must also be quantified. Data interpretation is most commonly accomplished through analysis of correlation of the various soil properties with the biological activity or through regression analysis of the results.

Problems in data interpretation with these field survey experiments are encountered in that statistically significant relationships may be obtained, that are irrelevant. For example, the relationship may be fortuitous in that the variables being measured could have correlated with a third factor that was not considered. A potential solution to this difficulty is that once significant correlations are determined in the field, the data can be verified under more controlled conditions in the laboratory. A further means of confirmation would be to use the field-collected relationships to construct a mathematical model of the biological activity of the soil ecosystems. Field validation of the model may provide evidence in support of the conclusion that appropriate relationships have been elucidated and quantified.

5.2 Impact of Individual Soil Properties on Microbial Activity

From a somewhat naive point of view, soil appears to present a reasonably accommodating habit for microbial growth, Mineral nutrients, organic matter (including fresh plant material), and growth factors necessary to meet microbial nutrient requirements generally exist within appropriate temperature, moisture, and pH ranges for microbial growth. Also, most soils contain microbial populations with the necessary genotypic capacity to catalyze all requisite reactions for ecosystem development. Unfortunately for the microbe, soils also contain a variety of growth inhibitors — for instance, heavy metals, as well as organic and inorganic acids. Therefore, to fully understand biological activity in soil, the needs of the organisms must be considered in relation to the form and distribution of chemical and physical prerequisites for microbial growth.

5.2.1 Availability of Nutrients

Microbes require an energy source, electron acceptors, trace minerals, and several macro-nutrients plus, frequently, a variety of growth factors (i.e., vitamins and amino acids that are incorporated into cellular substituents intact) to proliferate. Although these substances are generally present in soil, the extent of microbial community development is determined by their distribution and concentrations within the microsites where the microbes reside. Therefore, the following discussion is presented with the objective of evaluating the biological requirements for growth and reproduction and the availability of these nutrients to soil communities. Details of microbial mechanisms for energy recovery and incorporation of micro- and macro-nutrients into biomass are presented in Chapter 4.

Soil Microbial Nutrient Resources: Energy Source: The primary energy supply for all living systems is solar energy. Photosynthetic higher plants and microbes (photoautotrophs) convert this energy into the chemical energy used by nonphotosynthetic (chemoautotrophs) organisms to support cellular growth and replication. Because of the diversity of metabolic capacities of microorganisms, a small portion of the soil microbial community is also capable of taking advantage of nonsolar-derived, energy sources available in soil including reduced minerals (e.g., S^0, Fe^{2+}) and products of cellular respiration (e.g., NH_4^+, H_2). Typical energy yielding transformations of these substances are

$$NH_4^+ \rightarrow NO_2^- \rightarrow NO_3^-$$

$$H_2 \rightarrow H_2O$$

$$S \rightarrow SO_4^{2-}$$

$$Fe^{2+} \rightarrow Fe^{3+}$$

Although there are many opportunities for chemolithotrophic (energy recovery from oxidation of inorganic chemicals) metabolism in soil, the bulk of the soil community is driven by the oxidation of organic substances. Heterotrophic organisms (chemoorganotrophs, heterotrophs) primarily use photosynthetically synthesized organic carbon, most commonly entering the soil ecosystem as root exudates, dead or decaying plant biomass, or soluble organic substances leached into the soil from surface litter. Plant biomass can also become available to soil microbes through anthropogenic intervention such as through soil amendment with manure or composts or with a variety of anthropogenically modified and industrially produced waste materials (e.g., petroleum by-products and industrial wastes). Because the energy contained in newly synthesized plant biomass is easily recovered by soil microbes, sites receiving these substances could be likened to oases within a desert of biological activity. This conclusion is based on the fact that in the absence of the reasonably constantly supplied source of carbon and energy produced by growing plants, soil microbes generally are left to metabolize the components of soil humus. This latter nutrient source is of lesser quality than the fresh, green plant biomass. Previous microbial activity has reduced the concentrations of easily metabolized organic matter. Thus, the substance remaining, the humus, is enriched in more complex organic compounds and microbial products.

Terminal Electron Acceptors: Recall that a primary chemical maxim is that all oxidative reactions must be balanced by reductions (Chapter 4). The biological reactions catalyzed within soil are not an exception to this rule. The most common electron acceptor in soil is molecular oxygen, which is reduced to water. For facultative anaerobic organisms, oxygen or organic carbon can serve as electron acceptors, whereas for strictly anaerobic bacteria, organic carbon acceptors such as organic acids are reduced. Some aerobic organisms growing under anoxic conditions reduce nitrate or nitrite (denitrification) or sulfate (dissimilatory sulfate reduction). (See Chapter 4 for further details of these processes.)

Macronutrient Sources: Following energy sources, the next-most abundant requirement for cellular growth and proliferation in soil is a variety of macronutrients (i.e., carbon, nitrogen, phosphorus, and sulfur supplies). These substances are the major building blocks for cellular biomass. Carbon and nitrogen are generally derived from the mineralization of plant biomass, whereas phosphorus and sulfur are not only components of biomass but are also present in high concentrations in soil minerals. Thus, pathways of carbon and nitrogen mineralization are frequently interlinked whereas phosphorus and sulfur supplies and transformation in soil is generally independent of organic matter availability and mineralization rate.

Determination of *in situ* Limiting Nutrients: Evaluation of nutrient limitations in soil communities is usually focused on energy or macronutrient supplies. Typically, documentation of such limitations entails amendment of soil samples with a nutrient that is suspected to control microbial metabolic activity and quantification of its effect on soil processes. If the soil amendment is utilized by the soil microbes, a variety of responses of the microbial community may be observed. An idealized depiction of the type of data that could be recovered from such studies is presented in Fig. 5.2. If the microbial population was not controlled by the soil amendment or an alternative factor limited biological activity, curves of type A are observed. For example, assume that the objective of an experiment is to determine the primary limitation to aerobic microbial respiration in a soil collected from a swamp. Observation that metabolizable fixed carbon compounds occur at extremely low levels in the soil leads to development of the hypothesis that microbial populations are carbon limited. A typical experiment could be to amend the soil with an easily metabolized carbon source, such as glucose, glycerol, or acetate, under the conditions occurring in the field site and observe the effect of the amendment on microbial activity. When this is done, no change in microbial growth of culturable aerobic bacteria occurs. Apparently, the carbon source was not limiting to microbial growth. In this situation, the absence of the electron acceptor, free oxygen, precluded the rapid catabolism of available carbon and resultant multiplication of the aerobic organisms. Thus, prior to soil amendment, sufficient carbon and energy existed in the low levels of organic substrate

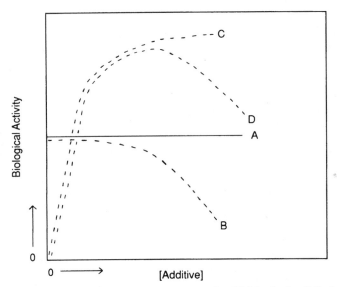

Fig. 5.2 Hypothetical curves describing response of soil biological activity to amendment with proposed growth limiting nutrients. A: The added substance is not controlling cell replication. B: The additive is nonlimiting and toxic to the biological community. C: The substance was present in the soil sample in growth-controlling concentrations. D: The additive is growth limiting, but it is also toxic in elevated concentrations.

to support the degree of microbial respiration occurring therein at a maximal level. Oxygenation of the system or provision of an alternative electron acceptor (e.g., nitrate) could result in development of a carbon-limited system. The primary delimiter of microbial activity (a terminal electron acceptor), in that case, would be relieved by addition of free oxygen or nitrate. Then, the available energy source could become becomes a prime candidate for primary obstacle to maximal microbial activity.

An alternative to this neutral response of soil amendment is the situation described by Curve B. These data reflect the condition where increases in soil concentrations of a hypothetical controlling nutrient results in inhibition of overall microbial respiration; that is, the amended substance is toxic. An example of this effect of soil amendment is provided by the frequently observed result of addition of a variety of aromatic compounds to a soil sample. In low concentrations [typically less than 0.1 percent (wt/wt)], microbial respiration may be stimulated by such compounds, such as catechol or benzoic acid, but as their concentration is increased, microbial respiration is inhibited. Thus, if maximally tolerated levels of these substances already exist in a soil site, further amendment with them inhibits cellular respiration.

More easily interpreted examples of relief of metabolic limitations are exemplified by Curves C and D. In these situations, the test compound controls microbial activity. As its concentration is increased, microbial activity or respiration increases until a plateau is reached. No further increase in activity is detected. At this point, another soil factor has become limiting to the microbial community or population studied. At maximum concentration, the amendment may have no further impact on soil activity (Curve C) or it may become inhibitory (Curve D), as was described above by the model Curve B.

Fluctuations in soil microbial populations following clear-cutting of a forest provide an excellent example of the impact of nutrient availability on biological activity *in situ* (e.g., Lundgren, 1982). Generally, clear-cutting of the forest initially results in an increase in soil microbial biomass. This microbial response results from the decomposition of excess organic matter, both living and that previously accumulated in soil as plant residues. These substances become available to the microbial community through destruction of the aboveground portions of the plant biomass. Roots die and carbon compounds from the debris produced by the cutting of the trees leach into the soil profile. Furthermore, machinery and logging traffic on the soil surface results in disruption of soil structure, which enhances availability of any organic matter occluded in soil aggregates. Three microbial communities can be described for this system in transition: the native forest soil community existent prior to anthropogenic intervention, that supported by the release of plant debris during logging (a disturbed soil community), and one that develops in response to the newly established ecosystem properties (bare soil). In a nutrient-controlled situation, the disturbed soil community would exhibit maximal population density, whereas the bare soil reflects a nutrient and energy limited community structure reflective of the denuded, minimally productive aboveground ecosystem.

Carbon Limitation of Microbial Activity: Examination of the occurrence of macronutrients in a soil sample (carbon, nitrogen, sulfur, and phosphorus) and their ratios in soil microbial communities reveals that of these substances, carbon resources are generally the primary controller in soil microbial community development. Soil microbial biomass contents of soils usually correlate with soil organic carbon contents and are generally stimulated by fixed-carbon amendments (e.g., Schnürer et al., 1985, Knapp et al., 1983). The capacity for soil microbes to utilize efficiently and effectively the pools of metabolizable carbon contained in soil and to be easily stimulated by soil amendment with such materials has been adequately and frequently documented. Similarly, it is commonly noted that nitrogen is rarely limiting to soil microbes, but the active incorporation of nitrogen into soil microbial biomass may cause nitrogen to become limiting to the plants growing in the ecosystem (see Chapter 11 for a further discussion of soil nitrogen transformations). The low concentrations of sulfur and phosphorus required by soil microbes may be supplied through mineralization of plant biomass or through solubolization of soil minerals. Again, although these macronutrients are commonly present in growth-controlling concentrations for higher plants, soil generally contains adequate sulfur and phosphorus pools for microbial development.

Growth Factors: Growth factors are organic compounds required by the cell in small quantities—for example, vitamins and amino acids, purines, and pyrimidines. They are incorporated into the cell intact. It must be noted that if sufficient concentrations of these materials exist in soil, they may also be used as carbon, nitrogen, and/or energy sources. For example, a large proportion of the bacteria growing in the rhizosphere are capable of mineralizing amino acids to meet these nutrient and energy requirements.

Species requiring growth factors are more common in soil than are those capable of synthesizing them *de novo*. This implies that common sources of growth factors must be present in soil. Four reservoirs of microbial growth factors can be readily identified:

- *Other organisms growing in soil*: Each microbe in soil is capable of synthesizing a portion of the growth factors that are necessary for its biomass production (prototrophs). A microorganism may produce more than is required for internal consumption. Excess growth factors thus may leak from the cell into the suspending milieu. This cross-feeding of nutrients (utilization of growth factors leaking from a producer cell to cohabitants of the habitat) is frequently observed when viable cells are grown on petri plates containing vitamin- or amino acid-free media that support development of heterotrophic bacterial colonies. Small satellite colonies develop around the prototrophic bacterial colonies. Alternatively, supplies of the growth factor become available to auxotrophs (those requiring the growth factor) upon death and lysis of the neighboring cells.

- *Root exudates and sloughed cells*: Plants commonly leak a mixture of growth factors (including vitamins and amino acids) from their roots. Furthermore, as the root extends through soil, packets of nutrients as sloughed cells are released from the root surface and consumed by the soil microbial community.

- *Anthropogenically supplied organic substances*: Organic matter incorporated into the soil (waste disposal, sludges, composts, and accidental spills) also contains a variety of vitamins as well as macronutrients. These substances are found within the residual cells of the composted vegetative materials as well as in the associated microbial community. Lysis of the plant cells and death and lysis of the microbial cells makes the nutrients available to the soil microbes.

- *Whole plant and animal cells*: Predators and parasites receive their growth factors directly from the host cells.

The dominant organisms present in an ecosystem are usually those requiring a few growth factors, at most. More fastidious metabolic needs are usually associated with the flora of living plants or animals (e.g., rumen or intestinal organisms, and root symbionts). In these latter situations, an organism capable of scavenging growth factors from its environment clearly has an advantage over one that must expend energy to synthesize the factor, but once the available supply of the vitamin or amino acid becomes limiting, the roles are reversed. In a soil system barren of growth factor resources, prototrophs are better competitors than are auxototrophs. An exception to this generalization is associated with those microbes that are capable of forming resting structures, such as endospores. For these creatures, once the easily recoverable supply of growth factors is depleted and their competitive advantage is lost, they form a resting structure wherein they persist until growth conditions improve.

Trace Metal Supplies in Soil: A variety of metallic ions are required for microbial growth. Examples of universally required cations are potassium, iron, and magnesium, whereas most living cells need calcium, copper, and zinc. Cobalt is likely required by most organisms for synthesis of vitamin B_{12}. Molybdenum is required for nitrogen fixation and nitrate metabolism. As would be anticipated by the predominance and variety of mineral components of soils, except for molybdenum, these metals are generally not limiting to soil microbes.

5.2.2 Moisture

Water is an essential contributor to the growth of soil microbes. Water not only provides an essential medium for growth of microbial populations, but it is also a primary participant in a variety of cell processes. The major roles for water in soil can be summarized as follows:

- *Serves as an essential material for flora and fauna*: A number of soil enzymes catalyze the incorporation of portions of the water molecule into biological compounds. Water is a direct participant in hydrolysis reactions (for example, cleavage of cellulose into glucose and proteins into amino acids) as well as hydroxylations, such as the hydroxylation of picolinic acid and other pyridine ring-containing compounds (e.g., see Dagley and Johnson, 1963; Hirschberg and Ensign, 1972; Hochstein and Dalton, 1965; Tate and Ensign, 1974.).

- *Affects gas exchange in soil*: Soil pores are filled with air and water. If all pores are water-filled, the soil is waterlogged or saturated. Since gaseous diffusion is greater in air than in water, the greater the water content of a soil, the higher the probability of its becoming oxygen-limited or even anoxic.

- *Affects microbial nutrient supply*: Water is the transport medium of microbial nutrients to the cell and of waste materials away from the cell. For nonwater-soluble nutrients, water serves as the transport medium for the extracellular enzymes or solubolizing agent to reach the substrate and convert it to water-soluble products or forms. Extracellular enzymes are generally involved in the cleavage of complex polymers into water-soluble subunits, whereas surfactants increase the availability of hydrophobic materials (such as lipoidal substances) to microbial cells.

- *Affects soil temperature*: The higher the soil water content, the more resistant is that soil ecosystem to temperature fluctuations. Hence, soil temperatures rise more slowly during daily cycles in moist or water-saturated soils during the Spring than occurs in the desiccated, or air-dried soils of the mid-Summer.

- *Serves as a growth medium for microbial colonies*: Microbes function in the water layer on soil particles or within soil pores. Even in a seemingly dry soil sample, a layer of water must exist surrounding microbial microcolonies and soil particles to support a viable soil community. In reality, although soil is predominantly a particulate system, microbial life forms contained therein live in an aquatic world. This water-saturated system may consist of a microfilm a few microns thick or it may extend throughout a soil micro- or macropore. In either case, for the microorganism to function properly, it must be totally bathed in water.

Measurement of Soil Moisture: Standard techniques for quantification and expression of soil moisture levels are a necessity to the development of predictive mathematical models of the impact of soil moisture variation on soil processes and to the provision of a basis for comparison of moisture effects on biological activity in contrasting soil types. The procedure should be easily conducted and reproducible, and should assess an aspect of soil moisture that directly affects soil microbial activity.

Traditionally, soil moisture has been reported as gravimetric water content [ω, g H_2O/g dry soil (usually dried to a constant weight at 105°C)], and volumetric water content (θ, $cm^3 H_2O$/cm^3 soil). These two means of expressing soil moisture are related to the soil bulk density (ρ_b, g solids/cm^3) by the following relationship:

$$\theta = \omega \rho_b \qquad (5.1)$$

A third term encountered on a perhaps all-too-frequent basis is g H_2O/g wet soil. Use of the latter term is particularly distressing. It provides some means for comparison of soils data when a single soil type is considered, but no commonalty for evaluation of results between soils types exists.

Although gravimetric and volumetric water content are probably the most frequently used representations of soil moisture by soil microbiologists, their use is somewhat problematic in that neither value describes the actual availability of water to the soil microbial community. Furthermore, these values do not vary independent of soil type. Thus, comparable gravimetric or volumetric moisture values in two different soil types do not represent comparable water availability to the resident microbial communities. Although a relationship between soil moisture and a specific microbial activity can be derived for a single soil type using these values, it is not possible to determine the generalized principles behind the relationship.

A partial solution to this difficulty is to use water potential as a measure of water availability in soil pores. Water potentials—measurements with a long history of use by soil scientists in general—are governed by surface tension forces and the radius of the menisci of the water. Thus, values can be assessed by principles of capillary rise. The mathematical expression of this relationship is as follows:

$$\tau = 2\gamma/r \qquad (5.2)$$

where r is the equivalent or effective capillary radius, τ is the suction necessary to absorb the water, and γ is the surface tension of water. Note that with this relationship, as the equivalent capillary radius becomes smaller, less water is present and the suction necessary to absorb the water increases. Equivalent expressions of suction are as follows:

$$1 \text{ bar} = 10^6 \text{ dyn cm}^2 = 100 \text{ kPa} = 1022 \text{ cm H}_2\text{O} = 75 \text{ cm Hg} = 0.987 \text{ atm}$$

Common water potential values found in the soil microbiology literature are the moisture content of soil at -0.3 bar (field capacity).

Another useful expression of soil moisture is water-filled pore space (Linn and Doran, 1984). Soil microbes exist in a mixed soil air-water system. In reality, it is the degree of displacement of air with water within the soil pore that combines with the rate of microbial respiration to produce anoxic or anaerobic microsites within the soil profile [see Skopp et al. (1991) for a detailed study of this process]. Complete displacement of air with water restricts diffusion and thereby hastens imposition of oxygen-limited conditions. When moisture approaches or exceeds field capacity, the percent soil pore space filled with water or air becomes a better indicator of aerobic vs. anaerobic microbial activity than water content or water potential. Use of percent water-filled pore space to express soil moisture levels is practical since it only requires knowledge of the soil bulk density and the gravimetric soil water content. (The particle density of mineral soils is assumed to be 2.65 Mg/m^3.) The percent water-filled porosity is defined by the following relationship:

$$\%\text{WFP} = (\theta_v/\text{TP})100 \qquad (5.3)$$

where θ_v is equivalent to the product of the gravimetric water content (ω) and the soil bulk density (ρ_b) and TP is the percent total porosity $(1 - \rho_b/\rho_p)(100)$, where ρ_b is the soil bulk density and ρ_p is the soil particle density. For aerobic processes, maximal activity is detected at a percent water-filled porosity of about 60 percent. Below 60 percent, water limits activity (insufficient moisture exists to provide the contiguous surface film required by the soil microbes), whereas above 60 percent aerobic microbial activity decreases (apparently due to reduced oxygen availability). Caution is necessary in use of these generalized figures. Neilson and Pepper (1990) found discrepancies in use of percent saturation or water-filled pore space to characterize soil aeration, especially when soils of differing bulk densities or those receiving particulate organic matter amendments are compared.

The effect of water in a heterogeneous system can also be expressed as water activity. This measurement of moisture is most commonly used in food storage to describe water availability. Water activity is related to the equilibrium relative humidity (ERH) as follows:

$$A_W = \text{ERH}/100 \qquad (5.4)$$

The difficulties with using water activity in soil microbiological research are the complications associated with its measurement. Long incubation times in a chamber where the moisture level of the test sample is allowed to equilibrate with the chamber atmosphere are usually required. Data can be extremely variable. Water availability affects species diversity, survival, movement, and activity of microbes. Of all of the expressions that may be used to define the moisture conditions under which the soil microbe is abiding, water activity is probably the most useful. Unfortunately, it is also the most difficult and time-consuming (i.e., impractical) measure of soil moisture.

Impact of Moisture Variation on Soil Microbial Activity: Total microbial activity varies from nearly nonexistent levels at low water availability to a maximal level under optimal soil air/moisture mixtures (Fig. 5.3). Typically, soil bacteria are capable of functioning under seemingly severe soil moisture tensions. For example, Wilson and Griffith (1975) found that respiration of a mixed bacterial population in soil was maintained at a reasonably high level between -8 and -30 bar, but was negligible at -50 bar. Similar results have been reported by Wildung et al. (1975) and Knight and Skujins (1981). As soil moisture levels are saturated and free oxygen supplies are depleted, overall activity again declines. This generalized curve is applicable to the summation of microbial respiratory activities and to a certain degree individual processes. In the latter situation, the points of minimal or optimal activity are characteristics of the trait being studied. Obligatorily anaerobic processes clearly are optimized under saturation conditions.

Tolerance to restrictive soil moisture levels varies with major microbial groups as well as with individual microbial species. Generally, bacteria are more exacting than are actinomycetes and fungi. Hence, the latter organisms tend to predominate in the drier soils. This observation results at least in part from the capacity of the latter groups of organisms to form resting structures. Robertson and Firestone (1992) postulated that bacteria may use extracellular polysaccharide to enhance survival under moisture-limiting conditions. A purified extracellular polysaccharide from a *Pseudomonas* species was shown to contain severalfold its weight in water at low moisture tension. Within the realm of the

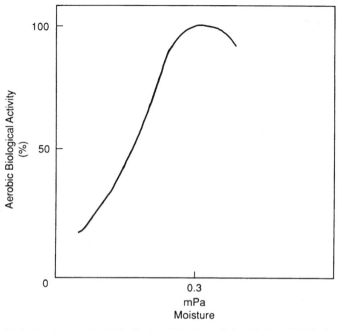

Fig. 5.3 Variation in aerobic biological activity in soil. In this hypothetical example, optimal moisture is approximately 0.3 mPa.

bacteria, a gradient of tolerance is noted to occur. For example, ammonium tends to accumulate in dry soils due to the ammonifying microorganisms being less sensitive to the low water activities then are nitrifiers.

Selection of microbial populations with increased resistance to moisture limitations in chronically arid soils was demonstrated by Knight and Skujins (1981). These workers used adenosine triphosphate and soil respiration measurements as indicators of microbial biomass and activity. The impact of variation of soil moisture between -2 and -100 bar tension on microbial activity in two arid soils and a subalpine forest soils was evaluated. The microbial activity increased in arid soils between -2 and -20 bar soil moisture, whereas the optimal moisture in the chronically wetter forest soils was -2 bar and activity continually decreased below that level. Thus, it was concluded that a characteristic microbial flora capable of managing the prevailing environmental conditions had developed in each of the extreme environmentally different soils studied.

It is important for the student of microbiology to understand clearly what these observations reveal about the function of the soil microbial community and at least as importantly, what is not stated. The positive side of this observation is that it follows that as long as the soils are not so dry as to preclude biological function, an appropriate microbial population develops.

The negative aspect is that it is not implied that this microbial population selected for growth under restricted soil moisture is as active as a comparable population growing in less stringent conditions. That is, were microbial respiration measured under limited soil moisture and at optimal soil moisture, even if the organisms in each situation were absolutely optimized, less respiration would be found in the low-moisture soil. Practically, this means that one should not anticipate as much organic matter to be decomposed in desert soils without irrigation as could occur in an irrigated field of the same soil.

The impact of excessive soil moisture (flooded or chronically waterlogged conditions) on soil microbial activities is more predictable than situations observed in habitats experiencing a shortage of free water. Interactions of soil moisture levels with availability of molecular oxygen become evident in the flood or chronically waterlogged soil sites. A traditional division of the microbial realm has been separation by oxygen tolerance (anaerobes, facultative anaerobes, and aerobes). Flooded soils are not necessarily oxygenfree, but should available free oxygen supplies be depleted, anaerobic and facultative microbial populations are stimulated while obligate aerobes die or enter resting stages. The most dramatic effect of institution of anaerobic conditions in soil is the induction of facultative (e.g., fermentations), anoxic aerobic metabolism (e.g., denitrification), or strictly anaerobic processes (e.g., methanogenesis). Organic acids may accumulate, nitrogen oxides could be evolved, and ultimately, methane generation or sulfate reduction may occur. Practically, anoxic soils may be recognized by precipitation of metal sulfides, or evolution of an amine or sulfide odor. Characteristics of the microbial community are dependent upon the duration of the flood period and the nature of electron acceptors present. Methanogenic bacteria require development of highly reducing conditions, whereas fermentative organisms are less stringent (see Chapter 4).

As stated above, a flooded soil is not necessarily either anoxic or highly reducing. Water movement through the environment as well as diffusion of atmospheric gases into the system may prevent exhaustion of free oxygen. This phenomenon was shown in a study of the muck soils of south Florida (Tate, 1979). With imposition of flooding of the soils, an increase in aerobic bacterial populations was initially observed (Fig. 5.4). The soils were flooded to maximum depths of about 31 cm. As the free oxygen and easily metabolized carbon source levels declined, a parallel reduction of aerobic bacterial populations occurred. Thus, were the experimenter assuming that flooded and anoxic conditions were equivalent, during the initial portion of the study, a behavior of the soil microbes contrary to what occurred *in situ* would have been expected.

Soil Systems with Fluctuating Soil Moisture: Further underlying assumptions or perhaps attitudes that must be developed in relationship to understanding the effect of soil moisture on the microbial community are (1) the moisture level of a particular site tends to be continually changing

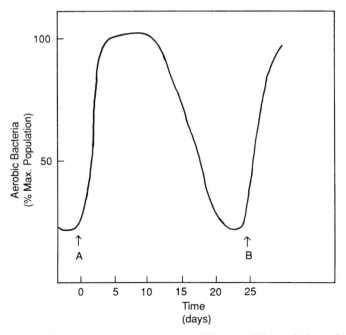

Fig. 5.4 **Effect of flooding a dry histosol on aerobic bacterial populations. Soils were flooded to a maximum depth of approximately 31 cm commencing on day 0 (Arrow A). Arrow B indicates the time of drainage of the flood waters. (See Tate, 1979 for further details).**

and (2) moisture effects on biological activity are interactive with other system properties.

Moisture fluctuation may be on a short-term, localized basis as would be found (1) with daily impact of temperature variation and water evaporation on the top centimeter of an exposed soil surface, (2) following a heavy rainfall, or (3) due to seasonal flooding, as occurs in a wetland rice soil. For example, microbial biomass fluctuations can reveal the short-term effect of soil moisture levels induced by precipitation (Clarholm and Rosswall, 1980). Precipitation caused increases in microbial biomass lasting one to two days, even in nonmoisture-limited soil. This observation supports the conclusion that although data descriptive of macrosite soil moisture tensions support a conclusion that the microbial populations are not limited by this soil property, it is microsite discontinuity in moisture films covering soil particles that determines the rate of change in and ultimate level of microbial population densities. In flooded rice field soils, anoxic conditions induce anaerobic microbial populations during the somewhat long term of flooding, but draining of the site following harvest or for establishment of seedlings induces aerobic and facultative microbial populations. Strict anaerobes may be further limited by the flow of oxygenated water through the system. In that situation, a mosaic develops in soil, with pockets of anoxic soil interspersed with oxygenated soil.

Note that soil systems are described as containing a mixture of aerobic and anaerobic micro- or macrosites, *not* "a partially aerobic system." By definition, each individual soil site is either anaerobic (anoxic) or aerobic. "Partially aerobic" is an inappropriate description in that "aerobic" indicates the presence of oxygen in the system, which is an all-or-none state. Even if only a few molecules of oxygen are present per cubic centimeter of soil, the site is still described as being aerobic.

The interactive nature of soil moisture with other soil properties is documented by examining organic matter availability and osmotic pressure variation in soil. Variation of soil moisture with time not only affects the nature of the microbial population present in the soil, but it may also alter the quantity of nutrients or the physical structure of the microbes' environment. Cycling of soil moisture liberates occluded organic matter through such processes as disruption of soil aggregate structure or by death of the more stringent microbial populations. Therefore, microbial oxidation of soil organic matter is greater in soils with fluctuating moisture than in constantly wet or dry conditions (e.g., see Lund and Goksoyr, 1980; Orchard and Cook, 1983; Reddy and Patrick, 1975; and Sørensen, 1974). Jager and Bruins (1975) noted that the loss of carbon due to microbial respiration in a soil experiencing approximately 60 wet/dry cycles was also proportional to drying temperature, with greatest loss (31.2 percent) in the soil dried at 85°C and least (18 percent) in soil dried at 35°C. Drying resulted in an ultimate reduction in microbial population densities, but the highest population densities were found in soil dried at 85°C, suggesting that more organic matter was made available to support microbial growth by the more drastic treatment. A practical implication of this observation relates to maintenance of soil organic matter levels in mineral soils and the minimization of subsidence of histosols (see Tate, 1980). In both situations, maintenance of the native organic matter levels is desirable for preservation of the ecosystem and/or encouragement of optimal soil structure (see Tate, 1987).

An indirect effect of declining soil moisture is the impact of osmotic pressure on biological activity. Generally, the osmotic potential of water is less important than water availability, but in some localized or transient situations, variations in osmotic potential emerge as significant controllers of microbial growth. Potentially lethal or inhibitory salt concentrations may be the product of microbial metabolism (e.g., mineral dissolution through acidification of microsites) and chemical equilibria (e.g., ionization of water-soluble salts), as well as physical phenomena (evaporation). Localized problems are associated with areas of salt encrustation due to water evaporation; saline waters such as are found in some irrigation systems, around saltwater bodies; or in association with saline soils. Transient difficulties may be associated with severe desiccation of soil.

Major impacts of osmotic pressure are likely averted in soil because of the fact that reduction of soil moisture results in induction of cellular survival adaptations (for example, sporulation). Also, in regions with chronically saline soils, selection of organisms with increased resistance to pressure changes may occur. For a discussion of microbial adaptations to osmotic changes, see Csonka and Hanson (1991).

5.2.3 Aeration

The primary role of molecular oxygen in soil is as a terminal electron acceptor. As was indicated in Chapter 4, energy is produced during electron transport to oxygen. In fact, oxygen is the highest energy-producing electron acceptor. Thus, the greatest cell yields are produced by the total oxidation of an organic carbon-containing substance to carbon dioxide and water. This important function of molecular oxygen in aerobic metabolism is augmented by its direct incorporation into molecular structures of such compounds as aromatic ring-containing substances and alkanes by oxygenases. The significance of aerobic metabolism to the soil microbial community is underscored by the observation that most microbes in surface soils are aerobes. Exceptions are soils that are consistently flooded, such as bog soils.

Soil oxygen tension is controlled by gaseous diffusion rates as well as by the respiratory activity of the soil biota. Since diffusion is a relatively slow process, it is not unusual for the rate of consumption of this essential gas to be greater than its rate of replacement. Thus, free oxygen tensions are rarely too high for soil aerobes, but clearly soil oxygen levels may be too low for obligate aerobes.

Indirect limitors of soil oxygen are soil moisture and the concentrations of metabolizable organic matter. As was indicated above, moisture affects soil oxygenation in that the greater the soil moisture, the less air-filled pores exist, and the slower the diffusion rate. Decomposable organic carbon availability controls the rate of oxygen consumption indirectly in that as this fixed carbon is oxidized, free oxygen may be depleted. With an influx of easily metabolized organic matter, it is not unusual for anoxic conditions to develop. Examples of practical consequences of oxygen depletion due to soil microbial respiration are as follows:

- Oxygen depletion in cropped soils after a heavy rainfall event can result in crop damage.

- After a heavy rainfall, with the presence of nitrate and suitable carbon sources, significant losses of fixed nitrogen from soil can result from induction of denitrifiers.

- Spills of easily decomposable organic matter in soil or incorporation of readily metabolizable organic matter can result in depletion of free oxygen supplies thereby resulting in ecosystem disruption through oxygen starvation of plant roots, death of small animals, earthworms, nematodes, mites, and so forth.

As the free oxygen in soil is depleted, a number of predictable changes in the microbial activity occur. When the soil oxygen tension has been reduced to less than 1 percent (v/v), the microbial population appears to shift from being predominantly aerobic to anaerobic. With the development of a reducing atmosphere, growth yields decline since the energy yielded per mole of fixed carbon oxidized anaerobically is far less than that produced from aerobic respiration (see Chapter 4). [Note that dissimilatory respiration of aerobic bacteria (denitrification and sulfate reduction) cell yields approximate those with oxygen as the terminal electron acceptor since a cytochrome based electron transport processes occurs and the carbon substrates are oxidized completely to carbon dioxide, water, and ammonia.] Fermentation results in accumulation of incompletely oxidized carbonaceous substrates (organic acids, amines, etc.). When highly reducing conditions develop so that methanogenesis and sulfate reduction occur, complete mineralization of fixed carbon substrates again occurs.

5.2.4 Redox Potential

It is difficult to separate the impact of redox potential from oxygen tension effects experimentally. To artificially raise the redox potential in the absence of free oxygen, it is necessary to add oxidizing agents, which are generally toxic to microorganisms. Thus, it must be noted that changes in soil redox potential are related to changes in oxygen levels. If organic matter is added to soil, oxygen is depleted and the potential drops — at times quite precipitously. This is a microbial reaction since inhibitors of microbial activity prevent both oxygen depletion and development of reducing conditions — which is somewhat logical since in an actively respiring community, oxygen is used as an electron acceptor.

As has been alluded to above, the occurrence of a variety of microbial processes is related to specific redox potentials. Some of these are as follows:

Aerobic Carbon Oxidation	>0.2 V
Denitrification	0.15 to 0.2 V
Methanogenesis	-0.2 to -0.1 V
Sulfur reduction	-0.2 to -0.1 V

Although these data suggest somewhat of a mutually exclusive nature of microbial activity in soil, it is not unusual to observe these processes occurring concurrently across a landscape of limited size. Again, this results from the heterogeneity of the soil system. Two examples of this situation (variation of redox potential in a flooded soil and within the profile of a drained histosol) are provided in Fig. 5.5. In both situations, an oxygen-rich, high redox potential region overlays an anaerobic portion of the soil profile. A region of variable conditions generally occurs between these two extreme environments. Similarly, a mosaic of activities can occur in a well-aggregated soil where

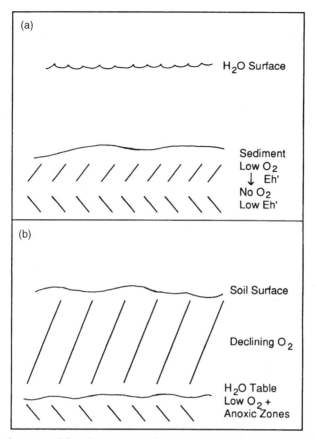

Fig. 5.5 Redox potential and oxygen tension patterns in a hypothetical flooded soil (A) and a soil with an elevated water table (B).

aerobic processes predominate on the surfaces of the aggregates and anoxic processes occur in the interior.

5.2.5 pH

The microbiologist's bias tends to favor conducting experiments with soils with approximately neutral pH and the study of the microbes capable of functioning therein. Our laboratory media developed for the culture of microorganisms are so constituted, and the majority of the soil systems studied tend to fall in a pH range of approximately 5.5 to 7.5. Yet acidic soils constitute a major portion of the world's soil resources, especially as exemplified by soils of tropical regions and temperate forests, and acidification is a major reclamation problem in the normally near-neutral soils of the temperate region. For example, with many mining processes, metal-sulfide–bearing slags are produced. Biological activity in these waste materials results in oxidation of the sulfide to sulfate. Subsequent drainage of these acid-bearing waters presents major problems for downstream soils and receiving waters. Hence, it is imperative to understand the effect of soil pH on soil microbial community development and function.

Acidic conditions present a particularly stressful situation to the microbial cell. Organisms existing in acidic habitats could be described as being in a somewhat schizophrenic state. The internal cell pH must be highly buffered at or near neutrality, the pH optimum for internal enzymes. Thus, the effect of variation of the external pH necessitates development of cell-surface and membrane-associated mechanisms to control the internal pH. Furthermore, metabolic processes occurring at the cell surface must be adapted to cope with the acidic conditions.

A variety of pH-associated metabolic problems must be surmounted by the cell. For example, cellular surface charge varies with environmental pH. This controls the interaction with soil particles as well as with charged nutrients. Furthermore, surface-bound enzymes as well as extracellular enzymes (see Chapter 6) must also be capable of functioning at the elevated hydrogen ion concentration. Maximal activity must either be expressed at the predominant pH, or enzyme synthesis rate must be modified to compensate for the reduced enzymatic activity occurring under the acidic conditions. Also, adjustment of enzyme synthesis rates may be necessary for those enzymes that are denatured under acidic conditions. If the enzyme product is essential for cell function, then the organism must adjust the enzyme synthesis rate so that it is always present at effective levels.

Indirect Mechanisms for pH Limitation of Biological Activity in Soil: Nutrient concentrations and toxicity of environmental substituents are both affected indirectly by soil acidity. Trace mineral availability provides an excellent example of pH interactions with nutrient availability. Both iron and manganese are more water soluble at low pH, whereas molybdenum is precipitated at low pH.

Similarly, capacity to transport charged carbonaceous substances such as fatty acids and amino acids into the cell is dependent upon the charge of the molecules. Reduction of the soil pH below the pKa of acids results in conversion of the negatively charged entity to a neutral compound. Zwitter ions may have an overall negative charge or positive charge, or be uncharged depending upon the pH of the suspending medium. For example, glycine ($^+H_3NCH_2COO^-$) has both + and − charges at neutral pH. As the pH is varied, the positive and negative charges are neutralized independently. Each of these states would result in alteration of the capacity of the cell or its catabolic enzymes to catalyze transformation of these materials.

Under acidic conditions, organic acids are toxic to microbial growth. This is not a pH effect, but is an impact of the presence of the organic acid itself in an uncharged or neutral state. Thus, accumulation of organic acids under anaerobic conditions in acidic environments could be a mechanism for limiting microbial activity. This interaction of pH and anaerobic conditions may contribute to organic matter accumulation in acid swamps.

A final indirect effect of acidic soil pH conditions on biological activity relates to the potential for solubolization of toxic compounds or elements. Aluminum is more available under acidic conditions. It is not uncommon for toxic levels of this cation to be detected in acidic tropical soils.

Direct Modifications of Biological Activity by pH Variation: The direct effect of pH relates to the fact that each microbial strain has an intrinsic range of pH within which it is capable of functioning. The diversity of soil life is exemplified by the observation that microorganisms grow from about pH 1 to 11. Similarly, since it is to a large degree the property of organism's enzymes that determine the diversity of processes catalyzed, a comparable reaction to environmental pH is found for these cell components. Although microbial life occurs at essentially all naturally occurring soil pH levels, each individual species (and to some extent, strain) has a characteristic pH range and optimum pH for growth. For example, *Thiobacillus thiooxidans*, a key organism in oxidation of reduced sulfur-containing compounds, grows only at extremes of acidity (pH 2 to 4), whereas *Nitrosomonas* sp., a key player in biological nitrification, has an optimum pH in the alkaline range and is not active below about pH 5.5. Similarly, another sulfur-oxidizing bacterium, *Thiobacillus thioparus,* grows only at near-neutral pH.

Some selection of overall contributors to biological activity results from variation of soil pH. In acid soils, the dominant activity is generally the result of the fungal community. This results from the poor growth of most heterotrophic bacteria at acid pH. An exception is with waterlogged acid soils. Here, the aerobic fungi are inhibited by oxygen limitations, so the acid-tolerant bacteria are better able to compete.

Some microbially catalyzed processes occur in soils with a specific pH range. For example, nitrification (oxidation of ammonium through nitrite to nitrate) is suggested to occur primarily in soils with a pH greater than 5.5. This observation results from the fact that nitrification is catalyzed primarily by a small group of autotrophic bacterial species that are pH sensitive (Dommergues et al., 1978). Exceptions to this pH limitation have been observed to occur in forest soils, but an unequivocal microbiological explanation of the data has not been presented.

It can be generalized that if a process occurs over the entire pH range conducive to development of living cells, the transformations result from the activity of a variety of microbial species or strains with varying pH optima. For example, carbon and nitrogen mineralization occur in essentially any soil ecosystem where life is possible. Thus, a measure of the diversity of the microbial community is revealed by the breadth of conditions under which the organisms are able to function. This maxim relates not only to soil pH but also to other properties, such as moisture and aeration.

Conflicts Between Laboratory Observations and Actual Field Activities: Some anomalous situations have been encountered. These are associated with conditions where laboratory experiments with axenic microbial cultures yield data suggesting that the only known microbes capable of catalyzing a particular process cannot function at the prevailing soil pH of the site under study, yet the products of the reaction are detected. For example, as indicated above, acidophilic autotrophic nitrifiers have not been detected in acidic forest soils, yet nitrate is commonly found in these soils. A variety of explanations may be proposed and should be investigated when such situations arise (Dommergues et al., 1978). The microbes may be growing in soil microsites with a favorable pH. That is, the acid-limited microbes may be constrained to islands of more moderate conditions contained within a sea of restrictive soil. This situation would be more likely to occur in soils where the overall pH is near the extremity of function of the microorganism. That is, an overall soil pH of approximately 5.0 could be measured, yet microsites 0.5 pH units above or below that value may exist.

An alternative explanation to the anomalous occurrence of a pH-limited process may involve our capability of culturing the responsible microbes. A previously unknown population of microbes with a pH optimum in the range existent in the ecosystem could be present and functioning in the site under study. Only a small portion of the microbial community in any soil can be cultured and studied in the laboratory. Techniques for the study of the vast majority of this diverse community remain to be developed. Thus, occurrence of processes in the field may be a better indicator of microbial capabilities than is the library of information collected from axenic cultures.

Last, a spontaneous chemical conversion of the reactants may occur outside of the pH range optimal for biological activity. This is exemplified by denitrification under acidic conditions. In this situation, bacteria can be cultured that catalyze reduction of nitrate to nitrous oxide and dinitrogen under acidic conditions, but their function in the native acidic soil is questioned. The kinetics of the spontaneous chemical reaction favor that process over the biologically catalyzed nitrogen oxide reduction.

In conclusion, a variety of interactions of soil hydrogen ion concentration and soil biological process have been elucidated. Life does occur in extremely acidic or alkaline soils. The participants may vary from those under less stringent conditions, and the range of processes occurring therein may be somewhat truncated. Stewardship of acid soils and reclamation of acid-affected soils requires consideration of the reactions occurring therein and the limitations of the system. Due to the intrinsic buffer capacity of soil, management of an acidophilic community may be necessitated by the inability to adjust the pH of many soils to a more neutral status (e.g., See Magdoff and Bartlett, 1985).

5.2.6 Temperature

Greatest resistance to temperature extremes among living entities is found in the microbial community. The typical growth range for soil inhabitants is more restrictive than the extremes associated with other ecosystems. The typical breadth of temperatures supportive of microbial growth in soil is from approximately 0°C to 70°C. This contrasts on the upper side of the range to growth of bacteria in hot springs. These latter organisms are capable of growth at 100°C (Heinen and Lauwers, 1981). The record for maximum growth temperature is attained by sulfate-reducing, barophilic thermophiles located in deep Pacific hydrothermal vents. These organisms have been shown to be capable of growth at temperatures over 250°C under pressure (Baross and Deming, 1983). Minimal growth temperatures have been shown to be slightly below 0°C in Antarctic soils. With either maximum or minimal growth temperatures, the limiting factor appears to be free water. As long as the conditions allow existence of free water (either through boiling temperature elevation by augmented pressure, or freezing point reduction by salt concentration), microorganisms appear to be capable of growth, albeit slow growth at the reduced temperatures.

Reduction of soil temperatures below freezing results in significant declines in the microbial populations (Biederbeck and Campbell, 1971). The cause of the cell death is not the freezing of the cell, in that freezing itself caused only a small reduction in microbial numbers. The lethal factor appears to be associated with slow thawing of the soil. As soils thaw, ice crystals form in the microbial cells and disrupt cellular integrity. Sensitivity also relates to the growth status of the microbial cells in that recently cropped soils exhibited greater population declines due to freezing than was detected in inactive soils (Biederbeck and Campbell, 1971). Furthermore a differential sensitivity of the microbial populations was noted in that bacterial population reduction was greater (92 percent) than that of fungi (55 percent), which was greater than that of actinomycetes (33 percent). Similar differential sensitivity of soil bacteria to freezing and slow thawing was noted by Nelson and Parkinson (1978) in a study of a *Pseudomonas* sp., *Bacillus* sp., and an *Arthrobacter* sp. Survival was dependent on soil moisture, storage time, and the thaw rate.

More typically for soil microbes, growth can be depicted as an activity range (Fig. 5.6). The growth rate increases until a maximal temperature, characteristic of the organism, is reached. The maximum temperature that supports growth is determined by the sensitivity of the most fastidious enzyme essential for cell replication. This temperature sensitivity curve is a good example of the *law of tolerance*. This law states that a microorganism will have an ecological minimum and maximum with a range in between that represents the limits of tolerance.

Tolerance to temperature can be used to divide microorganisms into three classes. Psychrophiles are defined to grow optimally at temperatures less than 20°C. Mesophiles grow best between 15 and 45°C, whereas thermophiles are active at temperatures greater than 45°C. Few if any psychrophiles are found in most soils. Exceptions are associated with Arctic and Antarctic sites. Mesophiles constitute the bulk of the soil microbial community, whereas most soils contain some thermophiles, even if the soils never achieve the high temperatures.

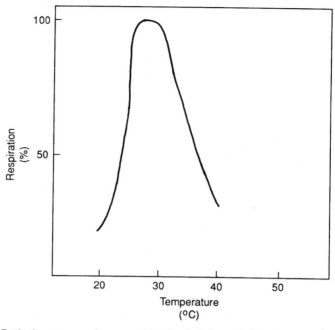

Fig. 5.6 Typical response of a mesophilic bacterial population to temperature variation. An optimum temperature of 25 to 30°C is depicted.

The presence of these classes of microbe in soil assures occurrence of metabolic activity over a wide temperature range. Typically, soil biological activity increases from a minimum at or near 0°C to a maximum and terminus point near 70°C. The rate increase can be represented by several mathematical relationships. (See Tate, 1987, for a discussion of modeling of microbial temperature responses.) Among these models are the Arhenius and Q_{10} relationships. Historically, the latter is likely the most commonly used parameter for soil microbiological studies. The Q_{10} is equal to the reaction rate at temperature one divided by the rate at temperature one minus 10°C, that is,

$$Q_{10} = K_T/K_{T-10} \tag{5.5}$$

Q_{10} values for biological processes generally range from 1.5 to 3.0, with a mode of approximately 2.0. Generally, if the Q_{10} is much greater than 3.0, an environmental parameter besides temperature is controlling the reaction rate. For example, Q_{10} values for petroleum decomposition at low temperatures (10 to 20°C) are commonly greater than 5. At the low temperatures, catabolism of the oil is controlled by its water solubility and viscosity. These parameters dramatically change as the temperature is raised, so a disproportional impact of temperature elevation of the microbial oxidation is observed.

5.3 **Microbial Adaptation to Abiotic Stress**

From an analysis of the diversity of soil properties affecting microbial activity, it almost appears that life in soil would be limited at best. Yet, nearly every site on earth — ranging from hot desert soils to the sediments receiving hot spring effluent — supports active microbial populations. Even in situations where a priori considerations would predict that two factors, such as moisture and temperature, exist at limits that preclude microbial growth, microorganisms have been shown to occur in extremely specialized niches. For example, in cold dry deserts of Antarctica, cyanobacteria-dominated communities have adapted to growth between the crystals of porous rocks (cryptoendolithic) where some modulation of the extreme environmental stress occurs (Friedmann, 1982; Johnston and Vestal, 1991). This portends a variety of mechanisms of community adaptation to these stresses. Three reactions to stressful conditions should be highlighted.

First, if the stress is extreme, there is no adaptation. This is exemplified by soils with temperatures greater than 70°C around hot springs or adjacent to nuclear reactor water outflows, some extremely acidic mineland reclamation sites, and soils containing excessively high metal concentrations. The existence of such ecosystems underscores the fragile nature of the biotic system and highlights the fact that at least some growth is necessary for implementation of the two following survival mechanisms.

Second, if growth is possible, microbial strains resistant or at least tolerant to the stress factor are selected. This has been observed with osmotic pressure, temperature, oxygen tension, and moisture stress. This adaptation mechanism may result in reduction of the species diversity, especially if the stress is severe, but a functional community develops. Reduction of the microbial diversity, thereby the genetic diversity, of the community could portend enhanced vulnerability to ecological insult. Soil presents a somewhat unique exception to this principle in that the potential always exists for invasion by propagules from adjacent soils. Thus, the genetic potential required to overcome chemical or physical stress in a particular site could exist in less stressed neighboring soils. In this case, passive or active transfer of propagules from the neighboring soils may result in establishment of a new community in a site that previously lacked the resources necessary for survival.

Lastly, in some situations, the microbial community may posses the capacity to alter the stress factor, thereby relieving the impediment to growth or survival. For example, microorganisms may modulate inhibitory acidic or alkaline conditions or decompose toxic substances. Organic acids or even antibiotics may be decomposed to carbon dioxide by microbial strains resistant to their toxicity. Even nonbiodecomposible toxins such as heavy metals may be transformed to nontoxic states. They may be complexed, precipitated, oxidized or reduced, or even chelated. The products are rendered nonwater soluble, converted to water-soluble but nontoxic forms, or incorporated into a chelate complex that can be leached from the microbial habitat. Inhibitory inorganic acids may also be rendered ineffective. For example, sulfate is reduced to hydrogen sulfide, which is precipitated by metal ions, such as iron, which are common in soil. In each of these examples, a homeostasis is returned by activity of resistant populations conferring benefits to the survivors of the susceptible organisms.

5.4 Concluding Comments

The nature, activity, and future of any soil microbial community is determined by the capacity of its individual members to adapt to or modify negative soil properties. Those members of the community best equipped to cope with the combination of ecological stresses are active, while more stringent organisms succumb or enter resting stages. Modification of the physical, biological, or chemical properties of the site results in induction of microbial processes to restore the homeostasis of the community. This adaptation may result in reemergence of an ecosystem resembling that existing prior to perturbation, development of a new community better suited to the newly developed conditions, or total destruction of the biological activity. The ultimate fate depends upon the genetic diversity of the populations and the severity of the stress imposed, as well as the time allowed for recovery. Some situations may allow recovery in hours to days, whereas in others, full recovery may not be evident for decades or centuries.

References

Baross, J. A., and J. W. Deming. 1983. Growth of 'black smoker' bacteria at temperatures of at least 250°C. Nature (London) 303:423–426.

Benoit, R. E., and C. L. Hall. 1970. The microbiology of some dry valley soils of Victoria Land Antarctica. *In* M. W. Holdgate (ed.), Antarctic Ecology 2:697–701. Academic Press, NY.

Biederbeck, V. A., and C. A. Campbell. 1971. Influence of simulated Fall and Spring conditions on the soil system. I. Effect on soil microflora. Soil Sci. Soc. Am. Proc. 35:474–479.

Cameron, R. E. 1972. Microbial and ecological investigations in Victoria Valley, Southern Victoria Land Antarctica *In* G. A. Llano (ed.), Antarctic Terrestrial Biology. Antarctic Research Series 20:195–260. Am. Geophys. Union, Washington, DC.

Clarholm, M., and T. Rosswall. 1980. Biomass and turnover of bacteria in a forest soil and a peat. Soil Biol. Biochem. 12:49–57.

Csonka, L. N., and A. D. Hanson. 1991. Prokaryotic osmoregulation: Genetics and physiology. Annu. Rev. Microbiol. 45:569–606.

Dagley, S., and P. A. Johnson. 1963. Microbial oxidation of kynurenic, xanthurenic, and picolinic acid. Biochem. Biophys. Acta 78:577–587.

Dommergues, Y. R., L. W. Belser, and E. L. Schmidt. 1978. Limiting factors for microbial growth and activity in soil. *In* M. Alexander (ed.), Advances in Microbial Ecology. 2:49–104. Plenum Press, NY.

Doran, J. W., L. N. Mielke, and S. Stamatiadis. 1988. Microbial activity and N cycling as regulated by soil water-filled pore space. Paper No. 132. Proc. Int. Soil Tillage Res. Org. ISTRO, 11th. 11–15 July 1988.

Friedmann, E. I. 1982. Endolithic microorganisms in the Antarctic cold desert. Science 215:1045–1053.

Heinen, W., and A. M. Lauwers. 1981. Growth of bacteria at 100°C and beyond. Arch. Microbiol. 129:127–128.

Hirschberg, R. L., and J. C. Ensign. 1972. Oxidation of nicotinic acid by a *Bacillus* species: Source of oxygen atoms for the hydroxylation of nicotinic acid to 6-hydroxynicotinic acid. J. Bacteriol. 108:757–759.

Hochstein, L. I., and B. P. Dalton. 1967. The purification and properties of nicotine oxidase. Biochem. Biophys. Acta 139:56–64.

Jager, G., and E. H. Bruins. 1975. Effect of repeated drying at different temperatures on soil organic matter decomposition and characteristics, and on the soil microflora. Soil Biol. Biochem. 7:153–159.

Johnston, C. G., and J. R. Vestal. 1991. Photosynthetic carbon incorporation and turnover in Antarctic cryptoendolithic microbial communities: Are they the slowest-growing communities on Earth? Appl. Environ. Microbiol. 57:2308–2311.

Knapp, E. B., L. F. Elliott, and G. S. Campbell. 1983. Microbial respiration and growth during the decomposition of wheat straw. Soil Biol. Biochem. 15:319–3232.

Knight, W. G., and J. Skujins. 1981. ATP concentration and soil respiration at reduced water potentials in arid soils. Soil Sci. Soc. Am. J. 45:657–660.

Linn, D. M., and J. W. Doran. 1984. Effect of water-filled pore space on carbon dioxide and nitrous oxide production in tilled and nontilled soils. Soil Sci. Soc. Am. J. 48:1267–1272.

Lund, V., and J. Goksoyr. 1980. Effects of water fluctuations on microbial mass and activity in soil. Microbial Ecol. 6:115–123.

Lundgren, B. 1982. Bacteria in a pine forest soil as affected by clear-cutting. Soil Biol. Biochem. 14:537–542.

Magdoff, F. R., and R. J. Bartlett. 1985. Soil pH buffering revisited. Soil Sci. Soc. Am. J. 49:145–148.

Neilson, J. W., and I. L. Pepper. 1990. Soil respiration as an index of soil aeration. Soil Sci. Soc. Am. J. 54:428–432.

Nelson, L. M. and D. Parkinson. 1978. Effect of freezing and thawing on survival of 3 bacterial isolates from an arctic soil. Can. J. Microbial. 24:1468–1474.

Orchard, V. A., and F. J. Cook. 1983. Relationship between soil respiration and soil moisture. Soil Biol. Biochem. 15:447–453.

Reddy, K. R., and W. H. Patrick, Jr. 1975. Effect of alternate aerobic and anaerobic conditions on redox potential, organic matter decomposition, and nitrogen loss in a flooded soil. Soil Biol. Biochem. 7:87–94.

Robertson, E. V., and M.K. Firestone. 1992. Relationship between desiccation and exopolysaccharide production in a soil *Pseudomonas* sp. Appl. Environ. Microbiol. 58:1284–1291.

Schnürer, J., M. Clarholm, and T. Rosswall. 1985. Microbial biomass and activity in an agricultural soil with different organic matter contents. Soil Biol. Biochem. 17:611–618.

Skopp, J., M. D. Jawson, and J. W. Doran. 1991. Steady-state aerobic microbial activity as a function of soil water content. Soil Sci. Soc. Am. J. 54:1619–1625.

Sørensen, L. H. 1974. Rate of decomposition of organic matter in soil as influenced by repeated airdrying — rewetting and repeated additions of organic matter. Soil Biol. Biochem. 6:287–292.

Tate, R. L., III. 1979. Effect of flooding on microbial activities in organic soils: Carbon metabolism. Soil Sci. 128:267–273.

Tate, R. L., III. 1980. Microbial oxidation of organic matter of histosols. *In* M. Alexander (ed.), Advances in Microbial Ecology 4:169–201. Plenum Press, NY.

Tate, R. L., III. 1987. Soil Organic Matter. Biological and Ecological Effects. John Wiley & Sons, NY. 291 pp.

Tate, R. L., and J. C. Ensign. 1974. Picolinic acid hydroxylase of *Arthrobacter picolinophilus*. Can. J. Microbiol. 20:695–702.

Wardle, D. A., and D. Parkinson. 1990. Interactions between microclimatic variables and the soil microbial biomass. Biol. Fert. Soils 9:273–280.

Wildung, R. E., T. R. Garland, and R. L. Buschbom. 1975. The interdependent effects of soil temperature and water content on soil respiration rate and plant root decomposition in arid grassland soils. Soil Biol. Biochem. 7:373–378.

Wilson, J. M., and D. M. Griffin. 1975. Water potential and the respiration of microorganisms in the soil. Soil Biol. Biochem. 7:199–204.

Soil Enzymes as Indicators of Ecosystem Status

Logic suggests that soil enzymatic activity should be a useful measure of soil biological potential. Enzymes are the mediators for biological transformations of native and foreign soil organic and mineral components. Thus, these catalysts could be predicted to provide a meaningful assessment of reaction rates of soil processes. In several instances, a variety of enzyme activities have been shown to relate to soil biological potential (e.g., Dick, 1984; Nannipieri et al., 1979; Tate et al., 1991), but frequently little or no correlation between apparently essential enzymatic activities and soil respiration, microbial biomass, microbial population density, or microbial activity is observed. In soil systems, synthesis of new enzyme molecules, total enzyme present in the soil, and substrate levels for specific enzyme activities are frequently uncoupled. This observation appears to be rather illogical, but when the heterogeneity of soil systems and the variety of forms in which an enzyme may occur are considered, logic is restored.

For example, a reaction may be catalyzed in a soil microsite yet provide no metabolic benefit for the soil biological community. This stems from the fact that for an enzymatically catalyzed process to benefit growing organisms, its product(s) must be available to the cells at the appropriate time and place. With cell-linked enzymes (those contained in the cytoplasm or associated with cell membranes and walls, including periplasmic enzymes), this concurrence of location of product and need is easily achieved. Yet with the large number of extracellular enzymes in soil, substrates may be transformed by these enzymes, which are not associated with cells, into products of potential utility for cellular function. Unfortunately for the cells, no real value may result because the products and needy cells may not encounter each other. In this case, the cell could synthesize enzyme *de novo* since no product from the existing extracellular enzyme is encountered. Thus, synthesis, existing enzyme, and substrate levels appear to be uncoupled.

Interpretation of soil enzymological data is further complicated by the fact that a particular soil enzymatic activity does not necessarily represent a homogenous entity. The rate of the enzyme catalyzed process is the result of cell-associated as well as cell-free enzyme molecules. Also, in contrast to pure culture analyses where the cell may produce a single enzyme type to catalyze a particular reaction, in soil, a large number of different species may be synthesizing enzymes with different capacities to catalyze the same reaction. Enzyme sources in soils include plant, animal, fungal, and bacterial cells. Thus, the quantity of any soil enzymatic activity detected in a particular soil sample is the sum of active and potentially active enzymes from any or all of these sources. Furthermore, due to the variety of different sources of enzymes, the basic properties of the specific enzymes catalyzing an individual reaction may differ (i.e., there may exist several isoenzymes for each enzymatic activity occurring in soil). The constants describing the kinetics of the catalysis can vary significantly, depending on the source of the enzymatic activity. Thus, not all the enzymes catalyzing the same reaction are assessed under their optimal conditions when a generalized procedure is used for their quantification, thereby resulting in an underestimation of the enzymatic activity of the soil sample.

Last, to understand the relationship of soil enzymatic activity to other biological processes in soil, our concept of enzymes must be expanded beyond that normally acquired in general microbiology or biochemistry courses. After completion of those basic science studies, enzymes are viewed as reasonably labile components of living cells, or if they are excreted from the cell, their association with the cell is still strong. In a soil system, enzymes synthesized by members of the soil community may become separated from the cell producing them both in location and time. They may be leached beyond the microsite where the microbe resides (locational displacement). Soil enzymes may also be covalently bound to soil humic substances (humified) and still retain their enzymatic capabilities (albeit with altered kinetic properties). Once humified, these enzymes assume degradation susceptibilities similar to those of the humic acid—for example, years, decades, even centuries. Thus, the bound enzyme can survive in soil long past the lifetime of the cell that produced it (separation in time). A specific term—*abiontic*—has been coined to describe this situation (Skujins, 1967). Recall that *abiotic* indicates without life, whereas biotic refers to the living systems. An abiontic substance is produced by the living system but has been separated from it. That is, the substance is a product of the living world but has become part of the nonliving world.

These introductory observations reveal the complexity of the ecosystem in which enzymes must function in soil. Soil enzymes are essential in assessing ecosystem function, yet the complexity of the soil system complicates interpretation of data involving these catalysts. Thus, this chapter is presented with the overall objective of developing a basic understanding of the nature of soil enzymes and their role in ecosystem function.

6.1 Philosophical Basis for the Study of Soil Enzymes

Soil enzymology has been an accepted scientific study for nearly one hundred years. A large body of published reports documents extensive scientific effort directed at (1) the evaluation of the variation in the quantities of numerous enzymatic activities with soil and ecosystem type as well as (2) the optimization of the procedures for measuring these activities. These data provide a basis for developing a conceptual model of the soil interactions affecting enzymatic activity. As a starting point in developing an understanding of the diversity of soil enzymatic activity, it is reasonable to assume that all enzymes necessary to support growth and replication of indigenous soil populations are present in any soil sample. The impact of soil particulates (e.g., adsorption), chemical properties (e.g., pH, salinity), or the presence of inhibitors (e.g., heavy metals) may preclude direct quantification or result in underestimation of the total quantity of a metabolic activity present in a soil sample, but the principles of comparative biochemistry predict that the enzymes must occur in the soil sample. Levels of enzyme activity detected in a soil sample and their utility is affected by the diversity of natural and xenobiotic organic compounds of the soil system, the chemical and physical complexity of the system, and the extreme heterogeneity of the microsite wherein the enzymes must be active.

Justification of research involving quantification of soil enzymes is exemplified by examining several examples of fundamental research areas. Soil enzyme research may involve:

- *Evaluation of basic properties of enzymes in soils*: This would include determination of the source of the enzymatic activity, the kinetics of the process catalyzed, physical limitations to enzymatic activity, stability of the enzymatic activity in soil, and the variation of activity with soil and ecosystem type (e.g., Duxbury and Tate, 1981; Sarkar et al., 1989; Tateno, 1988; Zantua and Bremner, 1975).

- *Enzymes as a measure of soil fertility*: Soil fertility relates to a composite of soil traits including biological and abiotic soil properties. Considerable research effort has been expended to elucidate the relationship between basic enzymatic activities and overall soil fertility (e.g., Moore and Russell, 1972; Verstraete and Voets, 1977). Contrasting conclusions have been reached in regard to the utility of soil enzymes in estimating soil fertility. For example, Moore and Russell (1972) found little linkage between dehydrogenase activity and soil fertility. Similar results were noted by Brendecke et al. (1993) in sewage sludge-amended desert soils. Verstraete and Voets (1977) noted that yields of winter wheat related positively to phosphatase activity, whereas a negative relationship of this enzymatic activity to sugarbeet yields was observed. Soil enzymatic activity has been shown also to reflect cropping and soil management practices (e.g., Chander et al., 1997; Curci et al., 1997; Ross et al., 1995; Serrawittling et al., 1995).

Thus, soil enzyme profiles may serve as indicators of change in soil structure and organic matter levels due to alteration of system management programs.

- *Enzymatic activity as an indirect measure of microbial biomass*: Considerable effort has been expended in determining the relationship between the levels of particular enzymatic activities and total microbial biomass or soil respiration. Theoretically, any metabolic activity — that is expressed in active cells, is present in concentrations proportional to the number of cells, and can be accurately quantified in soil samples could be used to measure microbial biomass. This activity must be directly associated with the microbial cells and rapidly inactivated or destroyed when cellular integrity is disrupted or with cell death. A common enzymatic activity used to estimate microbial activity is dehydrogenase (e.g., Casida, 1977). Dehydrogenase activity may be proportional to soil respiration but direct correlation with microbial population density is unusual (Tabatabai, 1982; Frankenberger and Dick, 1983). Frankenberger and Dick (1983) found that alkaline phosphatase, amidase, α-glucosidase, and dehydrogenase activities significantly correlated with microbial respiration in glucose-amended soils. No correlation was found in unamended soils. Alkaline phosphatase, amidase, and catalase activities did correlate with microbial biomass.

- *Indicator of capability to conduct biogeochemical cycling*: Phosphatases, sulfatases, and nitrogen oxide reductases have been used to estimate potential to catalyze essential portions of the carbon, phosphorus, sulfur, and nitrogen cycles, respectively (e.g., Häussling and Marschner, 1989; Tarafdar and Jungk, 1987; Tate, 1984).

- *Indicator of the negative effects of pollutants*: General overall ecosystem stability is related to the "general health" of the soil microbial community. Disruption of soil microbial activity as shown by changes in levels of metabolic enzymes can serve as an estimate of ecosystem disruption. This relationship is clearly shown when soils are polluted with heavy metals. The metal may combine with exposed enzymes through association with sulfhydral groups, thereby altering the tertiary structure of the protein sufficiently to reduce or destroy the enzymatic function (e.g., Cole, 1977; Doelman and Haanstra, 1979; Doelman and Haastra, 1989; Mathur and Sanderson, 1980; Tyler, 1981). For example, Doelman and Haanstra (1979) found increasing inhibition of soil respiration and dehydrogenase activities with increasing lead concentrations from 375 to 1500 μg Pb/g dry soil. Similarly, Cole (1977) noted inhibition of amylase, α-glucosidase, and invertase in lead-amended soils.

- *Enzymes as a predictor of bioremediation and potential success*: As the biological community is reestablished in damaged soils, associated enzyme activity increases. Similarly, in organic chemical-contaminated soils, an increase in enzymatic activities involved in mineralization of the pollutant serves as a predictor of capacity of native populations to return the soil to the previously existing conditions (e.g., Burns and Edwards, 1980; Dick et al., 1988; Klein et al, 1985; Stroo and Jencks, 1982). Also, a variety of enzymatic activities in soil have been used to monitor metal-polluted sites, such as ecosystems associated with metal smelters (e.g., Brookes, 1995; Kandeler et al., 1996; Kelly and Tate, 1998).

- *Indicator of soil quality*: Although soil is a renewable resource, restoration of damaged soil sites to their optimal functional level requires decades to centuries. (See Hillel, 1991, for a discussion of the decline in the quality of world soils and the resultant impact on society.) Normal soil usage for sustenance of society commonly results in declines in soil quality. Therefore, it is vitally important to manage soils in a way that maintains or optimizes soil quality. To accomplish this end, indicators are needed to assess changes in soil quality due to normal usage as well as determination of the success of reclamation-management procedures. (See Doran and Parkin, 1994; Karlen et al., 1997; Sims et al., 1997 for discussions of soil quality definitions, its assessment, and the importance of soil microbial communities in optimizing soil quality.) One such tool is soil quantification of soil enzymatic activity. Soil enzymes vary with management and are affected by soil pollution. Therefore, they could be anticipated to be an indicator of variations in soil quality. For example, soil enzymatic activity has been used to assess remediation effects on industrially disturbed soils (Roswell and Florence, 1993) and impacts of cropping history on soil quality parameters (Halvorson et al., 1996; Jordan et al., 1995; Miller and Dick, 1995).

With each of these examples, research results can be presented that demonstrate successful application of enzymological data to the achievement of the study objectives, but numerous failures have also been recorded. These conflicting results are frequently the product of a failure to consider fully the complexities of the soil environment. For a selected enzymatic activity to reflect the level of soil biological processes, the following criteria should be met:

- The enzymatic activity must be associated with living cells.

- The enzyme must be a participant in a biological reaction essential for the process of interest. (This point is particularly critical when attempting to relate enzymatic activity to soil fertility levels.)

- The activity measured must be proportional to the quantity of enzyme present. Active enzyme must not be obscured by large pools of soil-stabilized, abiontic enzyme or occurrence of spontaneous chemical transformations of the same or similar processes as are catalyzed by the enzyme.

■ The enzyme activity must be essential for the process of interest (i.e., alternate synthetic pathways must not be major contributors to the conversion, or the proportion of the reaction rate contributed by the alternate route must not change as the soil condition is altered).

These observations regarding the utility of enzymes as an indication of soil biological activity as well as the considerations regarding valid interpretation of the data reveal an underling conceptual viewpoint of the soil enzymologist. The philosophical understanding of the soil enzymologist resembles that of a biologist evaluating the activity of a tissue more closely than it does that of a bacteriologist studying axenic cultures. For the soil enzymologist, activity of the individual microbial species is less important than the sum of all metabolic activities occurring in the soil system. Each cell, enzyme, animal, or plant tissue piece is considered merely to be a part of the larger entity, the soil sample. From the view of a soil microbiologist (or perhaps a better term — *ecosystem microbiologist*), the function of the whole is more important than the intricacies of the parts. This is not to say that the identity of the individual has been trivialized. Indeed, to gain a complete understanding of the contributors to system homeostasis, as well as ecosystem potential, the potential and expressed communities of each individual life form must be elucidated and understood, but it is their interactions as a "team" rather than their "stellar performances" that are generally of utmost importance for the sustenance of the total ecosystem.

6.2 Basic Soil Enzyme Properties

At the outset, it must be noted that the basic biochemical concepts relating to enzyme function and kinetics are unchanged in the soil ecosystem. Enzymes are proteins whose activity is characterized by parameters reflective of the interaction between the enzyme, its substrates, and the products of the reaction. Synthesis rates of *de novo* enzymatic activity are controlled by cellular metabolic processes Some enzymes are always produced by the cells (constitutive). Others are produced only when their activities are required (inducible). A third group is intermediate between these two groups. In the latter situation, some enzymatic activities are synthesized in low levels when no inducer is present and at high levels when the activity is required. Apparent variation in enzyme properties as measured in soil, which appear to be contradictory to data collected in axenic cultures or with partially purified or characterized enzyme preparations, result from the interaction of the enzyme molecules and biological cells with the variety of chemical and physical components existent in the soil microsite. Thus, the following presentation is directed at delineating the common points between *in situ* function of enzymes in soil microsites and properties of these catalysts in purified or semipurified preparations studied in the laboratory, as well as the divergence associated with understanding these entities in the laboratory and field.

Soil Enzymes as Proteins: Enzymes are proteins that combine with their substrates in a stereospecific manner to catalyze biochemical reactions through the modification of stress on the molecular configuration around the chemical bonds at the site of catalysis, thereby lowering the activation energy of the reaction. This rather simple definition provides a basis for predicting modifications in enzymatic activity and limitations to the longevity of these activities that are imposed by soil properties.

Initial considerations relate to the proteinaceous nature of enzymes. These proteins are folded into a three-dimensional structure (tertiary structure) that optimizes interactions with the substrates for the reaction. Located within this three-dimensional structure of the protein is the enzyme's active site. The active site is the cleft in the molecule where the substrate(s) fit and are transformed into the product. The configuration of the active site is critical for the occurrence of the reaction. Minor changes in the structure of the protein in or around the active site can have major consequences on the rate of reaction. Thus, any soil properties that modify the protein tertiary structure can alter the enzymatic reaction rate. The tertiary structure of the protein can be affected by interaction with soil metals, by the ionic changes induced by varied soil pH, and changes in hydration due to desiccation, as well as by a covalent linkage to soil humic substances. The enzymatic reaction rate in soil is therefore dependent upon the substrate concentrations, pH, ionic strength of the interstitial water, and temperature as well as the presence or absence of a variety of enzyme inhibitors and activators. Some of these factors affect molecular configuration, whereas others involve the probability of interaction or collision of the substrate and the enzyme molecules.

Extreme disruption of the enzyme three-dimensional structure is termed *denaturation*. This could be likened to the changes that we observe when egg white proteins are heated to high temperatures (i.e., the egg white proteins are denatured by the high temperature into a substance that a major percentage of the population considers more palatable than the uncooked egg). Denaturation of enzyme proteins in soil may be induced by pH extremes and high temperatures, as well as by interaction with a variety of polyvalent metalic anions (such as mercury, copper, and cadmium).

Enzyme vs. Enzymatic Activity: A basic aspect of describing soil enzymes is the terminology used to designate that activity assayed in the soil sample. It is not unusual to hear soil enzymologists state that a particular enzyme (e.g., phosphatase), was measured in a soil sample. Yet, it must be understood that the only time that the name of an enzyme can be used is when it is clearly shown that the activity of a single enzyme type has been measured. For example, the phosphatase produced by a specific soil bacterial species can be quantified, say that of a *Pseudomonas* species. In that situation, the experimenter could state that *Pseudomonas* phosphatase was examined. But with a soil sample, there could exist a wide variety of bacteria and fungi, as well as higher animals and plants that produce distinct phosphatase enzymes. Should the conversion of organic phosphate to mineral phosphate be assayed in that soil sample, the technician would find it difficult to state which of the specific phosphatase enzymes were being quantified. Thus, the most accurate description of the procedure conducted would be to state that phosphatase *activity* was measured. This concept is important because the amount of activity detected in a soil sample is the sum of contributions of all of the individual enzyme types present. Thus, it is likely not characteristic of any specific individual enzyme. This requisite in terminology is even more important if the enzyme kinetic properties are examined. The kinetic constants so derived are necessarily average values more representative of the whole enzyme population than any specific enzyme.

A further qualification of the description of the soil process measured involves an indication of whether potential or actual field activity is being estimated. This point relates to enzyme kinetics parameters to be discussed later in this chapter and the specific assay conditions used to characterize the soil enzyme. Enzyme activities are generally measured in the laboratory under optimal chemical conditions with nonlimiting substrate levels, but in soil, rate-limiting substrate concentrations are generally available and the enzyme must cope with such difficulties as moisture, pH, or salinity limitations. Thus, the activity measured in a test tube is the maximum rate at which the enzyme activity could be expressed, something probably not often experienced in the soil sample. Thus, the value measured in soils is best referred to as the potential enzymatic activity

Role of Enzymes in Soil: The most clearly delineated from the view of functional importance enzymatic functions in soils are those cytoplasmic activities associated with intermediary metabolism. Less distinct is the contribution of extracellular enzymes to the sustenance of the biota. Extracellular enzymes are secreted by soil organisms to solubolize or metabolize substrates external to the cell in order to produce nutrients needed by the microbes, to detoxify environmental substituents, or to modify the microenvironment of the living cell in a manner to improve the probability of cellular survival and ultimately proliferation. These enzymes are highly valuable for total ecosystem function as long as the products of their activity are available to living cells. Note that a positive benefit for ecosystem function may be credited to extracellular enzymes as long as *any* living entity benefits. Thus, the cell producing the enzyme may be at a disadvantage, but the community may gain. Should these products or cells requiring the products of their activity not concurrently exist, the extracellular enzymes could be classed as having no ascribable role in the soil ecosystem. For stabilized enzymes—enzymes covalently linked to humic acid—involved in processes associated with soil nutrient cycles, a community benefit is more easily postulated. For example, phosphatase catalyzes the conversion of organic phosphates to mineral phosphate. This production of a required plant nutrient by a common soil enzyme clearly has a role in maintaining general soil fertility — in agricultural systems as well as in native, unmanaged ecosystems (i.e., those not receiving external phosphate sources).

Soil Enzyme Longevity: Since soil enzymes, be they within the living cell or existing free in soil interstitial water or bound to colloidal soil organic matter and clay particles, are proteins and since proteases (enzymes that hydrolyze proteins to amino acids) are ubiquitous in soil, enzymes are reasonably labile soil substituents. As indicated above, some enzymes may be stabilized in soil organic matter and therefore have a half-life of decades or even centuries (Skujins and McLaren, 1969), but most enzymatic activities are reasonably short lived. Those enzymes contained within cellular protoplasm — that is, associated with intermediary metabolism — turn over at rates dictated by normal cellular metabolism. The change in enzyme concentration within the cell with time (dE/dt) could be described by the following general equation:

$$\frac{dE}{dt} = \frac{DE_{synthesis}}{dt} - \frac{DE_{decomposition}}{dt} \tag{6.1}$$

Within the cell, these rates are controlled by the metabolic status of the cell. In a starving cell or even a resting cell, the decomposition rate may actually be much larger than the synthesis rate — especially of nonessential activities — so that the quantity of a particular enzyme activity rapidly declines.

External to the cell, some enzymes may be stabilized by covalent linkage to soil humic acids. The physical attachment of the enzyme to these substances limits access to the enzyme by proteolytic enzymes, thereby reducing the potential for proteolysis. This covalent linkage of the enzyme proteins to soil organic matter also stabilizes the enzyme tertiary structure in a manner to reduce the potential for denaturation of the protein. For enzyme activities that are a combination of endo- and extracellular activities, the apparent change in total activity in soil is modulated by the quantity of stabilized enzyme. That is, total enzyme activity (E_T) is described as follows:

$$E_T = E_{stabilized} + E_{cellular} \tag{6.2}$$

where the quantity of cellular enzyme and its turnover rate is described by equation 6.1. The practical result of this relationship is that in systems containing large quantities of a particular stabilized enzyme, such as phosphatase, the changes in newly synthesized enzyme can be obscured by the quantity of stabilized enzyme already present. In these situations, quantities of enzyme measured in soil samples will not reflect variation in activity of the living biomass.

Stabilization of Enzyme Activities in Soil: Since a significant portion of the enzymatic activity detected in soil samples may be the result of enzymes covalently bound to colloidal soil organic matter or associated with clay particles, consideration of the nature of these stabilized proteins and the processes leading to their long-lived state is worthwhile. There are two major fates of enzymes lost from cells due to cell lysis or to active secretion of the proteins into the extracellular milieu: The proteins may be decomposed, or they may be stabilized through linkage to soil humic materials or adsorption to clay minerals.

Decomposition of proteins in soil interstitial waters may be preceded by denaturation. Proteins whose native environment is the cell cytoplasm, the cell membrane, or the periplasm generally find the environment external to the cell harsh. These proteins are vulnerable to suboptimal pH and unfavorable ionic strength of soil interstitial water as well as the presence of denaturing agents, such as heavy metals. Sensitivity to the external milieu varies with the nature of the enzyme protein. For example, Frankenberger and Bingham (1982) found a differential sensitivity of a variety of intra- and extracellular enzyme activities to soil salinity. Dehydrogenase activity was severely limited by salinity whereas a variety of hydrolases were more resistant. The authors concluded that osmotic desiccation of the microbial cells may have resulted in release of the intracellular enzymes into the soil interstitial water and their subsequent proteolysis.

Adverse conditions tend to disrupt the protein tertiary structure in a manner to increase susceptibility to proteolytic activity. Within the cell, proteins may be protected by physical separation of the proteins and proteases as well as existence of the protein in a tertiary structure that limits or perhaps even precludes proteolytic attack. For example, should the terminal residues be protected by being embedded within the protein molecule, C- or N-terminal specific proteases would not be capable of initiating hydrolysis of the protein. This protection could be lost once the protein normally contained in the cytoplasm enters the soil environment. The conditions external to the cell encourage unraveling of the normally intracellular resident enzyme so that proteases may more easily attach to and cleave the protein.

Those enzymes that are synthesized by the cell and secreted for functions external to the cell wall or membrane (extracellular enzymes) generally have a tertiary structure that is more strongly held together by covalent, hydrogen bonding, and ionic linkages than is found to occur with intracellular enzymes. Thus, extracellular proteins are more resistant to denaturation and proteolytic activity than are enzymes whose prime site of function is the protected cove of the cellular protoplasm.

As indicated above, enzymes can be stabilized through ionic, hydrogen, or covalent bonding to soil humic materials. In situations where the linkages is through ionic or hydrogen bonds, the enzymes can be easily removed from the soil by washing in buffer suspensions. For example, peroxidases can frequently be easily extracted from soil suspensions (Bartha and Bordeleau, 1969). Few enzyme molecules are extractable from soil with buffer solutions, so it is logical to conclude that this type of association between enzyme proteins and soil organic matter is somewhat uncommon. Similarly, although there are numerous reports of separation of soil enzyme activities from mineral matter (e.g., Ladd, 1972; Nannipieri et al., 1980; Perez-Mateos et al., 1988), these purified enzyme preparations still contain humic substances associated with the proteins. Gosewinkel and Broadbent (1986) did separate about 3 percent of the phosphatase activity from a field soil with electron-donating substances, suggesting a disruption of the enzyme-humic acid covalent linkage. The proteins may be bound to humic acids through quinones by nucleophilic substitution, to sulfhydral groups, or to terminal and ε-amino acids. In these situations, enzymatic activity of the proteins is maintained if the protein configuration is not disrupted and the bonding does not obscure the active site. Should either of the conditions not be met, then the enzyme would enter the class of humic acid-bound proteins rather than being considered to be a soil-stabilized enzyme.

The quantity and type of enzymatic activities stabilized in particular soil samples may be distinctive for the individual soil site, such that their analysis could be of value for forensic investigations. Thornton et al. (1975) found that phosphatase, arylsulfatase, urease, invertase, and trypsinase activities in three soil series were sufficiently characteristic of the soils that soil samples collected within close proximity could be distinguished by their enzymatic patterns. The Michaelis constants of the enzymes could be used to characterize the soil samples in greater detail.

6.3 Principles of Enzyme Assays

To quantify a specific enzymatic activity in soil samples, knowledge of a variety of properties of the enzyme is essential, even if the activity of interest has been studied previously. In the latter situation, the enzyme assay must still be optimized for the particular soil of interest and perhaps modified to increase the probability of achievement of the experimental objectives. Contrasting goals for a study may be (1) to quantify total activity of a particular enzyme present in a soil sample or (2) to elucidate the amount of that activity expressed *in situ*. These values may differ. For example, the enzyme activity may be measured in a buffer with a pH that is optimal for a particular enzymatic activity. This may allow determination of the highest amount of substrate transformed by an enzyme in the soil sample. Unfortunately, few enzymes are functioning in soil at their optimal pH. Thus, activity values detected in buffered assays may show little relationship to real field catalysis rates. To best estimate field activity rates, the assay should approximate the conditions where the enzyme normally functions. The assay procedure may be modified so that the *in situ* soil pH and the native enzyme activity levels are more accurately reflected. Note that even if the assays are conducted at the prevailing soil pH, the level of activity detected still reflects only potential activity. This fact results again from the ideal conditions provided during the quantification of the enzymatic activity. Other variables generally optimized are substrate concentration, interaction of the substrate and enzyme (the assay mixtures are generally mixed while the enzyme is being quantified to maximize interaction between the enzyme and its products), and temperature.

To provide a reasonable and reproducible estimate of activity of an enzyme *in situ* in the soil sample, several properties of the enzyme and the reaction catalyzed must be understood in detail. These include (see Tabatabai, 1982, for a more detailed discussion of this point.).

- The reaction catalyzed should be understood in stoichiometric detail. This includes knowledge of the reactants and the products as well as an understanding of the chemical and physical properties of each of these chemicals. The latter traits are useful in developing enzyme assays in that they may provide a useful measure in their quantification. For example, a strong adsorption maximum could be exploited for a quick assay of product formation or substrate disappearance. Competing fates of the product, or of the substrate, also must be understood. These can lead to over- or underestimation of the actual enzymatic activity. For example, sorption of the substrate to soil particles (such as clay) could result in production of data that suggest greater activity than actually occurs (if substrate disappearance serves as the indicator of the reaction). Alternatively, if product appearance is quantified, utilization of the product in another reaction could result in an underestimation of the enzymatic activity.

- The requirement for chemicals besides the substrates for optimal enzyme reaction should be understood. This would include any necessary metal requirements for the proteins as well as electron acceptors. This is less of a problem for assay of enzymes in a crude milieu such as soil, but is becomes a greater difficulty as the enzyme activity is "purified."

- The kinetics of the reaction in relationship to substrate concentration must be elucidated. Enzyme assays are commonly conducted with saturation concentrations of the substrate. This procedure results from the necessity of having the assay proportional to the quantity of enzyme present and not limited by the substrate levels. An exception to this is encountered when saturation levels of the substrate are inhibitory to the enzyme. In this situation the V_{max} of the reaction can be used to estimate the quantity of the enzyme present. (See Section 6.4.)

- Even though the experimental objective may be to quantify the enzyme activity under field conditions, the optimum pH, temperature, and ionic strength of the enzyme activity should be understood. This is particularly important when a variety of soil types are to be assayed. Similar levels of enzyme protein may be present in each soil, yet the expressed activity could vary significantly due to occurrence of suboptimal chemical or physical conditions.

- A reasonably rapid means of monitoring changes in the substrate or product must be developed. This is particularly important in the soil system, since the altered conditions of the soil in the assay vessel could cause synthesis of new enzymatic activity or even, in the rare situation, result in a decline in enzyme levels. Generally, it is more expeditious to measure changes in product levels since it is generally more accurate to quantify increases in level of a compound from zero than to measure slight changes in a substrate that is present in high concentrations.

Methodological Difficulties of Assessing Soil Enzymatic Activity: Even after developing an apparently ideal assay for the enzyme activity of interest, application of the procedure to soil samples may be precluded or special considerations may be necessary in interpreting the results of the procedure. As has been stressed above, these difficulties arise primarily because of the complexity of the soil system and the reactions occurring therein. Four specific problems will be used to illustrate these difficulties: (1) biologically vs. chemically catalyzed processes, (2) complications due to microbial growth or enzyme synthesis during the assay time, (3) stability of the enzyme products in the reaction mixture, and (4) impact of sample treatment prior to enzyme assay.

Biologically vs. Chemically Catalyzed Processes: Within the reasonably defined confines of a microbial culture vessel, it is somewhat easy to assign cause to a particular process; that is, biological and chemical reactions are usually easily distinguished. In soil, because of the potential for catalysis of organic transformations by clay particles, abiontic enzymes, as well as occurrence of spontaneous chemical reactions, such distinctions are much more difficult. It is not unusual for a chemical conversion to be catalyzed both biologically and chemically within the same soil sample and, perhaps even, the individual microsite. Although it is catalyzed by intracellular enzymes, denitrification provides an excellent example of the overlap between biologically catalyzed and spontaneous chemical reactions at acidic pH values. Denitrification is predominantly the result of enzymological catalysis of the reduction of nitrogen oxides in agricultural soils with mildly acidic (5.5 to 7.0) pH levels. As the soil pH declines, nitrogen oxides may be denitrified through spontaneous chemical reactions. At the acid pH extremes (<3.5), although it has been shown that biological denitrification can occur (Muller et al., 1980), it is generally assumed that chemical reactions predominate. At the intermediate pH ranges, both chemical and biological reactions can occur.

The question thus becomes, "How can biological and chemical processes be distinguished in soil?" Three reaction properties that may separate abiotic or biotic contributions to a soil process are the impact of sterilization of the soil on the reaction, heat stability of the enzymes, and reaction kinetics.

The most commonly used procedure for distinguishing biological processes from spontaneous chemical reactions is to assess activity in sterile soil controls. Biological activity may be destroyed through heat sterilizing [generally steam sterilization at elevated temperatures and pressures (121°C, 15 psi), but a variety of chemical sterilants have been used] as well as radiation sterilization. The assumption underlying these methods is that only the biological aspects of the soil are altered by the sterilization procedure and that the chemical processes continue at unaltered rates in the treated soil.

Ideally, an all-or-none effect on the reaction rate by sterilization is observed. Unfortunately, most commonly such easily interpreted data are not produced. The reaction rate may be reduced by sterilization but not precluded. Contributions to the intermediate effect include the possibility that the soil samples may not have been rendered sterile by the sterilization procedure, the process may actually be occurring as a result of chemical processes, at least in part, or sterilization of the soil created conditions where the chemical reaction could occur when it would not necessarily be anticipated to occur in the native soil sample. Radiation-sterilized soil samples are particularly vulnerable to the latter problem. Radiation sterilization creates free radicals in the native soil organic matter. For transformations, such as polymerization of aromatic ring-containing compounds, formation of free radicals increases the probability of incorporation of aromatic ring-containing pesticides, for example, into native soil organic matter as well polymerization of the test compounds. Alternatively, the assumed sterility of the soil may not have been achieved or maintained during the analysis period. A sterility control must be conducted. Samples of the treated soil should be examined for surviving microbial populations, or any that developed subsequent to sterilization but prior to use in the experiments, by culturing soil samples in a variety of media. Testing for microbial growth in a single medium does not preclude the existence of viable microbes unable to grow in the test medium, but capable of growth in an alternate medium. Thus, several microbial growth media should be used.

Heat stability of the catalysts can also be used as a measure of enzyme-catalyzed reactions. Proteins are denatured quickly (within minutes) by temperatures above 55°C. Thus, heating of a soil sample at 55°C for approximately 30 minutes should result in destruction of an enzyme-catalyzed reaction. Again, interpretation of soils data must be tempered with the realization that proteins stabilized in native soil organic matter may become reasonably heat resistant. Thus, a portion of the enzyme activity may survive the heat shock.

A further consideration in differentiating chemical and biological soil processes relates to reaction kinetics. As is discussed below, saturation kinetics are observed with enzyme-catalyzed processes. As the substrate concentration is increased in the reaction mixture, a level is reached where no further increase

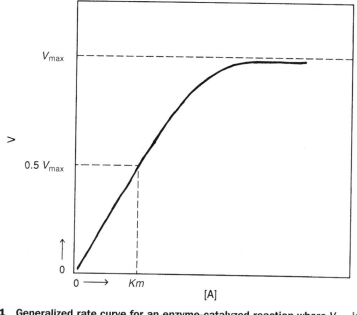

Fig. 6.1 Generalized rate curve for an enzyme-catalyzed reaction where V_{max} is the maximum reaction rate, v is the reaction rate, and [A] is a substrate for the enzyme.

in reaction rate is detected (see Fig. 6.1). The reaction becomes proportional to the quantity of catalyst present rather than the substrate concentration. In contrast, increased concentration of reactants in a chemically catalyzed mixture generally leads to increased product production until another rate limitation is imposed, such as substrate solubility. Therefore, an evaluation of the reaction kinetics can frequently be used as an indicator of biologically vs. chemically catalyzed reactions.

Complications due to Microbial Growth or Enzyme Synthesis: Another methodological difficulty is associated with the assay procedure itself. Again, with a defined mixture of enzyme substrates, cofactors, and buffer, it is reasonably easy to keep the principles (the enzyme concentration and substrates, including any requisite cofactors) at constant or saturating levels throughout the assay period. This is frequently difficult, if not impossible, in evaluating soil enzyme activities. Soil samples are a complex mixture of substances that may allow replication of enzyme-producing cells as well as *de novo* enzyme synthesis by preexisting microbial populations. Conditions are generally ideal for growth of the enzyme-producing microbial cells or synthesis of nascent enzyme during enzyme assays.

Since the objective with any assay procedure is to determine the quantity of enzyme activity present in the soil sample prior to assay, not that produced during the quantification procedure, enzyme synthesis must be prevented, or at least minimized. For assays of a few minutes' duration, enzyme synthesis is not a significant problem, but some procedures require incubation times of several hours for adequate product to be synthesized for accurate quantification. For example, phosphatase and some protease activity measurements require four to six hours or longer. A variety of microbial growth or protein synthesis inhibitors may be incorporated into the enzyme assay mixture. Inhibitors of microbial growth commonly used in soil enzyme assays are toluene, γ-irradiation, and antibiotics. Toluene [10 to 25 percent (volume/volume) concentrations] is the oldest, and is still a relatively commonly used procedure. The inhibitory mechanism of toluene is primarily the dissolution of cell membranes. Therefore toluene inhibition of cell growth is most useful for assessment of extracellular enzyme activity and for those activities where disruption of the cell membrane does not effect the activity. It has been used for phosphatase and invertase activities. Dehydrogenase activity — a nondescript enzyme activity that is associated with intact cell membranes — is inhibited by toluene inhibition.

In contrast, irradiation of cells with 5–10 Mev electron beam, hard X-rays or γ-rays is useful for assay of intracellular as well as extracellular enzymatic activity. Of consideration in using irradiation is the differential sensitivity of soil biomass and the potential to effect spontaneous chemical reactions. (See Skujins, 1967, for a more complete discussion of radiation sterilization applicability to enzymatic assays.)

Antibiotics have also been used to prevent *de novo* protein synthesis during soil enzyme and respiration measurements. Considerations in interpretation of data derived from use of antibiotic inhibitors relate to the differential sensitivity of soil biomass (bacteriocides vs. fungicides) and inactivation of the antibiotic due to adsorption to soil components, as well as biodegradation susceptibility of the inhibitors. The degree of inhibition and specificity of the antibiotic should be examined for each soil type to be used in the individual study.

Stability of Enzyme Products in the Assay Mixture: A further methodological complication to be considered in developing enzyme assays involves the fate of the reaction product. As indicated above, the most sensitive enzyme assays involve quantification of the reaction product. With defined systems, it is relatively easy to chose a substrate or modified substrate that yields products that are not further metabolized within the reaction mixture. This goal is frequently not possible with an ill-defined soil sample. The product of the enzyme activity may be an ideal substrate for other organisms or enzymes existent in the soil mixture. For example, it is common to measure carbohydrases by quantifying the monosaccharide production rate, but in soil, monosaccharides are highly degradable, and frequently short lived. Thus, a control for such assays must involve a determination of the longevity of the monosaccharide yielded by the enzyme.

Impact of Sample Treatment Prior to Enzyme Activity Assessment: The final methodological problem to be analyzed herein relates to soil sample storage techniques. Few enzymatic activities can be measured immediately upon selection of representative soil samples. Soil samples must be collected, transported to the laboratory, generally stored for some period of time, and then assayed. Procedural details that may affect the accuracy of the enzyme assay include conditions during sample storage [temperature (-21 or $4°C$, or room temperature) or moisture (field moisture or air dried)], sterilization procedure for the assay, reaction conditions (pH, temperature, etc.), and static vs. agitated incubation of the reaction vessel. Controls must be recorded to estimate changes in field enzyme levels by each of these procedural variations.

Along with the controls indicated above, any enzyme assay procedure must include a no-soil control (to demonstrate stability of the reaction mixture and with colorimetric assays, to indicate background absorbance of the reaction mixture) plus a no-substrate control. With the chemical complexity of soils, the possibility exists for compounds to occur in the soil sample that mimic the product to be assayed, or sufficient substrate may preexist in the soil sample to allow the synthesis of the reaction product in the absence of amended substrate.

6.4 Enzyme Kinetics

Evaluation of the impact of substrate concentration on rate of product formation of an enzyme-catalyzed reaction (i.e., measurement of the kinetics of the enzymatic catalyzed reaction) is a valuable means for elucidating such enzyme and ecosystem properties as the following:

- The properties of the soil microsite in which the enzyme is functioning

- The number of enzymes catalyzing the reaction of interest that are present in the soil sample

- The total amount of enzyme in a soil sample, and the impact of soil properties on expression of the enzyme activity

This type of assessment generally involves determination of process reaction rates with increasing substrate concentration, under otherwise constant assay conditions. Rates of soil biological processes are generally either zero-order (the reaction rate is constant and independent of the substrate concentration) or first-order (the rate is proportional to the substrate concentration). Modification of incubation temperature, pH, and ionic strength of the incubation solution, plus alteration in concentration of minor elements or cofactors required for the reaction may have a significant impact on the reaction rates.

Zero-Order Biological Processes in Soil: The disappearance of substrate (dS/dt) for zero-order reactions is described by the following simple equation:

$$\frac{dS}{dt} = k \qquad (6.3)$$

The constant k is independent of substrate concentration and thus, it is proportional to total enzyme. Although first-order kinetics are more commonly recorded with soil enzyme-catalyzed reactions, zero-order kinetics are frequently encountered, especially in situations where the substrate concentrations used in the assay procedure are considerably higher than would normally be existent in the soil site.

Observations of zero-order kinetics for soil processes commonly reflect availability of the rate-limiting substrate for the transformation rather than a basic property of the enzyme protein. Note with the first-order reaction curve depicted in Fig. 6.1, that in the presence of high substrate concentrations the reaction rate is independent of increased substrate amendment. That is, the enzyme present in the reaction mixture is saturated and no further increase in the reaction rate due to increases in the amount of substrate in the mixture is possible. Thus, any enzyme reaction may appear to be zero-order if rate-saturating quantities of substrate are (1) used in the reaction mixture to quantify the enzymatic activity or (2) already existent in the soil sample due to the presence of high indigenous levels of reactants. (This enzyme saturation will not be apparent with enzymes that are inhibited by high substrate concentrations, since the reaction rate declines at high substrate levels.) Thus, it is highly probable that some reactions considered to be zero-order in native soil samples are in reality first-order processes functioning at saturating levels of substrate.

Alternatively, zero-order kinetics are also observed when the access to the enzyme by the substrate is limited. In this situation, the reaction rate is controlled by the rate of encounter of the enzyme and protein. This situation occurs when the substrate must diffuse to the active enzyme or where the substrate availability is controlled by its limited water solubility. In the latter situation, the effective concentration of the substrate is controlled by its dissolution rate.

First-Order Processes in Soil: In contrast, rates of first-order processes are substrate dependent. The most commonly used mathematical representation for first-order reactions is the Michaelis-Menten equation. This relationship and some common transformations are presented in Fig. 6.2. Other mathematical models have been developed for describing first-order reaction kinetics, but most soil enzymology is still based on the Michaelis-Menten equation. Note that the Michaelis constant (K_m) is proportional to three rate constants in the general enzyme reaction depicted, and the maximum reaction velocity (V_{max}) is proportional to the total quantity of enzyme in the reaction

$$A + E \underset{K_2}{\overset{K_1}{\rightleftharpoons}} X \underset{K_4}{\overset{K_3}{\rightleftharpoons}} P + E$$

Michaelis - Menten Relationship:

$$v = V_{max} \frac{[A]}{K_m + [A]}$$

where $V_{max} = K_3 E_t$

$$K_m = \frac{K_2 + K_3}{K_1}$$

Lineweaver Burk Transformation:

$$1/v = 1/V_{max} + (K_m / V_{max})(1/[A])$$

plot $1/v$ vs $1/[A]$

slope $= K_m / V_{max}$; intercept $= 1/ V_{max}$

Hanes - Wolf Transformation:

$$[A] / v = K_m / V_{max} + (1/ V_{max}) [A]$$

Eadie - Hofstee Transformation:

$$v = V_{max} - K_m \, v/[A]$$

Fig. 6.2   **Michaelis-Menton equation and some conversions that are useful in calculating kinetic constants.**

mixture. (Those unfamiliar with the derivation of the basic Michaelis-Menten relationship should consult an elementary biochemistry text.)

Several interesting relationships are revealed by plotting data from a reaction showing Michaelis kinetics (Fig. 6.1). Note that the Michaelis constant is that concentration of substrate that allows the reaction to proceed at one-half the maximum velocity ($0.5 \, V_{max} = K_m$). whereas $0.1 \, K_m$ and $10 \, K_m$ are approximately equivalent to 0.1 and $0.9 \, V_{max}$.

It is difficult to derive rate constant values from a hyperbola. The maximum velocity is an asymptote, and it is reasonably difficult to distinguish with precision the substrate concentration yielding half the maximum velocity for determination of the Michaelis constant. Therefore, a variety of mathematical transformations have been derived for the reaction that result in straight lines when plotted (Fig. 6.2). With the Lineweaver Burk Transformation, the slope of the line (m) is equal to the Michaelis constant divided by the maximum velocity (K_m/V_{max}), whereas the x-intercept is the reciprocal of the maximum velocity. These values can thus be easily determined by linear regression of the reciprocals of the reaction velocity and substrate concentration data.

Compromises Associated with Determination of Michaelis Constants in Whole Soils: Michaelis constants derived from study of whole soils rather than purified extracts in defined mixtures are best termed apparent Michaelis constants (K_{app}). This conclusion is based on the following enzymological principles. Michaelis constants are a parameter characteristic of purified enzymes. The reaction kinetics are determined using carefully defined conditions. Therefore, values derived for enzyme activities assessed in suspensions of soil particulates may not reflect the same properties that would be derived from study of the same enzyme molecules exclusive of the undefined soil matrix. Difficulties in interpreting and assigning Michaelis constants to soil enzymatic activities result from the following:

- The existence of multiple enzymes capable of catalyzing the reaction in the soil system

- Competing reactions for the substrate (resulting in a lower concentration of substrate available to the enzyme than assumed by the experimenter)

- The existence of a rate-limiting step before the substrate reaches the enzyme of interest [e.g., permease activity (i.e., the enzymatic activity involved with transfer of the substrate into the cell where catalysis can occur) or diffusion limitations of the substrate into the microsite of enzyme activity.]

Abiotic difficulties include complications associated with substrate sorption (whereby becomes unavailable to the enzyme); in addition, suboptimal conditions for the occurrence of the reaction (e.g., presence of inhibitors) may also cause inaccuracies in assessing Michaelis constants of soil enzymes.

The importance of site variation in substrate concentration on enzyme reaction rates can be seen in data published by Tabatabai and Bremner (1971). The arylsulfatase and phosphatase activities in a variety of Iowa surface soils were assessed. These workers found that the Michaelis constant varied with soil type (as would be expected) as well as with incubation procedure. Continuous mixing of the soil samples resulted in lower Michaelis constants than were detected in static samples. The elevated Michaelis constant in the static samples reflected the reduced availability of the substrates to the enzyme. Soil particles containing the enzyme activities of interest settled to the base of the test tube during the incubation period in the static soil samples. The reaction rate was dependent upon the quantity of substrate present, therefore, in the base of the tube and not the total concentration in the reaction vessel. Furthermore, the quantity of substrate in the settled soil was controlled by diffusion, a slow process. Once the localized substrate in the soil mass was exhausted, its restoration was slow. Thus, the Michaelis constants derived from the static systems were a product of the assay procedure and not necessarily related to the actual enzyme capability of the soil samples.

Application of K_m Values to Detection of Occurrences of Isoenzymes in Soil: Apparent Michaelis constants can be used to demonstrate multiple enzyme forms in soil. Nannipieri et al. (1982) measured a variety of enzyme activities in soil and plotted the results using Eadie-Scatchard plots. Generic examples of the types of data from this analytical procedure are shown in Fig. 6.3. With a single enzyme, a linear relationship is observed (Fig. 6.3A), whereas with two distinct enzymes, a break in the kinetics curve is detected. The data actually appear to be nonlinear, but in reality they fit two straight lines (Fig.

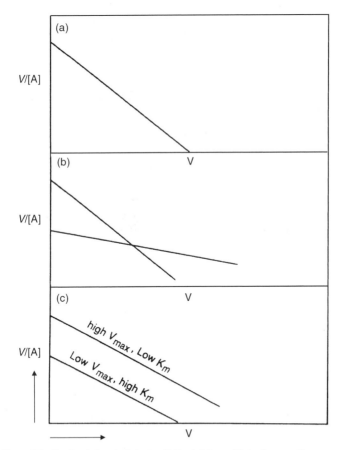

Fig. 6.3 Use of Eadie-Scatchard plots to distinguish multiple forms of enzymes in soil. (A) example of plot of data resulting from activity of a single enzyme; (B) plots demonstrating occurrence of two distinct enzymatic enzyme activities; (C) situation where multiple forms of an enzymatic activity are obscured by divergent kinetics of the two isoenzymes.

6.3B). In the situation for activity resulting from two enzymes, the reaction rate is described by the following equation:

$$v = \left(\frac{V_{\max 1} S}{K_{m1} + S} \right) + \left(\frac{V_{\max 2} S}{K_{m2} + S} \right) \tag{6.4}$$

As the number of different types of enzyme proteins increases, clearly the complexity of the reaction increases. The procedure cannot be used for all combinations of enzymes. Problems are encountered with isoenzymes having highly divergent kinetics constants (Fig. 6C). The rate curve for an enzyme with a high maximum velocity and low apparent Michaelis constant would totally overshadow the curve descriptive of an enzyme with a low maximum velocity and a high apparent Michaelis constant. In this situation, the reaction would appear to result from the activity of a single enzyme. This is an excellent means of analyzing multiple forms of enzymes in soil under the simple conditions described, but frequently the data are too complicated for such an analysis.

6.5   Distribution of Enzymes in Soil Organic Components

Experimental design for study of soil enzymological activities and interpretation of data derived from such studies must be based on a thorough understanding of the complexities of distribution of the enzyme proteins within the soil matrix. As indicated above, enzyme reaction kinetics are controlled by the degree of exposure to a variety of inhibitors and limiting chemical conditions in soil. Differential protection is derived from containment in cellular structures, suspension in soil interstitial water, or linkage to soil colloids. A conceptual model of enzyme pools in soil and the movement of these materials between the various organic matter pools is presented in Fig. 6.4. Central to the concept of enzyme function in soil is the living cell. From a teleological viewpoint, it is for the function of the living cells that enzymes are initially synthesized. Hence, the cell must occupy the central position of the model. Two classes of enzyme molecules are linked to the living cell (for this discussion the classic definition of a living cell will be used; i.e., a living cell is one that is capable of reproduction). The cell contains intracellular enzymes and is the primary source of extracellular enzymes.

Intracellular enzyme activities are those found in living microbial, plant, and animal cells. In most studies of soil enzymes, those activities associated with animal and plant cells are of less significance than those of fungal, actinomycete, and bacterial cells simply because of the mass of each cell type and their contribution to total soil respiratory and metabolic activity. Recall that in preparation of soil samples for analysis, recognizable plant material is generally removed. Most intracellular enzymes are involved in the various aspects of cellular metabolism — for example, glycolysis, Krebbs cycle, and so on. These enzymes cannot function outside of the cell due to cofactor needs

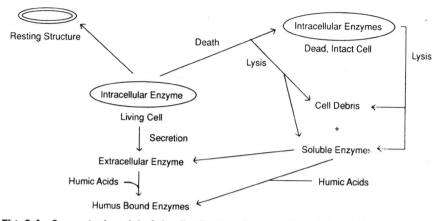

Fig. 6.4   Conceptual model of the distribution of enzymatic activity within major soil enzyme reservoirs.

and sensitivity to variation in pH, redox potential, heavy metals, and other inhibitory physical conditions. For example, a variety of polyphenyl oxidases contain ferrous iron in their active center. Exposure to molecular oxygen results in oxidation of the ferrous ion to ferric ion and inactivation of the enzyme. Some intracellular enzymes are capable of maintaining their activity upon cell lysis (e.g., some proteases), but generally intracellular enzymes are short lived outside of the protective environment of the cell.

Other enzymes that are directly associated with respiring microbial cells and therefore are not classed as extracellular enzymes per se are periplasmic (the enzymes contained between the cell wall and outer membrane of gram negative cells). The activity of periplasmic enzymes extends into the ambient medium. In culture, periplasmic activities are distinguished from cytoplasmic enzymes by conversion of the cell to spheroplasts. This determination is clearly difficult to accomplish with soil samples, but it can be concluded that enzymes that exist in the periplasmic space in culture would also so exist in soil. Examples of this class of enzymes include alkaline phosphatase and penicillinase. Leakage of periplasmic enzymes into the external milieu would result in their being classified as extracellular enzymes.

Enzymes attached to the outer surface of viable cells are usually defined as extracellular since their activity is directed toward external functions. These enzyme activities are exemplified by a variety of bacterial polysaccharidases, but enzymes embedded in the extracellular gum of plant roots (mucigel) may also be included in this category. These enzymes catalyze the conversion of molecules whose structure precludes transport through the cell membrane into the cell into small permeable molecules.

True extracellular enzymes are those enzymes that are secreted by living cells during normal cell growth and division. These activities are found in soil interstitial water. Most extracellular enzymes have low molecular weights (20,000 to 40,000) and are produced abundantly in soil by gram positive bacteria, fungi, and plant roots. An easy means of differentiating extracellular enzymes is by their function. If their substrates cannot enter the cell, then it may be concluded a priori that the enzymes catalyzing their hydrolysis must function external to the cell membrane. Examples of such enzymes include those responsible for hydrolysis of high molecular weight or water-insoluble polymers (cellulose, hemicellulose, etc.), those involved in destruction of toxins, and those associated with dissolution of host tissue prior to invasion by a pathogen.

Other pools for activities normally classed as being intracellular are those enzymes contained in nonproliferating cells (fungal spores, protozoan cysts, plant seeds, bacterial endospores), attached to or contained within entire dead cells, and those linked to cell debris. Enzymes of nonproliferating cells are generally surrounded by thick, biodegradation-resistant walls. Thus, these activities are not detected in enzyme assays as commonly conducted unless a procedure is included that disrupts the cell structure. This task can be accomplished through sonic oscillation. Cell debris-associated enzymes are those linked to cell membranes, cell walls, plus internal structures of eucaryotic cells. These structures are rapidly decomposed, but the continuing turnover of microbial biomass assures a constant, low level of enzymes existent in this class. These groups of enzymatic activities mostly includes hydrolytic activities. Longevity of enzymes in this grouping is dependent upon the degree of protection afforded by and the lifetime of the cellular structure in which the enzyme is contained (cell membrane or wall), as well as the sensitivity of the enzyme to the soil environment.

Enzymes that have become stabilized in humic matter and associated with clays (abiontic enzymes) are more problematic in developing enzyme study protocols and in interpretation of experimental results than any of the enzyme pools outlined above. Synthesis of these enzymes was induced in the cells to meet an existing need, but the enzyme activities were uncoupled from the soil biomass through association with abiotic soil components. Thus, although these enzymatic activities may be significant contributors to total activity detected in a soil sample, their presence in the soil sample may not relate to actual biological processes occurring therein. Soil stabilized enzyme activities are a major contributor to weak or negative correlation of enzymatic activity with soil biogeochemical processes. They provide a background from which it may be difficult, if not impossible, to distinguish the modulations in cell-associated enzymatic activity. The stability of extracellular enzymes relates primarily to the longevity of humic polymers, and any association with cellular function may be fortuitous. Benefit to soil biomass from these activities results entirely from the potential for enzymatic catalyzed products to diffuse to active cells.

An example of a practical implication of this dynamic distribution of soil enzymes among biotic and abiontic soil fractions is the interpretation of data derived from the analysis of air-dried soils. Drying of the soil may result in redistribution of enzyme proteins among these intracellular, extracellular, and abiontic pools. Desiccation may result in death of some viable cells and lysis of viable cells and whole dead cells, as well as disruption of soil aggregates where humic or clay-associated enzymatic activity may have been occluded and therefore not able to be an active participant in soil biochemical activities. Hence, although an enzymatic activity level can be measured in air-dried soils, extrapolation of the results to actual field conditions may be difficult, if not impossible. In reality, drying of the soils creates a new ecosystem with some relationship to that native site from which the soil was collected but with new and perhaps unique properties resulting from the altered physical state of the soil.

6.6 Ecology of Extracellular Enzymes

If we may take the liberty of personifying the soil microbe, we may note that the microorganism residing in soil is faced with a major dilemma. Considerable potential energy resources, that are not water soluble exist in the soil. These are not easily used as carbon or energy sources. The insoluble organic substances cannot be consumed by microbes until they are converted to a form that moves to the microbial cell and is readily transported across the cell membrane; that is, a water-soluble form must be produced. A survival advantage is derived by any cell capable of using these "less available" substances as a carbon and energy source. The difficulty encountered is that extracellular enzymes must be used in transforming the external energy sources into units that can be consumed. These enzymes are an inefficient but essential means of procuring this energy since the probability of their function in soil and the subsequent recovery of the product of their action by the enzyme producing cell is limited.

As indicated above, soil is an unfavorable environment for any enzyme. Even if the protein is not denatured, adsorbed, or inactivated, it may not encounter its substrate because of the heterogeneity of soil and the tendency of soil organic components to be occluded by soil mineral particulates. Furthermore, if the substrate and enzyme are found in the same microsite, the substrate concentrations may not be sufficient for optimal enzyme activity. Even if the enzyme encounters substrate in sufficient quantities to produce beneficial levels of the product, the transformation may be hindered by soil physical or chemical conditions. For example, enzyme reactions occur in a water matrix. Drying of the soil may result in too little moisture for continued enzyme activity. Even if all impediments to expression of an extracellular enzyme activity are avoided, one final difficulty must be overcome before the cell that originally synthesized the enzyme benefits from expenditure of energy resources to produce the extracellular enzyme protein. The product of the reaction must reach the cell. Since soil is a three-dimensional entity, the water-soluble product of extracellular enzyme catalysis may diffuse or be leached in paths that lead it away from the "hungry" cell. The product of the extracellular enzyme may also be consumed by cells other than the extracellular enzyme-originating cell. Clearly, the impediments for recovery of benefits from production of extracellular enzymes are many, but so are the rewards.

A variety of adaptations increase the probability of success of extracellular enzymes. Synthesis of the extracellular enzyme may be dependent upon the presence of the substrate in the vicinity of the cell. This phenomenon reduces the probability of synthesis and excretion of enzymes with no possibility of energy gain. One means for this adaptation to be effective is for the cell to synthesize continuously a low level of the enzyme. Full enzyme production occurs only when the enzyme's substrate is present in the region of the cell. When the extracellular enzyme synthesized during the noninduced period encounters its substrate, a product is yielded that induces full enzyme synthesis. The reduced enzyme production in the absence of the enzyme substrate reduces energy expenditure and is hence of survival advantage to the microbe. This mechanism would be advantageous to the growing cell only in sites where there is a reasonable probability of encountering the substrate. Otherwise, the synthesis of even low levels of the enzyme would be a waste of metabolic energy.

A variation on the limited induction mechanism involves organisms for which the synthesis of the extracellular enzymatic activity is totally inducible — that is, no extracellular enzyme is synthesized in the absence of its substrate. This condition is advantageous in that no energy would be wasted in synthesis of unnecessary enzyme molecules. A price is extracted in that the response to an influx of metabolizable substrate is slow since *de novo* enzyme synthesis must be induced. Enzyme induction would involve one of the following mechanisms:

1. Production of soluble inducer molecules could be the result of activities of coresident microbes in the microsite. This would be an example of commensalism.

2. Inducers could be yielded by nonbiological catalysis of the substrate. Some substrates may be decomposed chemically into intermediates that could stimulate enzyme synthesis.

3. The inducer could be produced by death and lysis of a proportion of the population. In this case either the inducer or enzymes necessary to yield it are released from the cell.

4. As a final alternative, the inducer could be produced through action of the humic acid fraction — that is, catabolism of the enzyme substrate by the stabilized enzymes present in the soil microsite.

Other potential adaptive mechanisms involve physical limitation of the distance the enzyme could migrate from the cell. This adaptation includes enzymes that by definition are not classified as extracellular enzymes, but they certainly are synthesized for external activity. Activities grouped in this category include periplasmic enzymes as well as those linked to cell walls or retained in cell slime layers. Linkage to the cell of externally active enzymes requires cellular adaptations to optimize the potential for the cell to encounter the substrate. For this mechanism to be successful, the microbe must be in direct contact with the substrate. Generally, this mechanism would be advantageous to the motile organism capable of a chemotactic response to available substrates. This adaptation is commonly observed with plant cell wall decomposers—for example, colonies of cellulolytic bacteria growing on the surface of plant debris.

6.7 Concluding Comments

Soil enzymes have been shown to be useful, if not essential, tools in evaluating soil life processes. When properly applied, enzyme assays reveal the nature of the biological processes occurring in a particular soil and the impact of external forces on these transformations. Interpretation of data is limited by our capacity to relate activity measured in the laboratory-manipulated soil sample to conditions at the actual field site, yet with a proper appreciation for soil heterogeneity and the limitations to biological processes occurring therein, such analyses provide a real depiction of the soil system. Such data can be useful in evaluating biogeochemical process rates *in situ*, determining the impact of reclamation management on microbial community recovery, and estimating soil quality.

References

Bartha, R., and L. Bordeleau. 1969. Cell-free peroxidases in soil. Soil Biol. Biochem. 1:139–143.

Brendecke, J. W., R. D. Axelson, and I. L. Pepper. 1993. Soil microbial activity as an indictor of soil fertility—long-term effects of municipal sewage sludge on an arid soil. Soil Biol. Biochem. 25:751–758.

Brookes, P. C. 1995. The use of microbial parameters in monitoring soil pollution by heavy metals. Biol. Fert. Soils 19:269–279.

Burnes, R. G., and J. A. Edwards. 1980. Pesticide breakdown by soil enzymes. Pestic. Sci. 11:506–512.

Casida, L. E. Jr. 1977. Microbial activity in soil as measured by dehydrogenase determinations. Appl. Environ. Microbial. 34:630–636.

Chander, K., S. Goyal, M. C. Mundra, and K. K. Kapoor. 1997. Organic matter, microbial biomass and enzyme activity of soils under different crop rotations in the tropics. Biol. Fert. Soils 24:306–310.

Cole, M. A. 1977. Lead inhibition of enzyme synthesis in soil. Appl. Environ. Microbiol. 33:262–268.

Curci, M., M. D. R. Pizzigallo, C. Crecchio, R. Mininni, and P. Ruggiero. 1997. Effects of conventional tillage on biochemical properties of soils. Biol. Fert. Soils 25:1–6.

Dick, W. A. 1984. Influence of long-term tillage and crop rotation combinations on soil enzyme activities. Soil Sci. Soc. Am. J. 48:569–574.

Dick, R. P., D. D. Myrold, and E. A. Kerle. 1988. Microbial biomass and soil enzyme activities in compacted and rehabilitated skid trail soils. Soil Sci. Soc. Am. J. 52:512–516.

Doelman, P., and L. Haanstra. 1979. Effect of lead on soil respiration and dehydrogenase activity. Soil Biol. Biochem. 11:475–479.

Doelman, P., and L. Haanstra. 1989. Short- and long-term effects of heavy metals on phosphatase activity in soils: An ecological dose-response model approach. Biol. Fert. Soils 8:235–241.

Doran, J. W., and T. B. Parkin. 1994. Defining and assessing soil quality. Pp. 3–21. *In* J. W. Doran et al. (eds.), Defining Soil Quality for a Sustainable Environment. SSSA Spec. Publ. 35. Soil Science Society of America. Madison, WI.

Duxbury, J. M., and R. L. Tate III. 1981. Enzyme activities in cultivated Histosols. Soil Sci. Soc. Am. J. 45:322–328.

Frankenberger, W. T., Jr., and F. T. Bingham. 1982. Influence of salinity on soil enzyme activities. Soil Sci. Soc. Am. J. 46:1173–1177.

Frankenberger, W. T., Jr., and W. A. Dick. 1983. Relationship between enzyme activities and microbial growth and activity indices in soil. Soil Sci. Soc. Am. J. 47:945–941.

Gosewinkel, U., and F. E. Broadbent. 1986. Decomplexation of phosphatase from extracted soil humic substances with electron donating reagents. Soil Sci. 141:261–267.

Halvorson, J. J., J. L. Smith, and R. I. Papendick. 1996. Integration of multiple soil parameters to elevate soil quality—a field example. Biol. Fert. Soils 21:207–214.

Häussling, M. and H. Marschner. 1989. Organic and inorganic soil phosphates and acid phosphatase activity in rhizosphere of 80-year-old Norway spruce [*Picea abies* (L.) Karst.] trees. Biol. Fert. Soils 8:128–133.

Hillel, D. 1991. Out of the Earth: Civilization and the Life of the Soil. Univ. of California Press, Los Angeles.

Jordan, D., R. J. Kremer, W. A. Bergfield, K. Y. Kim, and V. N. Cacnio. 1995. Evaluation of microbial methods as potential indictors of soil quality in historical agricultural fields. Biol. Fert. Soils 19:297–302.

Kandeler, E., C. Kampichler, and O. Horak. 1996. Influence of heavy metals on the functional diversity of soil microbial communities. Biol. Fert. Soils 23:299–306.

Karlen, D. L., M. J. Mausbach, J. W. Doran, R. G. Cline, R. F. Harris, and G. E. Schuman. 1997. Soil quality—a concept, definition, and framework for evaluation. Soil Sci. Soc. Am. J. 61:4–10.

Kelly, J. J., and R. L. Tate III. 1980. Effects of heavy metal contamination and remediation on soil microbial communities in the vicinity of a zinc smelter. J. Environ. Qual. 27:609–617.

Klein, D. A., D. L. Sorensen, and E. F. Redente. 1985. Soil enzymes: A predictor of reclamation potential and progress. Pp. 141–171. *In* R. L. Tate III and D. A. Klein (eds.), Soil Reclamation Processes. Microbiological Analyses and Application. Dekker, N.Y.

Ladd, J. N. 1972. Properties of proteolytic enzymes extracted from soil. Soil Biol. Biochem. 4:227–237.

Mathur, S. P., and R. B. Sanderson. 1980. The partial inactivation of degradative soil enzymes by residual fertilizer copper in Histosols. Soil Sci. Soc. Am. J. 44:750–755.

Miller, M., and R. P. Dick. 1995. Thermal stability and activities of soil enzymes as influenced by crop rotations. Soil Biol. Biochem. 27:1161–1166.

Moore, A. W., and J. S. Russell. 1972. Factors affecting dehydrogenase activity as an index of soil fertility. Plant Soil 37:675–682.

Muller, M. M., V. Sundman, and J. Skujins. 1980. Denitrification in low pH spodosols and peats determined by the acetylene inhibition method. Appl. Environ. Microbiol. 40:235–239.

Nannipieri, P., B. Ceccanti, S. Cervelli, and E. Matarese. 1980. Extraction of phosphatase, urease, proteases, organic carbon, and nitrogen from soil. Soil Sci. Soc. Am. J. 44:1011–1016.

Nannipieri, P., B. Ceccanti, S. Cervelli, and C. Conti. 1982. Hydrolases extracted from soil: Kinetic parameters of several enzymes catalyzing the same reaction. Soil Biol. Biochem. 14:429–432.

Nannipieri, P., F. Pedrazzini, P. G. Arcara, and C. Piovanelli. 1979. Changes in amino acids, enzyme activities, and biomass during soil microbial growth. Soil Sci. 127:26–34.

Perez-Mateos, M., S. Gonzalez-Carcedo, and M. D. Busto Nuez. 1988. Extraction of catalase from soil. Soil Sci. Soc. Am. J. 52:408–411.

Ross, D. J., T. W. Speir, H. A. Kettles, K. R. Tate, and A. D. Mackay. 1995. Soil microbial biomass, C and N mineralization, and enzyme activities in a hill pasture — influence of grazing management. Aust. J. Soil Res. 33:943–959.

Rowell, M. J., and L. Z. Florence. 1993. Characteristics associated with differences between undisturbed and industrially-disturbed soils. Soil Biol. Biochem. 25:1499–1511.

Sarkar, J. M., A. Leonowicz, and J.-M. Bollag. 1989. Immobilization of enzymes on clays and soils. Soil Biol. Biochem. 21:223–230.

Serrawittling, C., S. Houot, and E. Barriuso. 1995. Soil enzymatic response to addition of municipal solid-waste compost. Biol. Fert. Soils 20:226–236.

Sims, J. t., S. D. Cunningham, and M. E. Sumner. 1997. Assessing soil quality for environmental purposes — roles and challenges for soil scientists. J. Environ. Qual. 26:20–25.

Skujins, J. J. 1967. Enzymes in Soil. *In* A. D. McLaren and G. A. Peterson (eds), Soil Biochemistry 1:371–414. Dekker, NY.

Skujins, J. J. and A. D. McLaren. 1969. Assay of urease activity using ^{14}C-urea in stored, geologically preserved, and in irradiated soils. Soil Biol. Biochem. 1:89–99.

Stroo, H. F., and E. M. Jencks. 1982. Enzyme activity and respiration in minesoils. Soil Sci. Soc. Am. J. 46:548–553.

Tabatabai, M. A. 1982. Soil enzymes. *In* A. L. Page (ed), Methods of Soil Analysis. Part 2. Chemical and Microbiological Properties. American Society of Agronomy, Madison, WI.

Tabatabai, M. A., and J. M. Bremner. 1971. Michaelis constants of soil enzymes. Soil Biol. Biochem. 3:317–323.

Tarafdar, J. C., and A. Jungk. 1987. Phosphatase activity in the rhizosphere and its relation to the depletion of soil organic phosphorus. Biol. Fert. Soil 3:199–204.

Tate, R. L. III. 1984. Function of protease and phosphatase activities in subsidence of Histosols. Soil Sci. 138:271–278.

Tate, R. L., III, R. W. Parmelee, J. G. Ehrenfeld, and L. O'Reilly. 1991. Enzymatic and microbial interactions in response to pitch pine root growth. Soil Sci. Soc. Am. J. 55:998–1004.

Tateno, M. 1988. Limitations of available substrates for the expression of cellulase and protease activities in soil. Soil Biol. Biochem. 20:117–118.

Thornton, J. I., D. Crim, and A. D. McLaren. 1975. Enzymic characterization of soil evidence. J. Forensic Sci. 20:674–692.

Tyler, G. 1981. Heavy metals in soil biology and biochemistry. *In* E. A. Paul and J. N. Ladd (eds.), Soil Biochemistry 5:371–414. Dekker, NY.

Verstraete, W., and J. P. Voets. 1977. Soil microbial and biochemical characteristics in relation to management and fertility. Soil Biol. Biochem. 9:253–258.

Zantua, M. I., and J. M. Bremner. 1975. Comparison of methods of assaying urease activity in soils. Soil Biol. Biochem. 2:291–296.

Chapter 7

Microbial Interactions and Community Development and Resilience

Soil is a harsh habitat for microbial community development. A variety of common soil physical and chemical properties exist that singly or in combination could easily preclude or minimize the extent of microbial growth and activity. These physical and chemical conditions of the soil ecosystem may appear to provide a sufficient barrier to invading microbes to prevent their colonization of soil ecosystem, and indeed, that could easily be envisioned to be the true explanation for exclusion of invaders from the soil system. Interestingly, as long as life-precluding extremes in physical and chemical properties are not encountered by invading microbes, it is the biological community that frequently determines the ultimate destiny of the alien organism. That is, the invader may be able to adapt to the physical and chemical environment, but dealing with the competitive and defensive mechanisms of the indigenous community is more problematic. This conclusion is based on the long-understood fact that microbes do replicate when introduced into sterile soil samples (e.g., Conn and Bright, 1919; Katznelson, 1940). Although the microbial population may be functioning at the margins of its acceptable physical and chemical limitations in a particular sterilized soil sample, if all the metabolic and physical requirements of the microorganism are met, growth occurs.

Typical data exemplifying this principle are provided by an evaluation of the behavior of *Escherichia coli* in a sterile organic soil, Pahokee muck (Fig. 7.1) (Tate and Terry, 1980). With initial populations of approximately 1.0×10^5 propagules g^{-1} dry soil, more than one hundredfold increase in population density resulted. Note that when large populations of bacteria were added to the soil, growth was limited or prevented, as was shown when bacterial populations of greater than 3.0×10^7 propagules g^{-1} dry soil were added to muck samples. In this case, the population density of the added *E. coli* in the

189

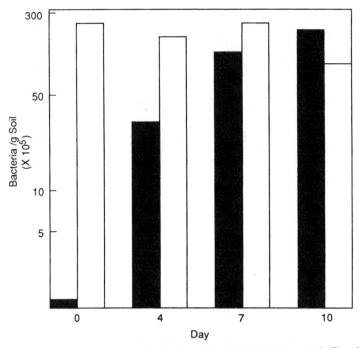

Fig. 7.1   **Survival and growth of *Escherichia coli* in sterile Pahokee muck. The shaded bars represent soil samples amended with approximately 1.1×10^5 bacterial propagules g^{-1} soil, whereas the open bars depict soil samples receiving approximately 3.0×10^7 cells g^{-1} soil.**

sterile soil remained stable over the 10-day incubation period. The limitation to cellular replication when excessive population densities are added to soils samples may result from space limitations (i.e., the physical space for cellular replication is not available) or from biological interactions between the soil-amended bacterial propagules themselves inhibiting or preventing growth. Furthermore, it should be recorded that over the duration of the study (10 days) a significant decline in the microbial populations did not occur in sterilized soil.

These laboratory observations should not be surprising, since colonization of sterile soil habitats or sites with low population densities must be assumed a priori to occur naturally. The most obviously related situation — that is, colonization of a newly established, essentially sterile soil site — is the invasion and colonization of nascent soils, such as volcanic ashes. Conditions may be extremely stressful, especially in newly formed volcanic soils, but as the physical and chemical limitations to microbial growth are ameliorated (e.g., surface area increased through fragmentation of rocky structure, incorporation of fixed carbon as the result of growth and death of pioneer microbial communities), extensive microbial population densities and the intricate microbial communities characteristic of mature soil ecosystems do develop.

These somewhat simplistic laboratory observations and rare-case field situations do not represent the most common experience involving behavior of invader populations in native field soils. When mature field soils are amended with foreign microbes (such as through application of sewage effluent), the introduced organisms generally die quickly. Data exemplifying this natural phenomenon are easily recovered from an examination of the public health literature. For example, Reddy et al. (1981) noted half-lives for fecal coliforms amended to a variety of soil samples of 2 to 150 hours, for *Salmonella* sp. of 2 to 185 hours, and for poliovirus half-lives of 7 to 416 hours. This short-term survival of microbes foreign to the soil community is fortuitous for purification of pathogens from societal waste materials and has provided the scientific justification for approval of land-based sewage and similar waste material disposal methods.

In contrast, the failure of soil-amended microbial populations to survive for extended periods is more troublesome in situations where development of new microbe-based catabolic activity in a soil site is desirable. This situation can be exemplified by the current efforts to develop genetically engineered or laboratory-selected bacterial strains for use in soil renovation procedures. The limited probability of survival and/or activity of the laboratory-derived or selected microbial strains when inoculated into the contaminated soil site raises a significant concern regarding the practicality of soil bioremediation with use of laboratory-produced microbial strains.

Since alien organisms readily survive when added to sterile soil samples, it is logical that the primary barriers to establishment of exogenous microbial populations in soil relate to a limited capacity for the introduced organisms to cope with the preexisting soil populations. This minimal survival of alien propagules in soil may involve a variety of biological exclusionary mechanisms, including the inability of the invader organisms to compete successfully with indigenous populations, sensitivity to biologically produced toxicants existing in soil, and susceptibility to common soil predators or parasites.

These contrasting observations between native and sterile soil samples underscore the importance of biological interactions in the determination of the nature of the soil microbial community composition and function. Soil communities are logically in a constant state of flux. As nutrient supplies are exhausted or reduced by microbial community respiration and growth, or alternatively expanded through influxes of plant biomass, the stresses on microbial activity and selective forces on microbial communities are continually changing. Thus, situations are readily envisioned where minorities or newly imported members of the community may become more competitive due to the changing ecosystem properties and therefore become major players in community dynamics. The soil dynamic forces controlling these changing community dynamics are the same delimiters determining successful soil inoculation for bioremediation or purification of sewage effluent. Thus, to understand situations where insufficient die-off of undesirable microbes occurs in soil or the contrasting inability to establish desirable microbial species in a soil site, a clear delineation of the interactions within the total soil community (i.e., viruses, bacteria, fungi, protozoa, nematodes, and higher animals) is required. Thus, this chapter is presented with the objective of analyzing the general traits or types of microbial interaction in soil and determination of the impact of these interactions on community adaptation and resilience following ecological insult.

7.1   Common Concepts of Microbial Community Interaction

To elucidate the basic principles of the interactive nature of the components of the soil biological community, the nature of the soil ecosystem must be appreciated. Thus, a consideration of the terms delimiting an ecosystem and the life contained therein is necessitated. Foremost among these relationships is emergence of a common concept regarding the meaning of the phase *ecology of the organism.* The word *ecology* is commonly misused by both the general public as well as the scientific community. By definition, **ecology is the study of the relationship of organisms to their environment.** This environment includes all abiotic and biotic factors affecting the cell. Biotic properties of an ecosystem include not only those traits commonly classified by a society as being "natural," but also any consequences of anthropogenic intervention or interaction with the site. Although the effects of human activities on a particular soil ecosystem are by definition natural, or are a normal portion of the driving forces in an ecosystem (since people are commonly a component of the aboveground ecosystem), they are usually distinguished in evaluation of ecosystem function. Perhaps this mental separation of anthropogenic factors from what is generally considered "natural" results from an impression that they can be eliminated, minimized, or at least to some degree managed.

Anthropogenic interactions do alter the properties of the soil biological community and the nature of the interactions between members of that community when they modify a controlling property of the soil ecosystem in sufficient magnitude to shift the equilibrium of the site. This alteration of system descriptors can be exemplified by examining the conditions that must be met for amendment of a chemical to a soil site to alter the biological equilibrium. Three preconditions for effect may be highlighted:

- *The chemical must be added in concentrations that significantly change the levels preexisting in the site.* For example, nitrogen contained in sewage effluent could be a major nitrogen source to plant communities in land disposal systems. There are notable situations where the nitrogen contained in sewage effluent would be an insignificant contributor to the total soil mineral nitrogen pools. Drained histosols provide an example where effluent nitrogen would not significantly change the *in situ* soil nitrogen concentration. The soils of the Everglades agricultural area are composed of approximately 85 percent organic matter, which is in part biodegradable. Decomposition of this organic soil results in the mineralization of about 1400 kg N ha^{-1} annually (see Tate, 1980). Thus, addition of a few kg of mineral nitrogen to these soils through sewage effluent disposal would be not make a meaningful change in the soil nitrogen concentrations.

- *The chemical must be added in sufficient quantity to alter the steady-state equilibrium.* For example, there is considerable concern regarding the impact of acid rain on environmental systems. In well-buffered, calcareous soils, the impact of the acid content of the rain water is modulated by the buffering action of the calcium carbonate and soil organic matter.

- *The chemical may be toxic to the soil organisms.* The toxic impact may be of short duration (e.g., as might be observed with a readily mineralized organic toxicant) or longer lived (e.g., a biodegradation-resistant toxicant). Biodegradable toxicants temporarily inhibit microbial processes. Examples of biodegradation susceptible compounds commonly encountered by soil communities are pesticides and petroleum components contained in accidental spills or used as carriers during pesticide application. In contrast, a nonbiodegradable compound would have a more lasting impact on ecosystem function if toxicant-resistant microbial communities are not developed.

Due to the interactive nature of soil biological communities, soil alterations that might be expected to stimulate only a few species (e.g., those capable of mineralizing a specific pesticide added to the soil) may cause reverberations throughout the entire biological hierarchy, due to the simple change in population density of a single group of organisms. For example, significant increases in easily metabolized carbon will increase microbial competition for the new nutrient and energy source. The augmented bacterial population density resulting from the increased energy and nutrient supply will stimulate the activity of secondary feeders, such as bacterial-feeding nematodes and protozoa. If sufficient carbon resources are introduced into the soil system, competition for mineral nitrogen or space could be accentuated. Thus, the anthropogenic activity associated with the chemical spill or even the purposeful amendment of the soil ecosystem with biologically decomposable organic matter alter the biological interactions at all trophic levels.

The impact of an environmental insult on the basic properties of an undisturbed site are commonly of interest at the ecosystem level, but a full explanation of the processes occurring therein requires specific targeting of a particular portion of the soil biota. Interactions may be evaluated at the individual organism, population, community, or total ecosystem levels. For most soil inhabitants, definition of the individual is reasonably simply accomplished—the single bacterial, algal, or protozoan cell. This entity is more difficult to define for filamentous organisms, fungi, and some actinomycetes. For most of these organisms, the total mycelial structure developing from a single spore could be considered to be the basic entity of this ecological hierarchy, the individual.

Above the individual level, which is studied on a limited frequency with soil microbiological research, is the more easily evaluated population. **A population is defined as being constituted of all the individuals of the same species or function.** More rigorous definitions require a species definition, but in soil microbiology it is not unusual to discuss populations of nitrifiers, denitrifiers, or diazotrophs as well as populations of *Bacillus* or *Pseudomonas* spp. The summation of the populations in a particular ecosystem constitutes the community. As indicated above, the ecosystem is the combination of both the living and nonliving components of the environment.

7.2   Classes of Biological Interactions

In simple situations — for example, an axenic culture growing in a laboratory — the microbial population could be said to be controlled by its nutrient supply, moisture (for colonies growing on solid media), energy source, and a variety of other reasonably defined chemical and physical traits defining the ecosystem. But, even in a simple growth medium, the determinants of cultural development extend beyond the more easily described physical and chemical conditions to the more ill-understood biological interactions. The microbes are not alone in their ecosystem. Each individual must deal with the physical presence and activity of its progeny. Even if a microbe enters a new habitat as a single spore and no other organisms are present — a highly unlikely situation — the primary product of the pioneer is its progeny. That is, for successful colonization of the site to occur, the growing, respiring microbe must replicate. At the least, the presence of this new cell results in competition for nutrients and space as well as an increase in waste materials accumulated in the vicinity of the developing microcolony. These parent-progeny interactions are simple in that the phenotypic and genotypic differences between the cells are minimal. Except for any mutations that may have resulted during the division cycle, each cell is essentially identical. Thus, both may be said to have equal capability and opportunity in the competition for available resources.

As the complexity of the community increases, this simplicity is replaced by a variety of more complex positive and negative interactions. There are few if any pure cultures in terrestrial ecosystems. Microbes must cope with actions of a variety of individuals with greater or lesser capability to deal with cohabitants of their microsite. A summary of the classical designation of the types of interaction affecting soil microbial community development is presented in Table 7.1. Each of these processes will be analyzed individually from the view of its importance to soil microbial community as well as to total ecosystem development.

Table 7.1 Biological Interactions Occurring in Soil Ecosystems

Type of Interaction	Species A	Species B
Positive Interactions		
Neutralism	0	0
Commensalism	+	0
Mutualisms:		
Protocooporation	+	+
Symbiosis	+	+
Negative Interactions		
Competition	−	−
Amensalism	0	−
Parasitism and Predation	+	−

7.2.1 Neutralism

Neutralism is included on the list of interactions of soil microbes for completeness, but it is unlikely that such minimal impacts of an organism on its neighbors occur in a soil ecosystem. By definition, in a neutralistic association the organisms living in the same habitat would have absolutely no effect on their neighbors. The rather simple situation of pioneer communities described above belies the possibility of occurrence of neutralistic interactions of microbes existing in the same habitat. Due to limitations to growth afforded by soil physical structure (space limitation) as well as restrictions resulting from supplies of growth factors, carbon resources, and the generally harsh conditions of soil, such a degree of noninteraction of adjacent populations is highly unlikely.

7.2.2 Positive Biological Interactions

Commensalism: In this opportunistic relationship, one organism is able to grow or function as the result of the action of a second organism. The second organism gains no benefit from the relationship. Commensalism is most likely the most common soil microbial interaction. Examples include such critical soil processes as existence and function of anaerobes, organic matter decomposition, cross-feeding of nutrients, provision of growth factors, and toxin inactivation.

Obligatorily anaerobic bacteria cannot function in the presence of free oxygen. These organisms are highly sensitive to even low levels of molecular oxygen and die quickly in its presence. Clearly, adaptations are necessary for these organisms to exist in an aerobic world. The cooccurrence of aerobes in the mosaic of aerobic and anaerobic microsites of the surface of most soil ecosystems facilitates the development of colonies of strict anaerobes. This enhancement of anaerobic processes is demonstrated by evaluation of microcolony development in and about a soil aggregate. Soil aggregates occur in soil with a predominance of aerobes growing on their surface and with anaerobes replicating internally. Facultative organisms may exist and function throughout the soil aggregate. For a benefit to be accrued by the internal, oxygen-sensitive anaerobes, populations of aerobes on the aggregate surface must utilize molecular oxygen at rates faster than it can diffuse into the interior of the aggregate. Thus, the anaerobes functioning within the soil aggregate do so at the behest of the rapidly respiring surface-residing populations. Those organisms on the aggregate surface do not necessarily gain any benefit from the internal populations. (The qualification "necessarily" does not pertain specifically to the matter under discussion directly. Conceivable benefits could be derived from the potential for products of the anaerobic metabolism to diffuse to the aggregate surface, where they could contribute to the metabolism of the aerobes. Such benefits would only be incidental to the commensal interaction involving molecular oxygen.)

Similar beneficial, gratuitous associations can be noted with the decomposition of plant tissue in soil. Many soil microbes are capable of decomposing the easily metabolized substituents of cellular cytoplasm (e.g., amino acids, proteins, amylose, and simple sugars), but those materials are separated from the general soil microbial community by the cellulosic cell wall. Cellulose-degrading microbes disrupt the cell wall, thereby rupturing the cell and making the internal contents available to noncellulytic microbial populations. Since, in general, cellulytic microbes are poor competitors for easily metabolized substrates, the cellulose degraders gain no or minimal benefit or detriment from the activity of those populations stimulated by the release of internal cellular substituents.

Similar metabolically based commensal relationships are associated with production of metabolic by-products and growth factors. A variety of waste products are yielded by the respiring microbes. These substances may include ammonium, produced from the dissimilatory reduction of nitrate, as well as a variety of partially oxidized carbonaceous substances, such as a variety of organic acids. The organic acids are of no use to the producer but they could be further metabolized by coresident organisms. For example, methanogens use the organic acids produced from catabolism of such complex organic substances as cellulose by clostridia to produce methane, a greenhouse gas. Additionally, the organic acids may be used by other soil organisms as carbon and energy sources.

Bacteria-requiring vitamins (auxotrophs) are commonly isolated from soil. In a study of 499 soil bacterial isolates, Lochhead and Burton (1957) found that 27.1 percent of the organisms required one or more vitamins for growth, with 63 percent of these more fastidious organisms needing more than one vitamin. The most common essential vitamins required by soil bacteria were thiamine and biotin (Table 7.2). By definition any organism that can grow without an external vitamin source must be synthesizing that vitamin (prototrophs). Thus, the latter organisms provide the essential substances to the

Table 7.2 **Vitamin-Producing and Vitamin-Requiring Populations in Soil**

Vitamin	Require[a] (Percent of Isolates)	Produce[b] (Percent of Isolates)
Thiamine	19.4	35.5–56.7
Biotin	16.4	15.2–32.7
Vitamin B_{12}	7.2	29.9–34.6
Pantothenic Acid	4.6	—
Folic Acid	3.0	—
Nicotinic Acid	2.0	—
Riboflavin	0.6	35.2–67.3

[a]From Lochhead and Burton (1957).
[b]From Lochhead (1957).

auxotrophic population. Most organisms synthesize only the quantities of vitamins required to meet their own metabolic needs. These vitamins become available to auxotrophic members of the community upon death and lysis of the auxotrophic cell. Perhaps of more significance to the active soil community is the fact that many prototrophs can excrete vitamins. More than 50 percent of the isolates studied by Lochhead (1957) excreted one or more vitamins. These latter organisms encourage development of commensal relationships with other soil inhabitants.

A variety of toxic substances entering an ecosystem through anthropogenic intervention or which are produced internally as by-products of microbial metabolism may have a generalized inhibitory effect on microbial respiration. For example acidification of a soil site by acid mine drainage or the internal production of sulfate from elemental sulfur or sulfide oxidation (see Chapter 15) results in the inhibition of all acid sensitive organisms. Reduction of sulfate by sulfate-reducing bacteria causes a generalized improvement of ecosystem conditions for the members of the soil community.

Mutualism: Mutualism involves an association in which both organisms benefit. Special cases or related terms include *protocooporation* and *symbiosis*. In protocooperation, any microbe that can catalyze the requisite reaction(s) can function in the combination. Thus, the combination is neither obligatory or specific. In contrast, symbiotic associations are both obligatory for the function to occur and only specific microbes are involved. A generalized example of protocooporation can be a process that could be postulated to occur in any environment. For this example, a soil system is proposed within which neither growth factors A nor B are present. Microbes requiring either of these growth factors would not be expected to grow in the system, yet both groups of organisms can be found to be active in the absence of their prerequisites for replication. For this combination of organisms to exist in the absence of the requisite nutrients for their growth, the microbial population that requires substance A must produce compound B. Similarly, the substance B-requiring strain must produce substance A. Neither of these organisms would be capable of growth in the absence of the coexisting population, but both can function in combination. The combination of organisms existing in this bimember community is determined by metabolic capacity, not species designation. That is, a third species with the requisite traits could just as easily participate in the development of the community. Specifically, such a system could be envisioned to occur in soils lacking both fixed nitrogen and carbon (newly formed volcanic soils). Successful biological associates in the pioneer community could include a green alga (a photoautotroph) and a nitrogen-fixing bacterial species (a diazotroph).

Symbiosis: To some, symbiosis is the most restrictive form of mutualism, whereas for others, it is considered to constitute a separate class of biological associations. Symbiosis is an obligatory, nontransitory, mutually beneficial interaction. This association is essential either for the existence of the organism in the ecosystem or for the occurrence of a particular processes, such as nitrogen fixation. The nature of this association does not mean that either of the partners cannot occur singly, under different circumstances. For independent existence, all of the requirements for growth must be supplied. For example, *Rhizobium* sp. and legumes combine in a symbiotic relationship to fix nitrogen (see Chapter 13). In soil, the association is obligatory for nitrogen fixation to occur, but both the legumes and rhizobial strains are capable of growth separately as long as all of their nutrient requirements are met. In soil, association of the bacteria with legumes is obligatory for nitrogen fixation to occur. In laboratory culture, the rhizobial strain can be induced to fix nitrogen exclusive of the symbiotic association, but this independence is not expressed outside of laboratory cultures.

Commonly encountered symbiotic associations in soil systems are the *Rhizobium* sp.-legume interaction, an actinomycete-higher plant nitrogen-fixing interaction (see Chapter 13), and mycorrhizal associations involving plant roots and a variety of fungi (see Chapter 8).

7.2.3 Negative Biological Interactions

Negative interactions or antagonisms are of interest from a variety of viewpoints. These detrimental associations may explain the failure to successfully inoculate soil communities with propagules of beneficial organisms (including genetically engineered microbes). Also, the principles of antagonistic associations underlie the use of soils as purifying agents for pathogen-containing waste materials and may be exploited in development of biological control methods for control of plant pathogens.

Competition: Competition, rivalry over a limiting factor, is generally described in relationship to a less active species being suppressed by a second, more vigorous, species. In theory it is considered that one species derives full benefit of the available resources, whereas survival or growth of the second population (the weaker competitor) is reduced. Should the competition be sufficiently intense, the poorer competitor is eliminated (**competitive displacement**). Generally, with competition for limiting nutrients, if the supply is continuous, the poorer competitor is eventually displaced.

This all-or-none effect of the competition is clearly not representative of the true field situation. If the weaker population replicates at all, it will exhaust at least a small portion of the object of competition (e.g., carbon, nitrogen, molecular oxygen, phosphorus, space). Thus, the ultimate population density of the more successful as well as its growth rate must be reduced. Furthermore, in natural ecosystems, competition based on availability of a single resource is unusual. A variety of ecosystem properties tend to mitigate the intensity of competitive interactions, thereby allowing coexistence of both competitors (Fredrickson and Stephanopoulos, 1981).

An obvious question regarding competition is: What makes a microorganism a good competitor? An understanding of these factors would be useful in such environmentally important projects as developing genetically engineered organisms for bioremediation. An organism with exceptional ability to mitigate environmental hazards is of limited or no value should it not be capable of competing with indigenous soil microbes. Thus, selection of desirable genetic traits in bioremediation inoculants should include survival as well as the desired metabolic capacities.

Several reasons why certain cell properties provide a competitive advantage have been proposed, but none fully explains the outcome of population interactions and community development in all situations. Most likely, there is no single phenotypic microbial trait essential for successful domination of a habitat. Rather a variety of capabilities contribute and are expressed depending upon the properties of the competitors, the habitat in which they are growing, and the population density-controlling condition. Four properties that clearly are beneficial for successful competition are (1) growth rate, (2) efficiency of nutrient utilization, (3) capacity to catabolize a variety of metabolic substrates, and (4) motility.

Growth rate can be proposed to be effective in competition for nutrients and space. Succinctly stated, the organism with the shortest generation time, all else being equal, would be the most effective competitor. This property is actually linked to the second mechanism (efficient nutrient utilization) in that efficient energy management is a necessary contributor to an augmented growth rate. The lesser the quantity of energy that must be expended for growth, the higher the growth rate. An excellent means of preserving energy for the microorganism is to avoid the unnecessary synthesis of enzymes. For example, an organism that synthesizes its own growth factors even in the presence of adequate levels of those substances would necessarily have its growth yields reduced by the quantity of energy used in synthesizing the growth factor and the requisite enzymes.

Substrate versatility and motility both offer contrasting advantages to the competing microbes. Less fastidious members of a community escape competitive limitations by catabolizing alternative carbon and energy resources. This capacity to utilize a second carbon or energy source once the original substrate has been exhausted would provide a competitive advantage for molecular oxygen and space competition. Motile organisms reduce their competitive pressures by migrating to another microsite.

Environmental factors can determine the outcome of a competitive interactions (for example, see Rosenzweig and Stotzky, 1979). In a competition between the fungus *Aspergillus niger* and the bacterium *Serratia marcescens*, amendment of soil with 3 percent kaolinite reduced the antagonism, and higher concentrations of the clay totally eliminated it. The antagonism was also influenced by soil pH and carbon availability (Rosenzweig and Stotzky, 1980). The data suggest a direct correlation between the degree of inhibition and the rate of glucose utilization by the bacteria, indicating that the antagonism resulted from competition for carbon. This competition was influenced by the clay content (kaolinite) of the soil and the soil pH.

Amensalism is the suppression of the growth of one organism by products of growth of a second. This growth inhibition may result from as simple a situation as alteration of the soil pH, or it may be derived from production of a growth-inhibiting or lethal biological product (e.g., antibiotics). For example, *Thiobacillus* spp. commonly reduce soil pH through the oxidation of sulfide to sulfate. Since the pH may reach values as low as 2, the growth of any pH-sensitive microbes is inhibited.

Two major types of biological inhibitor or toxin are produced by soil microbes: those effective at high concentrations (organic acids, chelators) and those inhibitory at low concentrations (antibiotics). The growth-controlling impact of the former compounds in soil has been reasonably well accepted, since the substances can be easily quantified in soil samples and their interactions with soil microbial populations easily shown. The role of antibiotics within the soil ecosystem is more problematic.

A conceptual barrier exists that has, for a considerable time, prevented appreciation of the impact of antibiotics in complex, heterogeneous systems such as soil. When a soil system is examined as a whole, it is difficult to demonstrate either the presence or activity of any antibiotic *in situ*. Although the antibiotic may be present in inhibitory concentrations in a soil microsite, averaging this localized concentration with that portion of the soil sample containing little or no antibiotic leads to an underestimation of the potential impact of these microbial products *in situ*.

Further conceptual difficulties are encountered when it is noted that in culture, antibiotics are commonly produced by microbes in a resting stage following a period of rapid growth. Since one could reasonably hypothesize that the primary benefit of antibiotics would derive during active growth, it is difficult to conceive of a role of these substances in native soil sites. This hypothesis is developed from the consideration that during growth the microbes would be involved in intensive competition for space and nutrients. There are other opportunities for microbes to benefit from antibiotic production. Resting cells could gain protection from predators. Also, during outgrowth of spores, the antibiotic could suppress susceptible members of the community sufficiently to allow the nascent colony to become established. That is, the antibiotic could provide a release from competitive pressures. This possibility has been shown for streptomycin production by *Streptomyces griseus* (Szabo et al., 1985). Streptomycin was synthesized by vegetative mycelia and bound by spores. Release of the antibiotic from the spores during germination was proposed as a competitive advantage for the young hyphae in the microenvironment.

A simple argument in favor of the role of these substances in soil relates to the fact that a major portion of the soil community is capable of producing antibiotics. The complexity of the molecules suggests a priori that the substance would not be synthesized without some selective advantage being conferred on the producer. The logic of this hypothesis is clear, but its truth in complex soil systems is difficult to demonstrate or disprove. At this point, it must suffice to say that gains from antibiotic production may be accrued at the point of cell-to-cell contact. Furthermore, although no obvious macrosite effects can be documented, the effects of community function observed at the macrosite level are clearly the product of the intensity of competition at the microcolony level. That is, although direct effects of antibiotics at the total system level cannot be demonstrated, any inputs of these chemicals to system function would have at least some effect on the totality of the properties of the system as a whole.

The actual importance of antibiotics in native ecosystems is still conjectural, but their potential has been clearly shown in laboratory, greenhouse, and field studies. For example, *Streptomyces olivocinereus*, which produces the antibiotic heliomycin, was shown to be antagonistic to *Arthrobacter crystallopoites* in laboratory media and in sterile soil (Polyanskaya et al., 1983). Also, several antibiotic-producing bacterial strains tentatively identified as *Bacillus subtilis* and a fungus, *Penicillium nigricans*, protected onions from infection with *Sclerotium cepivorum* in native muck soil in a controlled environmental chamber. These organisms also reduced the occurrence of the resulting disease, onion white rot, when used as a seed treatment in the field (Utkhede and Rahe, 1980). Similarly, amendment of the antibiotic streptomycin or streptomycin-producers to the *Rhizobium* inoculum for soybeans (*Glycine max* L. Merrill) and alfalfa (*Medicago sativa* L.) increased nodulation and plant biomass (Hossain and Alexander, 1984; Li and Alexander, 1986; Li and Alexander, 1988). Thomashow et al. (1990) found that inoculation of wheat roots with antibiotic-producing (phenazine-1-carboxylic acid) *Pseudomonas fluorescens* resulted in antibiotic synthesis in the rhizosphere of wheat plants and a reduction in the occurrence of take-all disease caused by *Gaeumannomyces graminis*.

Another group of biologically synthesized compounds that appear to be useful in reducing plant disease through antagonism of pathogens is siderophores (see Leong, 1986, for a review of this topic.) These substances appear to be active at higher concentrations than is characteristic of antibiotics, but when they result in suppression of microbial growth at low concentrations, they can be classed as antibiotics. Siderophores are extracellular, low-molecular weight (500 to 1000 dalton), iron-transporting compounds synthesized by a variety of microorganisms growing under low-iron conditions. These substances selectively bind ferric ion with a high affinity, thereby reducing iron availability to competing organisms. Most commonly studied siderophore synthesizing microbes from the view of controlling plant pathogens are members of the *Pseudomonas fluorescens-Pseudomonas putida* group. Inoculation with a variety of *Pseudomonas* strains appears to be useful in controlling *Gaeumannomyces graminis* var. *Tritici* (Hamdan et al., 1991, Thomashow and Weller, 1990) and *Rhizoctonia solani* (De Freitas and Germida, 1991) on wheat. Siderophores may also be instrumental in reduction of pathology in disease-suppressive soils (see Kloepper et al., 1980, as an example of these studies and Hornby, 1983, as a review of suppressive soils).

Parasitism and Predation: Predators and parasites, organisms that feed upon living biomass, also play a key role in the soil ecosystem. Soil bacterial and fungal populations are capable of producing resting structures or of entering resting stages (metabolic states where respiratory and metabolic activity are reduced to the minimal level necessary to maintain cellular integrity). Thus, without a means of encouraging the recycling of these nutrients contained in the resting cell biomass through the biogeochemical cycles, essential nutrients (e.g., carbon, nitrogen, and phosphorus) as well as space could be retained in or by these structures at the detriment of more active soil organisms. Facultative predators encourage the cycling of growth substances and liberation of space through parasitic and saprobic consumption of microbial biomass. Parasites and predators maintain the soil bacterial and fungal populations in an active state and enhance nutrient cycling between soil reservoirs through consumption of microbial biomass. This feeding activity of predators and infectivity of parasites maintains a younger, more active, soil microbial population.

Essentially all types of predators or parasites are present in the soil ecosystems. Bdellovibrios, bacteria that prey on other bacteria (e.g., Casida, 1980; Casida, 1983); bacteriophages, protozoa, and nematodes are all active in soil ecosystems. These organisms may ingest their nutrients by consuming intact cells (**holozoic feeding**) as is commonly described for protozoa or by producing extracellular enzymes, which lyse other bacteria, fungi, or algae. This latter process is exemplified by the predatory bacterial populations.

A key consideration in evaluation of predator or parasite behavior in any ecosystem relates to the observation that both the host and parasites or prey and predators coexist in the same ecosystem. Thus, it may be asked, "Why do predators and parasites not totally eliminate their prey and hosts?" A variety of laboratory studies have been conducted to elucidate the causes for coexistence of prey and predators. These studies generally involve incubation of a suscept with a predator (e.g., Danso and Alexander, 1975) or a parasite (e.g., Wiggins and Alexander, 1985) in a liquid culture. Most of the analyses of the specific survival adaptations outlined below were derived from study of this type of simplified microcosm-type interactions. It is commonly noted in these studies that the prey populations achieve an equilibrium density of approximately 10^6 propagules/ml of culture. [Furthermore, in the Wiggins and Alexander (1985) study, a minimal host population of approximately 10^4 colony forming units/ml was required for bacteriophage replication.] If prey population densities significantly greater than 10^6 propagules/ml are amended to the test system, they decline to about 10^6 propagules/ml. In studies where low prey densities were incubated in growth-supporting media in the presence of a predator protozoan population, the prey population increased to the limiting population density of about 10^6 propagules/ml culture, which was subsequently maintained.

Clearly, some mechanism must preclude total elimination of susceptible populations. Otherwise, eventually both groups of organisms would cease to exist. Extinction of prey would result in starvation of the parasite or predator. Some exceptions to this observation are noted. The continued survival of a "food" reservoir is essential for obligate predators, but facultative organisms are freed of this limitation in that they can use living prey or consume organic matter saphrobically. Some relief from this limitation to parasitic or predatory behavior is also gained by the organism that can consume a variety of host or prey species. Once one, perhaps even the prime, food source becomes limiting, consumption of the alternative food source not only spares the original prey population but also reduces the energy expended by the predator in search of food, since it is reasonable to assume that selection of the host is driven to at least a significant degree by the probability of encounter.

A variety of mechanisms have been proposed to explain the balanced attack by parasites or predators (for a detailed review of this subject, see Alexander, 1981). Examples include:

1. *Interactions among predators:* Increased predator density due to localized consumption of the prey may result in competition among the protozoa. This interference with grazing due to frequent contact among the animals could actually result in cannibalism of the predator population. This means of population control may occur in native sites, but it is not the primary mechanism in that in laboratory studies, amendment of stabilized bacterial-protozoan systems with more bacteria causes further increases in the protozoan population densities.

2. *Predators under biological control:* With biological control of predator/ prey or parasite/host associations, the predator and parasite populations may be controlled by populations that feed upon them directly or by biologically formed toxins. Because of the complexity the biological community in soil, it is likely that organisms that feed

upon bacterial predators or parasites would commonly occur. Again, since stabilization of predator-prey populations occurs readily in model systems with one protozoan and bacterial species present, feeding on the bacteriovorus organisms by higher organisms cannot be the sole explanation of population stabilization.

Toxic interactions is an attractive hypothesis, but insufficient data have been accumulated to allow its evaluation.

3. *Genetic feedback:* The primary assumption behind this mechanism is that a spontaneous mutation and population selection can occur. These mutations would result in changes in the prey cells that would reduced their susceptibility (desirability per se) to the predator. This mechanism is certainly a plausible adaptation of the prey population. Yet, at least in short-term laboratory cultures, such selection appears not to occur. In studies of the interactions of *Rhizobium meliloti* and the protozoan *Naegleria*, the surviving population appeared to be as susceptible to protozoan attack as the original population encountering the predator (Danso and Alexander, 1975). In contrast, development of bacteriophage-resistant cultures through modification of receptors on the bacterial cells is commonly observed. Thus, it is logical to assume that modification of the bacterial cell wall or even production of chemo-attractants could reduce interactions with predator populations, but this has yet to be demonstrated in laboratory studies of predator/prey interactions.

4. *Refuge:* Neither soil nor aquatic ecosystems are homogeneous. Both systems are composed of mixtures of particulates (detritus in aquatic systems) intermixed with a somewhat contiguous aquatic system or water layer. Thus, predators could be physically screened from the presence of prey. Pores between the particulate matter can easily be envisioned to be of a size through which the predator cell could not pass. Thus, even though its food source may be within a few microns of its cell, the predator may die. The more heterogeneous the environment, the greater the probability of the occurrence of refuges. This is particularly true of soils since a vast diversity of pores of varying sizes interconnect the various microsites suitable for microbial growth. Unfortunately, since survival of bacteria occurs in both soils and liquid cultures, refuges are not the universal mechanism for prey survival in soil, but it is certainly logical to propose some importance for this mechanism.

5. *Switching:* Most predators are capable of consuming a variety of prey species, although one or a few may appear to be more desirable. For example, Casida (1989) noted that nonsoil bacteria (*Escherichia coli* and *Bacillus mycoides*) amendment to soil samples resulted in protozoan population density increases. In contrast, no response in the predator population was detected following amendment of soil samples with

the native bacterial species *Arthrobacter globiformis*. Should the predator exist in a complex community, the random attack of a variety of bacterial species could result in less stress of each individual species or strain. No volition of the predator is implied in this process. The statistical probability of random encounter with the various species present in the system results in a reduction in destruction of any single population.

An exception to this probabilistic evaluation of the occurrence of this switching prey concept was suggested by Mallory et al. (1983). They observed that if alternative prey population densities were above the threshold for active predation, and the growth rate of the prey was less than the predation rate, then elimination of the prey species could occur in their microcosm study.

Again, the data to support switching as a mechanism for survival of prey cells are scant. Logic suggests that in a soil ecosystem, there may at least be a contribution of this probabilistic sparing of microbial cells from attack. Contributing factors in soil are low microbial growth and predation rates, as well as limited population densities of all species.

6. *Density dependence:* As with the above sparing mechanism, density dependence relies on the probability of the encounter of the predator and the prey. The higher the density of each population, the greater the probability of feeding. Danso and Alexander (1975) noted that at high population densities of *Rhizobium meliloti* cells in defined cultures, final densities of the predators (*Hartmanella* sp., *Naegleria* sp., and *Vahlkampfia* sp.) were proportional to the initial prey populations. The rate of elimination of the prey cells was proportional to initial abundance of the predators, but the final populations of both predators and prey were independent of this number. This observation leads to the conclusion that although many parameters of the feeding curve are determined by participant population densities, the population density of surviving bacteria is not. Similarly, in a study of *Bdellovibrio* feeding on a bioluminescent bacterial prey, Varon and Zeigler (1978) found that a minimal population of 3.0×10^6 prey/ml culture was required for *Bdellovibrio* cells to have at least a 50 percent chance of survival. A prey density for development of population equilibrium in their system was 7×10^5 cells/ml culture. It appears that once the prey density has been reduced to the point that energy expenditure in food acquisition exceeds that returned by prey consumption, sparing of the prey population occurs. That is, the prey population density must be sufficient to provide the requisite energy for the predator or parasite to replicate and search for new food supplies. A population of 10^6 cells per ml of liquid culture apparently approaches that minimal density

for adequate energy return to the feeding organisms. Density-dependent sparing of prey cells likely contributes at least a portion of the survival of suscepts in native environments.

7. *Replication to compensate for killing:* With high population densities of host cells, the probability of encounter between them and predators or parasites is reasonably high. Thus, a reasonably rapid population decline is anticipated to occur. But as the host populations recede, the rate of destruction is reduced simply because of the lower probability of encounter of feeders and host or prey cells. Should the latter populations be replicating, a point is reached where the rate of predation or parasitism is equivalent to the rate of growth of the prey. Considering the low bacterial growth rates commonly occurring in soil, replication to compensate for killing would be significant only in situations with low predator or parasite stress, such as would be associated with low feeder populations.

It is clear from analysis of these hypothesized means of controlling suscept population densities in soil that no single explanation is adequate to answer our initial question, "Why do predators and parasites not totally destroy their prey and hosts?" All the mechanisms proposed are logical. Although specific ecosystems can be sited where they are not functional, environmental situations can be proposed wherein they would be important. Data invalidating an all-inclusive role of a particular survival adaptation in sparing of suscept populations has generally been derived from study of two or three member liquid cultures. These microcosms are special cases per se. In themselves they lack the complexity of a soil microsite. Thus, a particular survival adaptation dismissed in the liquid culture may be crucial for continued existence of individual, predator-stressed soil microbial populations. Thus, the final conclusion regarding host or prey survival in soil is that a variety of properties of both the feeding populations and their food source contribute to the equilibrium population densities reached. The exact mechanisms contributing to this process vary with species involved and the nature and properties of the individual microsites wherein the interactions are occurring.

7.3 Trophic Interactions and Nutrient Cycling

To this point, this analysis of microbial interactions has been concentrated at the somewhat simplistic level of single or bilevel trophic interactions — that is, bacterial population-bacterial population or bacterial populations-protozoan community interactions. Clearly, all trophic levels of the soil biological community contribute to ecosystem development. This loosely associated cooperation is best illustrated by examining the trophic interactions involved with oxidation of plant biomass and some of the factors limiting the process rates and population densities. [For a more comprehensive evaluation of trophic level interactions in soil, see Anderson, 1988, and Anderson et al., 1985 (invertebrates); Bamforth, 1988 (protozoa); Freckman, 1988 (nematodes); Visser, 1985 (invertebrates); and Usher, 1989 (arthropods).]

7.3.1 Soil Flora and Fauna

Although microbiologists tend to concentrate their attention on the biochemical contributions of bacterial and fungal species (and to some degree actinomycete populations) to organic matter decomposition in soil, the overall rate of this process is dependent upon viability of the protozoan, nematode, microarthropod, and animal populations indigenous to soil. Although considerations of specific biochemical transformations and their rates are generally emphasized in evaluation of organic matter decomposition in soil, without the contributions (many of which are related only indirectly to organic matter transformation) of higher organisms, nutrient cycling in most ecosystems would be severely limited. This is best exemplified by listing and evaluating some specific roles for protozoa, nematodes, mites, and other animals in soil ecosystems.

For decomposition of plant debris to occur at rates essential for the return of mineral nutrients contained therein to the aboveground biomass, it must be populated with soil microbes. This occurs through mixing of the debris within the soil profile as well as by inoculation of the plant litter retained on the soil surface. Mixing of organic matter within soil profile is accelerated by the activity of earthworms, ants, termites, and even burrowing animals. For example, by collecting surface litter and storing it within their nests, termites and ants may create islands of plant debris within a sea of nearly litter-free soil. The biodecomposition of this termite- and ant-sequestered organic matter alters plant nutrient availability, soil structure, and soil humic and fulvic acid composition (e.g., see Anderson and Wood, 1984; Gupta et al., 1981; Pomeroy, 1983; Wood, 1988; and Wood et al., 1983). This localized improvement of soil properties is revealed by studies of tropical savanna soils (Lee, 1974; Ofer et al., 1982). Ofer et al. (1982) found that harvester ants in semiarid pastures enhanced plant growth in the vicinity of the mound, increased soil humic and fulvic acids, and altered the distribution of organic nitrogen within the various colloidal soil organic matter fractions. For a more detailed evaluation of the role of ant and termite activities in tropical soils, see Anderson and Flanagan (1989).

A contrasting situation is found in ecosystems where the plant litter remains distributed in a surface organic layer, as is commonly observed with temperate forest soils. Since the primary decomposers of plant debris are soil microbes, stimulation of the decomposition rate results from inoculation of the litter by animal movement between the A horizon soil and the litter layer. This transfer of microbial propagules may be accomplished by a variety of organisms including earthworms, mites, and ants. Generally, the inoculation of the plant debris results from passive transport of the microbial propagules on the surface of the animal body.

Soil animals may also modify the soil structure in a manner that stimulates organic matter decomposition. The activity of aerobic soil microbes may be controlled in part by the rate of oxygen diffusion into soil pores or by saturation of the pores with moisture due to low water infiltration rates. Each of these limitations may be reduced by earthworm tunneling and ant hill construction. Soil structural modification by earthworms and the resulting impact on plant residue decomposition are exemplified by studies of Zachmann and Linden (1989). Treatment of surface-applied corn residues with earthworms (*Lumbricus rubellus* Hoffmeister) resulted in more rapid residue decomposition and altered water balance of the soils.

As was indicated in Chapter 5, surface area can be a primary controller of biological decomposition. This is particularly acute with plant biomass. A primary indirect means of stimulating microbial decomposition of leaf tissue and fras is to increase its surface area. Soil animals both directly (through their own feeding) and indirectly (through physically breaking the plant structure by their passage through the litter layer) accelerate breakup of the biomass physical structure. Clearly, even higher animals, including human activity, accelerate breakdown of the plant physical structure.

A variety of soil animals directly catabolize soil organic matter. The contribution of these organisms to total organic matter respiration in soil is generally minuscule compared to the activity of soil bacterial and fungal populations, but it may be significant. Elliott et al. (1988) suggest that protozoa, nematodes, and microarthropods may in concert contribute up to 40 percent of the nitrogen mineralized in North American grasslands. Populations instrumental in this process include earthworms and isopods. For example, the woodlouse *Porcellio scaber* feeds on decaying pine needles. Their activity is primarily limited by the structural toughness of the needles themselves (Soma and Saito, 1983).

Probably, the major contribution of trophic levels above soil bacteria, fungal, and actinomycete populations to soil organic matter cycling results from their use of the primary feeders (bacteria, fungi, actinomycetes) as food sources. These secondary feeders include protozoa and nematodes. This feeding on the primary feeders has two effects on soil microbial activity:

- *Augmentation of nutrient cycling:* As exemplified by nitrogen mineralization (Woods et al., 1982), nutrients yielded from mineralization of organic nitrogen sources in soil are usually incorporated into the microbial biomass first. These sequestered nutrients are not made available to other participants of the biological community until the microbial biomass is mineralized. This is accomplished by the action of soil predators.

- *Stimulation of microbial activity:* The secondary feeders stimulate microbial growth, thereby keeping the primary organisms in a more active state (e.g., see Anderson et al., 1981). The activity of the secondary feeders can increase the rate of metabolism of soil organic matter by keeping the bacterial population young and active, but it must be remembered that the total amount of carbon mineralized does not change. All that is altered is the time-frame over which the organic matter is decomposed. The quantity mineralized is controlled more by its degradation susceptibility and the quantities present.

Alteration of the activity of secondary feeders can also result in a reduction in the rate of organic matter decomposition. In this situation, the primary population is reduced sufficiently that mineralization is reduced. Santo et al. (1981a, 1981b) examined the metabolism of creosote litter in desert soils by a community composed of microarthropods (tydeid mites), bacteriophagic nematodes, and bacteria. Destruction of the mite population resulted in an increase in their food source, the nematodes. This increased nematode density created greater stress on the bacterial population upon which they were feeding. The resultant decline in bacterial density was expressed as a reduction in litter decomposition. In contrast, elimination of both secondary feeders, nematodes and mites, increased bacterial numbers and litter decomposition. Therefore, it could be said that the mites were effective in controlling litter decomposition by indirectly stressing the soil bacterial population. Higher mite populations lead to reductions in the nematode density, which in turn increased bacterial populations densities and organic matter mineralization. Similarly, Grant et al. (1983) found that destruction of the invertebrate ostracod populations in wetland rice fields resulted in a tenfold increase in nitrogen fixation rates, which results from the parallel threefold increase in blue-green algal populations upon which the ostracods were feeding. In previous work by Grant and Alexander (1981), data were presented indicating that the selective feeding traits of the ostracod could be exploited to increase nitrogen fixation in flood soils. They found that animal size and temperature affected feeding rates of the animals, as did cell age of the cyanobacteria. (See Ingham et al., 1985 for a further analysis of bacterial, fungal, and nematode interactions in the soil community.)

7.3.2 Earthworms: Mediators of Multilevel Mutualism

To conclude that the summation of the effects of biological interactions in microbial communities is strictly limited to population control and invigoration plus nutrient supply interactions would be to totally miss the beauty of the intricacy of interations of the macro-biological world. Not only do higher plants and animals provide nutrients to microbes, but they engineer the soil structure in a manner that enhances life function in both realms and optimizes the mutualistic interactions at all trophic levels. An appreciation of the interdependent nature of plant roots and microbes in the rhizosphere is gained through study of Chapter 8. The discussion here will concentrate on elucidation of the impact of earthworms as ecosystem engineers and as major controllers of total community function.

An initial insight into the complexity of earthworm biology in soil is gained when it is noted that earthworm species are grouped into three major categories based on their site of activity. Endogeic species are active in mineral soil layers; epigeic species, in surface litter; and anecic move between deeper soil layers and soil surface litter (Coleman and Crossley, 1996). Thus, the direct impact of each of earthworm species on ecosystem function depends on its site of residence and activity. Endogeic species are major soil or ecosystem engineers because of their production of casts, galleries, burrows, and chambers (see Fragoso et al., 1997). A less obvious effect of eathworm activity on soil structure is the potential to increase the quantity of macroaggregates in soil. For example, addition of low densities of *Pontoscolex corethruus* in Peruvian Amazonian soils caused an increase in macroaggregates and a decreases in the proportion of small aggregates (Alegre et al., 1996; Blanchart et al., 1997; Gilotvillenave et al., 1996; Pashanasi et al., 1996). The region of the soil where soil is penetrated by earthworm borrow walls is commonly referred to as the drilosphere.

Soil community benefits easily ascribable to modification of soil structural elements — both aggregation and increased macropores — by earthworms in the drilosphere include increased water infiltration and gas exchange between the soil and atmosphere. Enhancement of aerobic microbial activity by the greater potential for import of molecular oxygen and export of carbon dioxide are clear. Also the gains in water movement could both reduce the incidence of water-saturated conditions (resulting in waterlogged soils) as well as increase the amount of water stored in the soil during drying conditions. Less obvious are the effect of the earthworm behavior on organic matter mineralization and community interactions in the drilosphere. Evidence that earthworms are increasing the rates of plant nutrient generation through enhancement of organic matter mineralization is seen through increased total phosphates in earthworm casts (Bossard et al., 1996), and increased maize production in the presence of endogeic earthworms (Gilotvillenave et al., 1996). Görres et al. (1997) examined soil directly affected by the anecic earthworm (*Lumbricus terristris*) and found that carbon mineralization was enhanced. Interestingly, microbial biomass carbon was decreased and nematode populations were augmented. These authors suggested that the indirect effect of the earthworms was to stimulate nematode populations and their resultant interactions with the microbial biomass within the drilosphere. The feeding of the nematodes would reduce the bacterial populations, thereby allowing for active growth of the latter populations. The actively growing bacteria would be expected to mineralize more organic matter than would the less active populations — that is, those not affected by nematode feeding.

7.4   Importance of Microbial Interactions to Overall Biological Community Development

In the above analysis, a dissection of system function has occurred. Truly, these essentials of the soil system cannot exist, and must not be studied, in isolation, or even as isolated pieces of a puzzle per se. They must be assembled into a conceptualization of the working whole. Indeed the ultimate importance of any biotic or abiotic trait or interaction is its contribution to community stability and resilience — that is, the capacity of the ecosystem to be maintained and its resistance to disruptive perturbations. Classically, this stability of the soil community is defined as **homeostasis** (the tendency of a biological system to resist change and to remain in a state of equilibrium).

Maintenance of the equilibrium that defines the existing ecosystem conditions is most commonly accomplished by **negative feedback** (a change in one or more populations induced by a modification of some environmental variable that brings about a response in other populations in such a manner that the original fluctuation is opposed or damped). Negative feedback could be said to be a biological equivalent to the physics action-reaction principle. Applicability of the principle of negative feedback to maintenance of community stability can be exemplified by a hypothetical situation wherein a biodegradable toxin is added to a soil community (Fig. 7.2). The system is defined to be composed of three biological groups — toxin-sensitive bacteria, protozoa capable of feeding upon the bacterial populations, and bacteria that decompose the toxin. For this example, it is not necessary that the toxin be mineralized, but rather a simple detoxification would be sufficient to allow return to the steady-state condition. Note that the initial reaction to the toxic situation is a decline in population densities of those organisms that are sensitive to the toxin. Because of the time necessary for induction of the enzymes associated with detoxifying the aggravating amendment, a delay in toxin destruction is commonly noted.

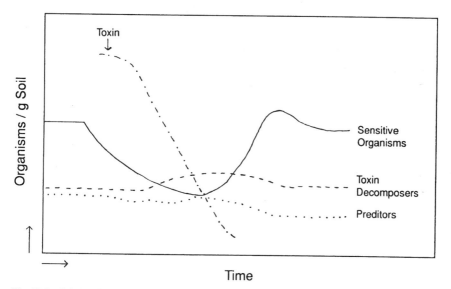

Fig. 7.2   Adaptations in a generalized microbial community to reestablish conditions existent prior to input of a biodegradable toxicant.

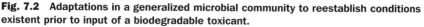

Increases in the population of toxin degraders can result from utilization of the toxin for carbon or energy. Similar increases in toxin decomposer population densities may result if they derive their energy from the biomass of the killed toxin-sensitive populations. In either situation, a finite quantity of carbon and energy is available for growth of the toxin degrader populations. Once the toxin has been removed from the system, the toxin-sensitive populations may again reproduce. At this point, conditions favor return to preinsult equilibrium population densities, but a further induction or negative feedback process must occur — protozoan populations increase to "assist" in the return of toxin degrader population density to preamendment levels. With the loss of the nutrient source (exhaustion of the toxin or depletion of the biomass of cells killed by the toxin), the toxin decomposers cease to grow. Protozoan activity facilities the return to the steady-state levels by consumption of the excess biomass. With time, the preamendment equilibrium can be reestablished. This is in reality an idealized situation in that it relies upon (1) the toxin being totally destroyed, (2) sensitive populations not being totally eliminated by the toxic material, and (3) no permanent change in the controllers of population densities. The overall manifestation of soil microbial populations may appear to represent a complete return to the preexisting equilibrium, but in reality, it would be more common for the system to move to a new equilibrium condition approximating the pretoxin insult conditions. A visible manifestation of the interaction at the microlevel on the total ecosystem would not occur unless the newly established microbial community differed from what preceded it sufficiently to alter overall system properties.

7.5 Management of Soil Microbial Populations

With anthropogenic intrusions into native sites as well as intensive management of developed soil systems, alteration or control of the nature of the soil microbial community is becoming more desirable, if not obligatory, for system survival. Common soil microbial population modification considerations range from alteration of dominant microbial strains (e.g., replacement of indigenous *Rhizobium* sp. strains with varieties possessing more efficient nitrogen-fixation capacities) to enhancement of the activity of specific biological decomposition capabilities (e.g., petroleum metabolizers in oil-polluted soils). In either of these situations, the underlying question regarding success of the inoculation or site management techniques rests at least as much if not more on the competitive abilities of the alien organism as on possession of the phenotypic capacity to achieve the desired ends of the project.

Most commonly, generalized soil inoculation has been a failure. Whether it is the use of free-living nitrogen-fixing microbes (e.g., *Azotobacter* sp.) or of bacteria highly efficient in the transformation of organic phosphorus to inorganic phosphorus (*Bacillus megaterium*), negative results predominate in the literature. A common explanations for stimulation of crop yields by soil inoculation with either of these organisms (i.e., the reported successes of soil inoculation) is that sufficient population densities of the alien organism were added to the soil that the death and decay of their biomass significantly increased soil-fixed nitrogen or phosphorus reserves.

Even with the scattered successes regarding generalized soil inoculation, the preceding evaluation of microbial interactions in soil suggests an explanation for why colonization by alien organisms should not be expected. Soil is a complex ecosystem in which a vast array of microbial populations have developed. It is reasonable to assume that in a mature (climax) system, if a niche exists for a microbe to function, organisms will have most probably already have been selected to fill it. The occupant of the niche would be the most efficient organism to fulfill that function. Thus, to alter the composition of the soil microbial community through inoculation, the added organism must be able to survive in the ecosystem sufficiently long to become established and either (1) to fill a niche that is not occupied in the soil ecosystem, or (2) to be capable of dislodging preexisting populations.

These prerequisites underscore the situations where soil inoculation appears or has been documented to be successful. In most cases, a new niche is created in the soil for the inoculated organism to occupy. This is seen in inoculation of root tissue with *Rhizobium* strains (the new niche is provided by the invasion of the nascent root tissue), addition of nitrogen-fixing blue-green algae to rice paddy fields (the new niche is provided by flooding of a previously drained soil), or addition of a specialized degrader population to a site to be renovated (the new niche is provided by the necessity of decomposing the biodegradable pollutant). In the latter situation, the probability of enhanced pollutant mineralization due to inoculation still rests partially on the capability of the amended organism to out-compete any indigenous populations with similar metabolic capabilities. Future expansion of the lists of inoculation success will be predicated upon better prediction of the nature of the unoccupied niche in soil and an improved understanding of phenotypic microbial traits that allows for augmented capacity to compete with indigenous microbes, thereby usurping their position in the soil community. (See Chapter 16 for a more detailed analysis of the use of soil inoculation for bioremediation of soils.)

7.6 Concluding Comments: Implications of Soil Microbial Interactions

It is reasonable that the vast majority of soil microbiology oriented projects have involved evaluation of the native soil procaryotic populations and to some degree fungal activity in isolation of consideration of interactions with other soil populations. This emphasis is understandable since the primary organisms responsible for decomposition of native biomass and xenobiotic compounds are bacteria, fungi, and to some degree actinomycetes. Yet it is mandatory to realize that the rate of decomposition processes in soil is controlled by the interactions of these primary feeders with a vast array of secondary feeders. The indirect activities of mixing of organic debris with soil, inoculation of this biomass with degrader populations, and the mincing of large plant structures have a significant and obvious effect on decomposition rates. The rate is further controlled by the feeding activity of parasites and predators. Hence, not only must any evaluation of ecosystem dynamics include a consideration of the metabolic capacity of the soil microbial community and their physical and chemical needs, but these studies must be conducted with an appreciation of the impact of higher organisms on the vitality and survival of these front-line organic matter mineralizers.

References

Alegre, J. C., B. Pashanasi, and P. Lavelle. 1996. Dynamics of soil physical properties in Amazonian agroecosystems inoculated with earthworms. Soil Sci. Soc. Am. J. 60:1522–1529.

Alexander, M. 1981. Why microbial predators and parasites do not eliminate their prey and hosts. Annu. Rev. Microbiol. 35:113–153

Anderson, J. M. 1988. Invertebrate-mediated transport processes in soils. Agric. Ecosystems Environ. 24:5–19.

Anderson, J. M., and P. W. Flanagan. 1989. Biological processes regulating organic matter dynamics in tropical soils. *In* D. C. Coleman, J. M. Oades, and G. Uehara (eds.), Dynamics of Soil Organic Matter in Tropical Ecosystems. University of Hawaii Press, Honolulu.

Anderson, J. M., S. A. Huish, P. Ineson, M. A. Leonard, and P. R. Splatt. 1985. Interactions of invertebrates, micro-organisms and tree roots in nitrogen and mineral element fluxes in deciduous woodland soils. Pp. 377–392. *In* A. H. Fitter, D. Atkinson, D. J. Reed, and M. B. Usher (eds.), Ecological Interactions in Soil: Plants, Microbes and Animals. Blackwell Scientific Publications, Oxford.

Anderson, J. M., and T. G. Wood. 1984. Mound composition and soil modification by two soil-feeding termites (Termitinae, Termitidae). Pedobiologia 26:77–82.

Anderson, R. V., D. C. Coleman, C. V. Cole, and E. T. Elliott. 1981. Effect of the nematodes *Acrobeloides* sp., and *Mesodiplogaster iheritieri* on substrate utilization and nitrogen and phosphorus mineralization in soil. Ecology 62:549–555.

Bamforth, S. S. 1988. Interactions between protozoa and other organisms. Agric. Ecosystems Environ. 24:229–234.

Blanchart, E., P. Lavelle, E. Braudeau, Y. Lebissonnais, and C. Valentin. 1997. Regulation of soil structure by geophagous earthworm activities in humid savannas of Cote D'Ivory. Soil Biol. Biochem. 29:431–439.

Bossard, M., P. Lavelle, and J. Y. Laurent. 1996. Digestion of a vertisol by the endogeic earthworm *Polypheretima elongata, Magascolecidae,* increases soil phosphate extractability. European J. Soil Biol. 32:107–111.

Casida, L. E., Jr. 1980. Death of *Micrococcus luteus* in soil. Appl. Environ. Microbiol. 39:1031–1034.

Casida, L. E., Jr. 1983. Interaction of *Agromyces ramosus* with other bacteria in soil. Appl. Environ. Microbiol. 46:881–888.

Casida, L. E., Jr. 1989. Protozoan response to the addition of bacterial predators and other bacteria to soil. Appl. Environ. Microbiol. 55:1857–1859.

Coleman, D. C., and D. A. Crossley, Jr. 1996. Fundamentals of Soil Ecology. Academic Press, NY. 205 pp.

Conn, H. J., and J.W. Bright. 1919. Ammonification of manure in soil. J. Agr. Res. 16:313–350.

Danso, S. K. A., and M. Alexander. 1975. Regulation of predation by prey density: The protozoan-*Rhizobium* relationship. Appl. Microbiol. 29:515–521.

De Freitas, J. R., and J. J. Germida. 1991. *Pseudomonas cepacia* and *Pseudomonas putida* as winter wheat inoculants for biocontrol of *Rhizoctonia solani*. Can. J. Microbiol. 37:780–784.

Elliott, E. T., H. W. Hunt, and D. E. Walter. 1988. Detrital foodweb interactions in North American grassland ecosystems. Agric. Ecosystems Environ. 24:41–56.

Fragoso, C., G. C. Brown, J. C. Patron, E. Blanchart, P. Lavelle, B. Pashanasi, B. Senapati, and T. Kumar. 1997. Agricultural intensification, soil biodiversity and agroecosystem function in the tropics — The role of earthworms. Appl. Soil Ecol. 6:17–35.

Freckman, D. W. 1988. Bacteriovorous nematodes and organic-matter decomposition. Agric. Ecosystems Environ. 24:195–217.

Fredrickson, A. G., and G. Stephanopoulos. 1981. Microbial competition. Science (Washington, DC) 213:972–979.

Gilotvillenave, C., P. Lavelle, and F. Ganry. 1996. Effects of a tropical geophagous earthworm, *Millsonia anomala*, on some soil characteristics, on maize-residue decomposition and on maize production in Ivory Coast. Appl. Soil Ecol 4:201–211.

Görres, J. H., M. C. Savin, and J. A. Amador. 1997. Dynamics of carbon and nitrogen mineralization, microbial biomass, and nematode abundance within and outside the borrow walls of anecic earthworms (*Lumbricus terristris*). Soil Sci. 162:66–671.

Grant, I. F., and M. Alexander. 1981. Grazing of blue-green algae (cyanobacteria) in flooded soils by *Cypris* sp. (ostracoda). Soil Sci. Soc. Am. J. 45:773–777.

Grant, I. F., A. C. Tirol, T. Aziz, and I. Watanabe. 1983. Regulation of invertebrate grazers as a means to enhance biomass and nitrogen fixation of cyanophyceae in wetland rice fields. Soil Sci. Soc. Am. J. 47:669–675.

Gupta, S. R., R. Rajvanshi, and J. S. Singh. 1981. The role of the termite *Odontotermes gudaspurensis* (Isoptera, Termitidae) in plant litter decomposition in a tropical grassland. Pedobiologia 22:254–261.

Hamdan, H., D. M. Weller, and L. S. Thomashow. 1991. Relative importance of fluorescent siderophores and other factors in biological control of *Gaeumannomyces graminis* var. *tritici* by *Pseudomonas fluroescens* 2-79 and M4-80R. Appl. Environ. Microbiol. 57:3270–3277.

Hornby, D. 1983. Suppressive soils. Annu. Rev. Phytopathol. 21:65–85.

Hossain, A. K. M., and M. Alexander. 1984. Enhancing soybean rhizosphere colonization by *Rhizobium japonicum*. Appl. Environ. Microbiol. 48:468–472.

Ingham, R. E., J. A. Trofymow, E. R. Ingham, and D. C. Coleman. 1985. Interactions of bacteria, fungi, and their nematode grazers: Effects on nutrient cycling and plant growth. Ecol. Monogr. 55:119–140.

Katznelson, H. 1940. Survival of microorganisms inoculated into sterilized soil. Soil Sci. 49:211–217.

Kloepper, J. W., J. Leong, M. Teintze, and M. N. Schroth. 1980. *Pseudomonas* siderophores: A mechanism explaining disease-suppressive soils. Curr. Microbiol. 4:317–320.

Lee, K. E. 1974. The significance of soil animals in organic matter decomposition and mineral cycling in tropical forest and savanna ecosystems. Trans. Int. Contr. Soil Sci. 10th. 3:43–51.

Leong, J. 1986. Siderophores: Their biochemistry and possible role in the biocontrol of plant pathogens. Annu. Rev. Phytopathol. 24:187–209.

Li, D.-M. and M. Alexander. 1986. Bacterial growth rates and competition affect nodulation and root colonization by *Rhizobium meliloti*. Appl. Environ. Microbiol. 52:807–811.

Li, D.-M., and M. Alexander. 1988. Co-inoculation with antibiotic-producing bacteria to increase colonization and nodulation by rhizobia. Plant Soil 108:211–219

Lochhead, A. G. 1957. Qualitative studies of soil microorganisms: XV. Capability of the predominant bacterial flora for synthesis of various growth factors. Soil Sci. 84:395–403.

Lochhead, A. G., and M. O. Burton. 1957. Quantitative studies of soil microorganisms. XIV. Specific vitamin requirements of the predominant bacterial flora. Can. J. Microbiol. 3:35–42.

Mallory, L. M., C.-S. Yuk, L.-N. Liang, and M. Alexander. 1983. Alternative prey: A mechanism for elimination of bacterial species by protozoa. Appl. Environ. Microbiol. 46:1073–1079.

Ofer, J., R. Ikan, and O Haber. 1982. Nitrogenous constituents in nest soils of harvester ants *Messor-Ebeninus* and their influence on plant growth. Commun. Soil Sci. Plant Anal. 13:737–748.

Pashanasi, B., P. Lavelle, J. Alegre, and F. Charpentier. 1996. Effect of the endogenic earthworm *Pontoscolex corethrusus* on soil chemical characteristics and plant growth in a low-input tropical ecosystem. Soil Biol. Biochem. 28:801–810.

Polyanskaya, L. M., P. A. Kozhevin, and D. G. Zvyagintsev. 1983. The dynamics of populations of an antagonist and an antibiotic sensitive microorganism in nonsterile soil. Mikrobiologiya 52:145–148 (Russian).

Pomeroy, D. E. 1983. Some effect of mound-building termites on the soils of a semi-arid area of Kenya. J. Soil Sci. 34:555–570.

Reddy, K. R., R. Khaleel, and M. R. Overcash. 1981. Behavior and transport of microbial pathogens and indicator organisms in soils treated with organic wastes. J. Environ. Qual. 10:255–266.

Rosenzweig, W. D., and G. Stotzky. 1979. Influence of environmental factors on antagonism of fungi by bacteria in soil: Clay minerals and pH. Appl. Environ. Microbiol. 38:1120–1126.

Rosenzweig, W. D., and G. Stotzky. 1980. Influence of environmental factors on antagonism of fungi by bacteria in soil: Nutrient levels. Appl. Environ. Microbiol. 39:354–360.

Santos, P. F., J. Phillips, and W. G. Whitford. 1981a. The role of mites and nematodes in early stages of buried litter decomposition in a desert. Ecology 62:664–669.

Santos, P. F., and W. G. Whitford. 1981b. The effect of microarthropods on litter decomposition in a Chihuahan desert ecosystem. Ecology 62:654–663.

Soma, K., and T. Saito. 1983. Ecological studies of soil organisms with reference to the decomposition of pine needles. II. Litter feeding and breakdown by the woodlouse, *Procellio scaber*. Plant Soil 75:139–151.

Szabo, I., A. Benedek, and G. Barabas. 1985. Possible role of streptomycin released from spore cell wall of *Streptomyces griseus*. Appl. Environ. Microbiol. 50:438–440.

Tate, R. L. III. 1980. Microbial oxidation of organic matter in histosols. *In* M. A. Alexander (ed.), Adv. in Microbial Ecol. 4:169–201.

Tate, R. L., III, and R. E. Terry. 1980. Effect of sewage effluent on microbial activities and coliform populations in Pahokee muck. J. Environ. Qual. 673–677.

Thomashow, L. S., and D. M. Weller. 1990. Role of antibiotics and siderophores in biocontrol of take-all disease of wheat. Plant Soil 129:93–99.

Thomashow, L. S., D. M. Weller, R. F. Bonsall, and L. S. Pierson, III. 1990. Production of the antibiotic phenazine-1-carboxylic acid by fluorescent *Pseudomonas* sp. in the rhizosphere of wheat. Appl. Environ. Microbiol. 56:908–912.

Usher, M. B. 1989. Population and community dynamics in the soil ecosystem. Pp. 243–265. *In* A. E. Fitter (ed.), Ecological Interactions in Soil: Plants, Microbes, and Animals. Blackwell Scientific Publications, Boston.

Utkhede, R. S., and J. E. Rahe. 1980. Biological control of onion white rot. Soil Biol. Biochem. 12:101–104.

Varon, M., and B. P. Zeigler. 1978. Bacterial predator-prey interaction at low prey density. Appl. Environ. Microbiol. 36:11–17.

Visser, S. 1985. Role of the soil invertebrates in determining the composition of soil microbial communities. Pp. 297–317. *In* A. E. Fitter (ed.), Ecological Interactions in Soil: Plants, Microbes, and Animals. Blackwell Scientific Publications. Boston.

Wiggins, B. A., and M. Alexander. 1985. Minimum bacterial density for bacteriophage replication: Implications for significance of bacteriophages in natural environments. Appl. Environ. Microbiol. 49:19–23.

Wood, T. G. 1988. Termites and the soil environment. Biol. Fert. Soils 6:228–236.

Wood, T. G., R. A. Johnson, and J. M. Anderson. 1983. Modification of soils in Nigerian savanna by soil-feeding *Cubitermes* (Isoptera, Termitidae). Soil Biol. Biochem. 15:575–579.

Woods, L. E., C. V. Cole, E. T. Elliott, R. V. Anderson, and D. C. Coleman. 1982. Nitrogen transformations in soil as affected by bacterial-microfaunal interactions. Soil Biol. Biochem. 14:93–98.

Zachmann, J. E., and D. R. Linden. 1989. Earthworm effects on corn residue breakdown and infiltration. Soil Sci. Soc. Am. J. 53:1846–1849.

Chapter **8**

The Rhizosphere/ Mycorrhizosphere

The soil microbial community is driven by solar power. The primary energy supply is photosynthetically fixed carbon that enters the soil ecosystem primarily as plant biomass and exudates. Hence, it is logical that a site of maximized biological activity is the portion of the ecosystem encompassed by the soil and root interface—the rhizosphere. The provision of energy to the microbial community by root exudates, dead roots, and sloughed cells results in intense microbial metabolic activity plus an enhancement of microbial interactions—for instance, competition, symbiosis, amensalism, and predation—over levels occurring in nonrhizosphere soil. Furthermore, since plants benefit from the mineralization capacity of the soil microbial community and the improved soil structure derived from the enhanced microbial activity, the rhizosphere microbial community represents that portion of the soil ecosystem with maximal effect on the aboveground community.

During the growing season, the rhizosphere is an expanding focus of biological energy. New microbial habitats are produced continually through root growth. Thus, scientific interest in elucidation of rhizosphere processes is derived from the desire to understand the basic biological activities occurring therein as well as from the fact that the rhizosphere interactions provide a model for successful modification of native microbial community dynamics (e.g., root inoculation with symbiotic nitrogen-fixing populations), as well as for study of the potential uses of genetically modified microbes. Opportunities for use of genetically modified microorganisms in the rhizosphere and adjacent soils include population manipulation to reduce problems with plant pathogenic microbes as well as to achieve soil renovation or other bioremediation objectives.

Therefore, this analysis of rhizosphere/mycorrhizosphere properties is presented with the overall objective of elucidation of the properties of the microbial processes occurring within the rhizosphere and those factors that make it a unique portion of the soil ecosystem. Traits of the total rhizosphere will be examined along with the microbe-plant interactions affecting plant growth and microbial community development. This analysis will be followed by an evaluation of mycorrhizal symbioses and the impact of these associations on properties of the rhizosphere.

8.1 The Rhizosphere

By definition, the **rhizosphere** is that portion of the soil under the direct influence of the roots of higher plants, whereas the **rhizoplane** encompasses the root surface and its adhering soil. As shown in Fig. 8.1, the soil microbes in the vicinity of the growing roots are stimulated by the provision of a surface upon which to grow as well as by the nutrients contained in leachates, sloughed cells, and decaying roots. Microbial colonies develop unevenly along the root surface, with maximal populations occurring where the exudates are lost from the root. Therefore, occurrence of microcolonies is a good indicator of carbon and energy source leakage from the root.

Characteristics that define the rhizosphere are as follows:

■ The properties of the plant are the predominant controllers of the ecosystem rather than those of the soil. This situation results from the

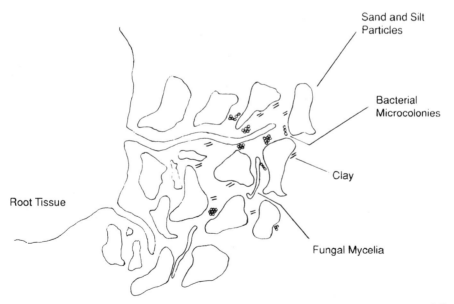

Fig. 8.1   Schematic diagram of the physical relationship of primary components of the rhizosphere (not drawn to scale).

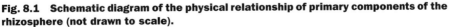

fact that the plant is supplying the nutrient and energy source, plus space for the microbes to grow, as well as altering the gaseous regime of the soil.

- The microorganisms prevailing in the rhizosphere originate in the soil and are characteristic of the soil microbial community. But, because of the variation in the nature of root exudates between plant species, the specific organisms constituting the rhizosphere microbial community do vary with plant species and soil type (e.g., Grayston et al., 1998; Latour et al., 1996; Maloney et al., 1997).

- Because of the combination of increased microbial respiration occurring in the root zone as compared to the adjacent nonrhizosphere soil and the root respiration, the gas regime around the root differs from that of the surrounding soil. Carbon dioxide levels are generally elevated and molecular oxygen tensions are reduced in comparison with the gaseous content of nonrhizosphere-affected soil.

- The influx of carbon and energy in the form of root exudates selects for microbes capable of using the nutrients most efficiently—the fast-growing microbes. In comparison, nonrhizosphere microbial communities typically exist in a "feast and famine" type of existence. Nutrients provided to nonrhizosphere populations primarily result from the influx of nutrients contained in infiltrated water (a typical source in a nonmanaged ecosystem is fixed carbon mobilized by the decay of surface litter), or are carried into the site through movement of animal life (including the nutrients contained in the animal body when the organism expires, as well as organic substances translocated by the activity of ants, termites, and earthworms). It is tempting to consider the rhizosphere community to be continuously bathed in nutrients. This concept may be of value in developing an overall model of the community, but it is far from the situation at the microsite. As is discussed in greater detail later in this chapter, quantities of root exudates as well as their chemical composition vary along the root length, change throughout the growing season, and are dependent upon root density and plant-plant interactions.

- Since many of the products of root metabolism are acids (including carbon dioxide, which forms a weak acid when dissolved in soil water), soil minerals are solubolized in the rhizosphere (e.g., Bolan et al., 1997; Hinsinger, 1997; Krishnamurti et al., 1997). Thus, osmotic impacts on microbial growth are accentuated in the rhizosphere, mineral nutrient availability is modified, and clay mineral transformation may occur. For example, linkage of potassium availability and loss of potassium ion from mica flakes plus transformation of phylogopite into vermiculite in inoculated pine root (*Pinus sylvestris* L.) rhizosphere soils was shown by Leyval and Berthelin (1971).

8.1.1 The Microbial Community

The rhizosphere effect or the degree of stimulation of an activity of population by the energy input of the plant root has classically been described by evaluation of the ratio of activity per unit weight of rhizosphere soil to the activity per unit weight of nonrhizosphere soil (the R/S value). Values greater then one indicate selective stimulation in the rhizosphere, equivalence of the two activities suggests no rhizosphere effect, whereas a R/S ratio of less than one reveals inhibition of the activity in the rhizosphere.

Rhizosphere Bacterial Populations: Bacterial populations are stimulated by root exudates. Populations greater than $10^9 \, g^{-1}$ rhizosphere soil are commonly detected. Selective stimulation of rapidly growing, gram negative, rod-shaped bacteria (*Pseudomonas* sp., *Flavobacterium* sp., and *Alcaligenes* sp. especially and occasionally *Agrobacterium* sp.) is observed. These bacterial species are found as microcolonies covering approximately 4 to 10 percent of the root surface. Competition, the primary selective, intermicrobial interaction, is based on growth rate, metabolic versatility, and growth factor requirements as follows:

- *Growth rate:* The conditions of the rhizosphere favor those bacteria that have short generation times. Clearly, organisms that convert the carbon and energy contained in the root exudates to biomass efficiently and rapidly will achieve greater population densities and occupy more of the space than the less vigorous and efficient members of the community.

- *Nutrients:* A variety of fixed carbon compounds are available to the microorganism within the rhizosphere. Stress of the competition may result in the exhaustion of a particular growth substrate, whereas abundant supplies of others remain. Thus, successful bacterial populations are those capable of using a variety of biochemical growth substrates. Data suggest that, in general, the organisms in the rhizosphere are not carbon limited (e.g., Cheng, 1996), but supplies of specific carbon sources may become more limiting due to intense competition.

- *Growth factors:* Root exudates commonly contain a variety of amino acids and vitamins. These compounds may be used as microbial nutrients or growth factors. Comparison of the growth factor synthesizing capacities of rhizosphere bacteria suggests that some advantage is conferred on organisms requiring amino acids and no other growth factors. Interestingly, the proportion of bacteria in the rhizosphere that require complex nutritional growth factors declines, but their actual number increases. This somewhat confusing observation results from the fact that the number of these organisms in the ecosystem increases but the total population increases at an even faster rate. Therefore, although there are more growth-factor–requiring bacteria in the rhizosphere than are found in nonrhizosphere soil, they constitute a smaller portion of the total population.

Two other factors affecting microbial growth in the rhizosphere are surface area available for colonial development and antibiotic production. Since the simple addition of nutrients to soil can also stimulate the same group of bacteria as are stimulated in the rhizosphere, nutrient supply has to be a primary selective factor in the rhizosphere, but as nutrient needs are met, the point of competition may be shifted to factors affected by the minute distances between individual cells and physical obstructions to growth—for example, surface area and toxicant production. Competition for surface area affects population development primarily at the microsite where cell-cell interactions are intensified. As was indicated in Chapter 7, antibiotic production appears to play a role in amensal interactions in the rhizosphere. With the close proximity of bacteria, the intense competitions occurring, and the positive impact of inoculation of plant roots with antibiotic-producing microbial strains, antibiotics could be of major importance in this ecosystem, but absolute elucidation of antibiotic production and importance in the root ecosystem has yet to be developed.

Fungi in the Rhizosphere: Quantification of the rhizosphere effect on soil fungal populations is equivocal at best. This difficulty is derived, to a large degree, from inability to separate active from inactive fungal propagules. Fungal population densities may be quantified through enumeration of colonial development on defined media or mycelial strands may be counted directly in soil samples. Colonies growing on defined media may result from outgrowth of spores contained in the soil sample or from mycelial strands. Direct observation of hyphal strands in soil provides a measure of total fungal biomass, but not species or genus identity. Colonies growing on the defined media can readily be classified to genus and species, but they may have developed from a quiescent spore serendipitously existing in the environs of the root or from an actively respiring fungal colony. Thus, data collected for the evaluation of rhizosphere effect on fungal populations are difficult to interpret. Total fungal counts appear not to be affected by association with the root, but some individual populations may be stimulated. Plant roots do produce compounds that stimulate outgrowth of fungal resting structures and hyphal development. Similarly, zoospores of phycomycetes are chemotactically attracted to the growing root tip. These observations suggests a positive rhizosphere effect on fungal populations.

Other Rhizosphere Inhabitants: Organisms whose populations tend not to react to influxes of readily decomposable organic matter are usually not affected by root growth. This grouping includes actinomycetes, protozoa, and algal populations. The actinomycetes generally derive their energy supply from decomposition of less readily decomposable soil organic matter components, whereas algal populations use solar energy. Protozoan populations are limited by the distribution and density of prey populations. Because of the high population density of prey required to support increases in protozoan cell numbers, and the limited precision of the methods used to quantify protozoa in soil, a slight increase of protozoa in rhizosphere soil could occur and escape detection.

8.1.2 Sampling Rhizosphere Soil

The basic definition of the rhizosphere (that portion of the soil profile under the direct influence of the plant root) provides an excellent conceptual resolution of the rhizosphere and nonrhizosphere domains, but from a practical viewpoint, it provides little upon which to base a soil-sampling procedure. Questions that may arise in preparation of rhizosphere soil samples pertain to the distance from the root that the plant-effect extends, the degree of the impact (for indeed we realize that there are minute changes in soil properties at considerable distance from the growing root resulting from the presence of the root), and how to avoid diluting true rhizosphere soil with nonrhizosphere soil. Even if it is considered that the effect in question will extend hypothetically only 1 cm from the root surface, that region of impact of the root exudates is not uniform along the length of the root. Greater intensity of interaction of the root with the soil community is found at the zone of active growth. Therefore, defining the rhizosphere as extending 1 cm from the root surface in all directions for the full extent of the root length will still necessarily result in dilution of more active soil with less active soil. Since in reality, the capacity to answer these questions involving the definition of the exact portion of the soil mass affected by the root is nonexistent, most studies of the rhizosphere have involved an arbitrary definition of the soil mass to be collected for field soils, in particular, and a variety of means of physical separation of roots and soil in laboratory and greenhouse studies.

The arbitrary nature of collection of rhizosphere soils in field plots is demonstrated by an examination of common soil-sampling techniques. The more time-consuming method is to dig a soil pit around a plant and aseptically remove soil within a predetermined distance of the plant root—commonly a 1 cm distance. For example, Ruark et al. (1991) compared the interaction of acid rain and ozone treatments on hydrogen, aluminum, calcium, and magnesium ion concentrations in root-adjacent (rhizosphere) and bulk (nonrhizosphere) soils for loblolly pine (*Pinus taeda* L.) seedlings growing in the field by this type of method. The procedure was adequate to demonstrate gross differences in anion concentration due to atmosphere problems and plant growth. Alternatively, the plant may be carefully removed from the soil. Nonadhering soil is allowed to fall to the ground whereas that soil adhering to the root system is gently shaken into a collection container. The latter soil is defined to be the rhizosphere soil. Soil in direct contact with the root that cannot be gently shaken from the root surface but must be washed from the root to be collected is arbitrarily defined as rhizoplane soil. The limitation associated with either of these procedures is that root-affected and -unaffected soils are inevitably mixed at unknown ratios. Thus, rhizosphere effects are at least underestimated and potentially may be totally overlooked should sufficient dilution with unaffected soil occur.

Techniques similar to those used to separate nonrhizosphere and rhizosphere soils in the field may be applied to microcosm studies incubated in the laboratory or in the greenhouse. For example, see Norton and Firestone (1991), where soil was sampled at small intervals from coarse and fine young roots of Ponderosa pine (*Pinus ponderosa* Laws.) seedlings for analysis of the metabolic status of the bacterial and fungal communities. Alternatively, rhizosphere and nonrhizosphere soils may be physically isolated in the microcosm (e.g., Hartel et al., 1989; Martens, 1982). Tate et al. (1991) isolated pitch pine (*Pinus rigida* Miller) seedling rhizosphere soils from nonrhizosphere soils with a 1-μm mesh fabric. Soluble exudates could pass through the membrane, but root tissue was retained. With either procedure, dilution of plant-affected soil by nonplant affected soil compromises data interpretation, as was indicated for field soil-sampling procedures.

It is reasonable to assume that results from the study of bulked soils underestimate the true stimulatory (or even inhibitory) effect of the plant root on soil microorganisms. This leveling of the data is further accentuated when the distribution of the active microbes along the root is considered (recall that root exudates are not released continuously along the root; hence, microbial colonies are found in "hot spots" in the rhizosphere). Thus, the bulked rhizosphere soil sample not only is diluted by nonrhizosphere soil but is also a composite of microsites containing highly active microbial communities plus microsites in which a more reduced stimulation or inhibition soil microbial activity has occurred.

8.1.3 Plant Contributions to the Rhizosphere Ecosystem

The role of the plant in the rhizosphere is to provide a surface for microcolony development and to produce the fixed carbon compounds used by the soil microbes for carbon and energy and as growth factors. Major questions regarding this process are associated with quantification of photosynthate entering the rhizosphere, identification of the organic components of root exudates, and elucidation of the variation in these parameters throughout the growing season and between plant species. A related concern in native systems and with dual cropping systems is determination of the impact of the interaction of root systems from different plant species on the quantities and composition of root exudates.

As a result of the observation that in native soils, photosynthate produced by higher plants (as compared to more limited quantities of carbon fixed by algal activity) generally provides the primary energy source for the soil microbial community, there has been a long-standing interest in quantifying the proportion of carbon fixed by the plant released into the rhizosphere community and the ultimate fate of the fixed carbon (i.e., incorporation into humic substances, microbial biomass, or oxidation to carbon dioxide by the soil biota). Microbiologists see this carbon/energy source emanating from the plant root structure as the major driving force for the soil ecosystem. In contrast, plant biologists may regard production of root exudates as both a benefit to the plant and a potential growth limitation. Benefits are accrued from the stimulation of soil microbial populations whose functions encourage plant development (e.g., through nitrogen mineralization, reduction of plant pathogenic interactions, and synthesis of growth-controlling plant hormones such as indol acetic acid and giberellic acid). Negative implications are associated with the potential for excessive loss of photosynthate, and therefore carbon and energy available for plant biomass production.

A variety of means of quantifying and tracing the movement of root exudates have been developed. Logically, it would seem that a useful procedure for assessing the nature and quantity of fixed carbon products exuded from growing roots would be to quantify the biochemicals accumulated in the growth medium of plants grown under axenic conditions in liquid culture either in a greenhouse or growth chamber culture [e.g., see study conducted by Griffin et al. (1977)]. The primary difficulty encountered with such studies is that the microbial community has a stimulatory effect on root exudate production and may alter the nature of the biochemicals contained in root exudates (Barber and Martin, 1976; Klein et al., 1988). Barber and Martin (1976) found that between 5 and 10 percent of the photosynthetically fixed carbon in wheat and barley plants was lost from the roots grown under axenic conditions, whereas 12 to 18 percent of the photosynthate was lost from roots growing in nonsterile soils. Thus, realistic studies of root exudation should be conducted in situations reflecting native soil conditions as closely as possible.

Utilization of radio-labeled carbon with plants growing in soils appears to provide an improved estimate of photosynthate incorporation into soil fractions. For this, root exudates are evaluated in greenhouse or growth chamber experiments where photosynthate can be labeled with ^{14}C-labeled carbon dioxide. Aboveground and belowground portions of the plant must be physically separated so that access of the soil microbial community to $^{14}CO_2$ or ^{14}C-labeled photosynthetically produced organic compounds can occur only through the root tissue. Movement of the ^{14}C-labeled photosynthate into the rhizosphere is determined by quantifying the $^{14}CO_2$ evolved from rhizosphere soil and the incorporation of ^{14}C into humic substances (e.g., using sodium hydroxide fractionation of soil organic matter) and microbial biomass (e.g., with the chloroform fumigation technique). An alternative to using ^{14}C-labeled substrates for quantifying root exudate production was exploited by Barber and Lynch (1977). These workers utilized the relationship between the quantity of nutrients available for microbial growth and the mass of cells synthesized (i.e., growth yields). By calibrating rhizosphere microbial biomass produced per unit of growth substrate, a growth yield constant was developed. This constant was used to convert the quantities of microbial biomass produced in the presence of roots growing in liquid culture in growth chambers into mass of photosynthate available for their growth.

A wide range of values for root exudate production are reported in the literature. Nutrients available to rhizosphere microbes vary with plant species, plant developmental stage, and soil temperature, as well as the species distribution of the plant community. Newman (1985) concluded that soluble root exudate ranges from 1 to 10 g per 100 g dry roots. Values found in the literature vary depending upon the inclusiveness of the data collected. Some reports of plant carbon incorporated into soil fractions include total plant carbon retained in soil organic fractions, whereas in other studies, root exudate production is evaluated exclusive of carbon contributions by root cell death and root death. Martin (1977) recorded a range of rhizodeposition of 14.3 to 44.4 percent of photosynthetically fixed carbon (for wheat plants), whereas Davenport and Thomas (1988) found 10 percent of $^{14}CO_2$ fixed by corn (*Zea mays* L.) in belowground plant components as compared to 40 percent for bromegrass (*Bromus inermis* Leyss). This differential separation of fixed carbon between aboveground and belowground plant biomass resulted in a rhizodeposition rate in the bromegrass twice that of the corn. During the last 31 days of the 55-day period from germination to seed set of blue grama [*Bouteloua gracilis* (H.B.K.) Lag.], 33 percent of the fixed carbon was found in the root biomass, 23 percent in the root-derived organic matter contained in the soil, and 22 percent was released as carbon dioxide (Dormaar and Sauerbeck, 1983). These values contrast to early studies, such as that by Martin (1977), where 0.8 to 1.3 percent of total organic carbon in root-free soils had been derived from plant photosynthate.

The rhizosphere microbial community can increase the rhizodeposition rate of photosynthetically fixed carbon. Microbes significantly increased root exudation of *Agropyron cristatum* and *Agropyron smithii* but had no effect on *Bouteloua gracilis* (Biondini et al., 1988). For the former grass strains, root exudates were reduced 40 and 83 percent, respectively, from native soil values when grown in axenic culture. The root biomass of *B. gracilis* was increased by the presence of an active rhizosphere community (Klein et al., 1988). This microbial stimulation of root exudation may result directly from synthesis of plant hormones by the microbes that are stimulatory to exudate production. An indirect stimulation of exudation could result from microbial utilization of the exudates, thereby preventing their accumulation. This would increase outward diffusion of exudate components from the root.

Merckx et al. (1986) compared the release of photosynthate by maize and wheat roots growing in a growth chamber. After 6 weeks' growth, 1.5 and 2.0 percent of the corn and wheat photosynthate, respectively, were found as soil organic carbon residues. Photosynthetically fixed carbon dioxide was also detected as soil microbial biomass and as carbon dioxide respired by the soil microbial community. For the wheat plants, approximately 20 percent of the fixed carbon was detected in each of these carbon pools. Distribution of photosynthetically fixed carbon within these various pools can vary with soil temperature (Meharg and Killham, 1989). Increasing the soil temperature from 5 to 25°C increased root-soil respiration from 5.7 to 24.15 percent. The proportion of the photosynthate retained in the plant root and soil was greater at 5 and 25°C , with a minimum at 15°C. The total fixed carbon released as root exudates by *A. cristatum, A. smithii*, and *B. gracilis* was estimated to be 8, 17, and 15 percent of photosynthetically fixed carbon.

Microbial growth in the rhizosphere can also be affected indirectly by the quantities of plant nutrients available for plant biomass production. Merckx et al. (1987) grew maize plants at high and low nutrient levels. Plant biomass production was limited at the low nutrient levels after 35 days' growth. In soils with the lower nutrient level, 2 percent of the total $^{14}CO_2$ fixed by the plant was retained in the soil at all harvest times (38, 35, and 42 days), whereas with the nongrowth-limiting nutrient levels, 4 percent of the plant photosynthate was retained in the soil. Maximal soil incorporation was found in the lower nutrient soils at 35 days' growth whereas the quantities retained in the high nutrient soils increased throughout the study period. Microbial biomass contained between 28 and 41 percent of the total soil ^{14}C in the lower nutrient soil, whereas in the higher nutrient soil, this value ranged from 20 to 30 percent of soil residual photosynthetically fixed carbon.

When evaluating plant contributions to soil microbial energy transformation, it is important to consider annual turnover of the root biomass itself. This fixed carbon contribution to microbial nutrient cycling may be considerable. For example, Whitford et al. (1988) found that in the Chihuahuan Desert, 85 to 90 percent of the mass of herbaceous annual roots and 40 percent of the mass of woody shrub roots was decomposed annually. These rates are comparable to or greater than similar processes reported for mesic ecosystems. For comparison, Dahlman and Kucera (1965) found an annual increment in the surface 25 cm of soil (the A horizon) of a prairie grassland in central Missouri (U.S.A.) of 429 g/m^2. This represented about 25 percent of the total root biomass in this horizon, suggesting an annual turnover rate of the root biomass of about 25 percent. For comparison, in a study of live root decomposition in four forests (a mixed deciduous forest in Virginia; a mixed deciduous forest, and a *Pinus resinosa* plantation in Massachusetts; and an *Acer saccharum-Quercus borealis* forest in Wisconsin, U.S.A.), losses of between 20 and 60 percent of their initial mass over a four-year period were detected for each forest (McClaugherty et al., 1984).

Other methods used to trace movement of carbon fixed by the plants through the soil microbial and organic matter pools include pulse-chase experiments with $^{14}CO_2$ (e.g., Rattray et al., 1995) and experiments assessing ^{13}C abundance in soils converted from growth of C-3 vegetation to C-4 plants (e.g., Qian and Doran, 1996). With the former studies (pulse-chase), plants are incubated in a $^{14}CO_2$- enriched atmosphere. After an appropriate incubation period in the enriched atmosphere, the ^{14}C-enriched air is replaced with normal ambient air. The change in the ^{14}C-enrichment of plant tissue and soil organic components is quantified with time. Initially, the bulk of the radioactivity is detected in plant tissue. After the labeled carbon dioxide is removed from the atmosphere, no further increase in labeled plant tissue occurs. Then, the movement of the labeled carbon into soil components is quantified. These studies provide an indication of the amount of plant carbon lost in root exudates and the time course of the process. The use of the stable, heavy isotope of carbon (^{13}C) is valuable in situations where plant communities can be changed (switched from C-3 to C-4 plants) and minimal alteration of soil structural properties is desired. In this case, the increased incorporation of ^{13}C into biomass of the C-4 plants provides the tracer for carbon movement in the soil system.

The composition of root exudate is a primary parameter in selecting for individual species active in the rhizosphere community and the diversity of that community. The composition of root exudates varies with growth condition of the plant and its developmental stage. The variety of compounds that may be contained in root exudates is indicated by a list of candidates compiled by Alexander (1977). These soluble, organic carbon sources accumulated by plants grown aseptically include:

- *Amino acids:* All naturally occurring amino acids

- *Organic acids:* Acetic, butyric, citric, fumaric, glycolic, lactic, malic, oxalic, propionic, succinic, tartaric, and valeric acids

- *Pentoses and hexoses:* Arabinose, deoxyribose, fructose, galactose, glucose, maltose, mannose, raffinose, rhamnose, ribose, sucrose, and xylose

- *Pyrimidines and puridines:* Adenine, cytidine, guanine, and uridine

- *Vitamins:* p-Aminobenzoate, biotin, choline, inositol, nicotinic acid, pantothenate, pyridoxine, and thiamine

- *Enzymes:* Amylase, invertase, phosphate, and protease

This smorgasbord of readily catabolized organic substances is supplemented by all components contained in sloughed root cells and decaying root biomass. As was indicated above, as much as one-fifth to one-third of the fine root biomass may turn over annually. The selective stimulation of microbial populations by the root exudates is readily demonstrated by the alteration of metabolic diversity (see Chapter 3 for a description of methods for determining the soil community property) compared to nonrhizosphere soil (e.g., see Garland, 1996a, b, for use with whole soils and Frey et al., 1997, for study of soil isolates). The method has been useful in assessing the effect of changes in root exudates due to elevated atmospheric carbon dioxide levels on rhizosphere community dynamics (e.g., Hodge et al., 1998; Rillig et al., 1997).

The activity of the rhizosphere population is determined not only by the composition of root exudate, sloughed cells, and decaying root tissue but also by the total quantities of fixed carbon produced, as was suggested above. From considering the relationship of these three fixed carbon pools to total plant productivity, it is clear that rhizosphere activity is dependent upon plant nutrient status and reaction of the plant to environmental stresses. Examples of data reflective of this microbe-plant growth state interaction is provided by studies of ryegrass (*Lolium perenne*) under a variety of fertilization strategies (Turner and Newman, 1984; Turner et al., 1985), and in studies of grassland management (Ingram et al., 1985). The ryegrass studies demonstrated increased root exudation under phosphorus deprivation. Evaluation of range plants showed an interaction between herbage removal and irrigation, with irrigation causing significant declines in general rhizosphere activity that was mitigated by herbage removal. Active fungal hyphae were decreased by both treatments, whereas overall microbial activity was stimulated by herbage removal, suggesting a shift in the relationship of active fungal and bacterial populations in the stressed plant rhizosphere.

A further question regarding the rhizosphere effect relates to the longevity of the plant impact on the soil community following demise of the plant. Death of the plant results in a decline in the rhizosphere community. Little if any microbiological effect is carried over in the soil until the next invasion by root tissue. Although there is an impact of adjacent plants on the nature of the rhizosphere community, preceding plant species existent in the ecosystem have little effect on rhizosphere community composition of succeeding plants. The new vegetation determines, to a large degree, its own rhizosphere community composition. Exceptions to this statement of no carryover of rhizosphere effects relates to soil properties that are altered by the presence of root surface or root-nodulating organisms. For example, fixed nitrogen accumulated in a soil environment due to the symbiotic association of *Rhizobium* sp. and legumes may provide a nitrogen nutrient source for succeeding plant communities. Similarly, physical or chemical modifications of the soil mineral environment necessarily persist.

8.1.4 Benefits to Plants Resulting from Rhizosphere Populations

Development of an active rhizosphere community has a variety of indirect and direct impacts on plant biomass production. Many of these benefits are proposed since it is difficult to quantify minute modifications of the plant root environment and their effect on overall plant development, but logic combined with sound data documents the fact that an active, dynamic rhizosphere microbial population is beneficial, if not essential, for optimal plant productivity. Rhizosphere property modifications beneficial to plant growth and development range from the more difficult-to-prove production of growth-promoting substances to the clearly documented gains of symbiotic root associations involved in nitrogen fixation.

A wide variety of rhizosphere microorganisms have been shown to produce plant growth hormones, including indolacetic acid (IAA), gibberellins, and cytokinins (e.g., see Nietko and Frankenberg, 1989; Tien et al., 1979). These substances could be proposed to stimulate root tissue development, thereby increasing the capacity of the root system to provide nutrients and water required for aboveground biomass function.

A more indirect but clearly beneficial result of stimulation of soil community by root invasion is the contribution of rhizosphere microbial communities to development of a stable soil structure conducive to plant community development (i.e., improved soil aggregation; e.g., Amellal et al., 1998; Andrade et al., 1998). (See Chapter 1 and Tate, 1987, for a further discussion of microbial involvement in soil structural development.) The rhizosphere community produces polysaccharide material such as capsules and slimes that cement soil mineral particulates into microaggregates. Furthermore, soil structural improvements are gained from fungal mycelial production as well as by association of soil particles with root tissue. This improvement in soil structure through increased soil aggregation results in improvements in soil aeration, water infiltration, and root penetration.

Perhaps the primary benefits of an active rhizosphere community to the plant community result from mineralization of plant biomass. Organic nitrogen, phosphorus, and sulfur compounds are oxidized, thereby liberating ammonium, phosphorus, and sulfate. Thus, the nonbioavailable organic forms of these essential nutrients are converted to plant-available mineral forms.

Two root-based symbiotic associations are instrumental in producing a thriving, stable plant community — nitrogen fixation associations (*Rhizobium*-legume and actinorrhizal associations; see Chapter 13) and mycorrhizal symbiosis (Section 8.2). These interactions increase fixed nitrogen resources and facilitate nutrient transfer to the higher plants.

Additionally, plant growth stimulation may result from solubilization of inorganic plant nutrients by production of organic or inorganic acids by rhizosphere organisms. Microbially produced organic acids may solubilize essential minerals simply through acidification of rhizosphere soil or by chelation (e.g., siderophore production) of the metal. The capacity of organic acids excreted by soil bacteria to solubolize soil minerals is exemplified by studies of 2-ketogluconic acid production by gram negative bacteria growing in glucose liquid media (Duff et al., 1963). Additionally, the acidification of the rhizosphere environs through metabolic production of hydrogen ions alters the pH sufficiently to mobilize soil minerals (e.g., see Gillespie and Pope 1990a, b).

Solubilization of a variety of metals contained in natural and synthetic insoluble salts and minerals has been demonstrated. Siderophores and other metal chelators are synthesized by a variety of bacteria, including common soil organisms (e.g., see Carrillo et al., 1992; Fekete et al., 1983; Jurkevitch et al., 1992; Hoefte et al., 1991; Kloepper et al., 1980a; Meyer et al., 1989; Pesmark et al., 1990) as well as mycorrhizal fungal associates (e.g., Cline et al., 1982; Federspiel et al., 1991). A number of examples of stimulation of plant biomass production by these microbially synthesized metal chelators have been published (e.g., Bar-Ness et al., 1991; Crowley et al., 1991; Derylo and Skorupska 1992; Kloepper et al., 1980b). Bosser and Verstraete (1986) have presented a model in which the rhizosphere is the most affected soil microsite for siderophore synthesis.

8.1.5 Plant Pathogens in the Rhizosphere

A common maxim in plant pathology is that a soil-born pathogen is more destructive to a plant growing in sterile medium than it would be to a plant developing in native soil, where an active rhizosphere community develops. This observation suggests that the rhizosphere microbial community directly or indirectly inhibits the invasion of the plant tissue by the pathogen. This disease reduction could result from direct competition between the pathogen and the root microbes, antagonism of the pathogen by root microbes, or alteration of root exudate diffusion into the soil environment in a manner that interferes with chemotaxis of the pathogen to the root. (See van Loon et al., 1998, for a review of this topic.)

Root-inhabiting microbes and plant pathogens may compete for space, nutrients, or even binding sites on the root surface. Space and nutrient competition could result in failure of the pathogen to develop critical population densities for disease initiation, whereas competition for specific binding sites would reduce the capability of the plant pathogen to initiate the infection process.

As indicated previously (Chapter 7), the rhizosphere is one soil site wherein antibiotic production may confer survival advantage upon microbial populations. Both antibiotics and siderophores produced by a variety of *Pseudomonas* strains have been shown to reduce or preclude infection of wheat roots by take-all disease (Hamdan et al., 1991; Kloepper et al., 1980b; Thomashow et al., 1990; Thomashow and Weller, 1990). Other suggested interactions of antibiotic-producing microbes in biocontrol of root disease are associated with rhizoctonia infections (De Freitas and Germida, 1991) and onion white rot (Utkhede and Rahe, 1980). Similarly, rhizobacteria may also be used to suppress growth of sensitive plants. For example, *Pseudomonas* spp. have been isolated from roots of winter wheat (*Triticum aestivum* L.) and downy brome (*Bromus tectorum* L.) that suppress growth of the downy brome weed (Kennedy et al., 1991).

An additional role of rhizosphere microbes in reducing root disease incidence is in interfering with chemotactic attraction of the pathogen to root receptor sites. Chemotactic interactions are more difficult to document in the field but are readily deduced to be of importance in plant disease production. A variety of compounds listed above as components of root exudates may serve as attractants for plant pathogens. Growth of root inhabitants (including mycorrhizal fungi) necessarily reduces both the quantity and diversity of organic compounds diffusing from the root, thereby diminishing the probability of encounter of a plant pathogen.

8.1.6   Manipulation of Rhizosphere Populations

Modification of the composition of any soil community structure through inoculation with exogenous microbes requires a thorough knowledge of microbial interactions as well as "a little bit of luck" if a reasonably predictable outcome of the procedure is desired. Historically, intervention into rhizosphere community dynamics has been a premier example of the microbiologists expertise—some might say utility. Biological nitrogen fixation has been enhanced through inoculation of legumes with *Rhizobium* sp. to augment symbiotic nitrogen fixation for many decades (Chapter 13). Long before the microbiology of the symbiosis was understood, legumes were exploited in crop rotations for their capacity to increase soil fertility. (Historically, in many agricultural systems, fallow cropping with legumes has been used to maintain soil quality. With these cropping systems, the effect of the legume was maximized by not harvesting the crop. Rather, the crop was plowed under to maximize the quantity of fixed nitrogen and organic matter retained in the system.) The long-established procedure of legume inoculation has been supplemented with methods to supply fungal inocula to stimulate mycorrhizal development plus utilization of a variety of plant-growth–stimulating bacteria, including the free-living nitrogen-fixing bacteria of the genus *Azospirillum* (e.g., see examples listed in Table 8.1).

Considering that the potential for modification of soil microbial community structure through inoculation with alien strains has long been viewed with pessimism, the question of why root inoculation is frequently successful must be examined. Although as many failures in inoculation for population alteration are reported as are successes, a cursory review of the literature could lead to the conclusion that the only limitations to the procedure are selection of an appropriate microbial strain for inoculation, development of a carrier to sustain the microbial population, and selection of a delivery system. Indeed a variety

Table 8.1   Examples of Reports of Modifying Root Communities by Amendment with Alien Microbial Populations

Microbe	Recipient	Reference
Rhizobium spp	legume seeds	See Chapter 13
Frankia spp.	higher plants	See Chapter 13
Mycorrhizal Fungi	higher plants	See text
Pseudomonas putida	wheat	Vrany et al., 1981
Bacillus polymyxa	lodgepole pine	Holl and Chanway, 1992
Pseudomonas fluorescens	wheat	Parke et al., 1986
Azospirllium spp.	spring and winter wheat	Harris et al., 1989
	maize	Fallik et al., 1988
Bacillus sp.	spring wheat	Chanway et al., 1988
Azospirillum brasilience	wheat	Bashan et al., 1987
Pseudomonas fluorescens	potato	Bahme et al., 1988
Pseudomonas serrata	wheat and peas	Astrom and Gerhardson, 1988

of strains, carriers, and inoculation procedures have been developed, but unanswered basic science questions still remain. Considerations that must be examined are why some inoculation studies are successful while others are not, and why some microbial strains are established in the community more readily than others. Complete understanding of these queries is not available, but some basic information exists to direct our attention to means of improving the probability of successful intervention into rhizosphere community dynamics.

A central maxim in soil ecology is that if a niche exists in an ecosystem for a microbe to function, it will be occupied by a member of the existing soil community. Additionally, displacement of an established, functional member of the community is difficult. Yet, such displacement is frequently desirable in that an improved function of the ecosystem may be achieved by establishment of a more efficient or active microbial strain.

Inoculation of the rhizosphere provides a good example of when this displacement or augmented activity is possible. The first consideration in inoculation of a soil site with an exogenously developed microbial strain relates to the occurrence of a niche in which the microbe has the capacity to function. Nascent root tissue can be a provider of both a habitat and a niche. The external structure of the root is a surface supportive of colonial development, whereas a variety of functional opportunities are derived from the invasion of the root tip into the soil (e.g., decomposition of components of root exudates, nitrogen mineralization, and nitrogen fixation) that at least initially have not been implemented by existing populations. Thus, the production of new root tissue overcomes the primary barrier to microbial community composition modification. Remaining obstacles are delivery of an appropriate inoculum to the habitat and the capacity of the alien organism to compete with the indigenous populations.

Clearly, inoculation of established root systems is difficult and results in a low probability of development of the inoculated populations. This problem of inoculation of established roots has been overcome by drenching of root systems (e.g., see Parke et al., 1986) or it has been avoided entirely by inoculation of seeds before planting. Typically, seeds may be drenched with liquid inocula or pelleted solid carriers of the microbial inoculum prior to planting [see Chapter 13 for discussion of inoculation with rhizobia and Bashan (1986a, b), for use with *Azospirillum* sp.]. The assumption underlying this procedure is that the growing root will carry the inoculum with it as the root mass expands. Considering that rhizobial inocula are at least partially established in the field using these methods, seed inoculation may be of less than perfect efficiency, but it still occurs. *Azospirillum brasilense* is also mixed throughout the soil and from plant to plant by growth of wheat roots (see Bashan and Levanony, 1987; 1989). Alternatives to root drenches and soil inoculation include banding of the microbial inoculum material into the soil adjacent to the expanding root mass. This procedure is underlain by the assumptions that the roots will growth through the band of inoculum, the inoculum will survive sufficiently long in the soil for this encounter to occur, and the inoculant will compete successfully with indigenous soil microbes for habitable surfaces on the root.

Competition of the alien organisms with indigenous soil microbial populations is most likely the least-understood aspect of any soil community inoculation procedure. The added organisms must be capable of overcoming the normal soil community defenses to invasion as well as coping with soil physical and chemical barriers. Many of the growth requirements for the microbe are met by the provision of nutrients and growth surface by the root; hence, dealing with biological competitors becomes the primary barrier to establishment.

Development of microbial strains that would be useful for field inoculation is generally initiated by isolation of an organism from the indigenous soil population that is amenable for selection of genetic variants. This isolate may then be modified either through selection of natural genetic variants (mutants) or by genetic engineering procedures where appropriate genes are inserted into the genome. In either case, the objective is to develop an organism that is more efficient in a given process (e.g., nitrogen fixation) or that possesses an improved metabolic capacity (e.g., ability to decompose an environmental contaminant).

By starting with a native member of the soil or root community, it may be assumed that most of the necessary traits for successful competition with the indigenous population preexisted in the parent strain. The question then becomes, "Did the genetic manipulation or laboratory culture reduce or eliminate this competitive capacity?" The query is frequently answered negatively. For example, legumes nodules are commonly formed by indigenous rhizobial strains in spite of the inoculation of seeds with more efficient varieties. Development of *Rhizobium* strains with improved nitrogen-fixation capabilities have been developed into a fine art, yet the art is commonly lost when the efficient strains do not enter the site of expression of their enhanced ability — the root nodule. Unfortunately, laboratory-selected rhizobia are not always the best competitors for nodule occupancy (e.g., Klubek et al., 1988; Materon and Hagedorn, 1982; Moawad and Bohlool, 1984; Robert and Schmidt, 1983; Smith and Wollum, 1989; van Rensberg and Strijdom, 1985). Variability in nodule occupancy by alien rhizobial strains is exemplified by data from van Rensberg and Strijdom (1985) where the range of nodule occupancy by inoculant strains was from 17.7 to 100 percent, with the highest values associated with legumes growing in soils with limited native rhizobial populations of appropriate specificity for the legume. As is discussed in Chapter 13, successful occupancy of the nodule by the inoculant depends upon its capacity to deal with the general rhizosphere population as well as possession of appropriate biochemical traits to efficiently interact with the root to form an effective nodule.

Thus, the rhizosphere microbial community structure has been manipulated through inoculation of the plant with alien organisms to improve plant development (i.e., in most cases, improve crop yields). Great potential exists for further exploitation of these skills to alter the microbial community to accelerate plant community development and to reduce soil pollutant loads. Plant growth improvement may be achieved by establishment of members of the rhizosphere community with the capacity to decompose plant toxicants or improve nutrient availability or augment soil granulation, whereas the rhizosphere community may also be utilized to couple the nutrient generation ability of higher plants (e.g., fixed carbon supplies) with the ability of an laboratory-generated rhizosphere inoculant bacterial species to degrade a troublesome pollutant (perhaps, even cometabolically). (See Chapter 16.)

8.2 **Mycorrhizal Associations**

Fungus-root or mycorrhizal associations have been known to exist since early descriptions by Frank in the early 1880s (cited in Allen, 1991). These associations are commonly studied and have been extensively described. The description, or definition, one provides for these fungal and higher plant symbioses, to a large degree, is linked to the definer's bias toward the consideration of the importance of the plant or the microbe to ecosystem function. This is best seen by comparing statements by Marx and Allen. Marx (1972) describes mycorrhizal associations as follows:

> The infection of feeder roots of most flowering plants by symbiotic fungi and
> the transformation of these roots into unique morphological structures
> called mycorrhizae (fungus-roots) undoubtedly constitutes one of nature's
> most widespread, persistent, and interesting examples of parasitism.
> Most plants of economic importance to man are actually dual organisms —
> part plant and part symbiotic root-inhabiting fungi.

This plant-oriented view stresses the utilization or perhaps even the "theft" of plant photosynthate by the controlled parasites in this symbiotic association. In contrast, Allen (1991) described the fungal-plant interaction from a more neutral or microbially oriented aspect:

> A mycorrhizae is a mutualistic symbiosis between plant and fungus localized in
> a root or root-like structure in which energy moves primarily from
> plant to fungus and inorganic resources move from fungus to plant.

In this definition, the mutualistic benefits of the interaction are stressed.

From an environmental or soil community consideration the fungal-plant associations in mycorrhizal symbioses can be viewed on a gradient of increasing association of the plant and soil community (Table 8.2). At the most extreme position lie the general soil microbial populations. These organisms are involved in mineralization of native soil organic matter, soil-incorporated organic components (plant and animal debris), and microbial biomass as well as soluble organic carbon components contained in infiltrated water. This community functions totally at the mercy of the nutrient supply, which may range from nearly total deprivation (i.e., a desert soil) to absolute luxury (e.g.,

Table 8.2   Comparison of Properties of Nonrhizosphere, Rhizosphere, and Mycorrhizosphere Soils

Nonrhizosphere	Rhizosphere	Mycorrhizosphere
All biogeochemical cycles—rates limited by organic matter (humus) availability	Maximal biogeochemical cycling	All biogeochemical cycles are supported—controlled by both plant inputs and soil humus contents
Nutrients may be leached to root tissue or groundwaters	Nutrients mineralized directly available for plant biomass as well as microbial biomass synthesis	Nutrients incorporated into fungal biomass and transported to plant tissue
Nutrients may be incorporated directly into microbial biomass	Microbial biomass synthesis is controlled by plant productivity and rate of root exudate production	Biomass controlled by plant productivity and transfer of photosynthate to the fungus
Microbial biomass limited by carbon and energy resources		
Note that nonrhizosphere soil metabolic activity is not affected directly by rhizosphere interactions, but mycorrhizal fungi may enter this soil region and catalyze biogeochemical processes		

soil receiving nutrient-laden industrial effluent). Of more direct association with the discussion at hand is the fact that the nutrients produced by organic compound mineralization by these general soil microbes may not be directly available to the aboveground community. Clearly, nutrient transfer to higher plants is dependent upon proximity of the microbial colony to active roots. In this case, a locational displacement as small as a few millimeters could limit the rate of mineral nutrient recovery by the plant.

An enhancement of ecosystem three-dimensional associations is derived from the transition from general soil populations to rhizosphere-linked communities. Rhizosphere communities catalyze all of the processes described above for nonrhizosphere soil, but their value to aboveground biomass sustainability is augmented by the improved efficiency for transfer of mineral nutrients to the plant community.

The ultimate association of soil microbes with growing plants is symbiotic fungal-root associations. This mycorrhizal symbiosis could be said to bridge these community types—nonrhizosphere, rhizosphere, and fungus root. Although there is an intimate association of the fungus with plant root cells (with some types of mycorrhizae, actual penetration of the root cells itself occurs), the fungal biomass extends well beyond the rhizosphere into nonroot-affected soil. Thus nutrients may be accumulated from nonrhizosphere soil by the fungus, and transported through the rhizosphere community into the plant root. Inversely, the photosynthate from the plant supports growth and development of the entire hyphal network. Furthermore, this mycelial network may directly link roots of plants of the same or different species in that the same fungal body may infect a number of individual plants.

The physical association of mycorrhizal fungi with plant roots has been extensively described and serves as the primary basis for classification of mycorrhizae. These associations may be classed as ectomycorrhizae, endomycorrhizae, or ectendomycorrhizae. In this grouping of mycorrhizae, the commonly encountered vesicular-arbuscular mycorrhizae are considered endomycorrhizae. A number of reviews of the biology of mycorrhizae have been published (e.g. see Allen, 1991; Mukerji, 1991). Thus, only basic descriptions of these classes of mycorrhizae will be presented herein.

Ectomycorrhizal fungi penetrate intracellularly and partially replace the middle lamellae between cortical cells of feeder roots. These fungi form a dense mycelial net around and between the plant cells termed a **Hartig net**. Ectomycorrhizal associations are also characterized by a dense, generally continuous hyphal network over the feeder root surface called a **fungal mantle**. This fungal mantle varies from one to two hyphal diameters thick to as many as 30 or 40 depending upon the fungal associate, the host, and the environmental conditions. Examples of plant species forming ectomycorrhizal associations are species in the families *Pinaceae, Salicaceae, Betulaceae,* and *Fugaceae*. Most ectomycorrhizal fungi are basidomycetes (primarily of the families *Amanitaceae, Boletaceae, Cortinariaceae, Russulaceae, Tricholomataceae, Rhizopogonaceae,* and *Sclerodermataceae*).

Endomycorrhizae are distinguished by the fact that the fungus penetrates the cortical cells of feeder roots and may form large vesicles and arbuscles (hence the term **vesicular-arbuscular mycorrhizae**). These fungi do not form dense fungal mantles, but do develop a loose, intermittent arrangement of mycelium on the root surface. Endomycorrhizae are formed by most agronomic, horticultural, and ornamental crops, as well as some forest tree species that do not form ectomycorrhizae. The fungal species are phycomyces—many of which are in the genus *Endogone*.

As with many, if not most biological classification systems, many organisms form associations that are not easily classified in any of the predominate groupings. For mycorrhizal associations, the **ectendomycorrhizae** constitute this grouping. These mycorrhizae resemble ectomycorrhizae in forming a Hartig net and a fungal mantle. A resemblance with the endomycorrhizae is associated with their penetration of cortical cells. This mycorrhizal grouping is the least studied, and the nature of the fungal symbionts has not been totally elucidated.

Although mycorrhizae provide intriguing subjects for biological and physiological analysis, the emphasis for this study is to delineate (1) the role of mycorrhizal symbioses in ecosystem function, (2) their importance in nutrient cycling in soil-plant interactions, and (3) their utility in bioremediation plans.

8.2.1 Mycorrhizae in the Soil Community

A variety of elegant studies of soil enzymes and respiratory activity in relationship to soil physical and chemical properties and biological interactions (both belowground and aboveground) have been conducted. Some of this research has been directed at quantifying fungal and bacterial contributions to these activities through incorporation of population-specific antibiotics (e.g., cycloheximide to inhibit fungal activity, and streptomycin to control bacterial activity) into the laboratory-incubated soil-plant samples, but evaluation of the role of mycorrhizal fungi in these soil functions is rarely considered. Standard methods for soil collection rarely preserve the soil-root-fungal associations. Hence, even direct counts of fungal mycelia in the soil would rarely, if ever, be linked to plant roots. This omission of a consideration of these important trophic interactions occurs in spite of the knowledge that mycorrhizal fungi constitute a major portion of soil biomass and that these fungi extend throughout the soil profile — far beyond the regions classified as rhizosphere soil. An example of the association of soil enzymatic activity with root and fungal development was provided by Spalding et al. (1975). In their study, cellulase, invertase, polygalacturonase, and peroxidase activities were found to increase in soil under lycopodium fairy rings, suggesting a mycorrhizal fungal association with organic matter metabolism.

The potential for mycorrhizal contribution to total soil metabolic activities can be appreciated by examining examples of the quantities of biomass contained in these structures. Fogel and Hunt (1983) in a study of a young, second-growth Douglas fir [*Pseudotsuga menziesii* (Mirb.) Franco] stand found that mycorrhizae constituted 6 percent of the total tree standing crop. Furthermore, they found that roots and mycorrhizae contained larger reserves of nitrogen, phosphorus, potassium, and magnesium than did the forest floor or soil fungi. Fine roots and mycorrhizae contributed between 84 and 78 percent of the total tree organic matter to the soil. This study suggests that although the mycorrhizal fungal biomass constitutes only a small portion of the standing crop, it may account for 50 percent of the organic matter throughput in the *Pseudotsuga menziesii* stand. Read (1984) suggests that mycorrhizal fungal biomass may be the largest microbial biomass component of many forest soils. Allen (1991) states that "mycorrhizal fungi may be the single largest consumer group of net primary production in many, if not most, terrestrial biomes." Vasicular-arbuscular fungal hyphae have been quantified at densities of up to $38\,\mathrm{m\,cm^{-3}}$ (Allen and Allen, 1986). It must be remembered that although a portion of the fungal biomass is intimately associated with the root, the fungal net extends throughout the soil surface (see Allen, 1991 for a review of this topic).

From these observations, a number of benefits of mycorrhizal development to the total soil microbial community can be delineated. Foremost among the contributions to the soil community is the capacity to cycle nutrients — that is, mineralize accumulated biomass. Although the bulk of the research efforts have been directed at quantifying mineral nutrient transfer from fungal biomass or soil to plant tissue, it is reasonable to assume that a portion of the metabolic products of this fungal activity will be released to the soil microbial community. For example, mycorrhizal fungi have been shown to mineralize soil organic phosphate through synthesis of phosphatase (e.g., see Dighton, 1983, and Pasqualini et al., 1992) and to solubolize mineral phosphorus through the acidification of the soil habitat through carbon dioxide production (see Knight et al., 1989).

Indirect benefits to the soil community by mycorrhizal fungal growth results from a generalized improvement of soil structure. Fungal hyphal development increases the association of soil particulates into aggregates (see Tisdall and Oades, 1982, and for a review of this topic, Tate, 1987). This improved soil structure has a direct impact on the indigenous microbial community through improved aeration and moisture infiltration, and an indirect effect via stimulation of plant root growth. Clearly, any augmentation of root development increases the quantities of fixed carbon reaching the soil microbial community.

8.2.2 Symbiont Benefits from Mycorrhizal Development

Mycorrhizal fungi are clearly instrumental in augmenting plant nutrient availability in nutrient-stressed ecosystems. A wide range of data (at times contradictory data) have been collected demonstrating nutritional benefits to plant communities in stressed communities through mycorrhizal symbiotic associations. Most research has shown improved nutrient transfer to the plant tissue through the augmentation of the adsorbing surface of roots by extension of the fungal mycelium into nonrhizosphere soil. Gains in phosphate, nitrogen, and water transfers are most commonly reported. The expansive literature documenting these benefits has been reviewed by Allen (1991), Bagyaraj (1991), Gupta (1991), and Mukerji et al. (1991).

Perhaps in a more intriguing vein, mycorrhizal fungi may not only enhance soil-plant transfer of nutrients, but they may also be instrumental in movement of nutrients between plants (e.g., see Camel et al., 1991; Eason et al., 1991; Hamel et al., 1991; and Read et al., 1989)). Read et al. (1989) demonstrated through the use of $^{14}CO_2$ that carbon moves freely between plants connected by mycorrhizal mycelium. This plant-plant bridging by the fungal hyphae occurs between host plants of the same or differing species. Similarly, Chiariello et al. (1982) found that ^{32}P-phosphate sprayed on leaves of *Plantago erect* in a grassland was transferred to the shoots of about 20 percent of the close neighbors. Vesicular-arbuscular mycorrhizae were noted to connect the root systems of neighbors of different species and considered to have mediated the nutrient transfer. Camel et al. (1991) found that the hyphal front may advance at a rate of 2.3 cm week^{-1} in soil sand mixtures and that bridges of at least 90 mm may be formed. These workers suggest a competitive advantage of this bridging in plant groupings where mycorrhizal inoculum is limited. Plants developing in a less desirable portion (e.g., a shaded site) of the ecosystem may gain benefits of the photosynthetic activity of the less stressed members of the community as well as the gains from nutrient production of the mycorrhizal function through an interconnected root system.

A further stimulus to plant biomass productivity is associated with the capacity of mycorrhizal associations to reduce or prevent plant disease development (see Jalali and Jalali, 1991, for a review of this topic). The association of the fungus with plant disease-susceptible root tissue may reduce pathogen vulnerability through modification of exudate composition or concentration, stimulate plant disease protective response through its own infection of the root tissue, inhibit competing microbial populations directly through synthesis of antibiotics, and limit access of the plant pathogen to root tissue by physically occupying the root surface.

Enhancement of fungal biomass by the symbiotic association with root tissue is generally attributed to augmented availability of photosynthate in the form of simple sugars for fungal energy and carbon needs, and the reduced competition with the general rhizosphere microflora. Kucey and Paul (1982) found that with in mycorrhizal, nodulated faba beans (*Vicia faba* L.), about 4 percent of the carbon fixed by the host was transferred to the fungal associate. [For a more detailed examination of this topic, see Allen (1991).] A competitive advantage for the fungal associate in dealing with soil microbial populations in general and more specifically rhizosphere microbes is derived from its close physical association with the host root. Intrusion into the interlameller space between cortical cells (as well as cell penetration by the endomycorrhizae) and intimate contact with the root surface facilitate nutrient recovery from the host tissue.

Of interest from the impact of the host plant on the behavior of mycorrhizal fungi in soil is the potential for selective stimulation of spore outgrowth. The published results regarding selective stimulation of fungal spore outgrowth by plant root products are mixed, yet a real microsite stimulation most probably does occur. Bechard and Piche (1989) found that in their study of carrots and the vesicular-arbuscular mycorrhizal fungus *Gigaspora margarita*, root volatiles provided little stimulation of spore germination and that the exudates alone had no effect. Considerable stimulation was detected from the synergistic interaction between the volatile and exudate root products. Elias and Safir (1987), from their study of the effect of root exudates of *Trifolium repans* on *Glomus fasiculatus* germination, conclude that the primary stimulation of spore outgrowth results from exudates from phosphorus-deficient clover seedlings when compared to the effect of exudates from phosphorus-sufficient seedlings, suggesting that it is the quality of the exudates altering the spore germination rates. Further studies (Nair et al., 1991) suggest that two isoflavonoids (7-hydroxy, 4″-methoxy isoflavone and 5,7,-dihydroxy,4″-methoxy isoflavone) may be signal molecules in vesicular-arbuscular mycorrhizal symbioses. Other workers (see Bowen, 1969) conclude that root exudates had no effect on spore outgrowth. It may be deduced from evaluation of these disparate observations that mycorrhizal fungal spores are sensitive to root exudates and that variability in composition (quality) of these plant products with host nutritional status greatly affects this interaction.

A further observation of the infection process in mycorrhizal development is a potential interaction of plant growth-promoting bacteria (e.g., Bowen and Theodorou, 1979; Garbaue and Bowen, 1989; Meyer and Linderman, 1986a,b). Bowen and Theodorou (1979) found that a variety of bacteria could either depress, have no effect, or stimulate mycorrhizal development in the rhizoplane of *Pinus radiata*, suggesting the necessity of fungal compatibility with the rhizosphere bacterial populations. Garbaue and Bowen (1989) actual found that "helper" microorganisms exist in the rhizosphere that improve the efficiency of ectomycorrhizal infection of *Pinus radiata* D. Don growing in a sandy podzol. Meyer and Linderman (1986a) found that a combination of plant-growth stimulating bacteria and mycorrhizal development of subterranean clover (*Trifolium subterraneum* L.) resulted in considerably greater plant biomass synthesis than with either microbial component taken singly.

8.2.3 Environmental Considerations

The statement by Allen (1991) that "mycorrhizae represent one of the least understood, most widespread, and most important biological symbioses on earth" cannot be more appreciated than when application of our understanding of this fungal-plant interaction to ecosystem problems is considered. Enlightenment about the status of successful exploiting of mycorrhizal associations is gained by a consideration of the current applications of mycorrhizal interaction management (development of functional plant communities on disturbed sites — land reclamation, and inoculation of nursery stock for field planting) and an interesting but as yet little-explored option (development of genetically engineered fungal associates for protection of host plants from chemical contaminants in their soil).

Although commonly considered from an industrially polluted soil viewpoint, reclamation management to improve soil quality is being more frequently related to "tired" agricultural soil systems (i.e., those cultivated in some cases for several centuries by methods designed to maximize crop yields, at times at the sacrifice of maintaining soil quality.) Traditionally, this soil management includes procedures that encourage development of a soil structure conducive to aboveground plant community development (see Tate, 1987, for a review of this topic) and establishment of stable populations of organisms involved in biogeochemical cycling (see Tate, 1985). Due to their ability to improve longevity and productivity of aboveground plant communities, mycorrhizal associations are a critical component in soil reclamation management of even these "tired" agricultural soils.

Plant community gains from management of mycorrhizal associations for soil quality improvement are derived from both the soil structural enhancements resulting from the fungal contributions to soil aggregate formation and the improved availability of essential plant nutrients. Rapid and somewhat long-lasting benefits are accrued from amendment of degraded soil with a variety of organic matter sources and fertilizers (e.g., see Fresquez et al., 1990; Stroo and Jencks, 1985; Tester, 1990; Wong and Ho, 1991), but for minimization of anthropogenic intervention, a stable soil microbe-plant interactive community must be developed. As was indicated above, mycorrhizal associations are developed best under stressed conditions — especially phosphate stress. Examples of additional limitations common to degraded soils that encourage mycorrhizal symbioses development include fertilizer application practice and cropping system (Ellis et al., 1992), soil erosion [i.e., loss of fungal propagules with surface soil losses (Day et al., 1987)], vegetation (Newman et al., 1981), and soil organic matter (Harvey et al., 1981). Hence, for rapid and enduring development of a reclaimed ecosystem, site management should include insurance of inclusion of fungal propagules in the reclamation target soil, implementation of plant nutrient management procedures that will not prevent mycorrhizal development, and where possible, utilization of inoculated transplants for establishment of aboveground plant communities.

Questions associated with inoculation of seedlings with mycorrhizal fungal propagules relate not to the benefit of such associations to plant survival and productivity — these have long been assumed to be major pluses — but more properly toward optimization of the inoculation processes and selection of the best fungal strains for the symbioses. For example, Marx et al. (1978) documented stimulation of loblolly pine seedlings by a variety of ectomycorrhizal fungi, whereas more recently, Theodorou and Bowen (1987) evaluated the selectivity of basidiospore germination and selection by pine seedlings and development of an effective method for inoculum conservation in the absence of host plants. Duponnois and Garbaye (1991) investigated the utilization of mycorrhization helper bacteria in establishing ectomycorrhizal fungal associations in forest nurseries. The unquestionable utility of such inoculation is exemplified by studies of Lindemann et al. (1984), where mycorrhizal interactions among other soil microbial activities were evaluated in mine spoil reclamation, and of Wilson et al. (1991), where mycorrhizae were utilized to reclaim portions of a dewatered flue-gas desulfurization sludge pond.

Future considerations of the utility of exploiting mycorrhizal associations for soil reclamation may also be based on extension of genetic engineering procedures to mycorrhizal fungi and modification of the carbon substrate metabolism capacity of these organisms. As was discussed above, mycorrhizal fungi extend from the growing root into the surrounding soil. The soil particle-associated mycorrhizal fungal hyphae catabolize soil organic carbon components. Clearly, a potential exists to develop mycorrhizal fungi with the capacity to decompose pollutants that may occur in the vicinity of the roots of both plants selected for soil reclamation (e.g., pine trees and associated shrubs) or even crop plants. The objective for the first situation would be to develop fungal associates that provide a benefit to the plant beyond those generally associated with mycorrhizal interactions — that is, removal of potential plant toxicants. In the case of the crop plant, the fungi could serve as a "first line of defense" per se, thereby preventing toxicants that may contaminate food plants from encountering and being accumulated in sensitive root tissue.

8.3 The Mycorrhizosphere

Traditionally, rhizosphere activity and mycorrhizal symbioses have been studied as if they represented separate ecosystem components, but a brief consideration of the preceding discussion of these ecosystem phenomena leads at least to the suspicion that they occur concurrently on many plant roots rather than as distinct entities. Indeed, the latter conditions is most likely the more common situation. In all soil systems a growing plant will develop some sort of microbial-plant interaction termed a *rhizosphere*. As our understanding of the occurrence of mycorrhizal symbioses expands, it is becoming more obvious that the vast majority of higher plants are mycorrhizal. Thus, we cannot conclude that all rhizospheres are mycorrhizospheres, but it is also evident that most of the data collected to date from field studies of rhizosphere populations must have resulted from evaluation of mycorrhizospheres — with the experimenter being guilty of ignorance of the manifestation of fungal mycelium in association with the roots, or of not possessing the requisite training to differentiate mycorrhizal root structures. This conceptual oversight is gradually being corrected.

In evaluating the quantities of exudates and their chemical substituents, quantification of the intensity of mycorrhizal association is necessary. With symbiotic fungal-plant associations, essentially the entire root surface may be covered by several layers of fungal mycelium (ectomycorrhizal associations), or a major portion of the root surface may be free of obvious fungal hyphal development with the fungus deriving its nutrient through intracellular penetration (endomycorrhizal associations). In either case, the heterotrophic fungi consume root exudates plus the organic carbon contained in sloughed cells, thereby altering both the quality and quantity of root exudates. Some components of the exudate may be decreased or totally depleted. A further complexity of carbon cycling in the more commonly occurring root soil complex, the mycorrhizosphere, is that one reservoir of fixed carbon in the soil is overlooked. Plant carbon is converted into the mycorrhizal fungal biomass. This latter carbon pool extends far beyond the rhizosphere, thus serving as a vehicle to translocate carbon from the plant to the soil community in general, following death of the fungal symbiont. (See Fogel, 1988, for further discussion of this topic.) Therefore, it must be observed that the total nature of the microbial populations in the region of the root is altered by the presence of the fungal symbiont. Direct and indirect impacts of this interaction between the fungus and root exudates on the rhizosphere and rhizoplane microbial populations include determination of the species that are active and alteration of the total microbial biomass produced (e.g., Meyer and Linderman, 1986b), as well as the control of plant pathogen interactions. These microbial consequences of fungal symbiont and rhizosphere bacteria interactions reflect back to the plant in that they may result in improved plant growth (Azcon, 1989; Will and Sylvia, 1990)

These limited evaluations of the impact of both rhizosphere populations and mycorrhizal associations with plant growth emphasize the need to expand future root community studies to include all levels of trophic interactions — be they the result of symbiotic fungi, nitrogen-fixation symbioses (*Rhizobium*-legume or actinorrhizal), or general rhizosphere bacterial, fungal, protozoan, or nematode population growth. Each of these entities withdraws nutrients from the common source — plant photosynthate — either directly or indirectly, produce biomass, and alter the productivity or lack thereof of their neighbors. Ultimately, total ecosystem dynamics are controlled by the complex interactions of each member of the soil community.

8.4 Conclusions

The rhizosphere is most certainly an oasis of biological activity within the soil ecosystem. This ecosystem is represented by a diversity of microbial populations, the complete range of plant-microbe and microbe-microbe interactions, and the relative inclusiveness of all essential biogeochemical processes for total ecosystem development. An overall view of the processes occurring in the rhizosphere/mycorrhizosphere may be summarized by three words — interactions, productivity, and protection. That is, the rhizosphere is populated by a vast variety of life forms each of which is affected both positively and negatively by all other creatures existing in the system; the productivity of the microbial community as well as the aboveground community is interlocked with the viability and stability of the microbial community (i.e., neither the plant community or the microbial components can function optimally singly); and last, the microbial community confers a resistance to the plant component of the system by competing with and in many cases destroying potentially destructive plant pathogens.

References

Alexander, M. 1977. Introduction to Soil Microbiology. John Wiley & Sons, NY. 467 pp.

Allen, E. B., and M. F. Allen. 1986. Water relations of xeric grasses in the field: Interactions of mycorrhizae and competition. New Phytol. 104:559–571.

Allen, M. F. 1991. The Ecology of Mycorrhizae. Cambridge University Press, NY. 184 pp.

Amellal, N., G. Burtin, F. Bartoli, and T. Heulin. 1998. Colonization of wheat roots by an exopolysaccharide-producing *Pantoea agglomerans* strain and its effect on rhizosphere soil aggregation. Appl. Environ. Microbiol. 64:3740–3747.

Andrade, G., K. L. Mihara, R. G. Linderman, and G. J. Bethlenfalvay. 1998. Soil aggregation status and rhizobacteria in the mycorrhizosphere. Plant Soil 202:89–96.

Astrom, B., and B. Gerhardson. 1988. Differential reactions of wheat and pea genotypes to root inoculation with growth-affecting rhizosphere bacteria. Plant Soil 109:263–270.

Azcon, R. 1989. Selective interaction between free-living rhizosphere bacteria and vesicular-arbuscular mycorrhizal fungi. Soil Biol. Biochem. 21:639–644.

Bahme, J. B., M. N. Schroth, S. D. Van Gundy, A. R. Weinhold, and D. M. Tolentino. 1988. Effect of inocula delivery systems on rhizobacterial colonization of underground organs of potato. Phytopathology 78:534–542.

Bagyaraj, D. J. 1991. Ecology of vesicular-arbuscular mycorrhizae. Pp. 3–34. *In* D. K. Arora, B. Rai, K. G. Mukerji, and G. R. Knudsen (eds.), Handbook of Applied Mycology. Volume 1. Soil and Plants. Dekker, NY.

Barber, D. A., and J. M. Lynch. 1977. Microbial growth in the rhizosphere. Soil Biol. Biochem. 9:305–308.

Barber, D. A., and J. K. Martin. 1976. The release of organic substances by cereal roots into soil. New Phytol. 76:69–80.

Bar-Ness, E., Y. Chen, Y. Hadar, H. Marschner, and V. Roemheld. 1991. Siderophores of *Pseudomonas putida* as an iron source for dicot and monocot plants. Plant Soil 130:231–241.

Bashan, Y. 1986a. Enhancement of wheat root colonization and plant development by *Azospirillum brasilense* following temporary depression of rhizosphere microflora. Appl. Environ. Microbiol. 51:1067–1071.

Bashan, Y. 1986b. Significance of timing and level of inoculation with rhizosphere bacteria of wheat plants. Soil Biol. Biochem. 18:297–302.

Bashan, Y. 1986c. Alginate beads as synthetic inoculant carriers for slow release of bacteria that affect plant growth. Appl. Environ. Microbiol. 51:1089–1098.

Bashan, Y., and H. Levanony. 1987. Horizontal and vertical movement of *Azospirillum brasilense* CD in the soil and along the rhizosphere of whet and weeds in controlled and field environments. J. Gen. Microbiol. 133:3473–3480.

Bashan, Y., and H. Levanony. 1989. Wheat root tips as a vector for passive vertical transfer of *Azospirillum brasilense* CD. J. Gen. Microbiol. 135:2899–2908.

Bashan, Y., H. Levanony, and O. Ziv-Vecht. 1987. The fate of field-inoculated *Azospirillum brasilense* CD in wheat rhizosphere during the growing season. Can. J. Microbiol. 33:1074–1079.

Bechard, G., and Y. Piche. 1989. Fungal growth stimulation by CO_2 and root exudates in vesicular-arbuscular mycorrhizal symbiosis. Appl. Environ. Microbiol. 55:2320–2325.

Biondini, M., D. A. Klein, and E. F. Redente. 1988. Carbon and nitrogen losses through root exudation by *Agropyron cristatum*, *A. Smithii*, and *Bouteloua gracilis*. Soil Biol. Biochem. 20:477–482.

Bolan, N. S. J. Elliott, P. E. H. Gregg, and S. Weil. 1997. Enhanced dissolution of phosphate rocks in the rhizosphere. Biol. Fertil. Soils 24:169–174.

Bossier, P., and W. Verstraete. 1986. Ecology of *Arthrobacter* JG-9-detectable hydoxamate siderophores in soils. Soil Biol. Biochem. 18:487–492.

Bowen, G. D. 1969. Nutrient status effects on loss of amides and amino acids from pine roots. Plant Soil 30:139–142.

Bowen, G. D., and G. Theodorou. 1979. Interactions between bacteria and ectomycorrhizal fungi. Soil Biol. Biochem. 11:119–126.

Camel, S. B., M. G. Reyes-Solis, R. Ferrera-Cerrato, R. L. Franson, M. S. Brown, and G. J. Bethlenfalvay. 1991. Growth of vesicular-arbuscular mycorrhizal mycelium through bulk soil. Soil Sci. Soc. Am. J. 55:389–393.

Carrillo, G. C., G. Del Rosario, and M. Vazquez. 1992. Comparative study of siderophore-like activity of *Rhizobium phaseoli* and *Rhizobium fluorescens*. J. Plant Nutr. 15:579–590.

Chanway, C. P., L. M. Nelson, and F. B. Holl. 1988. Cultivar-specific growth promotion of spring wheat by coexistent *Bacillus* spp. Can. J. Microbiol. 34:925–929.

Cheng, W. X. 1996. Is available carbon limiting microbial respiration in the rhizosphere. Soil Biol. Biochem. 28:1283–1288.

Chiariello, N., J. C. Hickman, and H. A. Mooney. 1982. Endomycorrhizal role for interspecific transfer of phosphorus in a community of annual plants. Science (Washington, DC) 217:941–943.

Cline, G. R., P. E. Powell, P. J. Szaniszlo, and C. P. P. Reid. 1982. Comparison of the abilities of hydroxamic, synthetic, and other natural organic acids to chelate iron and other ions in nutrient solution. Soil Sci. Soc. Am. J. 46:1158–1164.

Crowley, D. E., Y. C. Want, C. P. P. Reid, and P. J. Szaniszlo. 1991. Mechanisms of iron acquisition from siderophores by microorganisms and plants. Plant Soil 130:179–198.

Dahlman, R. C., and C. L. Kucera. 1965. Root productivity and turnover in native prairie. Ecology 46:84–89.

Davenport, J. R., and R. L. Thomas. 1988. Carbon partitioning and rhizodeposition in corn and bromegrass. Can. J. Soil Sci. 68:693–701.

Day, L. D., D. M. Sylvia, and M. E. Collins. 1987. Interactions among vesicular-arbuscular mycorrhizae, soil, and landscape position. Soil Sci. Soc. Am. J. 51:636–639.

De Freitas, J. R., and J. J. Germida. 1991. *Pseudomonas cepacia* and *Pseudomonas putida* as winter wheat inoculants for biocontrol of *Rhizoctonia solani*. Can. J. Microbiol. 37:780–784.

Derylo, M., and A. Skorupska. 1992. Rhizobial siderophore as an iron source for clover. Physiol. Plant. 85:549–553.

Dighton, J. 1983. Phosphatase production by mycorrhizal fungi. Plant Soil 71:455–462.

Dormaar, J. R., and D. R. Sauerbeck. 1983. Seasonal effects of photoassimilated carbon-14 in the root system of blue grama and associated soil organic matter. Soil Biol. Biochem. 15:475–479.

Duff, R. B., D. M. Webley, and R. O. Scott. 1963. Solubolization of minerals and related minerals by 2-ketogluconic-acid producing bacteria. Soil Sci. 95:105–114.

Duponnois, R., and J. Garbaye. 1991. Effect of dual inoculation of Douglas fir with the ectomycorrhizal fungus *Laccaria laccata* and mycorrhization helper bacteria (MHB) in two bare-root forest nurseries. Plant Soil 138:169–176.

Eason, W. R., E. I. Newman, and P. N. Chuba. 1991. Specificity of interplant cycling of phosphorus: The role of mycorrhizas. Plant Soil 137:267–274.

Elias, K. S., and G. R. Safir. 1987. Hyphal elongation of *Glomus fasciculatus* in response to root exudates. Appl. Environ. Microbiol. 53:1928–1933.

Ellis, J. R., W. Roder, and S. C. Mason. 1992. Grain-sorghum-soybean rotation and fertilization influence on vesicular-arbuscular mycorrhizal fungi. Soil Sci. Soc. Am. J. 56:789–794.

Fallik, E., Y. Okon, and M. Fisher. 1988. Growth response of maize roots to *Azospirillum* inoculation: Effect of soil organic matter content, number of rhizosphere bacteria and timing of inoculation. Soil Biol. Biochem. 10:45–50.

Federspiel, A., R. Schuler, and K. Haselwandter. 1991. Effect of pH, L-ornithine and L-proline on the hydroxamate siderophore production by *Hymenoscyphus ericae*, a typical ericoid mycorrhizal fungus. Plant Soil 130:259–261.

Fekete, F. A., J. T. Spence, and T. Emery. 1983. Siderophores produced by nitrogen-fixing *Azotobacter vinelandii* OP in iron-limited continuous culture. Appl. Environ. Microbiol. 46:1297–1300.

Fogel, R. 1988. Interactions among soil biota in coniferous ecosystems. Agric. Ecosystems Environ. 24:68–85.

Fogel, R., and G. Hunt. 1983. Contribution of mycorrhizae and soil fungi to nutrient cycling in a Douglas-fir ecosystem. Can. J. Forest Res. 13:219–232.

Fresquez, P. R., R. E. Francis, and G. L. Dennis. 1990. Sewage sludge effects on soil and plant quality in a degraded, semiarid grassland. J. Environ. Qual. 19:324–329.

Frey, P., P. Freyklett, J. Garbaye, O. Berge., and T. Heulin. 1997. Metabolic and genotypic fingerpriting of fluorescent pseudomonads associated with the Douglas fir *Laccaria bicolor* mycorrhizosphere. Appl. Environ. Microbiol. 63:1852–1860.

Garbaue, J., and G. D. Bowen. 1989. Stimulation of ectomycorrhizal infection of *Pinus radiata* by some microorganisms associated with the mantle of ectomycorrhizas. New Phytol. 112:383–388.

Garland, J. L. 1996a. Analytical approaches to the characterization of samples of microbial communities using patterns of potential C source utilization. Soil Biol. Biochem. 28:213–221.

Garland, J. L. 1996b. Patterns of potential C source utilization by rhizosphere communities. Soil Biol. Biochem. 28:223–230.

Gillespie, A. R., and P. E. Pope. 1990a. Rhizosphere acidification increases phosphorus recovery of black locust: I. Induced acidification and soil response. Soil Sci. Soc. Am. J. 54:533–537.

Gillespie, A. R., and P. E. Pope. 1990b. Rhizosphere acidification increases phosphorus recovery of black locust: II. Model predictions and measured recovery. Soil Sci. Soc. Am. J. 54:538–541.

Grayston, S. J., S. Q. Want, C. D. Campbell, and A. C. Edwards. 1998. Selective influence of plant species on microbial diversity in the rhizosphere. Soil Biol. Biochem. 30:369–378.

Griffin, G. J., M. G. Hale, and F. J. Shay. 1977. Nature and quantity of sloughed organic matter produced by roots of axenic peanut plants. Soil Biol. Biochem. 8:29–32.

Gupta, R. K. 1991. Drought response in fungi and mycorrhizal plants. Pp. 55–75. *In* D. K. Arora, B. Rai, K. G. Mukerji, and G. R. Knudsen (eds.), Handbook of Applied Mycology. Volume 1. Soil and Plants. Dekker, NY.

Hamdan, H., D. M. Weller, and L. S. Thomashow. 1991. Relative importance of fluorescent siderophores and other factors in biological control of *Gaeumannomyces graminis* var. *tritici* by *Pseudomonas fluorescens* 2-79 and M4-80R. Appl. Environ. Microbiol. 57:3270–3277.

Hamel, C., C. Nesser, U. Barrantes-Cartn, and D. L. Smith. 1991. Endomycorrhizal fungal species mediate ^{15}N transfer from soybean to maize in non-fumigated soil. Plant Soil 138:41–47.

Harris, J. M., J. A. Lucas, M. R. Davey, G. Lethbridge, and K. A. Powell. 1989. Establishment of *Azospirillum* inoculant in the rhizosphere of winter wheat. Soil Biol. Biochem. 21:59–64.

Hartel, P. G., J. W. Billingsley, and J. W. Williamson. 1989. Styrofoam cup-membrane assembly for studying microorganism-root interactions. Appl. Environ. Microbiol. 55:1291–1294.

Harvey, A. E., M. F. Jurgensen, and M. J. Larsen. 1981. Organic reserves: Importance to ectomycorrhizae in forest soils of Western Montana. Forest Sci. 27:442–445.

Hinsinger, P. 1997. Dissolution of phosphate rock in the rhizosphere of five plant species grown in an acid, P-fixing mineral substrate. Geoderma 75:231–249.

Hodge, A., E. Paterson, S. J. Grayston, C. D. Campbell, B. G. Ord, and K. Killham. 1998. Characterisation and microbial utilisation of exudate material from the rhizosphere of *Lolium perenne* grown under CO_2 enrichment. Soil Biol. Biochem. 30:1033–1043.

Hoefte, M., K. Y. Seong, E. Jurkevitch, and W. Verstraete. 1991. Pyoverdin production by the plant growth beneficial *Pseudomonas* strain 7NSK2: Ecological significance in soil. Plant Soil 130:249–257.

Holl, F. B., and C. P. Chanway. 1992. Rhizosphere colonization and seedling growth promotion of lodgepole pine by *Bacillus polymyxa*. Can. J. Microbiol. 38:303–308.

Holl, F. B., C. P. Chanway, R. Turkington, and R. A. Radley. 1988. Response of crested wheatgrass (*Agropyron cristatum* L.), perennial ryegrass (*Lolium perenne*) and white clover (*Trifolium repens* L.) to inoculation with *Bacillus polymyxa*. Soil Biol. Biochem. 20:19–24.

Ingram, R. R., D. A. Klein, and M. J. Trlica. 1985. Responses of microbial components of the rhizosphere to plant management strategies in a semiarid rangeland. Plant Soil 85:65–76.

Jalali, B. L., and I. Jalali. 1991. Mycorrhizae in Plant Disease Control. Pp. 131–154. *In* D. K. Arora, B. Rai, K. G. Mukerji, and G. R. Knudsen (eds.), Handbook of Applied Mycology. Volume 1. Soil and Plants. Dekker, NY.

Jurkevitch, E., Y. Haear, and Y. Chen. 1992. Differential siderophore utilization and iron uptake by soil and rhizosphere bacteria. Appl. Environ. Microbiol. 58:119–124.

Kennedy, A. C., L. F. Elliott, F. L. Young, and C. L. Douglas. 1991. Rhizobacteria suppressive to the weed downy brome. Soil Sci. Soc. Am. J. 55:722–727.

Klein, D. A., B. A. Frederick, M. Biondini, and M. J. Trlica. 1988. Rhizosphere microorganism effects on soluble amino acids, sugars and organic acids in the root zone of *Agropyron cristatum*, *A. Smithii* and *Bouteloua gracilis*. Plant Soil 110:19–25.

Kloepper, J. W., J. Leong, M. Teintze, and M. N. Schroth. 1980a. *Pseudomonas* siderophores: A mechanism explaining disease-suppressive soils. Curr. Microbiol. 4:317–320.

Kloepper, J. W., J. Leong, M. Teintze, and M. N. Schroth. 1980b. Enhanced plant growth by siderophores produced by plant growth-promoting rhizobacteria. Nature (London) 286:885–886.

Klubek, B. P., L. L. Hendrickson, R. M. Zablotowicz, J. E. Skwara, E. C. Varsa, S. Smith, T. G. Islieb, J. Maya, M. Valdes, F. B. Dazzo, R. L. Todd, and D. D. Walgenback. 1988. Competitiveness of selected *Bradyrhizobium japonicum* strains in Midwestern USA soils. Soil Sci. Soc. Am. J. 52:662–666.

Knight, W. G., M. F. Allen, J. J. Jurinak, and L. M. Dudley. 1989. Elevated carbon dioxide and solution phosphorus in soil with vesicular-arbuscular mycorrhizal western wheatgrass. Soil Sci. Soc. Am. J. 53:1075–1082.

Krishnamurti, G. S. R., G. Cieslinski, P. M. Huang, and K. C. J. Vanrees. 1997. Kinetics of cadmium release from soils as influenced by organic acids — implication in cadmium availability. J. Environ. Qual. 26:271–277.

Kucey, R. M. N., and E. A. Paul 1982. Carbon flow, photosynthesis, and N_2 fixation in mycorrhizal and nodulated faba beans (*Vicie faba* L.). Soil Biol. Biochem. 14:407–412.

Latour, X., T. S. Corberand, G. Laguerre, F. Allard, and P. Lemanceau. 1996. The composition of fluorescent pseudomonad populations associated with roots is influenced by plant and soil type. Appl. Environ. Microbiol. 62:2449–2456.

Leyval, C., and J. Berthelin. 1991. Weathering of a mica by roots and rhizospheric microorganisms of pine. Soil Sci. Soc. Am. J. 55:1009–1016.

Lindemann, W. C., D. L. Lindsey, and P. R. Fresquez. 1984. Amendment of mine spoil to increase the number and activity of microorganisms. Soil Sci. Soc. Am. J. 48:574–578.

Maloney, P. E., A. H. C. Vanbruggen, and S. Hu. 1997. Bacterial community structure in relation to the carbon environments in lettuce and tomato rhizospheres and in bulk soils. Microb. Ecol. 34:109–117.

Martens, R. 1982. Apparatus to study the quantitative relationships between root exudates and microbial populations in the rhizosphere. Soil Biol. Biochem. 14:315–317.

Martin, J. K. 1977. Factors influencing the loss of organic carbon from wheat roots. Soil Biol. Biochem. 9:1–7

Marx, D. H. 1972. Ectomycorrhizae as biological deterrents to pathogenic root infection. Ann. Rev. Phytopathol. 10:429–454.

Marx, D. W., W. G. Morris, and J. G. Meral. 1978. Growth and ectomycorrhizal development of loblolly pine seedlings on fumigated and nonfumigated nursery soils infested with different fungal symbionts. Forest Sci. 24:193–203.

Materon, L. A., and C. Hagedorn. 1982. Competitiveness of *Rhizobium trifolii* strains associated with red clover (*Trifolium pratense* L.) in Mississippi soils. Appl. Environ. Microbiol. 44:1096–1101.

McClaugherty, C. A., J. D. Aber, and J. M. Melillo. 1984. Decomposition dynamics of fine roots in forested ecosystem. Oikos 42:378–386.

Meharg, A. A., and K. Killham. 1989. Distribution of assimilated carbon within the plant and rhizosphere of *Lolium perenne*: Influence of temperature. Soil Biol. Biochem. 21:487–490.

Merckx, R., A. Dijkstra, A. den Hertog, and J. A. van Veen. 1987. Production of root-derived material and associated microbial growth in soil at different nutrient levels. Biol. Fert. Soils 5:126–132.

Merckx, R., J. A. van Ginkel, J. Sinnaeve, and A. Cremers. 1986. Plant induced changes in the rhizosphere of maize and wheat. I. Production and turnover of root-derived material in the rhizosphere of maize and wheat. Plant Soil 96:85–94.

Meyer, J.-M., D. Hohnadel, and F. Hallé. 1989. Cepabactin from *Pseudomonas cepacia*, a new type of siderophore. J. Gen. Microbiol. 135:1479–1487.

Meyer, J. R., and R. G. Linderman. 1986a. Response of subterranean clover to dual inoculation with vesicular-arbuscular mycorrhizal fungi and a plant growth-promoting bacterium. Soil Biol. Biochem. 18:185–190.

Meyer, J. R., and R. G. Linderman. 1986b. Selective influence on populations of rhizosphere or rhizoplane bacteria and actinomycetes by mycorrhizas formed by *Glomus fasiculatum*. Soil Biol. Biochem. 18:191–196.

Moawad, H., and B. B. Bohlool. 1984. Competition among *Rhizobium* spp. for nodulation of *Leucaena leucocephala* in two tropical soils. Appl. Environ. Microbiol. 48:5–9.

Mukerji, S., K. G. Mukerji, and D. K. Arora. 1991. Ectomycorrhizae. Pp. 187–215. *In* D. K. Arora, B. Rai, K. G. Mukerji, and G. R. Knudsen (eds.), Handbook of Applied Mycology. Volume 1. Soil and Plants. Dekker, NY.

Nair, M. G., G. R. Safir, and J. O. Siqueira. 1991. Isolation and identification of vesicular-arbuscular mycorrhiza-stimulatory compounds from clover (*Trifolium repens*) roots. Appl. Environ. Microbiol. 57:434–439.

Newman, E. I. 1985. The rhizosphere: Carbon sources and microbial populations. Pp. 107–121. *In* A. E. Fitter (ed.), Ecological Interactions in Soil: Plants, Microbes, and Animals. Blackwell Scientific Publications, Boston.

Newman, E. I., A. J. Heap, and R. A. Lawley. 1981. Abundance of mycorrhizas and root surface microorganisms of *Plantago lancelata* in relation to soil and vegetation. A multivariate approach. New Phytol. 89:95–108.

Nietko, K. F., and W. T. Frankenberg. Jr. 1989. Biosynthesis of cytokinins in soil. Soil Sci. Soc. Am. J. 53:735–740.

Norton, J. M., and M. Firestone. 1991. Metabolic status of bacteria and fungi in the rhizosphere of ponderosa pine seedlings. Appl. Environ. Microbiol. 57:1161–1167.

Parke, J. L., R. Moen, A. D. Rovira, and G. D. Bowen. 1986. Soil water flow affects the rhizosphere distribution of seed-borne biological control agent. Soil Biol. Biochem. 18:583–588.

Pasqualini, S., F. Panara, and M. Antonielli. 1992. Acid phosphatase activity in *Pinus pinea — Tuber albidum* ectomycorrhizal association. Can. J. Bot. 70:1377–1383.

Pesmark, M., T. Frejd, and B. Mattiasson. 1990. Purification, characterization, and structure of pseudobactin 589 A, a siderophore from a plant growth promoting *Pseudomonas*. Biochemistry 29:7348–7356.

Qian, J. H., and J. W. Doran. 1996. Available carbon released from crop roots during growth as determined by carbon-13 natural abundance. Soil Sci. Soc. Am. J. 60:828–831.

Rattray, E. A. S., E. Paterson, and K. Killham. 1995. Characterisation of the dynamics of C-partitioning within *Lolium perenne* and to the rhizosphere microibal biomass using C-14 pulse chase. Biol. Fertil. Soils 19:280–286.

Read, D. J. 1984. The structure and function of the vegetative mycelium of mycorrhizal roots. Pp. 215–240. *In* D. H. Jennings and A. D. M. Rayner (eds.), The Ecology and Physiology of the Fungal Mycelium. Cambridge University Press, NY.

Read, D. J., R. Francis, and R. D. Finlay. 1989. Mycorrhizal mycelia and nutrient cycling in plant communities. PP. 193–217. *In* A. E. Fitter (ed.), Ecological Interactions in Soil: Plants, Microbe, and Animals. Blackwell Scientific Publications, Boston.

Rillig, M. C., K. M. Scow, J. N. Klironomos, and M. F. Allen. 1997. Microbial carbon-substrate utilization in the rhizosphere of *Gutierrezia sarothrae* grown in elevated atmospheric carbon dioxide. Soil Biol. Biochem. 29:1387–1394.

Robert, F. M., and E. L. Schmidt. 1983. Population changes and persistence of *Rhizobium phaseoli* in soil and rhizospheres. Appl. Environ. Microbiol. 45:550–556.

Ruark, G. A., F. C. Thornton, A. E. Tiarks, B. G. Lockaby, A. H. Chappelka, and R. S. Meldahl. 1991. Exposing loblolly pine seedlings to acid precipitation and ozone: Effects on soil rhizosphere chemistry. J. Environ. Qual. 20:828–832.

Smith, G. B., and A. G. Wollum II. 1989. Nodulation of *Glycine max* by six *Bradyrhizobium japonicum* strains with different competitive abilities. Appl. Environ. Microbiol. 55:1957–1962.

Spalding, B., J. M. Duxbury, and E. L. Stone. 1975. Lycopodium fairy rings: Effect on soil respiration and enzymatic activities. Soil Sci. Soc. Am. Proc. 39:65–70.

Stroo, H. F., and E. M. Jencks. 1985. Effect of sewage sludge on microbial activity in an old, abandoned minesoil. J. Environ. Qual. 14:301–304.

Tate, R. L., III. 1985. Microorganisms, ecosystem disturbance and soil formation processes. Pp. 1–33. *In* R. L. Tate III and D. A. Klein (eds.), Soil Reclamation Processes. Microbiological Analyses and Application. Dekker, NY.

Tate, R. L., III. 1987. Soil Organic Matter. Biological and Ecological Effects. John Wiley & Sons. NY. 291 pp.

Tate, R. L., III, R. W. Parmelee, J. G. Ehrenfeld, and L. O'Reilly. 1991. Enzymatic and microbial interactions in response to pitch pine root growth. Soil Sci. Soc. Am. J. 55:998–1440.

Tester, C. F. 1990. Organic amendment effects on physical and chemical properties of a sandy soil. Soil Sci. Soc. Am. J.54:827–831.

Theodorou, C., and G. D. Bowen. 1987. Germination of basidiospores of mycorrhizal fungi in the rhizosphere of *Pinus radiata* D.Don. New Phytol. 106:217–224.

Thomashow, L. S., and D. M. Weller. 1990. Role of antibiotics and siderophores in biocontrol of take-all disease of wheat. Plant Soil 129:93–99.

Thomashow, L. S., D. M. Weller, R. B. Bonsall, and L. S. Pierson III. 1990. Production of the antibiotic phenazine-1-carboxylic acid by flurosecent *Pseudomonas* species in the rhizosphere of wheat. Appl. Environ. Microbiol. 56:908–912.

Tien, T. M., M. H. Gaskins, and D. H. Hubbell. 1979. Plant growth substances produced by *Azospirillum brasilense* and their effect on the growth of pearl millet (*Pennisetum americanum* L.). Appl. Environ. Microbiol. 37:1016–1024.

Tisdall, J. M., and J. M. Oades. 1982. Organic matter and water-stable aggregates in soils. J. Soil Sci. 33:141–163.

Turner, S. M., and E. I. Newman. 1984. Growth of bacteria on roots of grasses: Influence of mineral nutrient supply and interactions between species. J. Gen. Microbiol. 130:505–512.

Turner, S. M., E. I. Newman, and R. Campbell. 1985. Microbial population of ryegrass (*Lolium perenne*) root surfaces: Influence of nitrogen and phosphorus supply. Soil Biol. Biochem. 17:711–716.

Utkhede, R. S., and J. E. Rahe. 1980. Biological control of onion white rot. Soil Biol. Biochem. 12:101–104.

Van Loon, L. C., P. A. H. M. Bakker, and C. M. J. Pieterse. 1998. Systemic resistance induced by rhizosphere bacteria. Annu. Rev. Phytopath. 36:453–483.

van Rensburg, H. J., and B. W. Strijdom. 1985. Effectiveness of *Rhizobium* strains used in inoculants after their introduction into soil. Appl. Environ. Microbiol. 49:127–131.

Vrany, J., V. Vancura, and M. Stanek. 1981. Control of microorganisms in the rhizosphere of wheat by inoculation of seeds with *Pseudomonas putida* and by foliar application of urea. Folia Microbiol. 26:45–51.

Whitford, W. G., K. Stinnett, and J. Anderson. 1988. Decomposition of roots in a Chihuahuan desert ecosystem. Oecologia (Berlin) 75:8–11.

Will, M. E., and D. M. Silvia. 1990. Interaction of rhizosphere bacteria, fertilizer, and vesicular-arbuscular mycorrhizal fungi with sea oats. Appl. Environ. Microbiol. 56:2073–2079.

Wilson, G. W. T., B. A. D. Hetrick, and A. P. Schwab. 1991. Reclamation effects on mycorrhizae and productive capacity of flue gas desulfurization sludge. J. Environ. Qual 20:777–783.

Wong, J. W. C., and G. E. Ho. 1991. Effects of gypsum and sewage sludge amendment on physical properties of fine bauxite refining residue. Soil Sci. 152:326–332.

Chapter 9

Introduction to the Biogeochemical Cycles

The facts are clear. All living cells build their organic mass from carbon, hydrogen, oxygen, nitrogen, phosphorus, and sulfur. These elements are found in soil in a variety of organic and inorganic forms — some of which are usable (available to) by the microbes, others of which are not. Furthermore, there is a finite quantity of each element in the earth's systems. Thus, processes must exist to transform the various nutrients into appropriate life-available forms. Were these materials always usable by organisms once transformed from their unavailable states to forms that can be metabolized the situation would be simple, but it isn't. The very act of synthesizing living biomass makes the nutrients unavailable to other organisms. Thus, in order for the nutrients to again become biologically available after cell death, the organic substances must be decomposed by soil microbes. The process in this case is mineralization. Therefore, it is noted that all elements constituting microbial cells are cycled between organic or inorganic; water soluble or insoluble; gaseous, liquid, or solid forms in ecosystems. The rate of conversion from life-unavailable to -available states commonly determines the amount of biological activity in a given ecosystem. That is, it is the rate-limiting step for sustenance and stability of the soil biological community. These life-supporting, nutrient-generating, ecosystem-modifying processes constitute the various biogeochemical cycles; for instance, nitrogen, phosphorus, and sulfur cycles. Biogeochemical cycles are defined as the processes involved with cycling of a chemical element through its various biological, chemical, and geological forms in air, water, and soil systems. The various reactions are biologically and chemically catalyzed.

Conceptual models of biogeochemical cycles are a tool for evaluating and testing our understanding of nutrient movement and distribution in an ecosystem. The level of complexity of individual models necessarily varies with the objectives of the model developer. Rarely are all forms of the nutrient represented in a specific model. Rather, those processes most important to understanding a particular system function are detailed. For example, a model of carbon cycling can be as simple as a depiction of the size of the carbon dioxide and organic carbon pools in soil and assessment of the rates of movement of carbon between the two pools. Alternatively, movement of carbon among mineral (e.g., carbon dioxide and limestone) and soil organic matter, microbial biomass, and plant biomass pools may be evaluated. Should specific organic compounds be of interest, the model may be concentrated on chemical compounds (e.g., carbon dioxide, cellulose, lignin, and proteins). Each of these variations of a carbon cycle can answer specific questions about the nature of the cycling of carbon in an ecosystem.

The variability in the degree of complexity associated with consideration of biogeochemical cycles occurring in soil reflects the vast diversity of chemical and physical states of the elements (e.g., carbon, nitrogen, sulfur, phosphorus) that can be considered. Carbon is found as carbon dioxide in soil air, bicarbonate in soil water, and carbonate in soil minerals plus the variety of organic chemicals contained in living cells, soil humus, and any anthropogenic or xenobiotic compounds in the system. If this diversity occurs with an element that has only one valence state, consider the possible chemical forms of multivalent elements in soil. For example, valence states of nitrogen in compounds of importance to soil life include -3 (ammonium), $+1$ (nitrous oxide), $+3$ (nitrite), and $+5$ (nitrate). Comparable organic examples for these valence states of nitrogen can also be found in soil. (See Chapter 11 for further details on the various forms of nitrogen in soil.) Furthermore, the many chemical states present in a microbial habitat vary between ecosystems. Not all forms of a particular element are found in each soil. This situation is exemplified by the occurrence of carbonates (limestone) in neutral or alkaline soils. In acidic soils, the carbonate is converted to carbon dioxide. Also, a complete cycle of a particular nutrient may occur within an ecosystem of limited dimensions (e.g., a particular soil horizon of interest) or alternatively, integrity of the nutrient cycles may be reliant upon inputs from adjacent portions of a more extensively defined ecosystem (e.g., a soil community including aboveground populations plus atmospheric inputs).

It must be stressed that the realm of biogeochemical cycles is much larger than just the transitions of chemical forms necessary for sustenance of plant, animal, and microbial communities and the biological or chemical reactions involved in these processes. Should the sole interest in studying biogeochemical cycles be limited to an evaluation of the chemicals involved and these biological processes, then nutrient cycles are being examined. That is, nutrient cycles entail a limited portion of the subject area encompassed by biogeochemical cycles. A consideration of biogeochemical cycles should involve more of a total ecosystem outlook than may be included in evaluation of nutrient cycles. The definition of biogeochemical cycles includes a consideration of the totality of biological, chemical and physical interactions associated with the movement of nutrients between organic and inorganic pools and the variety of geochemical reservoirs existent within an ecosystem.

This definition of the complete realm of the biogeochemical cycles emphasizes the point that study of these chemical transformations has real implications beyond the world of plant science and general biology. Study of these cyclic transformations of life-sustaining nutrients is also essential for such broad-based disciplines as general ecology or environmental science. For example, consideration of biogeochemical cycles in disturbed environments is essential in optimizing anthropogenic activities for ecosystem sustenance, as well as in development of management plants for restoration or reclamation of damaged systems — that is, development of appropriate ecosystem stewardship procedures. Therefore, the topics are presented in this chapter with the overall objective of providing a basic understanding of biogeochemical cycles through an evaluation the general properties of the cycles and of elucidating the interconnectedness of these processes with the goal of maintaining an appreciation of the importance of these processes to overall ecosystem function and maintenance.

9.1 Conceptual and Mathematical Models of Biogeochemical Cycles

As suggested by the foregoing analysis, a study of biogeochemical cycles requires a means of assessing a wide array of complex interactions within a constantly varying environment. Some details may be studied in isolation, but it must be remembered that the primary objective of the exercise is to develop an understanding of the overall pattern of distribution and movement of nutrients in the ecosystem. A means of accomplishing the objective of developing an overarching framework for study of biogeochemical cycles is to assemble the data into models. These models may be used to elucidate gaps in knowledge and provide hypotheses for future study, as well as developing experimental plans for quantifying nutrient movement within the cycles and assessing the impact of environmental modification on these processes. Initial modeling activity, or perhaps modeling activity at its simplest, involves development of conceptual models. The substances of interest are grouped in boxes (commonly referred to as pools) with arrows between the boxes to designate direction of nutrient flow. For some models, size and shape of the arrows (also termed *vectors*) may be varied to represent major or minor pathways of flow. These models provide a basis for design of experiments that yield the data necessary for development of the more complex and predictive mathematical models.

9.1.1 Development and Utility of Conceptual Models

A primary objective of study of nutrient flow in an ecosystem is to elucidate (1) the chemical forms of a nutrient element (e.g., carbon as carbon dioxide and organic matter; nitrogen as mineral and organic forms; phosphorus in soil minerals, water-soluble, and organic forms) in an ecosystem of interest, (2) the quantity of nutrients retained in these compounds (i.e., the size of the nutrient pool), and (3) the rate of flux between these various nutrient pools. Once the identity of the various participants in biogeochemical cycles (even in lieu of determination of all of the pertinent intermediates), conceptual models may be assembled. These models are most easily presented as a series of boxes representative of the primary nutrient pools, with arrows connecting the boxes to indicate the directional transformation (i.e., nutrient flux) of the cycle components (e.g., see Fig. 9.1 for a depiction of the carbon cycle). As indicated above, to augment the information contained in a conceptual model, the type of arrow connecting the various reservoirs (e.g., dotted, size differential, or boldface) may be varied to suggest differences in the magnitude of the processes depicted. More commonly, a conceptual model is limited to indication of the nutrient reservoir and the direction of nutrient flow. Details such as distinction between relatively stable reservoirs, such as soil minerals or peat deposits, and the more ephemeral cycle components exemplified by simple organic compounds or mineral nitrogen forms may be minimal. That is, organic carbon may be considered as a single pool in the cycle or perhaps two pools, slowly and rapidly metabolized carbon compounds. A more quantitative representation of biogeochemical cycles or the individual reactions constituting them is entailed in extension of conceptual models into mathematical models, as described in Section 9.1.2.

Characteristically, conceptual models of biogeochemical cycles emphasize the distribution of an element in a variety of chemical forms. The elements of interest (i.e., carbon, nitrogen, phosphorus, or sulfur as well as a variety of trace metals) are found to move between a variety of chemical (organic vs. inorganic) and/or oxidation states at predictable rates. Basic chemical components of plant and animal biomass are transformed from biomass-unavailable forms to compounds that are readily incorporated into living cells. For example, the reasonably inert compound dinitrogen (N_2) is reduced by soil microbes to ammonium (nitrogen fixation), which, once it is released by the microbial cell that reduced the dinitrogen for its own use, can be incorporated into plant tissue. With the death of the plant tissue, the soil microbial community may return the nitrogen to the atmospheric pool of dinitrogen. This conceptualization of biogeochemical cycles is plant centered (or perhaps life centered, when microbial and animal biomass are also viewed as major nutrient recipients). Other models may be presented that could be centered on movement of elements and compounds to and from atmospheric pools, soil minerals, humus, or any other primary ecosystem component.

Conceptual models by their very design may support a biased view of biogeochemical cycles in that most representations are limited to designation of the basic chemical and biological forms of the cycle components. To provide a more complete model for soil microbiological or environmental science considerations, designation of the path for movement of the elements of interest through various ecosystem components (i.e., soil minerals, atmospheric

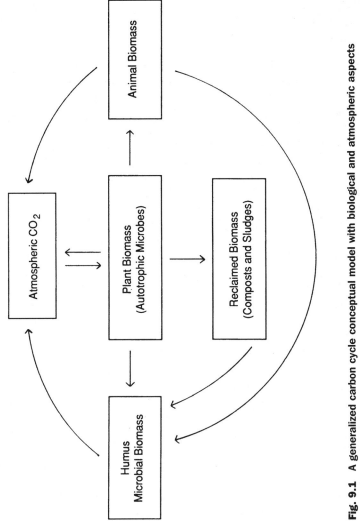

Fig. 9.1 A generalized carbon cycle conceptual model with biological and atmospheric aspects emphasized.

gases, and biotic entities) is useful. A sound ecological approach to study of biogeochemical cycles should include:

- An appreciation of interactions between the compounds transformed, the biological community catalyzing the processes, and the physical environment in which the reactions occur and which is itself modified during the process

- The contribution of spontaneous chemical reactions to the overall nutrient generation rates

The primary product of the physical, chemical, and biological processes constituting a biogeochemical cycle is a stable, sustainable ecosystem. Thus, although conceptual models of biogeochemical cycles are usually presented as "boxes connected by arrows," it must be remembered that the boxes represent "real-world" nutrient reservoirs, such as plant biomass. Furthermore, at least as important as the quantification of chemical form in a cycle is development of an understanding of the magnitude of the vectors (designated as arrows). The importance of a large pool of nutrients (such as atmospheric N_2), which in itself is unavailable to microbes, is determined not by its size but rather by the rate that it is converted into a form useable by microbes (i.e., ammonium). The rate of the transformations is controlled by not only biological processes themselves but also the properties of the environment in which the plant, microbe, or animal is located. Therefore, the occurrence of the biogeochemical cycles is imperative for long-term sustenance of the ecosystem, and the nature of the processes in the cycles is controlled directly by the physical and chemical attributes of the ecosystem.

Thus, collection of data for development of conceptual models should entail consideration of biotic as well as abiotic interactions. For example, mineral nitrogen availability determines productivity and nature of the aboveground community. This nitrogen can be derived through nitrogen fixation (i.e., the total amount of nitrogen available to the plants may increase through inputs from external sources), or the nitrogen contained in biomass may be recycled into nascent biomass through mineralization of deceased biomass and soil organic matter pools (i.e., the total amount of nitrogen contained within the soil system remains reasonably constant). In either of these situations, the productivity of the aboveground community is limited by the capacity of the soil microbial community (nitrogen fixers or nitrogen mineralizers) to provide a pool of nitrate and ammonium ions for incorporation into plant biomass. The activity of this soil microbial community is in turn controlled by the quantities of energy and nutrients provided by the aboveground community in the form of photosynthate, as well as by the soil physical and chemical environment. For example, in a high-moisture, high-clay soil, the limitation of plant productivity by the physical properties of the system will also curtail the soil microbial activity, which is dependent on the plants for their energy source. Also, anoxic conditions may favor loss of fixed nitrogen from the system through biological denitrification reactions, whereas chemical denitrification is accentuated in acidic soils. Furthermore, these life-limiting environmental properties are continually being modified by the physical presence and the biological products of the plant, animal, and microbial communities (see Tate, 1987). For example, improvements in the soil physical environment through enhancement of soil aggregate structure by growth of the microbial community augments plant biomass production, which may further stimulate microbial growth through augmented inputs of fixed carbon.

Therefore, essential steps in development and refinement of conceptual models involves elucidation of the forms of the nutrient element of interest in the system, perhaps tracing the movement of elements between these forms using such isotopes as ^{14}C or ^{15}N, and determination of the physical and chemical properties of the system controlling the relative magnitude of the various chemical pools and the rate of flux between these reservoirs. The process of model building is based primarily on characterization of the system and development of a reasonable understanding of the relationship between the various nutrient reservoirs occurring therein. These studies may entail determination of biological trophic interactions (e.g., Santos et al., 1981; Santos and Whitford, 1981; Santos et al., 1984), determination of the impact of management and climate changes on ecosystem function (e.g., Allen, 1990; Pastor and Post, 1988; Tabatabai and Chae, 1991; Walters et al., 1992), evaluation of impact of properties of ecosystem inputs on their mineralization (e.g., Meentemeyer, 1978; Fresquez et al., 1990), as well as examination of native ecosystem processes (e.g., Parker et al., 1983).

9.1.2 Mathematical Modeling of Biogeochemical Cycles

Commonly, the objectives of an evaluation of nutrient cycling in a soil ecosystem are twofold:

- Developing an understanding of the processes and their limitations in the particular site of interest. This type of experiment may involve development of a more clear understanding of the nuances of nutrient availability in a native forest ecosystem. Alternatively, these types of studies provide the basis for nutrient management for soil reclamation.

- Collecting a data pool that allows extension of the concepts learned at a few sites to a more extensive land area of soil ecosystems in general.

Conceptual and mathematical models aid in achievement of these types of objectives. A conceptual model provides an indication of the "design" of the biogeochemical cycles in an ecosystem. The components of the cycle and direction of flow of nutrients is described, but little is revealed about the size of the various pools or the magnitude of the fluxes between the pools. Frequently, a conceptual model can represent a hypothesis of how the "system works," thereby providing a conceptual basis for the experiment designed to test hypotheses. Extension of the data to other ecosystem usually requires development of mathematical models.

Initial steps for attaining an understanding of nutrient cycling processes include analysis of the quantities of various compounds present within the soil site of interest. Generally, it soon becomes clear from the data that a detailed understanding of the intricacies of the reactions is obscured by site heterogeneity and the shear number of potential soil properties that control the reaction rates and nutrient distribution. Thus, a mathematical approach at data summarization and interpretation is required.

First steps for developing more generally applicable mathematical relationships may stem from application of statistical analysis procedures. A simple model may be yielded through a linear or nonlinear regression analysis of the data. The resultant equations may be reasonably descriptive of the particular site under study, but applicability to other situations is limited. This conclusion stems from the fact that the equations derived are mathematical relationships fitted to specific data. That is, the relationship may be forced to fit a straight line, parabola, or hyperbola with the constants in the equations being representative of the conditions of the limited number of soil sites used in the experiment.

Equations more generally applied to biological, chemical, or physical reaction may be adapted to environmental data. For example, an enzyme-catalyzed reaction may be reasonably easily described by a first order equation or by the Michaelis-Menten relationship (see Chapter 6). As the complexity of the system increases (e.g., interest in the quantities of the reactant consumed within the profile of a conifer forest) and the variety of interactions with soil properties and climatic variations expands, inclusion of known relationships, such as the Q_{10} equation or Ahrenius relationship for estimating temperature impact on reaction rates, may be necessary to increase model utility. Furthermore, in the complex, heterogeneous environment of soil, effective chemical concentrations for a reaction may be defined by properties other than total substrate present. For example, reaction rates may be dependent upon soil parameters such as the rate of dissolution of the substrate (Stucki and Alexander, 1987) or diffusion of the enzyme substrate into the microsite wherein the organisms reside (Scow and Hutson, 1992; Scow and Alexander, 1992). Also, in an ecosystem adapting to a perturbation, the rate of occurrence of a process may be proportional to the growth or death of the microbial populations producing the requisite enzymes [i.e., the reaction may be best described by a Monod equation (see Chapter 4)]. Hence, a variety of mathematical relationships descriptive of not only biological but also chemical and physical interactions are combined to provide a mathematical relationship that at least in part is descriptive of the extent of a natural process and predictive of rate changes with perturbation of the system. Because of the general applicability of the mathematical relationships (e.g., Michaelis-Menton equations provide a reasonable model of the kinetics of the vast majority of enzymatic processes), the biogeochemical models based on them provide reasonable descriptors of process rates in a diversity of situations. It must be remembered that applicability of the mathematical relationship to alternative situations is still limited by the range of data used to validate the model. For example, a model descriptive of nitrogen transformation rates in well-drained agricultural soils would have to be proven to apply to swamp or wetland soils. This situation results from the fact that the mathematical relationships used to describe the soil processes are a mixture of variables and constants. Values for the variables are measured, estimated, or assumed from analysis of actual soil situations. The constants are based on or are commonly derived from these "real-world" studies. Thus, equations, or models, are only applicable within the range of variation of the data used to develop (validate) them. Their applicability to other soils with properties outside of the range of those used to validate the model must be proven. See Tate (1987) for a description of model development and application. A simple mathematical model of soil nitrogen mineralization rates and organic nitrogen pool size, which is based on first order reaction kinetics, is described in Section 9.5.4.

Applicability of mathematical models of varying complexity to biogeochemical cycling is exemplified by the following studies:

- *Assessment and elucidation of soil properties controlling specific portions of a biogeochemical cycle:* For example, study of the impact of distribution of aerobic and anaerobic microsites on denitrification rates (McConnaughey and Bouldin, 1985a, b; McConnaughey et al., 1985), as well as evaluation of nitrogen mineralization (Deans et al., 1986), sulfur and nitrogen mineralization (Ellert and Bettany, 1988), and carbon mineralization (Stroo et al., 1989) in a variety of ecosystems

- *Evaluation of ecosystem response to global changes:* For example, estimation of soil changes due to global warming (Jenkinson et al., 1991)

- *Development of a clear understanding of dynamics of a particular ecosystem:* For example, determination of the variation in decomposer populations and substrates in grasslands (Hunt, 1977)

- *Quantification of specific fluxes between nutrient pools:* For example, evaluation of the importance of nutrients contained in microbial biomass in nutrient cycling (van Veen et al., 1984)

- *Prediction of management changes on ecosystem dynamics:* For example, estimation of the effect of cultivation on carbon, nitrogen, phosphorus, and sulfur cycling in a grassland (Parton et al., 1988), or of erosion on plant phosphorus availability (Jones et al., 1984)

In each of these situations, varying combinations of mathematical relationships descriptive of biological transformations, pH variation, temperature effects on biological processes, adsorption processes, plant composition, and so on were combined to provide a mathematical representation of movement of nutrients between various biogeochemical nutrient pools.

9.2 Specific Conceptual Models of Biogeochemical Cycles and Their Application

Basic conceptual models of the primary nutrient cycles (carbon, nitrogen, sulfur, and phosphorus) are depicted in Figs. 9.1 through 9.4, respectively. These models are designed to emphasize the essential components of the cycles. The elements are transferred between organic and inorganic forms (all cycles) as well as through a variety of oxidation states (nitrogen and sulfur cycles). The oxidation state of nitrogen varies from -3 (ammonium ion) to $+5$ (nitrate ion), whereas in the sulfur cycle the oxidation state of the sulfur atom varies from -2 (sulfide) to $+6$ (sulfate). Similarly, a variety of soil and atmospheric inorganic compounds (e.g., nitrate, ammonium, dinitrogen, elemental sulfur, phosphate rock, sulfur-bearing minerals) and plant, animal, and microbial biomass organic components constitute primary nutrient pools in these models of the biogeochemical cycles. Note that geological materials are of major importance in both the sulfur and phosphorous cycles.

Traditionally, biogeochemical cycles are studied from the view of assessing the relationship of nutrient state to higher plant productivity. This limited purview could lead to an underappreciation of the impact of nutrient cycling on total ecosystem functions and properties. A number of intermediates in the various cycles have considerable impact on the integrity of the terrestrial ecosystem. Positive as well as negative implications are discerned, especially from the viewpoint of societal impact on system function. Perturbations of nutrient cycling could result in overproduction of substances that result in large-scale declines in system quality, altered biomass productivity, or modified terrestrial ecosystem sustainability. A well-known example is the decline in soil organic matter reserves due to soil cultivation (see Tate, 1987; Rosenzweig and Hillel, 1998) as well as total plant biomass productivity (e.g., see Adams et al., 1990; Allen, 1990; Dahlman et al., 1985; Reddy et al., 1989). Nitrous oxide, a key intermediate in denitrification, is also a greenhouse gas (e.g., see Firestone, 1982; Mosier et al., 1991). Nitrate and phosphate are often significant ground- and surface-water contaminants, whereas sulfate and sulfur dioxide are major components of acid rain and acid mine drainage problems.

In contrast to these associations of biogeochemical cycles with negative environmental situations, a number of benefits may also result from management of soils to optimize biogeochemical cycling. For example, soil organic nitrogen and carbon levels are generally enhanced when conservation tillage practices are implemented on soils previously managed with conventional tillage methods (e.g., see Blevins et al., 1984; Cambardella and Elliott, 1992; Doran, 1980). These increases in soil organic matter reserves could result in alteration in rates of accumulation of atmospheric carbon dioxide. Increased storage of carbon in soil organic matter pools moderates global warming due to greenhouse gas production. Furthermore, sulfate reduction to sulfide may

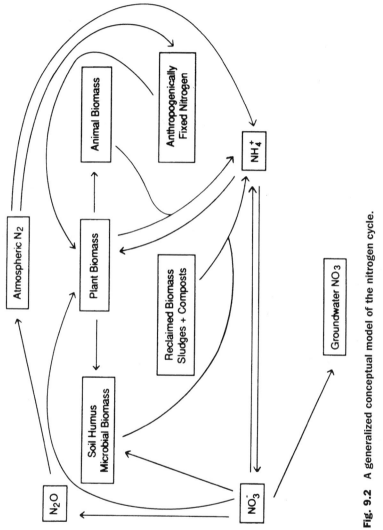

Fig. 9.2 A generalized conceptual model of the nitrogen cycle.

263

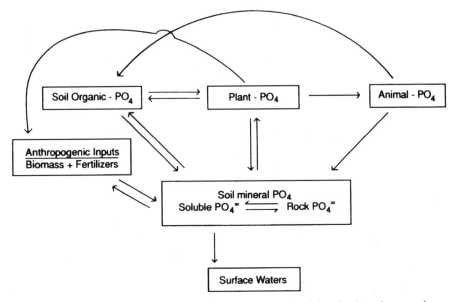

Fig. 9.3 A conceptual model of the soil phosphorus cycle with soil mineral reservoirs as the central component.

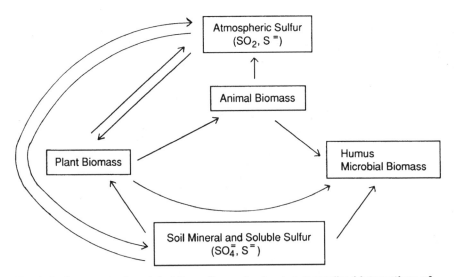

Fig. 9.4 A conceptual model of the sulfur cycle showing generalized interactions of biological resources with soil and atmospheric sources.

be exploited to reduce the difficulties associated with acid mine drainage-affected soils (see Chapter 15), and denitrification may be stimulated to reduce ground- and surface-water contamination with nitrate (see Chapter 14).

The ultimate impact of the soil microbial community on the total ecosystem is derived from its contributions to the orderly transition of the chemicals of life through its various oxidation states. Optimization of the ecosystem conditions for nutrient cycling results in development of a sustainable system in which nutrients are cycled with minimal losses of essential cycle components from the system or movement of potentially troublesome intermediates into ecosystems where their presence may be less appropriate (e.g., result in lake eutrophication or groundwater contamination). Microorganisms are instrumental in such essential ecosystem-sustaining activities as the conversion of plant-unavailable forms [e.g., dinitrogen (N_2)] to chemical entities that may be incorporated into plant biomass [nitrate (NO_3^-) or ammonium (NH_4^+)] as well as the return of organic components to inorganic forms that may be recycled into plant biomass (carbon or nitrogen mineralization). Loss of plant biomass nutrients phosphate or nitrate negatively affects the system by reducing its productivity and could be damaging to adjacent surface waters. Soil management to balance mineral nitrogen production and consumption minimizes surface- and groundwater quality decline.

9.2.1 The Environmental Connection

To truly understand the ramifications of any biogeochemical process, the implications of the synergism between the soil biota and their environment must be considered. Both the physical environment and the biological entities growing therein are modified by the products of biological growth. This synergism is absolutely essential for development of a sustainable ecosystem. In a managed site, such as a reclamation project, the sustainability of the developing ecosystem is reliant upon the capability of the microbial community to modify an initially hostile habitat to one more conducive to its own growth and sustenance as well as for the higher plant community. In highly damaged sites or in situations where rapid establishment of an aesthetic aboveground community is desirable or required, anthropogenic intervention to manage microbially mediated processes may be necessary. In the absence of such site management, the longevity of the biological populations essential for establishment of a viable aboveground and belowground community could easily be truncated. A typical scenario could include the following: To provide a rapid start to the soil microbial community and to sustain aboveground populations until the soil physical conditions are modified to those more conducive to plant development, the soils may be amended with an organic material (such as composts and sludges), nitrogenous fertilizers, and limestone to stimulate development of the indigenous microbial community. The organic amendments provide energy to the microbial communities as well as improve their growth conditions through complexing of heavy metals. The nitrogenous fertilizers are used to address imbalances in the carbon-to-nitrogen ratios of the organic components of the system. Liming of the soil may be necessary to adjust the soil pH to a range optimal for biological community development. Initially, this managed ecosystem development would be supported by the external nutrient inputs. A stable situation requires the slow evolution of soil populations and modification of the soil physical properties to a point wherein nutrients required for continued biomass productions are produced internally. This transition from artificially sustained to an internally sufficient system requires such physical-biological synergistic interactions as the enhancement of soil physical structure through aggregation (see Chapter 1), as well as microbial amelioration of soil acidity in sulfate-affected sites. (See Tate, 1985, for a more complete discussion of the importance of managing soil microbial populations for mineland reclamation.) This system development can be termed *quasi-synergistic* in that a feedback interaction develops between the developing microbial community and its physical environment. As the complexity of the microbial community increases, one product of its activity is an improvement in the physical and chemical conditions of its habitat. This alteration in turn supports further microbial growth. This cycle is repeated until ultimately a steady state develops where an optimized balance between the microbes and their physical and chemical environment has been attained.

Not only do these belowground processes optimize the soil microbial community activity and development, but conditions for plant population productivity and longevity are also improved. The development of soil aggregates through the action of soil microbes results in improved air diffusion and water infiltration rates. Therefore, not only is the soil microbial community stimulated by this habitat improvement, but plant root growth and aboveground biomass synthesis are also increased. The response of the plant community to the microbially modified soil structure allows for further enhancement of soil microbial activity. The microbial community is limited, to a large degree, by the fixed carbon supply. This fixed carbon in nonmanaged ecosystems is the direct product of plant photosynthesis. Therefore, the improvement of soil structure noted above results in better conditions for growth of the plants, which further stimulates the microbial community through the enhanced input of photosynthate into the soil. Thus, the biological aspect of that portion of the carbon cycle represented by microbial mineralization has modified the physical environment in a manner to increase carbon fixation by the plants, which in turn allows for further enhancement of the soil microbial populations dependent on generation of this plant carbon. This portion of the ecosystem development could be considered to be a true synergistic interaction.

The interdependence of higher plants, microbes, and their physical environment (not simply from a fixed environment, but more from the view that these interactions are resulting in a continuously evolving ecosystem) is demonstrated by the simple example of site management noted above. Similar interactive effects can be exemplified by evaluating the biological and chemical interactions involved with acid mine drainage-affected soil. For this situation, the pH of a soil has been lowered through the input of sulfate contained in acid mine drainage to a level that the indigenous acid-sensitive microbial population has ceased to function. Microbial activity may be manipulated to reduce the toxic conditions so that the concentrations of the toxic acid (in this case, sulfate) may be reduced sufficiently that biological activity may be resumed. Since biological sulfate reduction to sulfide requires anoxic conditions, site management must include flooding of the soil to encourage development of sulfate-reducing bacteria (see Chapter 15). The ameliorative processes eventually result in reestablishment of biogeochemical cycling of nutrients. [See Mills (1985) for a further discussion of management of acid mine drainage.]

These examples demonstrate the optimization of microbial community development results from modification of the physical and chemical environment due to the action of the biological components. Traditionally, this "optimization" is generally considered to entail improvements in community structure that translate, from an anthropogenic viewpoint, into an aesthetic aboveground community. But ecosystem evolution may involve alterations in the chemical environment that cause declines in aboveground productivity. For example, drainage of acid peats (either artificially or anthropogenically induced) stimulates the microbially catalyzed oxidation of the metallic sulfides contained therein into sulfate. The resultant declines in soil pH results in death of acid-sensitive plants in the aboveground community.

The biological cycling of nutrients in soils causes both permanent (long-term) and transient (seasonal) alterations of the physical environment. Solubilization of minerals and their subsequent movement into plant tissue or even leaching from the soil site results in an irreversible change in soil mineralogy. In contrast, more ephemeral changes in soil components are associated with the interaction between aboveground plant community and the soil microbes. For example, total system fixed nitrogen loads may be increased through the activity of nitrogen-fixing bacteria. This nutrient augmentation results in a general improvement in soil fertility. The effect may be limited chronologically should the newly incorporated fixed nitrogen be removed from the site (e.g., harvesting of crops — anthropogenically or by grazing animals), leached from the system in percolating waters, or denitrified.

In the preceding analyses, it was been shown that microbial and plant communities benefit from biogeochemical cycles. This gain by the biological community is enhanced by the associated modification of the physical environment by the soil microbes. It must be stressed that no volition on the part of the communities or individual organisms is involved in these adaptations. Beneficial alterations of the organism's surroundings are purely fortuitous. A developing situation is created wherein the microbial communities alter their habitat, which in turn results in selection of more effective microbial populations. This interaction produces a continuously evolving ecosystem.

9.2.2 Interconnectedness of Biogeochemical Cycle Processes

Conceptually, the most aesthetic means of considering biogeochemical cycles is to view them as independent entities, as has been the historical educational practice. Yet this to some degree presents an impediment to developing an absolute appreciation of the intricacies of nutrient cycling in the soil system. Central to all nutrient cycles are the various ecosystem components composed of organic matter—a variety of biomass pools plus intermediates in the mineralization of these components and soil humic substances. This organic matter-centered concept of biogeochemical cycling is depicted in Fig. 9.5,

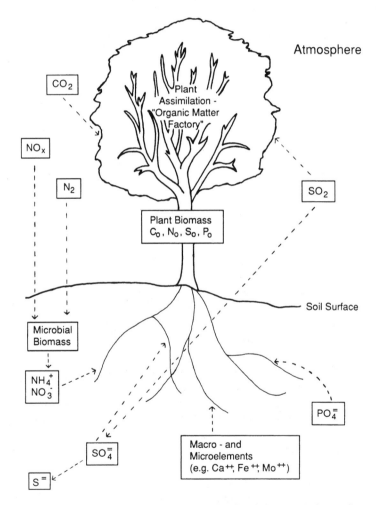

Fig. 9.5 Examples of the central role of plant biomass in sequestering various nutrient elements. (The subscript *o* refers to various organic pools of the individual element.)

where the assembly of the components of biomass is depicted, and Fig. 9.6, representing the disassembly or mineralization of organic substances to the primary mineral pools.

The pivotal role of organic matter in biogeochemical cycles results primarily from the fact that it is the primary reservoir of carbon, oxygen, nitrogen, phosphorus, and sulfur in soil. Thus, in an unmanaged system, the movement of nutrients into this organic matter reservoir and their subsequent mineralization are the primary controllers of plant nutrient availability. A conceptual model accentuating this fact is shown in Fig. 9.7. Movement of the primary components of biomass can be envisioned to be traveling along a highway that converges on a "square" consisting of living cells. Some organic substances are

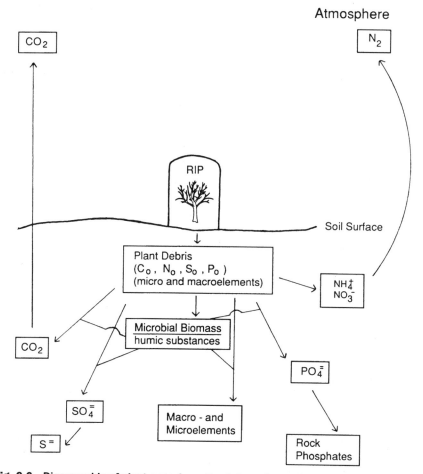

Fig. 9.6 Disassembly of plant organic matter into various primary inorganic nutrient pools. (The subscript *o* refers to various organic pools of the individual element.)

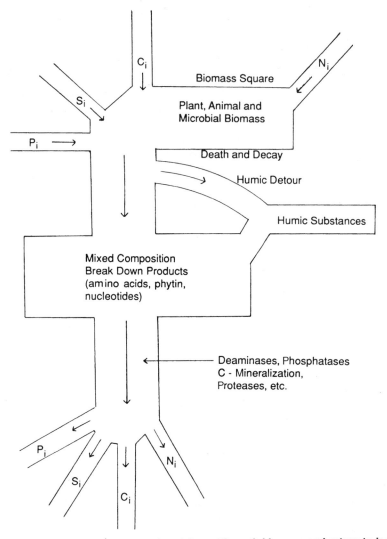

Fig. 9.7 The road from an inorganic existence through biomass and return to inorganic forms. (The subscript *i* refers to inorganic forms of the element.)

lost from the biomass and exit the square as biomass products (e.g., root exudates and extracellular enzymes), but most enter "decomposition lane" upon demise of the cell. Some of the organic substances take a detour through the neighborhood of humic substances, but eventually all enter a square consisting of mixed composition organic compounds—for example, amino acids, nucleotides, carbohydrates, and phytin. Continued mineralization of these products yield the original mineral components.

A practical difficulty resulting from this interaction of the biogeochemical cycles is the designation of a particular mineralization process as being driven by the needs of an individual nutrient cycle. For example, amino acid mineralization in soil can be easily quantified. Major products of the process are cellular energy, carbon dioxide, and ammonium. The question is whether the ammonium is produced because of a requirement of the decomposer population for ammonium to synthesize new biomass, or whether the mineral nitrogen was produced as by-product of the carbon and energy recovery processes. In fact mineralization of carbon and nitrogen is sufficiently linked in soil that assessment of carbon dioxide evolution rates has been suggested as a means of estimating net nitrogen mineralization (Gilmour et al., 1985). It is easy to produce models descriptive of various nutrient cycles, but difficult to designate particular processes as being primarily driven by needs of any specific nutrient cycle.

9.3 Biogeochemical Cycles as Sources of Plant Nutrients for Ecosystem Sustenance

A good way of gaining an appreciation of the interactive nature of the aboveground and belowground portions of the ecosystem is to examine the nutrient cycling interactions between the two regions. Nutrients enter soil through rainfall, dry deposition (e.g., sulfate, nitrate, and minerals contained in dusts), atmospheric gases (e.g., sulfur dioxide), anthropogenic sources (e.g., soil amendments), and fixation (dinitrogen, carbon dioxide). Primary sources of the basic building-blocks of organic matter are divided between the atmosphere (carbon dioxide and dinitrogen) and soil (phosphate- or sulfate-bearing soil minerals). Sulfur sources could be considered to be divided between soils (sulfur-bearing minerals) and atmospheric sources (e.g., sulfur dioxide generated by power generation and, to some extent, automobile exhausts). With the activities directed at alleviating air pollution, especially from automobiles and power plants, the latter source of sulfur compounds is being reduced, resulting in a need to consider soil mineral- and organic matter-derived sulfur inputs for plant biomass productivity.

As an ecosystem matures (approaches equilibrium), reliance on external nutrient inputs declines and the majority of the materials necessary for biomass productivity are produced by mineralization of resident organic matter pools. This situation is most clearly seen with nitrogen cycling in a mature forest. In a climax forest, the vast majority of the nitrogen is contained in soil organic matter and plant biomass nitrogen pools (e.g., Tate, 1987). In these situations, nutrient availability to the plant community is reliant nearly totally on the rate that the microbial community mineralizes soil humus.

In these ecosystems receiving minimal inputs of nutrients, a truncated nutrient cycle could be said to exist. The various nutrients are cycled between organic and mineral nutrient pools. Organic nutrient resources are composed of plant biomass and photosynthate entering the soil and native soil organic matter. These organic matter pools can be separated into (1) those substances separated on basis of solubility (humic acids, humin, and fulvic acids), and, (2) recognizable biochemical components of living cells — that is, polysaccharides, amino acids, nucleotides, and so forth. The latter substances are usually classed as readily available organic matter (see Chapter 10 for a discussion of the importance of this reservoir of organic material in ecosystem productivity). Readily available organic matter is that portion of the soil organic matter reservoir that is easily metabolized by the soil microbial community. It is distinguished by its short residence time (generally less than one to two years) and the fact that it provides essentially all of the metabolic energy for the soil microbial community.

Ecosystems may also derive a major portion of their energy from externally supplied organic substances. These situations are usually the product of anthropogenic inputs. Specific examples are intensively cultivated agricultural soils, where crop debris and anthropogenically produced fertilizers provide the bulk of plant nutrients, plus soils affected with varying inputs of sludges and composted materials. Soil amendment with these societal residues may be significant plant nutrient sources (e.g., see Boyle and Paul, 1989; Holland and Coleman, 1987; Geiger et al., 1992).

9.4 General Processes and Participants in Biogeochemical Cycles

The soil biological community mobilizes minerals, mineralizes organic substances, and immobilizes nutrients directly through a combination of oxidation and reduction reactions or indirectly through environmental modification. Direct reactions may involve alteration of the valence state of the particular element of interest as well as oxidation or reduction reactions. For example, energy recovery and mineralization of carbon-based compounds involves the oxidation (electron removal) of the carbon compounds to carbon dioxide. Similarly, ferrous iron may provide energy for microbial growth by removal of electrons to produce ferric iron. Similar alterations of the valence state as well as increases in the oxidation state of the total compound are associated with oxidation of ammonium to nitrate and sulfide to sulfate. In each of these instances, the energy yielded by the electron transfer produces increases in microbial biomass.

Lest a false impression be created regarding the direction of the oxidation-reduction processes in soil, it should also be noted that a variety of reductive transformations are essential for integrity of the biogeochemical cycles. For example, while organic compounds are being oxidized by denitrifying bacteria, the electrons so generated are utilized to reduce nitrate to dinitrogen. Similarly, sulfate may also serve as an electron acceptor — producing hydrogen sulfide among the products — in accompaniment with oxidation of a variety of organic carbon compounds.

Minerals may also be solubilized within the ecosystem by less direct means than direct alteration of their redox state by actively growing microorganisms.

For example, chelators solubolize metals and modification of the microsite pH may result in solubolization of a variety of cations. Acidification results in solubilization of metals such as aluminum as well as dissolution of carbonate rocks, thereby freeing associated cations.

The primary mediators of these processes are the soil bacterial, actinomycete, and fungal communities. Most research has historically been associated with these groups of organisms because they are responsible for catalyzing the bulk of the reactions involved in biogeochemical cycling, but recall from the presentations of Chapter 7 that the activity of these organisms in soil is controlled in part by their interactions with soil protozoa, nematodes, and microarthropods as well as a variety of soil animals (earthworms, ants, millipedes, and insects). Also, although the contributions of these latter biological groupings to nutrient cycling may be less than those of the primary decomposers, the higher members of the soil community also participate in mineralization of organic matter through the respiratory processes necessary to maintain their own biomass.

The interactions of protozoa and higher animals existent in soil are key to mineralization and cycling of nutrients through a variety of means:

- These organisms are instrumental in the mixing of organic matter within the soil profile, thereby augmenting the contact between the plant residues and the decomposer populations. Key participants in these processes include earthworms, ants, termites, and even to some extent burrowing animals.

- Soil animals are involved in inoculation of plant litter with propagules of decomposers. Movement of ants, termites, and other insects between soil nests and surface litter ensures contact of active microbial cells with plant debris.

- The soil animals indirectly stimulate decomposition by modifying soil properties. For example, increased aeration results from formation of earthworm galleries, borrows, and chambers and ant hills.

- The soil animals may increase the surface area of the organic debris indirectly through their movement through dry deposits of the plant biomass as well as directly via chewing and tearing of the plant parts. Increased surface area augments the potential contact with decomposer populations (see Chapter 7).

- As was indicated above, the organisms may mineralize the plant biomass directly. For example, the woodlouse (*Porcellio scaber*) feeds on decaying pine needles (Soma and Saito, 1983).

- Trophic interactions as described in Chapter 7 also stimulate decomposition. Interactions between the primary decomposers (bacteria and fungi) with population-controlling nematodes, protozoa, and mites are vital in maintaining active primary decomposer populations.

All soil decomposer populations, primary plus secondary decomposers, derive energy and reducing power (i.e., everything necessary for formation of a stable community) from the transformations of organic and mineral intermediates of the various nutrient cycles. The products of the decomposer populations benefit aboveground populations, plants directly and animals indirectly.

9.5 Measurement of Biogeochemical Processes: What Data Are Meaningful?

Design of field and laboratory experiments for the assessment of biogeochemical processes is complicated by the generally prevalent desire to extrapolate the data from the level of the soil sample contained within a test tube to an ecosystem-wide or even whole-earth view. Questions impinging on validity of data extrapolations include considerations of the inclusiveness (or appropriateness) of the data collected and the selection of representative soil samples. Furthermore, the manipulation of the soil sample (i.e., storage conditions, mixing, air drying) subsequent to collection may preserve or alter the relationship of process rates occurring in the sample from those occurring in the field.

9.5.1 Assessment of Biological Activities Associated with Biogeochemical Cycling

As was indicated in Chapter 2, a variety of methods are available for assessing biological activity in soil samples. Furthermore a multitude of assays of enzymatic activities, which are useful in the estimation of transformation rates associated with various biogeochemical processes (e.g., dehydrogenase and phosphatase, as well as assessment of the oxidation rates of a variety of carbonaceous substrates, such as glucose and aromatic compounds), have been adapted for use with soil samples. Applicability of the laboratory assessments of soil systems to the field situation is controlled by inherent limitations of the procedures themselves, as well as by difficulties associated with provision of a meaningful extrapolation of the results beyond the individual soil sample studied.

Problems with quantifying biological activities in soil for determination of nutrient-cycling potential and rates can be exemplified by evaluating applicability of various measures of soil microbial biomass. Several methods have been described for estimating microbial biomass levels and activity, for example, respiratory activity (oxygen uptake or carbon dioxide evolution plus dehydrogenase), chloroform fumigation, ATP measurement, pcr analysis, and direct examination. See Chapter 2 for a discussion of the basis for each of the methods. Considerations associated with utilization of each procedure involve the intrinsic limitations of the assay (i.e., what is and is not measured?) plus variability of microbial biomass densities between soil samples and within a soil site.

For any scientific procedure, the researcher must be cognizant of what the assay is truly measuring. For example, it is easy to postulate that carbon dioxide evolution rates from soil are proportional to the activity of the decomposer population in that soil sample, without realizing that carbon dioxide is also produced by those organisms feeding upon the primary decomposers plus any plant root tissue in the sample. Thus, such data could yield an overestimation of carbon mineralization in an ecosystem. Similarly, it is reasonable to consider that the total active soil microbial biomass would be related to carbon mineralization potential in a soil sample, but microbial biomass assessment techniques — chloroform fumigation, ATP measurement, and direct counts — all result in the quantification of both active and inactive microbial propagules.

9.5.2 Soil Sampling Aspects of Assessment of Biogeochemical Cycling Rates

Data interpretation is also complicated by the fact that the assays are generally conducted in the laboratory. Assuming that a representative field soil sample may be collected, data yielded from analysis of the soil sample in the laboratory at some distant time may not reflect actual population densities or activities in the field. A variety of changes in the soil properties generated by removal of the soil sample from its native site, its storage conditions, and preparation of the soil for study may result in alteration of the biological activity occurring therein. For example, collection of the soil sample as well as any sieving or mixing of the soil that occurs in the laboratory releases native soil organic matter trapped within soil aggregates. At least a portion of this organic matter can be mineralized by the microbial community. Thus, the biomass and the enzymes requisite for catabolism of the organic matter may be augmented above levels previously existing in the field. This complication is increased should storage of the sample be necessary. Even at refrigerator temperatures, significant growth of the microbial populations can occur. Some enzymatic activities are preserved at reduced temperatures, whereas other activities are not (see Tate, 1987, for a more complete discussion of this topic). Hence, for any measure of biological activity in a field sample, control experiments must be conducted to determine stability of the variable of interest during storage.

Extrapolation of laboratory-generated data to the field from which the sample was collected is reliant upon the soil sample being representative of the field situation. This prerequisite for data analysis is nowhere more essential than in studies of biogeochemical cycles. With few exceptions, laboratory assays are conducted with soil sample sizes ranging from less than 1 g to perhaps 100 g at most. As was indicated in Chapter 1, horizontal and vertical variation of microbial populations in soils as well as the containment of microbial populations within discrete microsites within the soil matrix make collection of manageable soil samples that are representative of an entire field or ecosystem site difficult. To produce data in the laboratory that are at least partially representative of the field situation, care must be taken to collect a series of soil samples that represent pertinent portions of the ecosystem of interest. This representative soil sample must then be preserved in a manner to minimize or preclude alteration of the biological activities prior to their quantification.

9.5.3 Environmental Impact of Nutrient Cycles

As was indicated above, biogeochemical cycles are composed of a variety of chemical compounds linked in a cyclic manner whose interconversion is accomplished by a combination of biological and chemical processes. Although a mass of data has been collected and published documenting the size of various nutrient pools within a number of ecosystems, considerable interest necessarily must be directed at consideration of rate of movement between these pools and the environmental impact of these transformations. Alteration of the distribution within the various pools or their rate of change will necessarily alter the ecosystem and adjacent system function. Frequently, complications arise from the failure to fully consider inputs or losses from the soil system. For example, fixed nitrogen may leave a system through leaching, denitrification, and volatilization (including ammonium volatilization), as well as being lost in harvested crop. The latter category is not limited to that biomass removed through agricultural processes. Wild and domestic animals may be prime movers of nutrients from an ecosystem through their grazing activities. Positive as well as negative system impacts are easily encountered.

This movement of nutrients between various reservoirs has been exploited for agricultural crop production. "Mining" of fixed nitrogen contained in soil humus has long been the basis for successful subsistence agriculture, but this enhancement of transfer of nutrients from organic to mineral reservoirs is not limited to low-input, limited-harvest cropping systems. Histosols, organic soils, are also so exploited. Peats and swamps represent side-tracks in nutrient cycles. Carbon, nitrogen, phosphorus, and sulfur compounds are incorporated into plant biomass that due to waterlogged conditions may be preserved for thousands of years. Removal of impediments to microbial mineralization of the organic matter results in rapid mineralization of the biomass. This produces sufficient nitrogen to support an intensive agricultural cropping system, as is exemplified by the Everglades Agricultural Area (South Florida, U.S.A.). Decomposition of the organic matter constituting the muck soils of that region produces as much as 3.3×10^4 kg C ha^{-1} year^{-1}. This is accompanied by production of approximately 1400 kg N ha^{-1} year^{-1} (Tate, 1980). For comparison, soils of a tropical rainforest are estimated to produce 405 to 2117 g C m^2 year (Schlesinger, 1977). The above carbon evolution rate for the histosols in southern Florida is equivalent to 3300 g C m^2 year^{-1}. For a more complete analysis of this topic, see Rosenzweig and Hillel, 1998), and Chapter 10.

In undisturbed ecosystems such as mature forests, mineralization and biomass synthesis rates are nearly balanced. For example, recovery of the mineral nitrogen produced by the soil microbial community through plant roots is sufficiently efficient that little nitrogen leaves the site through such processes as leaching to groundwater. The slight losses occurring through transfer of mineral nutrients to ground or surface waters are balanced by nitrogen inputs due to nitrogen fixation. Concerns arise from those systems, especially disturbed sites, where nitrogen leaching may reach ground or surface waters (such as the consideration of movement of nitrate and phosphates into Lake Okeechobee due to oxidation of this histosols of the Everglades Agricultural Area). Economic considerations arise when externally supplied nutrients (fertilizers) added to agricultural soils leach from the system rather being retained in the crop.

9.5.4 Example of Complications in Assessing Soil Nutrient Cycling: Nitrogen Mineralization

Nitrogen mineralization rate is a key indicator of the potential aboveground biomass production that can be supported by indigenous mineral nitrogen production. As indicated above, the vast majority of the nitrogen contained in soils is existent as organic nitrogen. Thus, barring any exogenous inputs, new biomass cannot be synthesized exclusive of mineralization of the resident organic nitrogen pools. The size of these nitrogen pools (organic and mineral) and flux rate between them are therefore meaningful values in characterization of an ecosystem.

Two procedures most commonly used to measure soil nitrogen mineralization potential are (1) the Stanford and Smith (1972) procedure, originally proposed for the study of soils contained in a column but also used for soils incubated as a batch sample in a beaker, plus (2) the buried-bag method (Eno, 1960; Pastor et al., 1984). With the Stanford and Smith method, soil samples are collected and transported to the laboratory, where they are sieved to increase homogeneity and to remove plant debris. The prepared soil samples are then dispensed into soil columns (leached method) or distributed into beakers (batch method). After an appropriate incubation period, mineral nitrogen is extracted from these samples with a dilute $CaCl_2$ solution. Nitrate plus nitrite and ammonium production are assayed over a time frame usually varying from 2 to 8 weeks. With the buried bag procedure, soils are collected in the field and placed either with minimal manipulation or following sieving into a polyethylene bag. The bag containing the field soil is returned to the soil horizon in a position as closely replicating the situation of the collected soil as much as possible. In theory, the conditions within the bag represent the field situation except that free exchange of soil water is prevented. That is, since air can diffuse through the thin plastic of the bag, it is usually assumed that gaseous conditions within and without the bag are equivalent. Data from Bremner's laboratory (Bremner and Douglas, 1971) indicate that free oxygen concentrations may decline inside the bag to levels that would reduce the activity of aerobic microbial populations. For an example of the use of these two procedures with forest soils, see Poovarodom and Tate (1988) and

Poovarodom et al. (1988). With each procedure, mineral nitrogen production rates are determined by quantifying ammonium and nitrate plus nitrite concentrations in the soil samples before and following incubation.

The primary objective of conducting these assays is to measure the quantities of organic nitrogen that can be mineralized (size of the nutrient pool) and how fast it is being mineralized (the flux rate). Note that these two values are not directly measured by the above procedures. Although direct assessment of the size and mineralization rate of labile soil organic nitrogen pool would be ideal, due to the complexity of the soil organic nitrogen pool, this is impossible. Soil organic nitrogen compounds are a highly diverse mixture of compounds, only a portion of which are easily metabolized by the soil microbial community. Furthermore, no soil organic nitrogen component has been found to be directly proportional to the mineralization rates. Therefore, these values must be derived indirectly. Both the quantity of mineralizable nitrogen contained within a soil sample and its mineralization rate are calculated from nitrogen mineralization data. Such data are predicated upon the assumptions that (1) soil organic nitrogen mineralization occurs in the laboratory-incubated or buried-bag samples at rates comparable to the field situation, (2) preparation of the soil samples for analysis does not alter the availability of the organic nitrogen to the microbial community, and (3) mineralization is a first order process occurring at the same rate throughout the incubation period. (See Section 11.3 for further discussion of the assumptions underlying the mathematical models used to describe nitrogen mineralization.)

It should also be noted that the Standford and Smith and the buried bag procedures yield an estimate of the net nitrogen mineralized in the soil sample — not the total nitrogen mineralized. As is discussed in Chapter 11, as nitrogen is mineralized, a portion can be detected as ammonium and nitrate, whereas some of the nitrogen is assimilated into new microbial biomass (immobilized). Therefore, the nitrogen detected by techniques based on quantification of ammonium and nitrate production in a soil sample is the difference between total nitrogen mineralized and the nitrogen immobilized (net nitrogen mineralized).

Determination of the nitrogen mineralization rate (Stanford and Smith procedure as well as the buried-bag technique) is usually predicated upon the observation that nitrogen mineralization follows first order kinetics:

$$\frac{dN}{dt} = -kN \tag{9.1}$$

where N is the concentration of mineralizable organic nitrogen, k is a rate constant, and t is time. Integration of this equation produces the following:

$$N_t = N_0 e^{-kt} \tag{9.2}$$

where N_t is organic nitrogen at time t and N_0 is organic nitrogen at time zero. Since the nitrogen at time t is equivalent to the initial nitrogen minus the nitrogen mineralized (N_m), the equation can be reduced to one containing a single unknown:

$$N_t = N_0 - N_m \tag{9.3}$$

$$N_0 - N_m = N_0 e^{-kt} \tag{9.4}$$

$$N_m = N_0(1 - e^{-kt}) \tag{9.5}$$

These calculations provide an estimate of the original organic nitrogen and its mineralization rates under the conditions where the data was collected (i.e., the laboratory or within the buried bag). Thus, the derived parameters are not necessarily applicable to true field situations (see Section 11.4.). For example, N_0 represents the initial nitrogen concentration in the disturbed soil sample — that is, the nitrogen available to the microbial population after the aggregates were disrupted by sampling and sieving of the soil sample. A significant proportion of that nitrogen could have been trapped in soil aggregates prior to sampling. Thus, the mineralization rate derived from the laboratory-incubated and buried-bag–retained soils may be an overestimation of that occurring in the field.

Further inaccuracies in the data derived by this method are added by the potential variation in nitrogen production due to diminished oxygen availability in the buried-bag procedure (and the potential for this to occur in batch-incubated samples), which would result in an underestimation of N_m. Complications associated with the Stanford and Smith leaching procedure involve the removal of mineralizable organic during leaching of the columns (Smith et al., 1980). With the latter nitrogen mineralization assay method, the soil columns are leached with a salt solution, prior to and at intervals during the incubation time. The resultant leachate contains the soil mineral nitrogen, which is the objective of the leaching procedure. It also includes a portion of the water-soluble organic nitrogen. Removal of this latter soil nitrogen fraction from the pool of organic nitrogen may reduce the actual mineralization observed during the incubation procedure. Each of these difficulties in incubation of the soils — modification of soil atmosphere and removal of mineralizable organic nitrogen — may alter the rate of ammonium, nitrate, and nitrite production by the soil microbial community. Thus, with all of these procedures, at best, an estimate of the field-available organic nitrogen pool and its mineralization rate is provided.

The model described above is based on observations that nitrogen mineralization in soil generally follows first order kinetics. Other mathematical models for deriving N_0 and K have been proposed that are based on alternative nitrogen mineralization kinetics (e.g., Bonde and Lindberg, 1988; Broadbent, 1986; Cabrera and Kissel, 1988; Deans et al. 1986, Juma et al., 1984), but the first order model described above appears to be the most commonly used for derivation of these nitrogen mineralization constants.

These observations demonstrate the degree of difficulties encountered in producing data representative of field nutrient pool size and the flux between these various pools. Such considerations of cycles where large mineral pools exist are complicated by difficulties in assessing small changes in mineral nutrient against a large background. This problem is commonly encountered with measurement of fluxes of organic phosphate and sulfate against backgrounds of soil-soluble and rock phosphate. These observations clearly reveal the basis for problems with ecosystem-wide and global extrapolation of such data.

9.6 Conclusions

An appreciation of the properties of the reactions occurring in biogeochemical cycles and the impact of variation in environmental factors on their reaction rates is essential for developing a complete understanding of a native ecosystem and in development of soil management or reclamation plans. These cycles are commonly appreciated solely for their contributions to aboveground plant productivity and their impact on ecosystem development, but it must be understand that biogeochemical processes are much more complex than the simple biological or chemical catalyzed transformation of an essential plant macro- or micronutrient. A complete depiction of biogeochemical cycles must entail an understanding of not only the biological entities involved but also of their synergistic interactions with the soil physical and chemical properties as they contribute to development of the native ecosystem. For reclaimed soils, this synergism affects development of a self-sustainable ecosystem.

References

Adams, R. M., C. Rosenzweig, R. M. Pert, J. T. Ritchie, B. A. McCarl, J. D. Glyer, R. B. Curry, J. W. Jones, K. J. Boote, and L. H. Allen, Jr. 1990. Global climate change and US agriculture. Nature (London) 345:219–224.

Allen, L. H., Jr. 1990. Plant responses to rising carbon dioxide and potential interactions with air pollutants. J. Environ. Qual. 19:15–34.

Blevins, R. L., M. S. Smith, and G. W. Thomas. 1984. Changes in soil properties under no-tillage. Pp. 190–230. *In* R. E. Phillips and S. H. Phillips (eds.), No-Tillage Agriculture. Van Nostrand Reinhold Co., NY.

Bonde, T. A., and T. Lindberg. 1988. Nitrogen mineralization kinetics in soil during long-term aerobic laboratory incubations: A case study. J. Environ. Qual. 17:414–417.

Boyle, M., and E. A. Paul. 1989. Nitrogen transformations in soils previously amended with sewage sludge. Soil Sci. Soc. Am. J. 53:740–744.

Bremner, J. M. and L. A. Douglas. 1971. Use of plastic films for aeration in soil incubation experiments. Soil Biol. Biochem. 3:289–296.

Broadbent, F. E. 1986. Empirical modeling of soil nitrogen mineralization. Soil Sci. 141:208–213.

Cabrera, M. L., and D. E. Kissel. 1988. Evaluation of a method to predict nitrogen mineralization from soil organic matter under field conditions. Soil Sci. Soc. Am. J. 52:1027–1031.

Cambardella, C. A., and E. T. Elliott. 1992. Particulate soil organic-matter changes across a grassland cultivation sequence. Soil Sci. Soc. Am. J. 56:777–783.

Dahlman, R. C., B. R. Strain, and H. H. Rogers. 1985. Research on the response of vegetation to elevated atmospheric carbon dioxide. J. Environ. Qual. 14:1–8.

Deans, J. R., J. A. E. Molina, and C. E. Clapp. 1986. Models for predicting potentially mineralizable N and decomposition rate constants. Soil Sci. Soc. Am. J. 50:323–326.

Doran, J. W. 1980. Soil microbial and biochemical changes associated with reduced tillage. Soil Sci. Soc. Am. J. 44:765–771.

Ellert, B. H., and J. R. Bettany. 1988. Comparison of kinetic models for describing net sulfur and nitrogen mineralization. Soil Sci. Soc. Am. J. 52:1692–1702.

Eno, C. F. 1960. Nitrate production in the field by incubation of soil in polyethylene bags. Soil Sci. Soc. Am. J. 24:277–279.

Firestone, M. K. 1982. Biological denitrification. Pp. 289–326. *In* F. J. Stevenson (ed.), Nitrogen in Agricultural Soils. Agron. Monogr. 22. Agronomy Society of America, Madison, WI.

Fresquez, P. R., R. E. Francis, and G. L. Dennis. 1990. Sewage sludge effects on soil and plant quality in a degraded, semiarid grassland. J. Environ. Qual. 19:324–329.

Geiger, S. C., A. Manu, and A. Bationo. 1992. Changes in a sandy Sahelian soil following crop residue and fertilizer additions. Soil Sci. Soc. Am. J. 56:172–177.

Gilmour, J. T., M. D. Clark, and G. C. Sigua. 1985. Estimating net nitrogen mineralization from carbon dioxide evolution. Soil Sci. Soc. Am. J. 49:1398–1402.

Holland, E. A., and D. C. Coleman. 1987. Litter placement effects on microbial and organic matter dynamics in an agroecosystem. Ecology 68:425–433.

Hunt, H. W. 1977. A simulation model for decomposition in grasslands. Ecology 58:469–484.

Jenkinson, D. S., D. E. Adams, and A. Wild. 1991. Model estimates of CO_2 emissions from soil in response to global warming. Nature (London) 351:304–306.

Jones, C. A., C. V. Cole, A. N. Sharpley, and J. R. Williams. 1984. A simplified soil and plant phosphorus model: I. Documentation. Soil Sci. Soc. Am. J. 48:800–805.

Juma, N. G., E. A. Paul, and B. Mary. 1984. Kinetic analysis of net nitrogen mineralization in soil. Soil Sci. Soc. Am. J. 48:753–757.

McConnaughey, P. K., and D. R. Bouldin. 1985a. Transient microsite models of denitrification: I. Model development. Soil Sci. Soc. Am. J. 49:886–891.

McConnaughey, P. K., and D. R. Bouldin. 1985b. Transient microsite models of denitrification. II. Model results. Soil Sci. Soc. Am. J. 49:891–895.

McConnaughey, P. K., D. R. Bouldin, and J. M. Duxbury. 1985. Transient microsite models of denitrification: III. Comparison of experimental and model results. Soil Sci. Soc. Am. J. 49:896–901.

Meentemeyer, V. 1978. Macroclimate and lignin control of litter decomposition rates. Ecology 59:465–472.

Mills, A. 1985. Acid mine waste drainage: Microbial impact on the recovery of soil and water ecosystems. Pp. 35–81. *In* R. L. Tate III and D. A. Klein (eds.), Soil Reclamation Processes. Microbiological Analyses and Applications. Dekker, NY.

Mosier, A., D. Schimel, D. Valentine, K. Bronson, and W. Parton. 1991. Methane and nitrous oxide fluxes in native, fertilized and cultivated grasslands. Nature (London) 350:330–332.

Parker, L. W., J. Miller, Y. Steinberger, and W. G. Whitford. 1983. Soil respiration in a Chihuahuan desert rangeland. Soil Biol. Biochem. 15:303–309.

Parton, W. J., J. W. B. Stewart, and C. V. Cole. 1988. Dynamics of C, N, P and S in grassland soils: A model. Biogeochem. 5:109–131.

Pastor, J., J. D. Aber, and C. A. McClaugherty. 1984. Above-ground production and N and P cycling along a nitrogen mineralization gradient in Black Hawk Island, Wisconsin. Ecology 65:256–268.

Pastor, J., and W. M. Post. 1988. Response of northern forests to CO_2 induced climate change. Nature (London) 334:55–58.

Poovarodom, S., and R. L. Tate III. 1988. Nitrogen mineralization rates of the acidic, xeric soils of the New Jersey Pinelands: Laboratory studies. Soil Sci. 145:337–344.

Poovarodom, S., R. L. Tate III, and R. A. Bloom. 1988. Nitrogen mineralization rates of the acidic, xeric soils of the New Jersey Pinelands: Field rates. Soil Sci. 145:257–263.

Reddy, V. R., D. N. Baker, and J. M. McKinion. 1989. Analysis of effects of atmospheric carbon dioxide and ozone on cotton yield trends. J. Environ. Qual. 18:427–432.

Rosenzweig, C., and D. Hillel. 1998. Climate change and the global harvest. Oxford University Press, NY. 324 pp.

Santos, P. F., N. Z. Elkins, Y. Steinberger, and W. G. Whitford. 1984. A comparison of surface and buried *Larrea tridentata* leaf litter decomposition in North American hot deserts. Ecology 65:278–284.

Santos, P. F., J. Phillips, and W. G. Whitford. 1981. The role of mites and nematodes in early stages of buried litter decomposition in a desert. Ecology 62:664–669.

Santos, P. F., and W. G. Whitford. 1981. The effects of microarthropods on litter decomposition in a Chihuahuan desert ecosystem. Ecology 62:654–663.

Schlesinger, W. H. 1977. Carbon balance in terrestrial detritas. Ann. Rev. Ecol. Syst 8:51–81.

Scow, K. M., and M. Alexander. 1992. Effect of diffusion on kinetics of biodegradation: Experimental results with synthetic aggregates. Soil Sci. Soc. Am. J. 56:128–134.

Scow, K. M., and J. Hutson. 1992. Effect of diffusion and sorption on kinetics of biodegradation: Theoretical considerations. Soil Sci. Soc. Am. J. 56:119–127.

Soma, K., and T. Saito. 1983. Ecological studies of soil organisms with reference to the decomposition of pine needles. II. Litter feeding and breakdown by the woodlouse, *Porcellio scaber*. Plant Soil 75:139–151.

Smith, J. L., R. R. Schnabel, B. L. McNeal, and G. S. Campbell. 1980. Potential errors in the first-order model for estimating soil nitrogen mineralization potentials. Soil Sci. Soc. Am. J. 44:996–1000.

Stanford, G., and J. J. Smith. 1972. Nitrogen mineralization potentials in soils. Soil Sci. Soc. A. J. 36:465–472.

Stroo, H. F., K. L. Bristow, L. F. Elliott, R. I. Papendick, and G. S. Campbell. 1989. Predicting rates of wheat residue decomposition. Soil Sci. Soc. Am. J. 53:91–99.

Stucki, G., and M. Alexander. 1987. Role of dissolution rate and solubility in Biodegradation of aromatic compounds. Appl. Environ. Microbiol. 53:292–297.

Tabatabai, M. A., and Y. M. Chae. 1991. Mineralization of sulfur in soils amended with organic wastes. J. Environ. Qual. 20:684–690.

Tate, R. L., III. 1980. Microbial oxidation of organic matter of histosols. *In* M. Alexander (ed.), Advances in Microbial Ecology 4:169–201.

Tate, R. L., III. 1985. Microorganisms, ecosystem disturbance and soil-formation processes. Pp. 1–33. *In* R. L. Tate III and D. A. Klein. (eds.), Soil Reclamation Processes. Microbiological Analyses and Applications. Dekker, NY.

Tate, R. L., III. 1987. Soil Organic Matter. Biological and Ecological Effects. John Wiley & Sons. NY. 291 pp.

van Veen, J. A., J. N. Ladd, and M. J. Frissel. 1984. Modelling C and N turnover through microbial biomass in soil. Plant Soil 76:257–274.

Walters, D. T., M. S. Aulakh, and J. W. Doran. 1992. Effects of soil aeration, legume residue, and soil texture on transformations of macro- and micronutrients in soils. Soil Sci. 153:100–107.

Chapter **10**

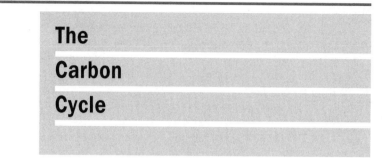

The

Carbon

Cycle

Conceptually, models of the carbon cycle can easily be reduced to representations of the processes associated with the transformation of carbon dioxide into organic matter and return of this primarily photosynthetically fixed carbon to its mineral form via biological mineralization (e.g., Fig. 9.1). This simplistic depiction of the carbon cycle processes provides a means for developing generalized principles, but amassing of the knowledge necessary for managing soil sites for ecosystem sustenance or reclamation is reliant upon separation of these two primary carbon reservoirs into pools designated by location in the ecosystem, residence time, and/or chemical form. Commonly studied soil organic carbon reservoirs include living and dead microbial biomass, animal and plant debris, and plant root exudates, as well as humic and fulvic substances (Fig. 10.1). Even this grouping of soil organic components must be expanded conceptually before a detailed understanding of the properties of both the individual compounds themselves and their interactions with the ecosystem components that limit the rate of flux between these organic reservoirs and the mineral pools can be presented. Quantification of biochemicals (e.g., protein, polysaccharide, lignin, fat, waxes, and resins) constituting each major soil organic reservoir that provides nutrients and energy to the soil microbial population, evaluation of the structural associations of these components in cells, and elucidation of the inclusion of these components in soil aggregates are essential for detailed modeling of the soil carbon cycle.

Some soil organic matter pools can be distinguished by their degradation resistance or vulnerability (e.g., humic substances vs. animal or plant biomass), but most of the organic matter pools designated in Fig. 10.1 are mixtures of biodegradation-susceptible and relatively stable organic components. Furthermore, due to the occurrence of these complexes of mineral and decomposable

284

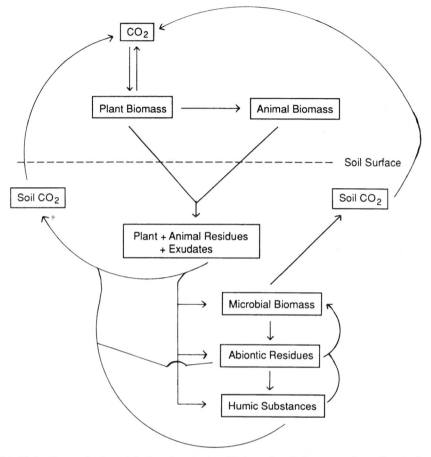

Fig. 10.1 Conceptual model of carbon cycle with transfers between major soil organic matter reservoirs emphasized.

organic matter and of the stability of plant cell structure, the decomposition rates of a portion of the easily metabolized substances reflects those of the more resistant materials. Consequences of interactions within the soil environment are exemplified by an analysis of plant biomass decomposition in soil. Accessibility of plant biomass components to the soil microbial population is determined by the rate of breaching of the plant cell wall. The cell wall protects the more easily decomposed substances contained in the cytoplasm. Biodegradation of readily metabolized organic compounds is also reduced by their occlusion in soil aggregates. Thus, estimates of the rates of mineralization of easily decomposable carbon pools must encompass the enhancement of their stability by inclusion within soil aggregates and stabilization of plant cellular structure. Similarly, a normally decomposable xenobiotic compound, such as a pesticide, may persist as the result of sorption to soil mineral particles or of noncovalent and covalent linkages to soil humic substances.

In conclusion, to develop an understanding of the complexity of the processes leading to increases or decreases in quantities of sequestered organic carbon in soil, the nature of the components of this soil fraction and the processes leading to their decomposition must be elucidated. Each of the major organic matter reservoirs in soil are a mixture of readily metabolized organic compounds plus relatively biodegradation-resistant substances, such as lignin and humic acids. Biodegradation-resistant humic substances constitute a major portion of the global carbon stores in soil. In contrast, the size of the pool of easily metabolized resources determines the quantity of energy available to drive the soil microbial population and "propel carbon through the carbon cycle." Ramifications of both processes relate to function and stability of individual soil systems as well as extending to global concerns, such as the role of soil ecosystem in mediation of greenhouse gas production and consumption. Hence, this chapter is presented with the objectives of (1) evaluating the nature of organic compounds retained within major soil organic matter reservoirs, (2) elucidating the mechanisms involved in the catabolism of these compounds, and (3) determining the properties of the carbon molecules, the microbial community, and the soil ecosystem itself, which are instrumental in controlling the return of soil organic matter to atmospheric carbon dioxide pools.

10.1 Environmental Implications of the Soil Carbon Cycle

It is reasonable to conclude that with the current state of knowledge, a plant community can be established on nearly any supportive matrix. This conclusion is based on the assumption that nothing toxic to the plants exists in the basic support material and that unlimited human intervention is feasible. The carbon cycle is an excellent example of the importance of organic matter transformations to the local community. Studies of organic matter fluxes through various soil reservoirs are frequently inspired by concern for such ecological and societal problems as demonstration of the potential for utilizing soil systems in recycling organic carbon based wastes, reclamation of sites affected by mismanagement — accidental or purposeful — and from a less direct viewpoint, the impact of soil carbon mineralization processes on global processes. Furthermore, historical interest has also centered on determining the best management procedures for enhancing soil structural quality and optimizing agricultural production. In summary, societal applications of studies of soil mineralization processes therefore include strong implications for decisions involving the management of greenhouse gas fluxes, maintenance of natural ecosystems, disposal of societal wastes, and reclamation of damaged soil systems as well as aspects of enhancement of agricultural production of food and fiber.

10.1.1 Soils as a Source or Sink for Carbon Dioxide

Soil ecosystems are both producers and consumers of the two carbon-containing mineralization products of organic substances, carbon dioxide and methane. Increases in soil organic matter levels result in the sequestering of excess carbon dioxide into soil organic matter (soils are a net pool for carbon dioxide). Conversely, management of the soil ecosystem in a manner causing declines in colloidal soil organic matter (e.g., intensive cultivation) converts the ecosystem into a source of mineral carbon. A good example of soil system management to increase organic matter resources is conservation tillage. Kern and Johnson (1993) estimate that maximization of the implementation of no-tillage soil management practices could result in sequestering of carbon in soil organic matter equivalent to 0.7 to 1.1 percent of the total projected U.S. fossil fuel carbon emissions of a 30-year period. A less direct but significant interaction between soil organic matter reservoirs and atmospheric mineral carbon concentrations results from management of soil properties for optimization of plant biomass production (i.e., ecosystem productivity). As will be discussed below, plant biomass is a primary carbon reservoir. Among the soil properties controlling biomass synthesis is soil aggregation, which is determined by interactions of soil minerals with humus (Chapter 1). Thus, improvements in soil structure resulting from increases in colloidal organic matter may also indirectly reduce atmospheric carbon dioxide loads by increasing terrestrial biomass pools. Thus, the benefits of the latter situation result from both increased soil organic matter levels due to improved soil structure and augmented production of organic matter by aboveground biomass.

Evidence for the role of soil organic matter in modulating increases in atmospheric carbon dioxide is provided by observation of an incongruity between known carbon dioxide sources and measurable atmospheric loads of this gas. That is, known sources of atmospheric carbon dioxide exceed the identifiable sinks. Soil organic matter and terrestrial biomass most likely account for at least a portion of this unknown fate of the carbon dioxide (e.g., Harden et al., 1992, and Sundquist, 1993). Since the terrestrial biosphere and associated soils contain far more carbon than the atmosphere, these carbon pools may serve as buffers or modulators of atmospheric carbon dioxide levels. A recent example of estimates of mass of carbon in various biogeochemical pools is provided by Sundquist (1993), where it is estimated that 750 Gt (1 Gt = 10^{15} g) of carbon was contained in the atmospheric pool in 1990 A.D. compared to 2160 Gt for terrestrial biosphere, with the soils component accounting for approximately two-thirds of this total. Eswaran et al. (1993) estimated that 1576 Gt of carbon is stored in soils, with about 32 percent of this carbon retained by tropical soils. The importance of soil organic matter as the primary carbon pool is exemplified by considering the distribution of carbon in a forest soil. It has been estimated that between two- and threefold the carbon retained in forest biomass is found in dead organic matter on the forest floor (Kögel-Knaber, 1993).

Conversely, the existence of the large biosphere carbon reservoir means that minor changes in biomass production or in retention of fixed carbon within soil can result in significant increases or decreases in atmospheric carbon dioxide levels. That is, biosphere carbon provides a buffer from changes in atmospheric carbon dioxide loads. Potentially world-affecting ecosystem modifications include (1) cultivation of virgin soils (see Tate, 1987), (2) drainage of peat soils (e.g., Armentano 1980; Tate, 1980) and (3) clear-cutting of forest biomass and its replacement with less productive plant species.

10.1.2 Diffusion of Soil Carbon Dioxide to the Atmosphere

A further consideration in modeling the exchange rate between soil organic carbon pools and atmospheric carbon dioxide is the diffusion rate of carbon dioxide between soil air and the atmosphere. Variation in the localized soil properties (i.e., moisture, structure, and temperature) control not only the gaseous exchange rates but also delimit the mineralization rate (i.e., the mass of carbon dioxide available to move from soil to atmospheric pools) characteristic of the soil site (see Oades, 1988). The direction of movement is generally from soil air to the overlying atmosphere in that concentrations of gaseous carbon dioxide existent in soil pores generally exceed atmospheric carbon dioxide loads. A variety of soil properties interact to facilitate or inhibit this gas exchange. Along with the increases in mass of carbon dioxide resulting from ecosystem modifications listed above, soil moisture, temperature, structure (aggregate formation and pore structure), biological activity (root development and earthworm activity), and mass movement of water (e.g., rainfall) also control carbon dioxide distribution. Less dramatic variability in ecosystem properties than those resulting from ecosystem-wide alterations, natural and anthropogenically induced, may determine the carbon dioxide exchange rate between soil and atmospheric reservoirs. For example, a minor alteration in soil moisture and temperature, such as those resulting from developing crop canopies, can alter the carbon dioxide flux rate. These interactions were presented in a model developed by Ouyang and Boersma (1992a; 1992b).

Not all carbon dioxide diffusing from the soil surface leaves the ecosystem where it was produced. A portion of it can be intercepted by the plant community and fixed into new plant biomass. Increased atmospheric carbon dioxide can thus result in an increase in total abovegound biomass production in established ecosystems (e.g., see McMurtrie et al., 1992; Polgase and Wang, 1992, Rogers et al., 1983). These increases in biomass production are in addition to the changes in atmospheric carbon dioxide loads due to deforestation and reforestation (e.g., Vitousek, 1991; Woodwell, 1983). In either situation, enhancement of plant productivity can also alter the equilibrium concentration of organic carbon retained in the underlying soil. Most of the photosynthetically fixed carbon entering soil is mineralized to carbon dioxide by the soil microbial community, but some carbon will be retained in soil in more biodegradation-resistant fractions — microbial biomass and humic substances. Increased fixed carbon inputs may therefore lead to increased soil microbial populations, and their products (e.g., extracellular polysaccharides), as well as greater accumulation of humified substances than existed prior to stimulation of aboveground plant community growth. Thus, stimulation of biomass productivity by increased carbon dioxide loads could lead to a reduction (albeit of variable magnitude) in atmospheric carbon dioxide and a resultant sequestering of carbon in soil organic matter pools.

10.1.3 Managing Soils to Augment Organic Matter Contents

Due to a historical predisposition for the study of soil processes affecting crop yields, most of the data useful for elucidating processes that control soil humus levels are associated with research with the primary objective of improving plant biomass productivity (crop yields). The results of decades of research regarding establishment of natural organic matter levels in soil could be succinctly summarized as follows (see Chapter 1 and Tate, 1987, for a more detailed review of this topic):

- The amount of soil organic matter retained in a particular soil system is a product of the balance between organic matter input (primarily aboveground biomass production) and decomposition (predominantly microbial mineralization of plant biomass).

- An equilibrium level of soil organic matter is attained in an undisturbed soil ecosystem that is determined by total ecosystem properties (e.g., the nature of such soil ecosystem properties as the plant community, soil texture, climate, topography, and anthropogenic intervention. See Amundson and Jenny, 1991; Jenny, 1980).

- Of ultimate importance in evaluation of the impact of soil organic matter transformations on terrestrial carbon cycling is the fact that disturbance of a soil, such as cultivation of virgin prairies or clear-cutting of a forest, results in establishment of a new equilibrium level of soil organic matter generally lower than that in the pristine or undisturbed ecosystem.

- Processes that reduce soil aggregation, aerating the soil, or reducing biomass incorporation rates result in reduced retention of organic matter.

A variety of examples of alteration of soil organic matter resources due to ecosystem disturbance (e.g., clear-cutting of forests, cultivation of forest or grassland soils) and of the relationship of subsequent soil management techniques on establishment of new steady-state organic matter percentages in the soil are available (e.g., see Aguilar et al., 1988; Alegre and Cassel, 1986; Mann, 1986; and Rasmussen and Rohde, 1988). The extent of change in soil organic matter levels in reaction to ecosystem disturbances tends to be proportional to (1) the amount of disruption of soil structure associated with the initial site clearing, (2) the extent of cultivation or related management of the soil site subsequent to the initial disturbance (i.e., continued degradation of soil structure and mixing of the soil results in greatest losses of native soil humus), and (3) the quantities of organic matter produced in or amended to the soil following disturbance. For example, in the study by Alegre and Cassel (1986), it was shown that in clearing forest soils in the Amazon jungle of Peru, slash-and-burn clearing caused greater retention of native soil organic matter than did mechanical clearing with a bulldozer. Similarly, maintenance of soil aggregate structure by bedding and lime application reduced organic matter losses compared to that observed in cleared soils that were managed by more structurally damaging procedures. Alteration of inputs of organic matter can be through maintenance of plant productivity (aboveground as well as root mass) and return of *in situ*-produced plant biomass to the soil ecosystem (Barber, 1979; Richter et al., 1990, Wood et al., 1992) as well as through soil amendment (e.g., for the long-term impact of manure amendments to cropped soils see Jenkinson and Rayner, 1977).

Reduction in site intervention, such as that associated with the conversion from intensive tillage to no-till management of soil, reduces susceptibility of organic matter to microbial decomposition through enhancement of soil aggregate development. Thus, the equilibrium level of organic matter on the soil surface horizon is increased. See Havlin et al. (1990) and Scott and Wood (1989) for examples of studies of the impact of variation in tillage practices on soil organic matter.

The overall environmental impact of each of these localized soil management procedures is alteration in the quantities of carbon dioxide released to the atmosphere or conversely, augmentation of the proportion of total terrestrial carbon residing in the soil organic matter pool. The latter situation is of greater overall benefit since it results in a reduced level of greenhouse gases and improves the stability and productivity of the ecosystem, a factor that can also reduce greenhouse gas production.

10.1.4 Carbon Recycling in Soil Systems

A major product of human societies is organic waste products. These substances include such products as industrial and residential garbage, and sludges, as well as a variety of petroleum waste products. Incineration of these substances results in conversion of essentially all the carbon to atmospheric carbon dioxide. Alternatively, utilization of the substances as soil amendments recycles plant nutrients contained therein into plant biomass and allows sequestering of portions of the organic matter into the various soil organic carbon pools. Thus, the fixed carbon components and associated nutrients are recycled through the soil biomass and associated carbonaceous components.

In the use of soils to recycle organic waste substances, the natural biological decomposers resident in soil are exploited as an economical means of returning bulky, waste vegetative material to less troublesome mineral carbon or stable soil organic carbon reservoirs. For example, see review by Boyle (1990) describing the land disposal of sludge or studies of landfarming of petroleum wastes (see Bossert et al., 1984; Dibble and Bartha, 1979a, b; Norris, 1980).

Questions regarding use of soils to recycle the organic waste products of society pertain not to the capacity of soils to decompose such materials. Indeed, the organic components contained in such materials as sludges and composts differ little from the organic substances entering soils as fresh plant biomass, litter, and thatch. Thus, it can reasonably be concluded that the biochemical processes involved in decomposition of these materials are the same as those associated with general biomass and products of biomass decomposition in soil. Due to differing ratios of decomposable and more biodegradation-resistant components in the composted materials compared to fresh plant biomass, the kinetics of decomposition will differ. Major concerns with disposal of organic wastes in soil systems commonly are related more (1) to the total quantities that can be added before the system is overloaded (i.e., either natural decomposition processes are reduced, ecosystem sustainability is threatened, or groundwater contamination is threatened), (2) to difficulties with con-taminants of the waste material that may impede soil biological activity or may be resistant to mineralization (e.g., heavy metal contaminants and synthetic organic chemicals), and (3) to the potential for spreading of pathogenic microbes (e.g., fecal coliforms, protozoa, and viruses). For examples of how these potential difficulties are avoided, see review by He et al. (1992) regarding the land disposal of composted municipal solid waste. Not only does societal benefit result from the disposal of troublesome waste substances, but if the process is properly conducted, soil properties can also be improved (e.g., Douglas and Magdoff, 1991; Epstein et al., 1976; Mays et al., 1973; Mitchel et al., 1978; Piccolo and Mbagwu, 1990).

These limited examples of the impact of soil carbon cycle processes on the total terrestrial ecosystem emphasize the necessity to develop a clear under-standing of the types of organic carbon compounds mineralized in soil, their distribution in major carbon cycle pools (Fig. 10.1), and their decomposition kinetics.

10.2   Biochemical Aspects of the Soil Carbon Cycle

The complexity of the biochemical processes associated with carbon cycling in soil is exemplified by the diversity of organic compounds mineralized and the array of microbes involved. The soil microbial community is adept at convert-ing native biomass components plus a variety of xenobiotic compounds to atmospheric carbon dioxide, albeit at varying rates dependent upon the substrate structure and ecosystem properties. This seemingly limitless capability of the soil microbial community to mineralize organic carbon compounds could be considered to be the ultimate in recycling. Indeed, at one time in the development of the concepts of soil microbiology, the soil microbial community was considered to be infallible. Failure to discern measurable mineraliz-ation of an organic compound was considered to relate to the inability of the researcher to provide conditions conducive to the activity of the requisite microbes, rather than to be an inherent genotypic limitation of the microbes. With the advent of the "chemical age," it has become apparent that the concept of microbial infallibility was at best optimistic from the view of microbial capacities and pessimistic in regards to the innovative abilities of chemists and the magnitude of the societal appetite for unique industrially-synthesized materials.

As a result of the expanding capacity to synthesize organic carbon-based compounds of little resemblance to those produced naturally, this fallibility/ infallibility concept of the soil microbial populations must be adjusted. More accurately, the soil microbial community can be considered to be infallible only at mineralizing biologically synthesized organic compounds. This conclusion is based on current concepts of genotypic diversity of indigenous soil microbes. Interestingly, the potential now exists to expand soil metabolic capabilities through bioengineering. With the increasing capability of developing "design-er genes," the enzymatic array of the soil microbes may come to include a variety of "unnatural" enzymes. These enzymes would be characterized by the fact that they result from modifications of cellular DNA to allow the produc-tion of enzymes with a specificity that includes the capacity to cleave molecular bonds not normally encountered in the soil system, thereby resulting in mineralization of the chemically synthesized organic compound. In that situation, perhaps, our concept of microbial infallibility may return to some-thing approaching the optimistic view of the prechemical age. The challenge is for soil microbiologists to become as adept at developing novel degraders of xenobiotic compounds as chemists are at producing such substances.

10.2.1   Individual Components of Soil Organic Carbon Pools

The organic components of soil amendments and native soil organic matter are relatively easily separated conceptually by chemical class; that is, polysacchar-ides, proteins, lignin, humic acids, and fulvic acids. Each of these groups of compounds is distinguished by specific chemical properties, which delimit their utility for carbon and energy metabolism of the soil microbial community. Since all but the humic and fulvic acids are sufficiently defined chemically that they can be studied in the test tube, the biochemical processes associated with their mineralization appear to be reasonably predictable in soil. Thus, a basic understanding of the biochemistry of the soil environment can be revealed through analysis of the organic compounds available as carbon and energy resources.

An understanding of the relative biodegradation susceptibility of each chemical group allows prediction of the relative activity of different soil ecosystems. For example, a site containing large quantities of polysaccharide would support a more active microbial community then one in which essentially all of the organic carbon is associated with humic and fulvic acids. The former situation is exemplified by a soil system receiving significant inputs of fresh biomass (most productive native ecosystems). Alternatively, the soil system characterized by a predominance of biodegradation-resistant humic substances would be one receiving little fixed carbon inputs from associated plant or animal activity. Thus, the indigenous microbial population would be expected to reduce the reserves of easily metabolized fixed carbon to a minimum. Then, the only remaining energy source would be the more difficult-to-catabolize organic substances. Such a site is exemplified by a fallow (bare) soil maintained with little or no plant growth for several seasons. Quantification of biologically decomposed substrates in a soil sample can only provide at best an estimate of the potential biological activity contained therein. This conclusion results from the fact that the organic compounds do not exist in isolation in soil, but instead they occur in complex mixtures of easily decomposed and more complex organic substances, not all of which are available to the soil microbial community. For example, proteins and starch are readily detected in soil. These substances may be decomposed within a matter of hours when amended as chemically pure preparations to soil, but both starch and proteins could have half-lives of days or longer in soil should they be retained within a reasonably biodegradation-resistant structure, such as a plant cell wall.

Polysaccharides are the primary energy source of the soil biological community. The predominant source of soil polysaccharides is plant tissue, although locally important contributions may be derived from influxes of animal biomass (e.g., amylose). Polysaccharides originating within the soil community are mainly the product of microbial synthesis. Between 5 and 25 percent of the carbon contained in native soil organic matter (i.e., that organic matter remaining in soil after all recognizable plant and animal debris has been removed) is composed of polysaccharides. Of this soil humus polysaccharide, that originating in newly synthesized biomass is most susceptible to biodegradation, whereas polysaccharides covalently bound to humic acids and polysaccharides coating soil minerals are more resistant. A diverse mixture of polysaccharides and their decomposition products have been isolated from soil (for greater detail on this subject, see Gupta, 1967 or Stevenson, 1994):

- Monosaccharides including hexoses (D-glucose, D-galactose, D-mannose) and pentoses (D-ribose, D-arabinose, D-xylose)

- Polysaccharides such as cellulose and a variety of hemicelluloses from plants, as well as amino sugar-containing polysaccharides including chitin, as found in some fungi and in insect exoskeletons, and the complex mucopeptides of bacterial cell walls

- Uronic acids and polyuronic acids such as glucuronic acid, galacturonic acid and polygalacturonic acid (i.e., pectin)

- Methylated sugars such as 2-O-methyl-D-xylose, 2-O-methyl-D-arabinose, 2-O-methyl rhamnose, 4-O-methyl galactose

These polysaccharides are distinguished by the diversity of linkages between the monosaccharide units (cellulose vs. amylose) and inclusion of nonsugar moieties (chitin and microbial cell walls), as well as combination of a variety of monosaccharides within a single polysaccharide molecule (hemicelluloses). Variation in each of these molecular properties effects decomposition susceptibility through alteration of the crystalline structure of the molecule and/or the number of enzymes necessary for its mineralization.

The effect of linkage of the sugar moieties on biological stability is best exemplified by comparing the structures of two glucopyranoses, amylose and cellulose (Fig. 10.2). Amylose, α-$(1 \rightarrow 4)$-D glucose, is usually decomposed in soil in a matter of hours, whereas cellulose, β-$(1 \rightarrow 4)$-D-glucose, has a half-life in soil that may be measured in years. Amylose is hydrolyzed by a single enzyme, amylase. In contrast, a complex of enzymes is required for disruption of the crystalline structure and cleavage of cellulose into monomers (e.g., Béquin, 1990). Enzymes are required to disrupt the crystalline structure of cellulose before it can be hydrolyzed. Chitin, β-$(1 \rightarrow 4)$-N-acetylglucosamine, also forms a rigid structure that is resistant to most soil bacteria.

In contrast to the reasonably simple structures of the above polysaccharides, the bacterial cell wall (Fig. 10.3) is composed of regular repeating units of N-acetylglucosamine and N-acetylmuramic acid interlinked by short peptide chains. A degree of biodegradation-resistance is necessary since these substances protect the easily metabolized substances constituting bacterial cytoplasm from degradation. Bacterial cell wall polymers are significant sources of mineral nitrogen following death and mineralization of the bacterial cell.

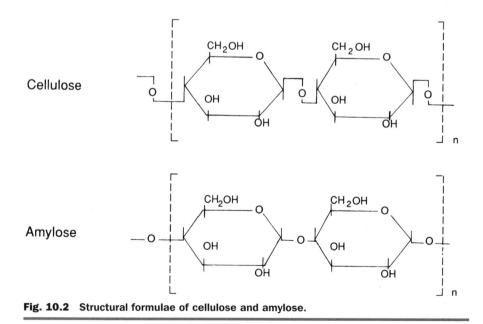

Fig. 10.2 Structural formulae of cellulose and amylose.

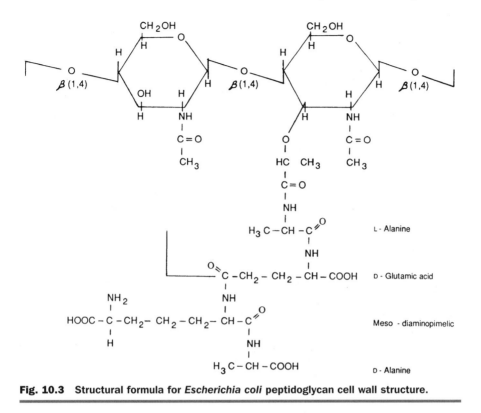

Fig. 10.3 Structural formula for *Escherichia coli* peptidoglycan cell wall structure.

Hemicellulose is second in abundance in plant tissue only to cellulose. It is found in proximity to cellulose in primary and secondary cell walls. Structurally, the only commonalty between cellulose and hemicellulose is that they are both polysaccharides. Two groupings of hemicellulose are introduced into soil — homoglucans and heteroglucans. Homoglucans, the less common grouping of plant hemicelluloses, consist of single monosaccharide units. Examples include xylan (polyxylose), mannan (polymannose), and galactan (polygalactose). Heteroglycans are composed of more than one monosaccharide or uronic acid. Most contain 2 to 4 different monosaccharides, but examples of 5 to 6 components are detected. The most abundant moiety is listed last in the name — for example, mannose for glucomannan and arabinomannan. Hemicelluloses as a group are structurally complex, containing 50 to 60 monosaccharide units with branching. A small group of monosaccharides (xylose, arabinose, mannose, glucose, galactose, glucuronic acid, and galacturonic acid) provide the bulk of the sugar moieties of hemicelluloses.

Lipids are a more heterogeneous group of soil organic compounds. This soil component is defined operationally. Soil lipids are those soil organic components that are soluble in organic solvents. They include fats, waxes, and resins. Chemical components include waxes (long-chain fatty acids and higher-chain aliphatic alcohols and some cyclic alcohols), organic acids (both short and longer chain), and normal parafins (C_{16} to C_{32}) as well as polycyclic hydrocarbons. The latter groups include polycyclic aromatic hydrocarbons, sterols, terpenoids, and chlorophyl. (See Stevenson, 1994, for a more complete discussion of these compounds.)

Lignin is a major component of plant biomass. Stems of woody angiosperms contain between 18 and 25 percent lignin (on a dry weight basis). For comparison, gymnosperms are composed of 25 to 35 percent lignin, whereas 10 to 30 percent of the weight of monocotyledons is lignin (Crawford, 1981). Lignin is a random polymer of sinapyl, coniferyl, and coumaryl alcohols containing a variety of complex organic linkages that are considerably less common in other plant substituents. Spruce (*Picea abies*), for example, contains 48 percent arylglycerol-β-aryl ethers, 6 to 8 percent noncyclic benzyl aryl ethers, 9.5 to 11 percent biphenyl, 7 percent 1,2-diarylpropane structures, 9 to 12 percent phenylcoumaran structures, and 3.5 to 4 percent diphenyl ethers (Crawford, 1981). Due to the complex aromatic structure of this plant component, lignin is decomposed slowly in soil (see review by Kirk and Farrell, 1987, for discussion of the biochemistry of lignin decomposition). The most labile components of this complex molecule are the side chains and methoxyl groups. For example, after a 6-month incubation period in a neutral sandy loam, about 23 percent of the ring carbons and 2-carbons and 39 percent of the methoxyl carbons of coniferyl alcohol units linked into model and cornstalk lignins were evolved as carbon dioxide (Martin and Haider, 1979). (See Tate, 1987 for a more detailed examination of the decomposition of this compound in soil.) Summary points regarding the decomposition in lignin of significance to the soil microbiologist and for evaluation of the impact of lignin on the rate of cycling of plant carbon through soil are:

- The complexity of the structure results in extended residence time for the plant carbon contained in lignin in soil.
- Although lignin contains a variety of chemical components that could yield energy to growing microbes, recovery of this energy by the soil microbial community is unlikely. The variety of components and their random assembly into the lignin molecule increases the quantities of energy that the microbes must expend to recover energy from the mineralization of lignin.
- The extended residence time of lignin results in incorporation of significant portions of the lignin molecule into soil humic acids. This extends further the longevity of these carbons in the soil ecosystem.

Humic substances are the most biodegradation-resistant components of soil humus. Although lignin presents a formidable barrier to microbes instrumental in cycling organic carbon entering or contained within the soil ecosystem back to mineral pools, an even more complex structure is encountered by soil microbial populations active in carbon mineralization in humic acids. Humic substances are commonly extracted from soil with dilute alkaline solutions (e.g., 0.5 M NaOH). Humic acid is that substance that is soluble in the alkaline solution but precipitates at pH less than 1. Fulvic acid is soluble in both the acidic and alkaline conditions. Humin is not soluble in acidic or alkaline solutions.

Humic acids are random polymers of aromatic and aliphatic moieties. Molecular weights for these substances have been shown to range from a few thousand to over one million daltons (e.g., see Stevenson, 1994). Malcolm (1990) specifies a molecular weight range for soil humic acids of 50,000 to 500,000 daltons. Although traditional concepts of the generalized structure of humic acids are based on the assumption that these substances are predominantly aromatic, recent research has shown that humic substances are characterized by a range of aromaticities (see Malcolm, 1990). Some humic acids are predominantly aliphatic whereas others (as traditionally believed for all humic acids) are composed of nearly all aromatic ring-containing substituents. For example, Hatcher et al. (1981) calculated aromaticities for humic acids from a variety of climatic zones ranging from 35 to 92 percent. Thus, the commonly encountered hypothetical structure of humic acids must be interpreted to contain representative aromatic and aliphatic moieties, but the ratio between these individual components differs from that shown in general models (see Stevenson, 1994, for examples of these molecular models).

A variety of proteins and polysaccharides may be covalently bound to the humic acid molecules. Furthermore, organic compounds entering soil or produced therein may become associated with humic acids through a variety of weak associations (e.g., hydrogen bonding and van der Waals forces, as well as ionic bonds). Individual components of the humic acid complex may be microbiologically synthesized aromatic compounds, xenobiotic compounds, such as pesticides, or even extracellular proteins and polysaccharides released into the environment by microbial and plant biomass either as a normal process of cellular metabolism or through death and decay of the biomass. Additionally, low molecular weight organic substances may become stabilized within the large humic acid molecule through noncovalent linkages (predominantly weak molecular interactions, such as hydrogen bonding and van der Waals interactions). These interactions may be nearly as stable as associations involving covalent bond formation (see Dec and Bollag, 1997). Also, stable, noncovalent interactions of humic acid and hydrophobic organic substances can occur in hydrophobic regions of the molecule (micelles or of smaller extent) of the humic acid molecule (see von Wandruszka, 1998).

The soil fulvic acid fraction contains a mixture of organic compounds that includes monosaccharides, oligosacchrides, polysaccharides, amino acids, and peptides, and organic acids plus fulvic acids. Removal of the defined organic compounds through ion exchange chromatographic procedures (Malcolm, 1990), leaves the more ill-defined soil organic component, fulvic acid. Fulvic acid is composed of a light-brown, low molecular weight organic material of only a postulated composition. These materials have a molecular weight range of about 1000 to 5000 daltons (Malcolm, 1990) and are more oxidized than humic acids (Stevenson, 1994). Their origin in soil is more problematic. They may be decomposition products of the larger, more complex humic acids, precursors of these molecules, or, more likely, both.

Humic and fulvic acids present a formidable but not insurmountable barrier to the microbes involved in mineralization of soil organic matter. A long list of papers has been published in which it is concluded that specific microbial strains with the capability of mineralizing these complex compounds have been isolated (e.g., see Andreyuk and Gordienko, 1978; Andriiuk et al., 1973; Blondeau, 1989; Federov and Il'ina, 1963). The fact that an equilibrium level of these substances is reached in soil further indicates that these substances are mineralized, albeit at a slow rate, in the soil environment. The latter conclusion is supported by the fact that ages of humic acids in soils have been estimated to range from a few decades to several thousand years (see review by Stout et al., 1981).

Examination of the structure of soil humic acids and of their distribution in soil indicates two sources of the resistance of these substances to microbial catabolism. Biochemical stability is derived from the diversity of the components of the molecule and their random polymerization. For a microbial species to utilize these complex molecules for a carbon and energy source, a large number of enzymes would have to be synthesized. The energy expended by the cell in producing the enzymes would most likely exceed what could be derived from catabolism of the complex aromatic/aliphatic structure. Further impediments to microbial decomposition of humic acids develop from the association of the large molecule with soil minerals, especially clay particles. This chemical and physical interaction would inhibit approach of the molecule by either cells or enzymes (see review by Theng et al., 1989).

Because of the diversity of the biological compounds associated with humic acids either through covalent linkage or a weaker association, the potential does exist for microbes to gain at least a limited amount of energy from catabolism of portions of the humic acid molecule. The most easily envisioned mechanism for energy production would be the mineralization of humified proteins and polysaccharides. Even the oxidation of these normally easily decomposed substances would be slow when they are associated with the large humic acid molecule. Not only would the proteins and polysaccharides be anticipated to be widely spaced on the humic acid molecules, but the humic acids are dispersed within the soil mineral components. Furthermore, the approach of the protease or hydrolase would be inhibited sterically by the obtrusive aliphatic/aromatic humic acid component. Thus, it can be proposed a priori that the organisms involved with mineralization of humified substances would be slow growing. This hypothesis appears to be true in that most reports of microbial growth on humic acids involve the activity of *Actinomyces* spp., *Nocardia* spp., and related microbes (e.g., Andreyuk and Gordienko, 1978; Andriiuk et al., 1973; Blondeau, 1989; Federov and Il'ina, 1963). Linkage of the observations that (1) humic acid molecules are decomposed biologically,

(2) oxidation of the random structure of the aliphatic/aromatic portion of the molecule would not yield sufficient energy to support microbial growth, and (3) humified polysaccharides and proteins could be oxidized to yield energy suggests that in the vast majority of the situations, the soil microbial community decomposes the aliphatic/aromatic humic acid core cometabolically. Energy for microbial growth could be supplied by oxidation of the humified proteins and polysaccharides as well as by nonhumified organic matter located in the same microsite as the humic acid molecule.

10.2.2 Analysis of Soil Organic Carbon Fractions

Difficulties in quantifying components of soil organic matter are not generally the result of shortage of analytical methods. An immense and diverse body of literature exists in which the quantification of xenobiotic chemicals or specific biologically synthesized compounds has been described. The primary difficulty results from attempts to apply the methods developed for analysis of reasonably homogenous or mildly complex samples to chemically complex soil samples. Association of the target organic compound with humic substances or soil minerals reduces recovery efficiencies. Furthermore, the reactivity of soil organic components may be enhanced during the extraction period, thereby creating a diverse array of procedural artifacts. Thus, implementation of any analytical procedure for use with soil systems requires modification in order to maximize extraction efficiency and to minimize artifact generation.

Analysis of a defined organic component in a heterogeneous matrix is complicated by the degree of interaction of various substituents with soil organic and mineral components. The intensity of the extraction procedure employed necessarily varies depending on the extent of association of the organic grouping of interest with the mineral and humic components, the complexity of the material to be extracted, plus the solubility of the substance to be analyzed. The least complex procedure would be that applied to analysis of substances dissolved in soil interstitial water or easily dissolved from particulate substances — for example, the extraction of free carbohydrates, water-soluble proteins, or simple organic compounds (e.g., acetate, amino acids) with water or ethanol. As the solubility of the organic component in water decreases or the association with soil minerals increases, use of more stringent extractants is necessitated. Differential solubility is commonly utilized to separate soil mineral and organic fractions. This situation is exemplified by the utilization of ether or methylene chloride to extract aliphatic compounds. Perhaps the most extreme extraction procedure is the separation of humic substances from mineral matter with strongly alkaline solutions (usually 0.5 M NaOH).

A clear understanding of the limitations of each fractionation procedure is necessary to determine its general applicability as well as to interpret the data yielded. Concerns range from the more mundane (how to remove the soil from the extraction suspension) to the highly critical determination of artifacts resulting from the extraction procedure. Some specific questions regarding soil organic matter fractionation follow:

- *What is the best method to reduce mineral contamination of the extract?* Soil extracts contain mixtures of the target compound plus dissolved soil minerals and suspended colloidal minerals such as clays. The degree of contamination with soil mineral colloids and salts is termed the **ash content** of the extract. Since these substances interfere with many of the analytical procedures used to quantify soil organic fractions subsequent to extraction, they must be removed or minimized. Ionic contaminants may be removed with ion exchange resins. Mineral inclusions are reduced by pretreatment of the soil with hydrofluoric acid prior to extraction.

- *Does the extraction procedure modify the organic carbon pool?* Strongly acidic or alkaline solutions are frequently utilized to dissociate soil organic components from soil minerals. This may result in oxidation, hydrolysis, or even total destruction of some compounds. Polysaccharides and proteins may be hydrolyzed under acidic conditions. The oxidation state of humic acids is increased under alkaline conditions in the presence of oxygen. Furthermore, some amino acids are destroyed by heating in the acidic solutions generally used for extraction of proteins from soil.

- *Are the extractions quantitative?* Two questions are generally considered in evaluating a procedure for extracting organic matter from soil: Is the procedure efficient? Is the array of compounds contained in the extract representative of that existing *in situ*? An indication of the proportion of material extracted is required for determination of *in situ* concentrations, but for many studies, a representative sample, not total extraction, is acceptable. As long as a representative, reproducible extraction of the organic matter is conducted, 100 percent efficiency is not required. Difficulties are encountered if not all components of a fraction are contained in an extract or if their ratio is altered by the extraction procedure. For example, should it be of interest to elucidate the diversity of polysaccharide monomers in an organic soil, all polysaccharide pools must be represented in the extract in proportion to their occurrence in soil. This mixture of sources includes the more easily extracted plant-associated materials as well as those less readily solubolized, bacterially synthesized heteropolysaccharides coating soil minerals.

- *How does the extraction procedure affect post-extraction analysis?* Due to the complexity of soil and the intensity of the extraction procedure necessitated by the association of organic materials with soil mineral and humic components, it is not uncommon to produce a soil extract containing high concentrations of substances that interfere with subsequent characterization of extract components. Probably the best example of this situation is the fulvic acid fraction. The high salt content in the fulvic acid fraction necessitates extensive purification of the extract before the individual organic components can be characterized.

Specific extraction procedures for analysis of soil organic matter typically involve dissolution of the organic matter in strongly alkaline or acidic solutions (e.g., Stevenson, 1994). All of the questions listed above pertain to these procedures. These methods are characterized by the intensity of the extractants (highly acidic or alkaline solutions), incomplete removal of the fraction from soil, modification of the substance of interest during the extraction, and production of an extraction solution that must be neutralized and desalted prior to analysis.

These limitations of extraction procedures create significant problems with data interpretation, particularly in comparing results attained with different soils or laboratories. Data must be interpreted with the understanding that extraction efficiencies may vary with soil type and that the degree of modification of the soil organic component during extraction may not be equivalent for different laboratories. The latter condition is particularly acute in comparing historical data in that many early experiments were conducted without an appreciation of the degree of artifacts created by the extraction procedure or of means of reducing these artifacts. For example, the alkaline extraction of soils must be conducted under a nitrogen atmosphere to reduce oxidation of the organic matter. See Stevenson (1994) for a more detailed discussion of specific procedures used in isolation of soil organic matter fractions, their applicability, and their limitations.

10.2.3 Structural versus Functional Analysis

As outlined in the previous section, procedures for analyzing organic components in soil are reasonably straightforward. For the soil microbiologist, the greater difficulty is derived from the fact that provision of a quantity and identity of a chemical entity in soil achieves only a minor portion of the objective. Experimental objectives usually require determination of a specific role of the organic materials in the soil ecosystem. None of the extraction procedures used for soil organic matter differentiate organic matter pools on a functional basis. More commonly, substances are extracted and quantified based on their solubility (e.g., humic substances), chemical class (e.g., protein, polysaccharide, lipid), or even distribution in specific biomass components (microbial, plant, or animal). Thus, the soil extracts contain a mixture of functionally active and inactive materials. For example, polysaccharide that is actively contributing to microbial respiration may be found in the fulvic acid, humic acid, and humin soil fractions. In contrast, protein, all of which can be quantified in an acid hydrolysate of soil, is a mixture of protein directly contributing to soil respiration plus biodegradation-resistant or chemically stabilized protein (i.e., that protein stabilized by covalent linkage into soil humic substances).

The capacity to analyze soil organic matter fractions based on their role in the ecosystem is limited. One approximate division of soil organic substances of some functional value — at least conceptually — is their separation based on its decomposition susceptibility. Organic components may be classified as being easily (readily) decomposable or relatively biodegradation resistant. This elementary division allows discernment of that portion of fixed carbon that provides energy for growth and development of the rapidly dividing microbial populations and that fuels those more ephemeral processes commonly associated with hourly, daily, or even seasonal changes in soil carbon, as opposed to the more constant processes driven by the gradual decomposition of the more stable organic fractions.

Readily metabolized soil organic matter provides a carbon and energy source for the soil microbial community with a minimal energy expenditure by the microbe. The microbial investment in energy acquisition involves synthesis of catabolic enzymes and direct substrate activation (as involved with the substrate-level phosphorylation of glucose in the hexose monophosphate pathway of glucose catabolism). Compounds that require the action of a few enzymes prior to producing energy to the microbe are more easily decomposed than complex substances requiring interaction with a variety of enzymes prior to imposition of an oxidative change that provides energy to the microbe. The greater the energy deficit created by the microbe in preparation for catabolism of a substance, the less likely that the substance will serve as a primary energy source for the soil microbial community. For example, glucose and other monosaccharides, disaccharides such as sucrose, amino acids, proteins, polypeptides, and simple aromatic compounds are all classified as easily metabolized compounds. Humic substances and lignin are extreme examples of more biodegradation-resistant substances.

A useful means of grouping compounds into readily decomposed and biodecomposition-resistant fractions is to evaluate the diversity of enzymes that must be synthesized to mineralize the substrate in relationship to the quantities of energy provided to the microbes. Simple organic compounds are easily decomposed and provide adequate energy to support viable microbial populations. Similarly, polysaccharides, although they may have extended longevity in soil as long as a few years, still yield growth-supporting energy with minimal expenditures by the microbes for enzyme synthesis. In contrast, considerable energy for microbial growth is contained in lignin and humic acids, but the energy expenditure involved in enzyme synthesis that would be required for the microbes to recover this energy in growth-supporting quantities is prohibitive.

Stabilization of Easily Decomposed Substances in Soil: A substance may be easily grouped in one of these two metabolic categories through analysis of its decomposition rate by axenic microbial cultures or even when added to soil samples. Again, the situation is complicated by the heterogeneous distribution of organic compounds and the diversity of their chemical associations in soil. An unquestionably easily metabolized substance, such as carbohydrate or proteinaceous material, may exhibit an extreme longevity in a soil ecosystem. This may result from its association with residual cell structures or association with soil humic or mineral matter.

Inherent properties of a chemical compound control its longevity in soil provided the substance exists in a free state and the microbial community (or its enzymes) enjoys unimpaired access to the material. Reductions in substrate concentrations decrease the probability of such access as does occlusion within soil mineral structures or retention within cell walls. All cytoplasmic components, irrespective of the mineralization potential, are occluded from microbial attack until the cell wall is breached. Similarly, access to these energy-yielding substances is also precluded or limited sterically by linkage to soil humic acids.

Carbon Dioxide Yields from Catabolism of Easily Metabolized Substances: The role of readily decomposable organic compounds in soil carbon cycling is unquestioned. These compounds are the primary sources of energy and nutrients in an active soil ecosystem. Acetate catabolism in soil provides an excellent example of the decomposition kinetics and fate of this chemical grouping. After 6 days of incubation in soil, typically greater than 70 percent of acetate carbon is usually evolved as carbon dioxide, with the remainder incorporated into microbial biomass components, such as carbohydrates and amino acids (Sørensen and Paul, 1971). As much as 90 percent of the carbon of readily decomposable compounds has been detected as carbon dioxide (e.g., Kassim et al., 1982,) with as much as 95 percent of the carbon contained in carbon dioxide plus microbial biomass pools. These decomposition efficiencies are sufficiently reproducible that if they are not achieved where anticipated, the procedure must be carefully reevaluated. If less than 70 percent of the carbon contained in a readily decomposable substance is evolved as carbon dioxide in a particular experiment, either an error in experimental design has occurred or some unanticipated soil property is inhibiting microbial activity. An example of the former situation is an incubation of a bulk soil sample where the soil moisture was not maintained throughout the experimental period. Declines of the soil water level to levels limiting microbial activity would result in an underestimation of the biodegradation potential of the soil microbial community. Alternatively, an environmentally meaningful result could have been occluded by a failure to account for all of the potential fates of the compound. An experimenter may assume that all of the amended substance should be mineralized to carbon dioxide. That reaction could occur, but insufficient carbon dioxide could be detected to account for total mineralization of the compound. This result could occur in a soil system where the pH of the soil favors conversion of carbon dioxide to bicarbonate and carbonate (pH > 7. The true quantity of carbon dioxide generated by carbon mineralization could be revealed by treatment of the soil sample following the incubation period with hydrochloric acid. Were the test substance radiolabled with ^{14}C, the $^{14}CO_2$ would be released to the gas phase where it could be quantified.

For comparison, a small portion of the carbon contained in a biodegradation-resistant compound is evolved as carbon dioxide, even after extended incubation periods. In a study of melanin decomposition, after a 1-year incubation period, between 5 and 13 percent of the carbon had been evolved as carbon dioxide with less than 0.7 percent of the carbon retained in microbial biomass (Stott et al., 1983). With this class of compounds, the bulk of the amended carbon is retained in unmodified or slightly changed parent molecules. The biodegradation-resistant compounds have an augmented probability of being incorporated into soil humic substances (see Tate, 1987).

10.2.4   Microbial Mediators of Soil Carbon Cycle Processes

The concept of microbial infallibility presented above was soundly based on an appreciation of the diversity of microbial species that function as primary decomposers in soils. Biologically synthesized organic carbonaceous compounds are mineralized in essentially all soil systems where biological life is supported by the chemical and physical traits of the site. Organic carbon compounds are oxidized in soil with a variety of compounds such as nitrogen oxides, molecular oxygen, organic substances, and sulfur compounds serving as terminal electron acceptors. Active populations include primary decomposers (bacteria, actinomycetes, and fungi) as well as secondary feeders.

The breadth of organisms involved in carbon mineralization is exemplified by listing a portion of the microbial world involved in cellulose catabolism. This primary soil community energy source is used as a carbon and energy source by fungi (including *Alternaria, Aspergillus, Chaetomium, Coprinus, Penicillium, Phanerochaete, Polyporus, Rhizopus, Trichothecium,* and *Zygorhynchus*), yeasts (including *Kluyveromyces* and *Candida*), actinomycetes (including *Micromonospora, Microbispora, Nocardia, Streptomyces,* and *Streptosporangium*), as well as a wide variety of commonly occurring bacterial genera (including *Bacillus, Cellulomonas, Clostridium, Corynebacterium, Cytophaga, Pseudomonas, Sporocytophaga,* and *Vibrio*) (Alexander, 1977; Béquin, 1990).

Considering the vast number of microbial genera that are involved in carbon-cycling processes, a functional division of cycle participants is commonly more useful in assessing metabolic potential and kinetics individual soil sites than is a detailed analysis of the microbial species present in the sample. Perhaps the most commonly encountered grouping of soil microbes is based on growth substrate and the rate of catabolism of the substrate. For example, soil bacteria may be grouped into such metabolic categories as amino acid oxidizers or cellulose degraders. The capacity of the carbon substrate to provide energy for growth and the capability of the microbial species to compete with other members of the microbial community result in their separation into two groups based on growth rate. Rapidly growing soil bacteria, which are primarily involved in catabolism of easily metabolized or fresh biomass, are termed **zymogenous** bacteria. The slow-growing organisms (i.e., those predominantly associated with the steady decomposition of native soil organic matter) are classed as **autochthonous** (Winogradsky, 1924).

This growth substrate/rate division of soil bacteria is useful to some degree, but as the metabolic diversity of the various species (especially members of the genera *Arthrobacter* and *Pseudomonas*) and their survival capacity in soil are elucidated, it becomes clear that an unequivocal separation of soil microbes into these defined groups is difficult. This delineation of soil bacteria is further complicated by the observation that growth rate of an organism is the result of the expression of genetically determined capabilities within the physical and chemical framework of the soil microsite. Thus, an organism that grows rapidly with an easily metabolized substrate as a source of carbon and energy source in the test tube may have generation times of days or longer in soil. An autochthonous or zymogenous designation becomes cloudy at best.

Other large-scale groupings that are commonly encountered, which are perhaps as tenuous as the autochthonous/zymogenous grouping, include division of the soil population into native and alien species and, a more modern apparition of this separation, natural vs. bioengineered. As with the bipartite division of autochthonous and zymogenous bacteria, these designations are useful in clear-cut cases, but become confusing or even convey an erroneous impression of the diversity of the soil system with the more cosmopolitan or even with little-studied microbes. For example, alien organisms to the soil system could be cynically defined as those organism that have not to date been isolated from soil. In such situations, future studies would result in "alien" becoming "native." Further confusion may be derived from varying degrees of inclusiveness associated with "native." Is a determination of native soil organisms related to a specific soil site, ecosystem, or all terrestrial soils? Difficulties with designations of bioengineered vs. natural will likely emerge as nonengineered organisms with the same or nearly identical genotypes to the laboratory-produced microbes are isolated from soil.

A caveat associated with analysis of data in which soil microbes are divided into any of these general groupings is that the subjectivity of the decision process must be evaluated to assure that a bias has not been introduced into the conclusions of the experimental study as a result of this artificial grouping of soil populations.

10.3 Kinetics of Soil Carbon Transformations

Identification and quantification of the proportion of the soil carbon distributed in various organic matter pools is both interesting and useful. But in reality, from the view of determining the impact of this soil component on our ecosystem, it is the movement of carbon between these various pools and their rate of return to the atmosphere that are better descriptors of the dynamics of carbon cycle processes occurring in any specific soil site. As discussed above, questions that are dependent upon carbon flux measurements (kinetics of the process) include evaluation of such terrestrially important processes as the capacity of the soil community to reduce or to modulate atmospheric carbon dioxide concentrations and its utility in recycling organic wastes.

A variety of mathematical models have been developed in which carbon mineralization processes in soil and the effects of soil environmental properties are described (see Tate, 1987; Alexander, 1999). Based on the observation that most carbon mineralization results from biological decomposition, it would be reasonable to assume that the carbon transformations are best described by Michaelis-Menten kinetics (see Chapter 6). The decomposition of simple biochemicals can easily be described by this simple descriptor of enzyme kinetics or similar equations describing microbial growth kinetics (the Monod equation).

More commonly, organic-carbon based substances decomposed in soil are a complex mixture of biochemicals that are decomposed by a variety of soil microbes as described above in a heterogeneous ecosystem. In this situation, carbon dioxide flux rates are a summation of the kinetics of individual biodegradation processes. Thus, any Michaelis-Menten–type relationships could be obscured by the variety of active microbes present, the diversity of carbon-based substrates catabolized, differences in individual growth rates and enzyme efficiencies, and variation in the impact of the physical environment on the activity of the organisms and the enzymes. That is, kinetics of carbon dioxide evolution become the sum of all sources of this gas. Therefore, the rate of decay of these substances is frequently more easily described by a simple first order equation with a series of terms for each biochemical group present in the complex substrate. A generic representation of the equation is as follows:

$$\Upsilon_t(\text{Biomass residue}) = C_1 e^{-k_1 t} + C_2 e^{-k_2 t} + \cdots \qquad (10.1)$$

Where C_1 and C_2 are concentrations of the individual organic carbon fraction, t is time, and k_1 and k_2 are decomposition rate constants. An individual term could be added to this equation for each biochemical component of the complex organic substance studied, but experience demonstrates that the decomposition rates can be grouped into readily metabolized and more biodegradation-resistant fractions. That is, the decomposition of complex organic substances in soil is biphasic (Fig. 10.4), with an initially rapid decomposition period (readily metabolized or labile organic carbon) followed by a slow decomposition phase (resistant and occluded organic matter). The magnitude of the rapid decomposition phase is proportional to the quantity of readily decomposed organic matter in the tissue and its accessibility. This phase could represent the majority of the carbon contained in the substrate (succulent green grass tissue mixed into soil) or a minority of the carbon (woody tissue). It must be noted that there is a certain amount of arbitrariness associated with the biodegradation susceptibilities connected with each portion of this biphasic curve. Some of the carbon dioxide produced during the rapid

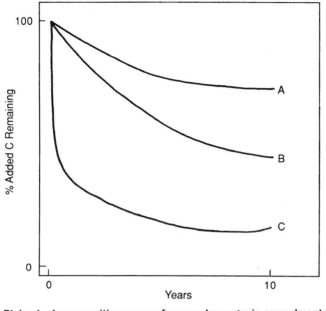

Fig. 10.4 **Biphasic decomposition curves for complex organic amendments to soil. The percent easily decomposable organic matter in A ≪ B ≪ C.**

decline phase (initial decomposition portion of the curve) is the result of mineralization of the more resistant biomass components. These substances are clearly mineralized throughout the incubation period, just at a much slower rate than is associated with the easily decomposable pool. Furthermore, some of the carbon dioxide produced late in the incubation period is the product of mineralization of those easily metabolized compounds whose decomposition kinetics reflect those of the more resistant substances. As indicated above, this phenomenon is the result of the fact that the degradable substances are protected by the resistant materials (e.g., located within a plant cell wall) and thus can only be mineralized after the barrier is breached.

Decomposition kinetics of rice and barley straw (Murayama, 1984), rye straw and maize (Jenkinson and Ayanaba, 1977), and water hyacinth [*Eichornia crissipes* (Mart.) Solmes] (Moorhead et al., 1986) provide examples of the applicability of this two-term equation in evaluating transformations of complex organic substances in soil. These materials are all constituted primarily of labile organic carbon. Therefore, most of their biomass is mineralized to carbon dioxide during the initial year following incorporation into soil. With these types of material, it is reasonable to anticipate that approximately two-thirds of the carbon will be evolved as carbon dioxide during the first year of incubation in soil. In contrast, first-year weight loss of leaf litter, which contains more biodegradation-resistant structures than do the succulent grass tissues and water hyacinths discussed above, have been shown to be considerably smaller (e.g., *Fagus grandifolia*, 21 percent; *Acer saccharum*, 32 percent; *Quercus alba*, 39 percent) (Shanks and Olson, 1961).

These data were collected in aerobically incubated soil samples, but similar results can be derived from anaerobically incubated soil samples, if all decomposition products are considered. Gale and Gilmour (1988) derived rate constants for decomposition of alfalfa (*Medicago sativa* L.) in aerobically and anaerobically incubated soil samples. The aerobically incubated samples were incubated at optimum soil moisture under a carbon dioxide-free atmosphere, whereas the anaerobic soils were flooded under a nitrogen atmosphere. Carbon dioxide, methane, and water-soluble organic carbon were quantified. Three decomposition phases were distinguished (rapid, intermediate, and slow). Aerobic rate constants were 0.123, 0.059, and 0.0095 day^{-1} for each phase. The constants for the anaerobic decomposition were slower (0.118 and 0.024 day^{-1} for the rapid and intermediate phases, respectively). No constant was derived in this study for the slow phase.

These rate constants in the mathematical model are system dependent in that they are the result of the summation of all environmental parameters affecting the decomposition process in the individual site studied at the time of the experiment. Thus, the specific equation derived from any individual study can be extrapolated only to a system with similar properties, such as temperature, pH, moisture, and soil mineral content. The magnitude of the variation in these constants is not only shown in the above study with variation in oxygen tensions, but was also revealed in a comparison of the impact of climate on ryegrass decomposition kinetics in England and Nigeria (Ayanaba and Jenkinson, 1990). Decomposition of ryegrass and maize in Rothamsted, England, was described by the equation:

$$\Upsilon(\text{percent retention}) = 70.9\,e^{-2.83t} + 29.1\,e^{-0.087t} \qquad (10.2)$$

where t represents time in years. The decomposition of these substances in Nigeria was described by the same equation, except that each decomposition coefficient was multiplied by four. Variation in the decomposition constants due to localized variation in seasonal soil moisture differences and soil type effects were also detected (Ayanaba and Jenkinson, 1990). Should the experimental objectives include derivation of an ecosystem-independent mathematical representation of biomass decomposition in soil, then more extensive models than the two-term model described herein must be derived (see Tate, 1987).

The data collected in the ·types of experiments described herein are commonly derived by assessing the evolution of ^{14}C-labeled carbon dioxide evolution from soil samples amended with ^{14}C-labeled plant biomass. Thus, strictly speaking, in such studies the longevity of the carbon atom in soil organic fractions is being assessed rather than the survival of the actual plant biomass. This distinction is critical in conceptualizing the nature of the organic carbon from which the carbon dioxide is evolved in the latter stages of such a study. Initially, the vast majority of the carbon dioxide collected results from mineralization of added labile plant carbon. During this time of rapid plant biomass decomposition, a portion of the carbon mineralized is incorporated into soil microbial biomass. As time passes, some of the ^{14}C-label is incorporated into other soil organic matter fractions, such as humic substances, *in situ*-synthesized polysaccharides, and stabilized proteins. Thus, during the slow phase of the decomposition curve (Fig. 10.4), the ^{14}C evolved from the soil sample could originate in residual (more biodegradation-resistant) plant components of the biomass initially added to the soil as well as from the mineralization of soil microbial biomass and humic substances synthesized *in situ* subsequent to amendment. For example, Sørensen (1987) found that after 8 to 15 years' incubation in soil, the portion of the carbon originally added to the soil sample as barley straw was distributed in the soil amino acid fraction (21 percent) and microbial biomass (2.7 percent).

10.4 Conclusions: Management of the Soil Carbon Cycle

The soil carbon cycle has regional as well as worldwide implications. For example, ecosystem productivity and stability relies to a large part on the activity of the soil microbial community, whose nutrients and energy is provided by mineralization of organic carbon compounds. The magnitude of soil organic matter reservoirs and their mineralization rate are modulators of atmospheric carbon dioxide loadings. In each of these situations, the rate and direction of change of the carbon forms are of greater importance than its mere presence. In soils at steady state, total soil organic carbon contents are relatively constant. Yet large quantities of carbon may be cycling between mineral and organic forms. The primary environmental effect of this nutrient cycling is in enhancement of soil properties conducive to plant community development, thereby resulting in increased system productivity. On a global scale, the primary soil impact on world carbon balances originates from ecosystems in transition, primarily disturbed or managed sites. Such perturbation of soil properties induces attainment of an altered equilibrium concentration of organic matter. A new steady-state carbon content (and therefore, carbon dioxide flux rate) is achieved. That is, there is an optimal level of organic matter attained by each soil ecosystem.

Soil organic matter contents may range from nearly zero (e.g., desert soils) to approaching 100 percent. Furthermore, mismanagement of a soil site (i.e., excessive cultivation to destroy soil structure) may result in a nearly total depletion of colloidal organic matter pools. In contrast, under swampy situations, soils (histosols) may be formed consisting of essentially 100 percent organic carbon. With most soils, a level of a few percent organic matter is usually attained and maintained, both in properly managed and native soil sites. The quantity of organic matter retained in a particular soil is the result of the achievement of a balance between inputs and mineralization as affected by the physical and chemical properties of the site.

Since organic matter synthesis and decomposition are primarily biologically controlled, they can be managed. Modulators of soil biological activity (see chapter 5) that may be anthropogenically controlled include pH, temperature, nutrient concentration, and moisture. Furthermore, as the capacity to develop genetically engineered microorganisms improves, inoculation of a particular soil site with microbes specifically designed for optimal function therein will become a management tool. The soil carbon cycle will be maintained to maximize societal benefits and stewardship of a renewable resource, our soil ecosystems.

References

Aguilar, R., E. F. Kelly, and R. D. Heil. 1988. Effects of cultivation on soils in Northern Great Plains rangeland. Soil Sci. Soc. Am. J. 52:1081–1085.

Alegre, J. C., and D. K. Cassel. 1986. Effect of land-clearing methods and post clearing management on aggregate stability and organic carbon content of a soil in the humid tropics. Soil Sci. 142:289–295.

Alexander, M. 1977. Introduction to soil microbiology. John Wiley & Sons, NY. 467 pp.

Alexander, M. 1999. Biodegradation and Bioremediation, 2nd edition. Academic Press, NY. 453 pp.

Amundson, R., and H. Jenny. 1991. The place of humans in the state factor theory of ecosystems and their soils. Soil Sci. 151:99–109.

Andreyuk, E. I., and S. A. Gordienko. 1978. Transformation of humic acids by soil actinomycetes. Mikrobiol. Zh. (Kiev) 40:690–697 (Russian).

Andriiuk, K. I., S. O. Hordienko, I. N. Havrysh, H. I. Konotop, and V. A. Martynenko. 91973. Decomposition of peat humic acids by associative cultures of microorganisms. Mikrobiol. Zh. 35:554–559 (Russian).

Armentano, R. V. 1980. Drainage of organic soils as a factor in the world carbon cycle. Bioscience 30:825–830.

Ayanaba, A., and D. S. Jenkinson. 1990. Decomposition of carbon-14 labeled ryegrass and maize under tropical conditions. Soil Sci. Soc. Am. J. 54:112–115.

Barber, S. A. 1979. Corn residue management and soil organic matter. Agron. J. 71:625–627.

Béquin. 1990. Molecular biology of cellulose degradation. Annu. Rev. Microbiol. 44:219–248.

Blondeau, R. 1989. Biodegradation of natural and synthetic humic acids by the white rot fungus *Phanerochaete chrysosporium*. Appl. Environ. Microbiol. 55:1282–1285.

Bossert, I., W. M. Kachel, and R. Bartha. 1984. Fate of hydrocarbons during oily sludge disposal in soil. Appl. Environ. Microbiol. 47:763–767.

Boyle, M. 1990. Biodegradation of land-applied sludge. J. Environ. Qual. 19:640–644.

Crawford, R. L. 1981. Lignin Biodegradation and Transformation. John Wiley & Sons, NY. 154 pp.

Dec, J., and J.-M. Bollag. 1997. Determination of covalent and noncovalent binding interactions between xenobiotic chemicals and soil. Soil Sci. 162:858–874.

Dibble, J. T., and R. Bartha. 1979a. Rehabilitation of oil-inundated agricultural land: A case history. Soil Sci. 128:56–60.

Dibble, J. T., and R. Bartha. 1979b. Effect of environmental parameters on the biodegradation of oil sludge. Appl. Environ. Microbiol. 37:729–739.

Douglas, B. F., and F. R. Magdoff. 1991. An evaluation of nitrogen mineralization indices for organic residues. J. Environ. Qual. 20:368–372.

Epstein, E., J. M. Taylor, and R. L. Chaney. 1976. Effects of sewage sludge compost applied to soil on some soil physical and chemical properties. J. Environ. Qual. 5:422–426.

Eswaran, H., E. van den Berg, and P. Reich. 1993. Organic carbon in soils of the world. Soil Sci. Soc. Am. J. 57:192–194.

Fedorov, M. V., and T. K. Il'ina. 1963. Utilization of humic acid by soil actinomycetes as a sole source of carbon and nitrogen. Microbiology 32:234–237.

Gale, P. M., and J. T. Gilmour. 1988. Net mineralization of carbon and nitrogen under aerobic and anaerobic conditions. Soil Sci. Soc. Am. J. 52:1006–1010.

Gupta, U. C. 1967. Carbohydrates. *In* A. D. McLaren and G. H. Peterson (eds.) Soil Biochemistry pp. 99–118. Dekker, NY.

Harden, J. W., E. T. Sundquist, R. F. Stallard, and R. K. Mark. 1992. Dynamics of soil carbon during deglaciation at the Laurentide ice sheet, Science 258:1921–1924.

Hatcher, P. G., M. Schnitzer, L. W. Dennis, and G. E. Marciel. 1981. Aromaticity of humic substances in soil. Soil Sci. Soc. Am. J. 45:1089–1094.

Havlin, J. L., D. E. Kissel, L. D. Maddux, M. M. Claassen, and J. H. Long. 1990. Crop rotation and tillage effects on soil organic carbon and nitrogen. Soil Sci. Soc. Am. J. 54:448–452.

He, X.-T., S. J. Traina, and T. J. Logan. 1992. Chemical properties of municipal solid waste composts. J. Environ. Qual. 21:318–329.

Jenkinson, D. S., and A. Ayanaba. 1977. Decomposition of carbon-14 labeled plant material under tropical conditions. Soil Sci. Soc. Am. J. 41:912–915.

Jenkinson, D. S., and J. H. Rayner. 1977. The turnover of soil organic matter in soil of the Rothamsted classical experiments. Soil Sci. 123:298–305.

Jenny, H. 1980. The Soil Resource, Origin and Behavior. McGraw-Hill Book Co., NY.

Kassim, G., J. P. Martin, and K. Haider. 1982. Incorporation of a wide variety of organic substrate carbons into soil biomass as estimated by fumigation procedure. Soil Sci. Soc. Am. J. 45:1106–1112.

Kern, J. S., and M. G. Johnson. 1993. Conservation tillage impacts on national soil and atmospheric carbon levels. Soil Sci. Soc. Am. J. 57:200–210.

Kirk, T. K., and R. L. Farrell. 1987. Enzymatic "combustion": The microbial degradation of lignin. Annu. Rev. Microbiol. 41:465–505.

Kögel-Knaber, I. 1993. Biodegradation and humification processes in forest soils. *In* J.-M. Bollag and G. Stotzky (eds.), Soil Biochemistry 6:101–135. Marcel Dekker, NY.

Malcolm, R. L. 1990. The uniqueness of humic substances in each of soil, stream and marine environments. Anal. Chim. Acta 232:19–30.

Mann, L. K. 1986. Changes in soil carbon storage after cultivation. Soil Sci. 142:279–288.

Martin, J. P., and K. Haider. 1979. Biodegradation of ^{14}C-labeled model and cornstalk lignins, phenols, model phenolase humic polymers, and fungal melanins as influenced by a readily available carbon source and soil. Appl. Environ. Microbiol. 38:283–289.

Mays, D. A., G. L. Terman, and J. C. Duggan. 1973. Municipal compost: Effects on crop yields and soil properties. J. Environ. Qual. 2:89–92.

McMurtrie, R. E., H. N. Comins, M. U. F. Kirschbaum, and Y.-P. Wang. 1992. Modifying existing forest growth models to take account of effects of elevated CO_2. Aust. J. Bot. 40:657–677.

Mitchell, M. J., R. Hartenstein, B. L. Swift, E. F. Neuhauser, B. I. Abrams, R. M. Mulligan, B. A. Brown, D. Craig, and D. Kaplan. 1978. Effects of sewage sludges on some chemical and biological characteristics of soil. J. Environ. Microbiol. 7:551–559.

Moorhead, K. K., D. A. Graetz, and K. R. Reddy. 1986. Decomposition of fresh and anaerobically digested plant biomass in soil. J. Environ. Qual. 16:25–28.

Murayama, S. 1984. Decomposition kinetics of straw saccharides and synthesis of microbial saccharides under field conditions. J. Soil Sci. 35:231–242.

Norris, D. J. 1980. Landspreading of oily and biological sludges in Canada. Proc. Ind. Waste Conf. 35:10–16.

Oades, J. M. 1988. The retention of organic matter in soils. Biogeochem. 5:33–70.

Ouyang, Y., and L. Boersma. 1992a. Dynamic oxygen and carbon dioxide exchange between soil and atmosphere: I. Model development. Soil Sci. Soc. Am. J. 56:1695–1702.

Ouyang, Y., and L. Boersma. 1992b. Dynamic oxygen and carbon dioxide exchange between soil atmosphere: II. Model simulations. Soil Sci. Soc. Am. J. 56:1702–1710.

Piccolo, A., and J. S. C. Mbagwu. 1990. Effects of different organic waste amendments on soil microaggregates stability and molecular sizes of humic substances. Plant Soil 123:27–37.

Polglase, P. J., and Y. P. Wang. 1992. Potential CO_2-enhancing carbon storage by the terrestrial biosphere. Aust. J. Bot. 40:641–656.

Rasmussen, P. E., and C. R. Rohde. 1988. Long-term tillage and nitrogen fertilization effects on organic nitrogen and carbon in a semiarid soil. Soil Sci. Soc. Am. J. 52:1114–1117.

Richter, D. D., L. I. Barbar. M. A. Huston, and M. Jaeger. 1990. Effects of annual tillage on organic carbon in a fine-textured udalf: The importance of root dynamics to soil carbon storage. Soil Sci. 149:78–83.

Rogers, H. H., J. F. Thomas, and G. E. Bingham. 1983. Response of agronomic and forest species to elevated atmospheric carbon dioxide. Science 220:428–429.

Scott, H. D., and L. S. Wood. 1989. Impact of crop production on the physical status of a Typic Albaqualf. Soil Sci. Soc. Am. J. 53:1819–1825.

Shanks, R. E., and J. S. Olson. 1961. First year breakdown of leaf litter in southern Appalachian forests. Science (Washington, DC) 134:194–195.

Sorensen, L. H. 1987. Organic matter and microbial biomass in a soil incubated in the field for 20 years with ^{14}C-labelled barley straw. Soil Biol. Biochem. 19:39–42.

Sørensen, L H., and E. A. Paul. 1971. Transformation of acetate carbon into carbohydrate and amino acid metabolites during decomposition in soil. Soil Biol. Biochem. 3:173–180.

Stevenson, F. 1994. Humus Chemistry: Genesis, Composition, Reactions. John Wiley & Sons, NY. 443 pp.

Stott, D. E., J. P. Martin, D. D. Focht., and K. Haider. 1983. Biodegradation, stabilization in humus, and incorporation into soil biomass of 2,4-D and catechol carbons. Soil Sci. Soc. Am. J. 47:66–70.

Stout, J. D., K. M. Goh, and T. A. Rafter. 1981. Chemistry and turnover of naturally occurring resistant organic compounds in soil. *In* E. A. Paul and J. N. Ladd (eds.), Soil Biochemistry 5:1–73. Dekker, NY.

Sundquist, E. T. 1993. The global carbon diodide badget, Science 259:936–941.

Tate, R. L. III. 1980. Microbial oxidation of organic matter of histosols. *In* M. Alexander (ed.), Adv. in Microbial Ecol. 4:169–201. Dekker, NY.

Tate, R. L., III. 1987. Soil Organic Matter: Biological and Ecological Effects. John Wiley & Sons. NY. 291 pp.

Theng, K. G., K. R. Tate, and P. Sollins. 1989. Constituents of organic matter in temperate and tropical soils. Pp. 5–32. *In* D. C. Coleman, J. M. Oades, and G. Uehara (eds.), Dynamics of soil organic matter in tropical ecosystems. University of Hawaii Press, Honolulu.

Vitousek, P. M. 1991. Can planted forests counteract increasing atmospheric carbon dioxide? J. Environ. Qual. 20:348–354.

Von Wandruszka, R. 1998. The micellar model of humic acid: Evidence from pyrene fluorescence measurements. Soil Sci. 163:921–930.

Winogradsky, S., 1924. Sur la microflora autochtone de la terre arable. Comptes rendus ebdomadaire des séances de l'Academie des Sciences (Paris) D, 178:1236–1239.

Wood, C. W., R. J. Mitchell, B. R. Zutter, and C. L. Linn. 1992. Loblolly pine plant community effects on soil carbon and nitrogen. Soil Sci. 154:410–419.

Woodwell, G. M., J. E. Hobbie, R. A. Houghton, J. M. Melillo, B. Moore, B. J. Peterson, and G. R. Shaver. 1983. Global deforestation: Contribution to atmospheric carbon dioxide. Science 222:1081–1086.

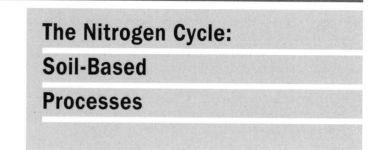

The Nitrogen Cycle:
Soil-Based
Processes

The nitrogen cycle consists of a series of oxidation and reduction transformations of nitrogen. The major processes occurring therein can be summarized as follows (Fig. 11.1): Relatively inert atmospheric nitrogen (dinitrogen, N_2) is converted into a biologically available form (fixed nitrogen in the biomass of the microbe catalyzing the process). The reduction of dinitrogen into organic

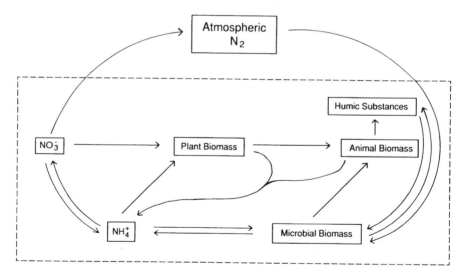

Fig. 11.1 Conceptual model of the nitrogen cycle. Nitrogen reservoirs enclosed in the dashed lines are resident in soil, whereas dinitrogen is replenished from atmospheric sources.

nitrogen is nitrogen fixation. This organic nitrogen is mineralized into ammonium (nitrogen mineralization) by decomposer populations. The oxidation of the ammonium to nitrate (nitrification) provides energy to the nitrifiers and produces a form of fixed nitrogen generally used by the plant community. Ammonium and nitrate may be assimilated into microbial biomass (immobilization). The most oxidized form of fixed nitrogen in soil, nitrate, is returned to the atmosphere as dinitrogen and nitrous oxide by organisms using the nitrogen as a terminal electron acceptor (denitrification).

For the practical objective of understanding the magnitude or relative importance of these processes in soil, the soil nitrogen cycle can easily be divided into two "minicycles": (1) a predominantly soil-resident portion consisting of those processes (mineralization, ammonification, nitrification, and immobilization) associated primarily with the conversion of organic nitrogenous compounds to plant-available mineral forms and their return to biomass (plant, microbial, and, indirectly, animal), and (2) the more inclusive (soil as well as the greater biosphere-resident processes) cycle entailing the conversion of atmospheric dinitrogen into forms available to the living components of the ecosystem (i.e., nitrogen fixation), the transformations within the soil described above, and the return of the fixed nitrogen to atmospheric dinitrogen (i.e., denitrification). The interrelationships of these two aspects of the nitrogen cycle are depicted in Fig. 11.1.

The significance of this positional differentiation of the nitrogen cycle in native environments is readily exemplified and contrasted by comparison of the situation in a climax forest ecosystem and a newly developing soil system, such as would occur with the reclamation of a mineland soil in which all naturally occurring organic matter has been destroyed. In the forest system, the use of fixed nitrogen sources from external sources (atmospheric deposition and nitrogen fixation) is minimal. As much as 85 to 90 percent of the fixed nitrogen incorporated into new plant biomass is derived from mineralization of soil humus and indigenous biomass. The approximately 10 to 15 percent of the fixed nitrogen lost from the system (via denitrification, leaching, runoff, etc.) is replaced by a low level of nitrogen fixation. In contrast, for a soil system with little accumulated organic matter, fixed nitrogen must be imported. This task is accomplished by nitrogen-fixing bacteria. For example, in a pioneer system (volcanic ash or even some mine spoils), indigenous fixed nitrogen resources are minimal. Plant biomass productivity is total reliant on external resources. Nitrogen can be provided symbiotically by lichens or free-living nitrogen-fixing bacteria (see Chapters 12 and 13) or through amendment with organic residues in managed systems.

In those situations where the bulk of the nitrogen utilized for plant biomass synthesis is the product of recycling soil-resident nitrogen compounds between organic and mineral forms, homeostasis of the ecosystem can be maintained only if nutrient losses are minimized. Thus, it is advantageous for the majority of the fixed nitrogen contained in soil reservoirs to be existent in nonmobile forms, such as organic or biomass nitrogenous compounds and, to some degree, ammonium ion. Fortunately, most of the fixed nitrogen in soil systems is retained in organic nitrogen pools. Typical data demonstrating the distribution of soil nitrogen in organic and mineral forms are presented in Table 11.1. Although soil nitrate and ammonium concentrations are highly variable with time and position in the soil profile, note that these compounds at most constitute a few percent of the total soil nitrogen. In the examples cited, quantities of nitrogen in soil organic matter ranged from about 10 to 500 fold the mass of nitrogen in mineral forms. (For further discussion of distribution of nitrogen in these various soil reservoirs see, Tate, 1987, or Stevenson, 1986 for a review of the topic or Bolton et al., 1990; Cabrera and Kissel, 1988a; Cassman and Munns, 1980; Clarholm et al., 1981, or Serna and Pomares, 1992, for examples of primary research involving nitrogen transformations involving these nutrient pools.)

This documentation of the predominance of organic nitrogen in soil ecosystems reveals a potential indicator of plant biomass production potential. Since plant biomass synthesis is frequently controlled by the availability of soil mineral nitrogen resources (other potential delimiters of plant productivity include soil moisture, phosphate or trace mineral levels, and a variety of soil physical and chemical properties), assessment of net nitrogen mineralization rates of the internal soil organic nitrogen pools and/or amendment rates of external fixed nitrogen resources can provide a valuable estimation of the fertility of an ecosystem. Nitrogen inputs from sources external to the soil biological community (fixation or soil amendment) are the primary determinants of ecosystem productivity in soils containing low levels of biologically decomposable organic matter or in cropped soils receiving inputs of industrially fixed nitrogen or organic residues.

A conceptual model of the dependence of plant community productivity and structure on nitrogen mineralization is depicted in Fig. 11.2. Nitrogen

Table 11.1 Examples of Nitrogen Distribution in Mineral and Organic Forms: Agricultural Soils from Valencia, Spain,[a] and Soils Beneath Shrub-Steppe Vegetation (Southeastern Washington, USA[b])

Soil	Organic N (mg N Kg^{-1} soil)	Mineral N (mg N kg^{-1} soil)	Mineral N (%)
Sandy soils (Spain) ($n = 9$)	300–2470	0.0–86	0.0–7.1
Loamy soils (Spain) ($n = 9$)	710–2400	3 –153	0.4–6.7
Clay soils (Spain) ($n = 9$)	720–3680	3 –85	0.2–1.8
Shrub-steppe soils ($n = 15$)	1500–3300[c]	3.71–7.00	0.2–0.3

[a]From Serna and Pomares (1992).
[b]From Bolton et al. (1990).
[c]Total nitrogen.

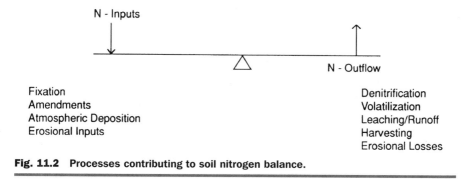

N - Inputs	N - Outflow
Fixation	Denitrification
Amendments	Volatilization
Atmospheric Deposition	Leaching/Runoff
Erosional Inputs	Harvesting
	Erosional Losses

Fig. 11.2 Processes contributing to soil nitrogen balance.

mineralization provides a pivotal base for total biomass production. In an ecosystem with large reservoirs of organic matter (microbial, plant and animal biomass plus colloidal soil organic matter), the quantity of fixed nitrogen available for plant biomass synthesis is controlled by the balance between nitrogen mineralization and immobilization ($N_{available} = N_{mineralized} - N_{immobilized}$). Due to the heterogeneity of soil and the fact that it is an open system, not all of this available nitrogen is retained in soil biomass. Some fixed nitrogen is lost through denitrification, volatilization, leaching, and runoff of soil water. The total quantity of fixed nitrogen within a soil system is also reduced by plant biomass removal (harvesting by animals in native sites and cropping in agricultural systems) and by soil erosion. [In reality, the fixed nitrogen in the latter reservoirs (harvested biomass and eroded soil) are not lost from the soil system in general. They are merely translocated to a different specific ecosystem.]

In a sustained ecosystem, nitrogen outputs must necessarily be balanced by inputs: nitrogen fixation, plant biomass inputs, atmospheric deposition, erosional deposition, or anthropogenic intervention (see Section 11.5.2). In a steady-state system, a balance exists between the rates of occurrence of mineral nitrogen-generating/removal processes and plant growth. Modification of these rates results in establishment of a new biomass productivity steady state wherein these factors would again be balanced. Continued losses of fixed nitrogen without concurrent replacement would eventually result in creation of a biomass desert.

An appreciation of the pivotal role of nitrogen mineralization on the nature and productivity of the plant community of an ecosystem is essential in managed systems. Both excessive nutrient inputs and losses have negative consequences in regards to system sustainability. Fixed nitrogen decrements in agricultural systems commonly arise from crop removal, enhanced leaching of mineral nitrogen, and augmented soil erosion. Conversely, system overloads may be encountered through excessive organic matter amendments, as may be encountered through disposal of sludge, composts, or animal waste materials or overuse of nitrogenous fertilizers.

Based on the essential nature of mineralization, immobilization, and nitrification on stability of soil ecosystems, the objective of this chapter is to evaluate the primary nitrogen-cycling processes in soil and the properties of soil that control these reactions. Biological nitrogen fixation will be evaluated in Chapters 12 and 13 and denitrification will be examined in Chapter 14.

11.1 Nitrogen Mineralization

Simply stated, **nitrogen mineralization** is the conversion of organic nitrogen to inorganic nitrogen. Since, in soil, ammonium is usually rapidly oxidized to nitrate by nitrifiers, net nitrogen mineralization in an ecosystem is estimated by quantifying ammonium and nitrate concentrations. A determination of the absolute amount of nitrogen mineralized requires quantification of the nitrogen immobilized (see Section 10.2). Since many studies involving assessment of nitrogen mineralization are concerned with determination of the quantity of fixed nitrogen available to the growing crop, most workers quantify net, not total, nitrogen mineralization. A more limited term, **ammonification**, is used to designate the conversion of organic nitrogen to ammonium. Mineralization is the sole, soil-resident, biological means of generating the nitrogen forms usable by green plants.

11.1.1 Soil Organic Nitrogen Resources

Soil organic nitrogen is distributed among plant, animal, and microbial biomass, litter, and humic substance pools. Thus, determination of the biochemicals mineralized by the soil microbial community entails elucidation of the nitrogenous substances contained in biomass plus the more exotic humified substances. All nitrogenous compounds commonly contained in living cells are found in the soil organic fraction. Humified nitrogen—that is, nitrogen linked to humic and fulvic acids, may be in recognizable biochemical forms (e.g., humified proteins and amino sugar-containing polysaccharides) or in heterocyclic ring compounds incorporated into the basic structure of humic or fulvic acids. (For more information on the chemistry of soil nitrogen associated with humic substances, see Stevenson, 1994, for a review of the topic or Bondietti et al., 1972; Malcolm, 1990; Piccolo et al., 1990, or Schnitzer and Kerndorff, 1980, for examples of primary research.) It is interesting that Schnitzer and Kerndorff (1980) found that humic and fulvic acids from industrially and environmentally polluted soils contained more nitrogen and sulfur and less oxygen than did these substances isolated from unpolluted soils.

Those complex heterocyclic aromatic compounds contained in humic substances are of interest for developing a complete picture of the soil organic nitrogen component, but they contribute minimally to soil nitrogen cycling. Their rate of mineralization would be equivalent to that of humic acid in general. Nitrogen mineralization rates are determined essentially in totality by the quantity and availability of the more readily metabolized organic nitrogen compounds found in soil. These substances are composed primarily of amino acids and amino sugars and their oligomers and polymers. Other contributors include nucleic acid components plus, in managed systems, such societal products as various heterocyclic ring-containing pesticides. Again, except in the more limited situations where these latter compounds enter the soil system in the high concentrations associated with accidental spills or mismanagement, they usually contribute minimally to overall soil nitrogen cycling.

This short list of classes of nitrogen-containing biochemicals leaves an implication of simplicity of nitrogen mineralization processes that does not represent reality in soil. Biochemically, the process of converting organic to mineral nitrogen is the result of catalysis of a limited number of compounds by a few relatively commonly occurring enzymes (e.g., proteases and amidases). The predominant nitrogenous compounds (amino acids and amino sugars — monomers, dimers, oligomers, and polymers) are all relatively rapidly decomposed by a vast variety of soil microbes. Yet it is reasonably easily demonstrated that nitrogen mineralization is not proportional to the quantities of these compounds present in the soil site, as could naively be predicted. For example, Fox and Piekielek (1978a; 1978b) reported that after a two-year period, a variety of chemical soil nitrogen parameters correlated with nitrogen mineralization in soils from corn fields, yet with a longer testing period (four years) and greater variability in soil and weather, none of the tests were found to be accurate predictors of soil nitrogen mineralization (Fox and Piekielek, 1983). Serna and Pomares (1992) found that prediction of soil nitrogen availability to plants in growth-chamber studies required analysis of initial mineral nitrogen compounds of the soils plus several chemical and biological assays. Similarly, Juma and Paul (1984) evaluated nitrogen mineralization dynamics using a ^{15}N-labeled Weirdale, loam soil. Their study of ^{15}N enrichment in mineral nitrogen products supported the conclusion that extraction of a highly labeled organic N pools from the soil only partly explained the substrates used for nitrogen mineralization in the soil. Their data supported the conclusion that a variety of nitrogen pools interact in nitrogen mineralization processes. Note that here, a nitrogen pool or reservoir is *not* equated or limited to a specific biochemical. For example, in this situation, nitrogen pools would include nitrogenous compounds in the relatively stable form of microbial biomass as well as the ephemeral pools of water-soluble amino acids, proteins, and amino sugars. Form or state of the biochemical are as important as chemical identity in determining or designating soil nitrogen pools. Nitrogen-containing biochemicals may be bound to soil organic matter or soil particles; contained in plant, animal, or microbial biomass or debris; or exist free in soil interstitial water, or simply, a biodecomposable organic nitrogen substance may be occluded within a soil aggregate. Each of these situations provides a unique impediment to access to the biochemical for liberation of the nitrogen contained therein. For example, humified compounds are protected by the inability of the enzyme to approach the substrate, whereas microbial cells present a reasonably resistant cell wall barrier to protect internal biodegradable compounds.

Another portion of the difficulty in using concentrations of decomposable organic nitrogen compounds as predictors of nitrogen mineralization potential is the fact that nitrogen mineralization may not be the primary reason for the conversion of the organic nitrogen to ammonium or nitrate. Organic nitrogenous compounds found in soil may also serve as sources of carbon and energy for the microbial community. Thus, any mineral nitrogen produced by these processes is a by-product of carbon and energy metabolism. For example, Learch et al. (1992) evaluated the mineralization of organic sewage sludge constituents in soils. The quantity of proteins extracted from the sludge was highly correlated with carbon mineralization but had limited relationship to the nitrogen mineralization rate. This observation supports the hypothesis that the soil microbial community was using the protein as a carbon rather than nitrogen source. In these situations, the carbon/nitrogen ratio tends to serve as a better indicator of nitrogen mineralization potential in that with a narrow ratio, significant production of mineral nitrogen can be expected to occur whenever the microbial community uses the substrate for a carbon/energy source (e.g., DeLuca and Keeney, 1993; Learch et al., 1992). (See discussion below regarding the interaction of carbon/nitrogen ratio of the substrates metabolized by soil microorganisms and the resultant impact of mineral nitrogen production.)

11.1.2 Assessment of Nitrogen Mineralization

Nitrogen mineralization in soil has traditionally been viewed from aspect of its contribution to plant biomass production — generally in agricultural systems. It must be stressed that not all the nitrogen mineralized in soil is available to the plant community. Mineral nitrogen is removed from soil through assimilation by plants and microbes, it may be denitrified, and it may be leached from the root zone. Hence, change in soil mineral nitrogen could be represented as follows:

$$\Delta N_i = N_m - (N_a + N_p + N_v + N_l + N_d) \tag{11.1}$$

where N_i represents inorganic nitrogen, N_a is nitrogen assimilated by microbes, N_p is nitrogen assimilated by plants, N_l is nitrogen leached from the soil, N_v is nitrogen volatilized, N_m is nitrogen mineralized, and N_d is nitrogen denitrified. The number of terms in this relationship provides an indication of the complexity of tracing the flux of nitrogen atoms between various soil nitrogen reservoirs in a field situation. The movement of this atom is simplified in laboratory-incubated soil samples, which are commonly used to estimate soil nitrogen mineralization potential.

Nitrogen mineralization potential, usually measured in laboratory-incubated soil samples, is most commonly estimated by assessing net mineral nitrogen production over a given time period in a closed system (batch soil samples or soil columns) in the absence of growing plants. Since such a system would be physically isolated, leaching, volatilization, and denitrification losses are also be minimized. Denitrification is generally reduced by maintaining aerobic soil conditions. Therefore, in these closed laboratory-managed, soil systems,

$$\text{Net } N_m = \Delta N_i = N_m - N_a \qquad (11.2)$$

Thus, it is apparent that to understand the dynamics of mineral nitrogen production for plant biomass synthesis, the nuances of nitrogen immobilization must be evaluated.

11.2   Nitrogen Immobilization

11.2.1   Process Definition and Organisms Involved

Nitrogen immobilization is the microbial assimilation of mineral nitrogen compounds. Any actively growing microbe in soil contributes to this incorporation of soil ammonium and nitrate into microbial biomass. If the environmental conditions favor microbial growth, nitrogen is immobilized. Thus, nitrogen immobilization occurs under such diverse environmental conditions as those associated with aerobic or anoxic soils, acid-affected sites, and soils receiving hot spring or even nuclear power plant waste waters. The microbes that convert mineral nitrogen to microbial biomass are aerobes, facultative anaerobes, anaerobes, thermophiles, mesophiles, psychrophiles, acidophiles, neutrophiles, and alkalinophiles. That is, nitrogen immobilization is catalyzed by all groups of soil microorganisms.

Historically, the primary justification for study of nitrogen immobilization has related to its role in the availability of plant nutrient nitrogen in agricultural soils. In unmanaged ecosystems, studies of nitrogen immobilization have mainly involved evaluation of basic nitrogen cycling processes, such as the quantification of total nitrogen mineralization and the proportion of soil mineral nitrogen reservoirs cycled through microbial biomass. In contrast, in agricultural soils or other managed sites, such as reclamation projects, interest in nitrogen immobilization kinetics acquires an economic aspect, since quite frequently the major justification of soil amendment with anthropogenically fixed nitrogen is to stimulate aboveground plant biomass synthesis. Excessive incorporation of the amended nitrogen into microbial biomass necessarily affects plant community development.

11.2.2   Impact of Nitrogen Immobilization Processes on Plant Communities

The quantity of mineral nitrogen incorporated into the soil microbial population depends on community activity, which in turn in most aerobic soil systems is controlled by the availability of carbon and energy resources. In anaerobic or oxygen-limited soils, moisture-limited sites, or soils of extreme acidity, micro-bial activity is usually sufficiently controlled by these factors that nitrogen immobilization is not a significant impediment to plant biomass productivity. In carbon-controlled communities, the impact of an influx of carbon rich material on mineral nitrogen resources (i.e., ammonium and nitrate) depends upon the carbon/nitrogen ratio of the soil amendment. If the carbonaceous compounds providing carbon and energy to the active portion of the soil microbial community contain sufficient nitrogen to meet the needs of the soil microbes decomposing them, then a nitrogen deficiency does not occur. Unfortunately with most commonly utilized carbon and energy sources in soil (plant biomass), this is not the situation. Plant biomass has carbon/nitrogen ratios as high as 60 to 80/1. Thus, since soil bacteria have carbon to nitrogen ratios of 4 to 6/1 and fungal ratios generally range from 10 to 12/1, synthesis of microbial biomass from these substrates necessarily results in a soil nitrogen deficit. That is, some soil nitrogen must be incorporated into microbial biomass to balance the carbon inputs. In soils where insufficient mineral nitrogen exists to compensate for the deficiencies of the carbon and energy source, the microbial growth rate is reduced by the nitrogen limitation. In either situation, a reduction in total soil mineral nitrogen resources by immobilization results in a loss of plant biomass productivity.

The differential fate of carbon and nitrogen contained in the organic matter mineralized by the soil microbial community is reflected in the change in the carbon/nitrogen ratio of the organic matter as it is transformed from plant biomass into soil organic matter. With the decomposition of biomass, the carbon contained therein is oxidized primarily to carbon dioxide with a small portion being incorporated into microbial biomass components. The nitrogen in the substance being decomposed is conserved in the soil site. Thus, the effective carbon/nitrogen ratio of soil humus decreases with time. Field data indicate that after the carbon/nitrogen ratio of the biomass residues amended to soil reaches about 30/1, net nitrogen immobilization is no longer a problem. At this point, instead of consuming mineral nitrogen, the catabolism of the plant biomass becomes a mineral nitrogen source; that is, there is net nitrogen mineralization within the system.

The obvious question in this situation relates to the apparent contradiction associated with the accumulation of mineral nitrogen in the field. When the carbon/nitrogen ratio of the decomposed substrate approaches the 30/1 carbon/nitrogen, there is a net accumulation of mineral nitrogen, yet the 10/1 mean carbon to nitrogen ratio of the microbial biomass predicts that a mineral nitrogen deficit should occur until the ratio approaches the latter value. If the microbes decomposing the plant residue have an average carbon/nitrogen ratio of 10/1, why is nitrogen not limiting until the metabolized substrate carbon/nitrogen ratio approaches 10/1 rather than 30/1? At least a partial answer to this question is derived from analysis of the basic biochemical composition of the plant biomass. It must be remembered that a portion of the plant biomass entering the soil system is of limited metabolic availability to the soil microbes. These carbonaceous substances include lignin and lignified substances, easily metabolized substances protected by the more slowly decomposed plant cell walls, plus substances that become associated with and thereby protected by soil mineral matter such as soil clays. Lignin has a high carbon to nitrogen ratio and is a major component of plant biomass. Thus, the effective C/N ratio of the substance actively metabolized by the soil microbial community may be much lower than the overall ratio detected by gross analysis of the decomposing organic matter. That is, the portion of the plant biomass mineralized by the soil microbial community is essentially the total plant biomass minus the high-carbon, low-nitrogen lignin component.

The rate of decomposition of even the more readily decomposed organic substance providing carbon and energy to the soil microbial community alters the impact of the amendment on nitrogen immobilization. This fact is exemplified by two studies, one using relatively simple carbon compounds, and the other with complex plant components. Kelley and Stevenson (1987) added glucose, a phenolic glycoside, or catechol with ^{15}N-labeled ammonium sulfate to soil and evaluated nitrogen immobilization. Most rapid immobilization occurred in glucose-amended soils (maximum incorporation at 6 days), compared to maximum incorporation times of 10 days with the phenolic glycoside and 20 days with catechol. It should be also noted that in their study, Kelly and Stevenson found that at maximum immobilization, the nitrogen incorporated into microbial biomass was 3.5- to 6-fold more susceptible to mineralization than was native soil nitrogen. Thus, more recently synthesized, most likely less humified organic nitrogen compounds are more metabolically available to the soil microbial community. Similarly, Mengel and Schmeer (1985) evaluated the effect of straw, cellulose, or lignin soil amendments on nitrogen immobilization, with the overall objective of the project being to increase the retention of nitrogen in soil as organic nitrogen, thereby reducing denitrification losses. These latter workers found that straw and cellulose stimulated greater nitrogen immobilization than did lignin in the 36-day incubation period. This observation relates to the biodegradation susceptibility of the substrates in that little microbial biomass synthesis resulted from amendment of the soil with lignin due to its inherent resistance to microbial catabolism. The complexity of the soil nitrogen cycle interactions was shown in this study in that straw and cellulose amendment also stimulated the denitrification rate in amended soils compared to the uncropped soil. Thus, incorporation of added nitrogen into the crop was higher in the unamended soils than occurred in those receiving exogenous carbon sources.

These observations suggest that the percent nitrogen of organic material entering soil declines or increases until it approaches the percent nitrogen content of the soil microbial biomass. In reality, this change can be predicted and is observed in soil systems receiving large influxes of organic material, but the final carbon/nitrogen ratio achieved in a soil system at equilibrium is considerably greater than that detected in axenic cultures of bacteria and fungi. This limitation results from the fact that in a biologically active mineral or organic soil, there is a significant concentration of soil humic substances (humic and fulvic acids plus humin). These humic substances have very wide (60 to 80/1) carbon to nitrogen ratios. Recall that these soil organic components are only minimally susceptible to catabolism by the biological community. The presence of these substances obscures changes induced by the amendment of soil with plant biomass.

The carbon/nitrogen ratio of native soil organic matter is a mean of the average 10/1 ratio of the microbial community and the elevated value of the soil humic substances. For example, Post et al. (1985) measured the carbon to nitrogen ratio of the humus from a variety of soil ecosystems. The ecosystems were from sites with varying climate and aboveground vegetation communities. They found that all but three samples had carbon to nitrogen ratios ranging from 10.2 to 30.2, with a mean of 16.0 for all soil types studied and a mean of 17.8 when the values were weighted for the variable number of samples per ecosystem type studied. The exceptions were a subtropical wet forest (carbon/nitrogen ratio = 3.3) and two soil samples from a tropical thorn wood land (carbon/nitrogen ratio = 9.2). Wet and dry ecosystems from a variety of climates were evaluated. Although none of these mean values for the carbon/nitrogen ratios of soils can be considered to be representative of all soils, they do provide an indication of the situation in steady-state native soil ecosystems.

11.2.3 Measurement of Soil Nitrogen Immobilization Rates

Measurement of total nitrogen mineralized requires quantification of the soil nitrogen immobilization rate. This assessment generally requires labeling of nitrogen atoms so that their movement between various soil nitrogen reservoirs can be traced. This task is accomplished through the use of ^{15}N. Mineral nitrogen compounds enriched in ^{15}N may be added to soil samples, and the incorporation of the labeled nitrogen atom into organic nitrogen pools can be quantified. Nitrogen mineralization can similarly be assessed by quantifying the enrichment of ^{15}N in mineral nitrogen pools subsequent to soil amendment with ^{15}N-enriched organic matter. Since nitrogen already occurs in all of the soil nitrogen pools, the value measured in these experiments is the rate of enrichment of the existing pool with the ^{15}N. Limitations of this procedure, aside from availability of the mass spectrometer necessary to detect the ^{15}N, are involved with mixing of the labeled substances into soil. Calculations of both nitrogen mineralized or immobilized by this technique are based on the assumption that a distinct nitrogen pool is not created by the soil amendment. That is, the added ^{15}N-labeled substance is of the same availability to the soil microbial community responsible for its metabolism as are the indigenous nitrogen reservoirs. Due to the heterogeneity of soil, this assumption is likely rarely met, but this procedure yields data that are a reasonable estimate of the rate of *in situ* processes.

There has been some interest in developing soil extraction procedures that are specific for recently immobilized nitrogen compounds. For example, He et al. (1988) and Kelley and Stevenson (1985) demonstrated that milder extraction procedures (e.g., hot water, hot 10 mM $CaCl_2$) removed more of the recently immobilized nitrogen than did the more intensive procedures (e.g., acidified permanganate). Although some enrichment of immobilized nitrogen was achieved with these procedures, the nitrogen contained in the extracts was still derived from a variety of soil pools. Serna and Pomares (1992) evaluated a combination of biological methods (plant uptake and aerobic soil incubations) and chemical extraction procedures (autoclave, 0.5 M $KMnO_4$, 6M HCl, or 0.01 M $NaHCO_3$) as predictors of nitrogen availability to maize in calcareous soils. They found that no extraction method was an effective indicator of the concentrations of organic nitrogen compounds that can be metabolized by the soil microbial community, but predictive capabilities were improved by combining results from assessment of initial mineral nitrogen contents of the soils plus several chemical and biological assays in regression analyses.

These observations document that nitrogen immobilization is an essential function in soil microbial communities in that it represents the proportion of nutrients incorporated into the microbial community. From the view of the total ecosystem, the nitrogen immobilization pathway provides an alternate fate (e.g., incorporation into plant biomass vs. soil microbial biomass) of mineral nitrogen. This concept provides a contrast to viewpoint of the agriculturist and, to some the degree, an ecologist or a reclamation specialist. To the former, optimizing aboveground crop production is of primary importance, whereas the latter individuals may be more concerned with ecosystem sustainability. For those optimizing or maximizing biomass production, immobilization represents a troublesome economic problem. With each nitrogen atom incorporated into the microbial biomass, additional external nitrogen inputs are necessitated to maximize the likelihood of achieving the desired ends of an attractive and/or productive aboveground community development.

11.3 Quantitative Description of Nitrogen Mineralization Kinetics

Net nitrogen mineralization in native soil systems is generally observed to follow first order kinetics (see Section 9.5.4). Its kinetics are approximated by the equation

$$\frac{dN}{dt} = -kN \qquad (11.3)$$

where N is the concentration of mineralizable nitrogen, k is a rate constant, and t is time. Thus, a commonly used integrated form of this equation used to calculate net nitrogen mineralization is as follows:

$$N_m = N_0(1 - e^{-kt}) \qquad (11.4)$$

where N_m represents the net nitrogen mineralized, N_0 is the organic nitrogen at zero time, and k and t are a rate constant and time, respectively. It must be noted that situations may also exist where zero order net nitrogen mineralization kinetics are detected. That is, nitrogen mineralization is independent of the organic nitrogen concentration. In this situation, nitrogen mineralization kinetics are described by the equation

$$\frac{dN}{dt} = k \tag{11.5}$$

where k is a rate constant. Occurrence of zero order kinetics can be predicted to occur in soils amended with high concentrations of nitrogen-rich organic substances (e.g., bacterial or fungal biomass waste substances from fermentative, industrial processes). In this situation, the capacities of the mineralizer population are saturated. Increased mineralization cannot occur without an increase in the enzymes or population densities of the organisms catalyzing the process.

The most commonly occurring alternative mathematical model used to describe nitrogen mineralization in soil samples is the double exponential model. As was indicated above, organic nitrogen occurs in a variety of forms and physical situations in native soil samples. Some organic nitrogen compounds are easily decomposed by the microbial community, whereas others are inherently biochemically resistant or physically occluded. In reality, each class or specific organic nitrogenous compound has a characteristic decay rate, but field data indicate that these substances can reasonably easily be grouped into two categories, those relatively rapidly metabolized and a more resistant reservoir. Thus, the above model can be expanded to include both organic nitrogen reservoirs as follows:

$$N_t = N_0 S(1 - e^{-ht}) + N_0(1 - S)(1 - e^{-kt}) \tag{11.6}$$

where N_t is the nitrogen mineralized in time t, N_0 is the potentially mineralizable nitrogen, S and $(1 - S)$ represent the labile and less accessible organic nitrogen fractions of the total organic nitrogen pool, and h and k are decomposition rate constants for the labile and less accessible pools, respectively. Deans et al. (1986) applied this double exponential model to a variety of historical studies and found that it provided a good description of the net nitrogen mineralization kinetics. Other mathematical representations of nitrogen mineralization include empirical models (e.g., see Broadbent, 1986) and a variety of double exponential and mixed-order representations (e.g., see Bonde and Rosswall, 1987, and Bonde and Lindberg, 1988). A caveat in deriving nitrogen mineralization constants from application of these mathematical expressions is that the values derived may vary depending on the methods used to fit the data to the model (e.g., see Ellert and Bettany, 1988; Juma et al., 1984; Talpaz et al., 1981).

Extrapolation of Laboratory-Derived Net Nitrogen Mineralization Rates to Field Situations: Independent of which mathematical model is used to analyze nitrogen mineralization data, application of the laboratory-derived results to field situations is dependent upon a number of assumptions. The primary assumptions that must be considered are (1) soil organic nitrogen mineralization occurs in the laboratory-incubated samples at rates comparable to those existing in the field, (2) preparation of soil samples for laboratory analysis does not alter the availability of various soil organic nitrogen reservoirs to the microbial community, and (3) (many workers assume that) nitrogen mineralization is a first order process; that is, alternative models are not evaluated in the data analysis process. In reality, all of these assumptions are of questionable validity, at best.

The difficulties with the assumption that laboratory nitrogen mineralization kinetics reflect field rates are readily exposed when it is considered that nitrogen mineralization kinetics vary widely with temperature and moisture (e.g., see Addiscott, 1983; Cabrera and Kissel, 1988b; Cassman and Munns, 1980; Kladivko and Keeney, 1987; Myers et al., 1982; Smith et al., 1977; and Stanford et al., 1975); whereas in the most commonly implemented laboratory incubation methods, these properties are controlled within a limited range. Cassman and Munns (1980) derived mathematical models to compensate for field variation in temperature (15, 20, 25, and 30C) and moisture (0.1, 0.3, 0.7, 2, 4, and 10 bars). Marion and Black (1987) used empirically derived models for the inclusion of the effect of time and temperature on nitrogen mineralization kinetics in Arctic tundra soils. Temperature variability may be compensated through use of the buried-bag procedure (Poovarodom et al. 1988; Westermann and Crothers, 1980), but field moisture variation is limited within the bag. Utilization of empirically derived relationships or even widely applied models, such as Q_{10} or the Arrhenius relationship, are useful in adapting laboratory parameters to the more variable field situation.

The second assumption — that collection and preparation of soils for laboratory incubation has minimal effect on nitrogen mineralization kinetics — is invalidated by a consideration of the soil properties controlling the availability of the various soil organic nitrogen pools to the microbial community. It has already been stated that a primary controller of nitrogen mineralization is accessibility of the substrates to the microbes and enzymes. Physical handling of soils as occurs during collection and sieving of the samples clearly alters the occlusion of organic compounds within soil aggregates and thereby their susceptibility to biological decomposition. For examples of studies of the impact of this liberation of soil organic matter in sample preparation on kinetics and the development of mathematical models to account for this laboratory generated artifact, see Beauchamp et al. (1986), Craswell and Waring (1972a, b), Cabrera (1993), Cabrera and Kissel (1988a, b). Since it is appreciated that even the minimal handling of soil samples associated with the buried-bag procedure results in alteration of nitrogen availability and thus the nitrogen mineralization parameters derived from the study, it can be concluded that currently no method exists to negate the impact of sample handling on estimation of soil nitrogen mineralization potential.

A final consideration in estimating field nitrogen mineralization parameters is their spatial and temporal variation. Among the many factors limiting laboratory analysis of this soil community property is the reality that only a limited number of soil samples can be analyzed and that the incubations require several weeks for completion. Insufficient numbers of soil samples are generally assayed to overcome the impact of variability of nitrogen mineralization rates in the field in space and time. Nitrogen mineralization is highly variable across the landscape and changes significantly on a microscale with variation of limiting physical and chemical parameters. Goovaerts and Chiang (1993) found that microscale nitrogen mineralization variability (<1m) was explained to a large degree by the distribution of soil-oxidizable carbon, which correlated with gravimetric water content. This microscale variation affects the nature of a representative soil sample. Goovaerts and Chiang estimated that 24 to 37 samples were needed to estimate nitrogen mineralization potential (with a precision of 0.1 at $p = 0.05$) for a 1600 m^2 plot. The impact of variation with time requires incorporation of data describing variability of primary soil properties controlling nitrogen mineralization into the analysis procedure. Similar results were presented by Starr et al. (1992) in determining soil sample size and number requirements for assessing field variation of soil nitrate concentrations.

The basis for this spatial variation in nitrogen mineralization potential relates to properties of both the macrosoil ecosystem as well as the microstructure of the system. The former is exemplified by studies of the impact of plant distribution on soil organic nitrogenous compounds and nitrogen mineralization potential (e.g., see Bolton et al., 1990). Typically in such studies, soil organic nitrogen as well as associated biological activities are concentrated within the portions of the soil profile that are under direct influence of the growing plant—that is, the rhizosphere and areas affected by plant litter. The rhizosphere effect may be direct through the addition of plant-derived organic matter to the soil system or indirect as a result of the improvement of soil properties conducive to microbial population development. The latter may include moisture preservation due to elevated soil organic matter levels, temperature modulation on hot summer days under plant canopies, and improved soil structure due to encouragement of granule formation by plant roots.

Variation in the distribution of soil organic nitrogen reservoirs is also associated with soil structural development. Separation of soil fractions by particle density or size results in fraction of soil organic matter into more readily decomposed and more biodegradation-resistant portions. Mineralizable carbon and nitrogen tend to be associated with smaller particle sizes (Anderson et al., 1981; Cameron and Posner, 1979; Schnitzer and Kodama, 1992). These fractions tend to have a lower carbon to nitrogen ratio, with most of the mineralizable soil nitrogen found in the fractions with smaller diameter particles. Similarly, separation of soil based on density fractionation results in enrichment of readily decomposable organic matter (hence, an enrichment for higher nitrogen mineralization potential) in light fractions (i.e., specific gravity greater than 1.2 g cm^{-3}) (e.g., see Sollins et al., 1984).

11.4 Microbiology of Mineralization

Nitrogen mineralization is essential for development of a sustainable ecosystem. Hence, it is reasonable to conclude that organic nitrogenous compounds are converted to mineral nitrogen in any situation where life is possible. Were this not a basic truth, the biotic community would require a continued input of mineral nitrogen and would have to adapt continuously to increasing levels of accumulated organic matter. This situation occurs rarely, if ever in natural soil systems. Plant biomass accumulates in swampy ecosystems, but aboveground plant production is balanced with the slow nitrogen mineralization rate occurring in waterlogged soils associated with the water-saturated conditions. Situations wherein soils receive constant inputs of mineral nitrogen are, in our society, either fertilized agricultural or landscaped situations (wherein the fertilization regime presumably balances plant-production needs) or pollution-affected situations (such as soils receiving nitrogen-laden industrial effluents).

Thus, based on occurrence of nitrogen mineralization in essentially any ecosystem where life can occur, it is predictable that an equally diverse microbial community is capable of catalyzing the process. Nitrogen is mineralized by a diverse group of bacteria, fungi, and actinomycetes. These organisms are aerobes, anaerobes, thermophiles, mesophiles, and psychrophiles, and are metabolic diverse in regards to their energy needs. Specific microbial species active in this process represent all microbial genera growing and reproducing in soil. Indeed, quantification of nitrogen-mineralizing populations in soil varies or is limited not by the number present in the soil sample, but rather by the ingenuity of the individual devising the growth medium for the microbial populations. Common nitrogen substrates used to isolate organisms involved in nitrogen mineralization include protein, chitin, and amino sugars.

11.5 Environmental Influences on Nitrogen Mineralization

In that the processes associated with nitrogen mineralization and immobilization are basic to microbial life in general, it is reasonable to assume a priori that these biologically catalyzed portions of the nitrogen cycle occur under essentially any environmental condition wherein microbial growth is possible. Simply stated, nitrogen immobilization is the incorporation of mineral nitrogen into cell biochemicals. Thus, nitrogen immobilization is easily concluded to be a major by-product of microbial growth. In contrast, the direct association of nitrogen mineralization with carbon mineralization provides an indirect link to biological energy generation. Thus, the question in regards to environmental influences on nitrogen mineralization and immobilization relates not to whether the processes occur but rather to the impact of modification of soil physical and chemical properties on the rate of these nitrogen conversions.

Primary ecosystem properties that are usually evaluated for their impact on nitrogen immobilization and mineralization rates are nitrogen concentration, soil moisture, pH, and temperature. Nitrogen concentration effects were considered above with the discussion of net nitrogen mineralization rates. The impact of moisture, temperature, and pH variation on these processes follows a pattern similar to that observed for most biologically catalyzed processes (see Fig. 5.3). There is a minimum and maximum level of the stress factor wherein no activity is possible. Between these limiting values, the biological activity tends to increase to a maximum level at the optimal adaptation point of the stress factor, after which the biological activity again declines. This generalized reaction of the microbial community is exemplified best by analyzing the impact of soil moisture on nitrogen mineralization. Again, a generalized description of the process is presented, since individual reaction rates at specific soil moisture levels vary with the nature of the microbial community existent in a particular soil ecosystem. With no other limitations, microbial growth, thereby nitrogen mineralization and immobilization, tends to increase as water availability increases. An optimal level of activity is generally reached at approximately water saturation. At this water level, adequate soil water is available for the microbes to interact with their growth substrates and for by-products of their metabolism to be removed from the cells' environs. As the soils become flooded, oxygen diffusion becomes a limiting factor to microbial respiration. Once the oxygen tension is reduced by an imbalance between consumption and influx of free oxygen, cellular processes become limited. Nitrogen mineralization and immobilization rates are reduced as the activity of soil microbes in general declines due to the anoxic conditions. Both nitrogen mineralization and immobilization do continue under anaerobic conditions, albeit at generally reduced rates compared to those occurring under oxygenated conditions. Optimal soil moisture conditions for aerobic transformations of nitrogen generally range between 0.3 bar and saturation (60 to 90 percent water-filled pore space) (e.g., see Walters et al., 1992), whereas the optima for pH and temperature are near neutrality and within the thermophillic range (40 to 60°C), respectively. Clearly, although most soils are not thermophillic and major active ecosystems have soils with pH values substantially below neutrality (e.g., tropical and temperate forest soils), these processes still occur at rates sufficient to sustain highly productive ecosystems under the less-than-optimal conditions of most soils.

11.6 Nitrification

Nitrification is the biological formation of nitrate and nitrite from compounds containing reduced nitrogen. These oxidative reactions are catalyzed by two

Table 11.2 Oxidation State of Nitrogen Compounds Associated with Soil Nitrogen Cycling

Compound	N Oxidation State
NH_4^+	-3
NH_2	-1
NO	$+2$
N_2O	$+1$
NO_2^-	$+3$
NO_3^-	$+5$

mutually exclusive groups of organisms, ammonium and nitrite oxidizers. The primary reactions involved are as follows:

$$2NH_4^+ + 3O_2 \rightarrow 2NO_2^- + 4H^+ + 2H_2O$$

$$2NO_2^- + O_2 \rightarrow 2NO_3^-$$

These processes involve the direct incorporation of molecular oxygen into the nitrogen oxide products. Note that hydrogen ions are produced by ammonium oxidation. Conversion of ammonium to nitrate results in a change in the oxidation state of the nitrogen from its most reduced state (-3) to the most oxidized form $(+5)$ (Table 11.2). As will be elucidated below, the primary implications of these observations are (1) nitrification is an obligatorily aerobic process and (2) oxidation of ammonium results in acidification of the soil environment.

11.6.1 Identity of Bacterial Species That Nitrify

All nitrifiers are closely related by their specialized biochemical reactions. All oxidize reduced nitrogenous compounds for energy and fix carbon dioxide for their carbon source (see Section 4.2.2). The primary taxonomic difference at the genus level is based on cellular morphology. Common ammonium oxidizers, nitrite oxidizers, their source and morphology are listed in Table 11.3. Note that the ammonium oxidizers are distinguished by the prefix *Nitroso-* for the genus name, whereas the nitrite oxidizers have the comparable prefix *Nitro-*.

Some confusion may develop in the study of classical literature due to the impact of a past dogma on our concept of the role of these various species in soil. Prior to the early 1980s, it was commonly held that of the nitrifiers, only *Nitrosomonas* spp. and *Nitrobacter* spp. occurred commonly in soil. This assumption led to the conclusion that these two genera were *undoubtedly* the primary nitrifying chemoautotrophs in soil ecosystems. Therefore, when population densities of ammonium oxidizers were estimated, the numbers were

Table 11.3 Common Species of Autotrophic Nitrifiers, Their Sources, and Morphologies

Bacterial Species	Source	Morphology
Ammonium oxidizers:		
Nitrosomonas europea	Soil, water, sewage	Ellipsoidal, short rods
Nitrosospira briensis	Soil	Spiral
Nitrosococcus oceanus	Marine	Spherical
Nitrosococcus mobilis	Marine	Spherical
Nitrosococcus nitrosus	Marine, soil	Spherical
Nitrosolubus multiformis	Soil	Lobate, pleomorphic
Nitrosovibrio tenuis	Soil	Comma shaped
Nitrite oxidizers:		
Nitrococcus mobilis	Marine	Spherical
Nitrobacter gracilis	Marine	Short rods
Nitrobacter winogradski	Soil	Short rods
Nitrobacter agilis	Soil, water	Short rods

presented as populations densities of *Nitrosomonas*. Similarly, the epithet *Nitrobacter* was used globally for nitrite oxidizers. With the advent of the use of fluorescent antibodies to quantify nitrifier populations in soil at the species level, it has become evident that a variety of different species were found in soil systems (e.g., see Belser and Schmidt, 1978a,b; Fliermans et al., 1974; Stanley and Schmidt, 1981). As more specific enumeration procedures such as polymerase chain reaction application (e.g., Navarro et al., 1992) and fluorescent antibodies specific for various genera of nitrifiers are applied to the study of the diversity of nitrifier populations in soil, an improved understanding of the variety of these organisms *in situ* will emerge. At this point, it is evident that sufficient variation in the identity of active nitrifiers in soils occurs to justify the use of the epithets *ammonium* and *nitrite oxidizers* in place of the more specific genera designations.

11.6.2 Benefits to the Microorganism from Nitrification

The primary organisms responsible for ammonium oxidation to nitrate in soil are chemolithotrophs. (See Section 11.6.6 for discussion of the possible role of heterotrophic nitrifiers in soil.) The organisms drive their energy from the oxidation of the nitrogen atom. Conversion of ammonium to nitrite yields 66 kcal per mole, whereas nitrite oxidation to nitrate produces 20 kcal. The recovery of this energy by the microbes ranges from about 5 to 15 percent. Thus, large quantities of ammonium or nitrite must be oxidized for each carbon fixed by the microbes. Ammonium oxidizers typically oxidize between 14 and 70 ammonium-nitrogen per carbon incorporated into cellular biomass, whereas between 76 to 135 nitrite-nitrogen must be oxidized for the comparable task.

11.6.3 Quantification of Nitrifiers in Soil Samples

Difficulties associated with culture of the bacterial species involved in nitrification have resulted in the development of indirect culture methods and procedures for direct observation of cells within the soil sample or in extracts. Historically, the two most commonly employed techniques for quantifying soil nitrifiers have been the most-probable-number method using nitrifier-specific growth media and direct enumeration of specific organisms using fluorescent antibodies. As marker genes for these organisms are identified, pcr-based procedures will provide a clearer estimation of these populations. Examples of studies applying 16S rDNA gene analysis to study of nitrifier population diversity include Hastings et al. (1997), Kowalchuk et al. (1997), and Stephen et al. (1996).

Two requirements of the most-probable-number method are (1) media and growth conditions must be used that allow any and all nitrifying cells present in the soil sample to grow to a population density that can be detected, and (2) the cells must be sufficiently separated from any particulate matter in soil that each cell will be individually and independently dispersed in the diluent (that is, nitrifier cells and not soil particles must be enumerated). No most-probable-number technique for quantification of nitrifiers meets either of these requirement. All media are selective. Only a portion of the nitrifier population is able to grow. This situation is the most likely reason for the development of the central dogma cited above in that the dominant species growing in the commonly used most-probable-number media are *Nitrosomonas* sp. A further complication associated with essentially all procedures based on selective media is that the more rapidly growing organisms limit or preclude development of populations of more fastidious organisms. Support for the role of this phenomenon with nitrifiers is seen in the work of Matulewich et al. (1975), where it was found that the much shorter incubation times commonly employed for nitrifier assays (e.g., Alexander and Clark, 1965) did not allow for maximal population development. Incubation times of up to 55 and 100 days were required for ammonium and nitrite oxidizers, respectively. Difficulties associated with using the longer incubation periods include delays in data acquisition as well as the generally encountered problems associated with the evaporation of the liquid growth media. With longer incubation times, the increased salt concentration in the growth media due to the loss of water may itself induce artifacts in population selection. Therefore, analysis of data procured through this procedure must always be predicated on an appreciation of the limitations imposed by the selectivity of the media, artifacts resulting from the long incubation times necessary for population development, as well as the failure of some investigators to use sufficient incubation periods for their studies. These inherent methodological problems are enhanced by the low statistical precision inherent in such procedures. The latter limitation precludes utilization of the procedure for studies where detection of slight changes in population densities is necessary.

In light of these methodological problems, the question may be asked, "Why is the most-probable-number procedure still the most commonly used procedure to estimate field populations of nitrifiers?" Perhaps, even more crucial is the question, "Why is its continued common usage likely?" The answer to both of these questions relates to the practicality of the procedure. Most-probable-number assays of nitrifiers are still the only methods available to most investigators in the field.

All of the problems inherent in the most-probable-number procedures are overcome with the use of fluorescent antibody techniques. Data are produced rapidly, specific populations are quantified, and small changes in populations of the nitrifiers are easily measured. Yet as with any scientific procedure, some methodological limitations of the fluorescent antibody technique govern the wide implementation of the procedure. A major barrier to its application is the availability of equipment necessary for the assays. Whereas most-probable-number incubations can be conducted with equipment available in essentially any field microbiology laboratory, fluorescent antibody procedures require at least the availability of a fluorescent microscope. Should in-house production of the species-specific antibodies be necessary, the investigator will also need to have the necessary animal-care facilities. Specific data interpretation difficulties associated with the use of fluorescent antibodies are (1) many serotypes of autotrophic nitrifiers occur (thus, large numbers of fluorescent antibody types may have to be used to quantify the total nitrifier population), and (2) it is difficult to isolate and purify nitrifiers, a prerequisite for antibody production. This latter limitation assures that noncultured nitrifiers will not be detected by this method.

11.6.4 Discrepancies Between Population Enumeration Data and Field Nitrification Rates

As indicated above, the energy yield from oxidation of ammonium and nitrite and the free-energy efficiency for the process are known. Thus, it is reasonable to assume that some relationship between the amount of nitrate produced in a soil system and the population of nitrifiers detected therein should occur. Unfortunately, the most common observation is that the population density of autotrophic nitrifiers in soil samples is insufficient to yield the nitrate concentrations detected. A number of reasons have been proposed to explain the difficulties:

- *Inherent inaccuracies of the most-probable number procedure:* These include the fact that in many situations, total nitrifier populations are underestimated because of their inability to grow in the selective media utilized as well as the imprecision of the procedure.

- *Noncoincidence of nitrifiers and nitrate production:* The nitrate present in the soil sample could have been produced in the past, and the population of nitrifiers responsible for its production subsequently declined.

- *Errors in estimating growth yields:* The growth yields in culture used to estimate population production per unit of ammonium or nitrite oxidized probably differ from those yielded in soil.

- *Imported nitrate:* The nitrate detected in a soil sample may not have been produced *in situ.* Nitrate moves freely in the soil profile. The nitrate detected in a particular sample of soil could have been produced elsewhere and transported into the soil sample with water infiltration.

Each of these explanations for the fact that the nitrifier population detected in soil samples is frequently less then 10 percent of that necessary to produce the nitrate present in the samples is plausible.

11.6.5 Sources of Ammonium and Nitrite for Nitrifiers

Ammonification is the primary *in situ* source of ammonium in nonfertilized soils. External ammonium inputs include point sources such as sewage or comparable waste substances amended to soil, as well as nonpoint sources such as nitrogenous fertilizers leaching from or applied to cropped lands. Continuous enrichment of nitrifiers may occur downstream from a sewage plant or intermittent population stimulation may occur from irregular flow from cropped soils due to variation in rainfall patterns.

Any nitrite found in soil is presumed to arise from ammonium oxidation. Under normal situations, nitrite does not accumulate in soils. Thus, the limiting factor in nitrite oxidation to nitrate in most soils is the conversion of ammonium to nitrite (low levels of ammonium are detected in most soils). The nitrite is usually converted to nitrate as rapidly as it is formed. Under alkaline conditions, nitrite may accumulate. This results from the fact that ammonia is toxic to *Nitrobacter* sp. An equilibrium between ammonium and ammonia normally exists in soils, with ammonium predominating in neutral and acidic conditions and ammonia predominating under alkaline conditions. Thus, under alkaline conditions, *Nitrobacter* spp. are inhibited but the ammonium oxidizers remain active. Hence, nitrite accumulates. This phenomenon also occurs in microsites when anhydrous ammonium is added to soil.

Nitrite oxidizers may also derive a portion of their growth substrate from the reduction of nitrate (e.g., see Belser, 1977). In a flooded system (Fig. 11.3), nitrate may be reduced to nitrite in anaerobic microenvironments. The nitrite may then diffuse to aerobic sites where the nitrite oxidizers can convert it to nitrate. In these situations, populations of nitrite oxidizers would be elevated in comparison to the ammonium oxidizers.

11.6.6 Environmental Properties Limiting Nitrification

In contrast to the situation with both carbon and nitrogen mineralizers, where a wide diversity of soil microbes are capable of catalyzing the process, the kinetics and occurrence of nitrification in soil reflect the fact that a limited

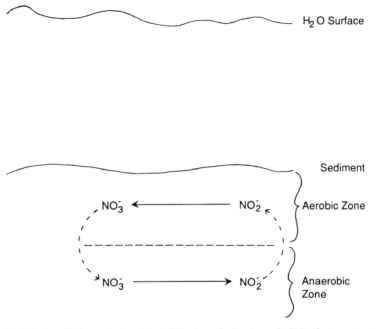

Fig. 11.3 Nitrate-nitrite cycle involving diffusion of nitrate and nitrite between aerobic and anaerobic zones in sediments.

number of bacterial species are involved in the process. Primary concerns in predicting the extent and rate of nitrification in an ecosystem are (1) substrate availability, (2) temperature, (3) pH, (4) oxygen tension, and (5) soil moisture.

Substrate Availability: As was indicated above, most soils contain low concentrations of nitrate and ammonium, with detectable nitrite levels occurring rarely (e.g., see Jones and Schwab, 1993). Thus, it can be concluded that the nitrite oxidizers are generally controlled by the presence of their growth substrate, which indirectly is determined by ammonium availability. Thus, the growth-limiting step for nitrification in general is nitrogen mineralization.

The kinetics of nitrification processes necessarily depend upon population density of nitrifying bacteria present and efficiency of nitrifying enzymes. Nitrifiers are highly efficient organisms at recovering ammonium and nitrite from their environment. That is, the enzymes responsible have low Michaelis constants. Properties of the nitrifier community that can be deduced from the reaction kinetics are as follows:

- Nitrate production may increase proportionally to ammonium concentration in the soil sample. In this situation, the ammonium-oxidizing enzymes in the existing nitrifier population are substrate limited. Thus, nitrate production in these systems is described by Michaelis-Menten kinetics (e.g., see Malhi and McGill, 1982).

- Nitrate production in some cases is described by zero-order kinetics. In this situation the indigenous ammonium production is adequate to provide all of the substrate necessary to support growth of the indigenous nitrifier population. Energy sources (NH_4^+) and energy consumers (ammonium oxidizers) are in balance. A relatively constant nitrification rate can be anticipated to occur unless this balance is shifted — that is, through an increase in NH_4^+ supply or inhibition of the nitrifier population. Augmentation of the energy supply can induce population growth. Then, the nitrifier population will increase until another environmental factor (e.g., space, oxygen, pH, species interactions) becomes limiting. Amendment of an ecosystem wherein the nitrifier population is controlled by one of these latter site properties rather than growth substrate availability will not induce population growth, and therefore deviation from zero-order kinetics will not occur.

- Nitrification may not occur in an ecosystem until several days following amendment. In this case, for some reason a functional nitrifier population has not developed. With amendment of the soil site with ammonium, a nitrifier population slowly develops. An example of this phenomenon was provided in a study of a pine forest soil by Vitousek and Matson (1985).

In conclusion, the *in situ* nitrification rate varies with the degree of saturation of the existent enzymes in the soil system and the environmental limitations controlling nitrifier population development. Changes in the nitrification rate are dependent upon the capacity of the nitrifier population density to increase.

Toxic concentrations of ammonium and nitrite can be reached in axenic cultures of nitrifiers. Generally, normally encountered environmental levels of the nitrification substrates are far below these inhibitory concentrations, but elevated ammonium concentrations have been shown to inhibit the process. For example, Malhi and McGill (1982) in a study of three Canadian soils found that concentrations of $200 \, \mu g \, g^{-1} \, NH_4^+$-N supported rapid nitrification, but elevation of the ammonium-nitrogen concentration to $300 \, \mu g \, g^{-1}$ inhibited the process. Malhi and McGill suggested that the inhibition could result from a combined effect of low pH and salt content.

Temperature: In the laboratory, ammonium oxidation to nitrite occurs at temperatures from near freezing to 65°C. Nitrite oxidation occurs from near freezing and is completely inhibited at about 40°C. The latter process has an optimum generally reported to be between 30 and 35°C. Since nitrification occurs in tropical and subtropical soils at temperatures far above the limits for nitrite oxidation, some compensation must occur in the microbe's native environment that is not provided under axenic culture.

pH: In culture, both the rate of nitrification and the number of nitrifiers declines below a pH of about 6.0. Negligible nitrification is detected below a pH of 5.0 to 5.5 The generally detected optimum for this process is between 6.6 and 8.0. This observation has led to the dogma that autotrophic nitrification cannot occur in acidic environments, yet it can readily be shown that nitrification does occur in acid soils, as is commonly encountered in forest soils. Explanations of this discrepancy between the laboratory and "real-world" observations include (1) microsite variation in soil pH; (2) invoking the possibility of the occurrence of nonculturable, acidophilic, autotrophic nitrifiers in these soils; (3) selection of heterotrophic nitrifier populations under these acidic conditions, and (4) adaptation of the growth properties of commonly encountered nitrifiers to accommodate the acidic conditions. Each of these hypothesized mechanisms for nitrification in the acid forest soils may be operative under some situations but none is totally satisfying. [See Dommergues et al. (1978) for a review of the pros and cons of these mechanisms.] For example, the possibility of nonculturable, acid-loving organisms is plausible, and some acid-loving autotrophic nitrifiers have been cultured (e.g., see Hankinson and Schmidt, 1988). The presence of humic acid has also been shown to provide some relief of pH inhibition of nitrification in soil columns (Bazin et al., 1991). The postulation of more neutral microsites in otherwise acidic soils is plausible in soils whose predominant pH is near the limit for autotrophic nitrifiers, but prediction of the occurrence of such islands of neutrality in extremely acidic soils is implausible. The potential for the formation of microaggregates of nitrifying bacteria in acidic soils is interesting and a potentially useful resolution of this laboratory-field anomaly (De Boer et al., 1991; De Beer et al., 1993). The size of the aggregates is limited by the penetration depth of the oxygen necessary for this obligatorily aerobic process.

The most commonly encountered (and perhaps debated) interpretation of the field data relating to nitrification in acidic soils involves the role of heterotrophic nitrifiers. Heterotrophic nitrification is the oxidation of reduced nitrogen forms to nitrite or nitrate by heterotrophic bacteria and fungi. The reduced nitrogen can be either organic or inorganic. Heterotrophic nitrification can be easily demonstrated in culture. Large numbers of soil bacteria, actinomycetes, and fungi have long been known to exist in soil that are capable of oxidizing ammonium to nitrite or nitrate (e.g., Doxtander and Alexander, 1966; Eylar and Schmidt, 1959; Schimel et al., 1984; Tate, 1977; Verstraete and Alexander, 1973), but their function in native soil ecosystems at rates sufficient to affect *in situ* nitrate concentrations is questioned. Ammonium or organic nitrogen oxidation by these organisms generally occurs after the active growth phase of the organism in media with a low carbon/nitrogen ratio. More nitrogen is usually present than the organisms need for growth. The quantities of nitrate or nitrite produced are small compared to those produced by the autotrophic population (less than 1 vs. 2000 or greater $\mu g \ ml^{-1}$ solution for the heterotrophs and autotrophs, respectively). Since the oxidation of the reduced nitrogen occurs subsequent to replication of the heterotrophic cells, it is unlikely that the organisms are recovering the energy from the oxidation. Although there are reports of existence of large populations of heterotrophic microbes in some ecosystems, their environmental significance is unknown.

Oxygen Tension: Nitrification is an obligatorily aerobic process. Molecular oxygen is directly incorporated into the final product. Therefore, both soil moisture and soil structure affect the nitrification rate indirectly through control of oxygen diffusion. Nitrification can occur at oxygen concentrations as low as 0.3 μg ml^{-1}; thus, the key in occurrence of this process in waterlogged or otherwise diffusion limited systems is the availability of a means for oxygen import. Influx of oxygen into soils where it would normally not be found can be through such processes as mass movement of oxygenated water into flooded soils or even transport of air through the plant into the rhizosphere, as occurs with rice plants.

Moisture: As was indicated above, limitations of nitrification in elevated moisture situations is derived primarily from the indirect control of availability of molecular oxygen by its diffusion rate in water. At the other extreme — desiccated soils — preclusion of nitrification and the direct proportionality of nitrification rate with moisture level between this totally inhibitory level and moisture saturation result from the fact that the soil microbial community is essentially aquatic. That is, microbes require a thin film of water on the soil particles within which to respire and grow. In the absence of this layer of water, no nitrification can occur. As the thickness of the coating of water on soil particles increases, the "aquatic" soil nitrifier bacterial population becomes active.

Other Limitations to Nitrification: Two site characteristics with potential to alter soil nitrification potential that should be considered in land reclamation or management plans are metal contamination and salinity. Both have been demonstrated to have a significant impact on nitrification rates.

Liang and Tabatabai (1978) assessed the inhibition of nitrification by a variety of elements including silver, mercury, cadmium, nickel, arsenic, chromium, boron, aluminum, selenium molybdenum, manganese, lead, cobalt, copper, tin, iron (II and III), and zinc. Soil pHs ranged from 5.8 to 7.8, clay contents from 39 to 50 percent, and organic matter from 2.58 to 5.45 percent. When added at a concentration of 5 μmoles g^{-1} soil, all 19 of the soil amendments inhibited nitrification, but the amount of inhibition varied with substance added and the soil. The potential exists for adaptation of the microbial population to the toxicant in situations where the active microbial population is not killed. This possibility was suggested by a report by Morrissey et al. (1974) when they recorded an initial inhibition of nitrification in a Cheshire fine sandy loam soil by cadmium, chromium, copper, lead, manganese, nickel, and zinc and a subsequent adaptation to the presence of the metal over a several-week period. Thus, these latter workers concluded that although with elevated metal concentrations a transient inhibition was observed, in the long run there was little significant inhibition of nitrification by the metals in their system.

McClung and Frankenberger (1985) noted that increased salinity generally decreased nitrification rate in three diverse soils. Inhibition of the nitrification rate varied from 8 to 83 percent, dependent upon the soil type and the nature of the salt added (Na_2SO_4, NaCl or $CaCl_2$). Generally the sulfate salt was less inhibitory than were the chloride salts. As with the metal-contaminated soils cited above, the potential exists for microbial adaptation to increased salt content. Somville (1984) found that freshwater nitrifiers were capable of adapting to increased salinity in an estuarine environment.

11.7 Concluding Observations: Control of the Internal Soil Nitrogen Cycle

In nonmanaged or minimally managed soil ecosystems (e.g., native forests, "wild lands"), nitrogen cycling could be described as being in balance. That is, organic nitrogen is mineralized to ammonium, a portion is nitrified to nitrate, and most of the mineral nitrogen is returned to biomass. This balance does not imply that no mineral nitrogen is lost from the site. Aside from denitrification processes (as discussed in chapter 14), some nitrate can be leached below the root zone through soil-saturating rain falls. In balanced systems, this small nitrogen loss is balanced by atmospheric inputs and nitrogen fixation (see Chapters 12 and 13).

Greater potential for disruption of the balance between nitrogen mineralization and incorporation into nascent biomass occurs in managed sites as exemplified by systems amended with external supplies of organic matter or receiving significant mineral fertilizer inputs. In these situations, the internal nitrogen cycle could be said to have been disrupted or skewed by the anthropogenic intervention. Rates of soil biological processes generally adjust in these situations (usually through enhanced biomass production, perhaps through increased denitrification), but the potential also exists for enhanced negative ecosystem impact. Since ammonium is relatively nonmobile in the soil profile (see Stevenson, 1986), most environmental contamination difficulties are derived from the synthesis of nitrate — that is, nitrification.

The process of nitrification in soil can be likened to a double-edged sword. To maximize crop production and plant community development in reclaimed sites, fertilization with nitrogenous fertilizers is desirable, if not essential. Nitrification of this added nitrogen is necessary to maximize plant incorporation, but this oxidation of ammonium to nitrate greatly increases the mobility of this essential nutrient. Hence, nitrate may be leach into groundwaters or be transported through overland flow into surface waters. A desirable soil substituent is transformed into a water contaminant.

The best means to reduce the environmental impact of nitrification in a managed system is to attempt to maintain nitrate availability to that level required by the growing plant. This control is generally achieved through the use of slow-release fertilizers or nitrification inhibitors.

Ammonium concentration is the primary nutrient limitation of nitrification in soil. Once the ammonium is formed, it is rapidly transformed to nitrate — a trait desirable in balanced systems where plant biomass synthesis is tightly linked to organic matter mineralization. Such is not the case in fertilized sites. To slow ammonium release from fertilizers, the fertilizer particles (for example, urea) can be coated with a substances that slowly dissolves in soil, thus gradually releasing the ammonium or ammonium precursor. The objective is commonly achieved through the use of slow-release fertilizers or timed mixing of plant growth nutrients with irrigation water (e.g., fertigation of turf or golf grasses or in drip irrigation systems for row crops). Dissolution of fertilizer nitrogen can be achieved by variation of fertilizer particle size, coating of the particles with slowly dissolved or metabolized substances such as sulfur-coated urea (e.g., Landschoot and Waddington, 1987; Severson and Mahler, 1988), or incorporation of the fertilizer in polymers (e.g., see Gandeza et al., 1991), which are commonly used in urban systems for yard and turf fertilization or in agricultural systems, as well as through the use of urease inhibitors (e.g., see Bremner and Douglas, 1971; Gould et al., 1978; Krogmeier et al., 1989). Urease is an enzyme that produces ammonium through the hydrolysis of urea. Similarly, a variety of compounds have been shown to be efficient and effective inhibitors of the conversion of ammonium, once it is produced, to nitrate (e.g., see Baldwin et al., 1983; Belser and Mays, 1980; Belser and Schmidt, 1981; Bremner and McCarty, 1993; Goring, 1962; McCarty and Bremner, 1986). It must be noted that the role of plant-derived toxicants in the inhibition of nitrification in soil is questionable, in that although these compounds may inhibit nitrifier activity in culture, there is little evidence supporting their function at concentrations naturally detected in soil systems (Bremner and McCarty, 1993).

References

Addiscott, T. M. 1983. Kinetics and temperature relationships of mineralization and nitrification in Rothamsted soils with differing histories. J. Soil Sci. 34:343–353.

Alexander, M., and F. E. Clark. 1965. Nitrifying bacteria. Pp. 1477–1483. *In* C. A. Black (ed.), Methods of Soil Analysis, 2. American Society of Agronomists, Madison, WI.

Anderson, D. W., S. Saggar, J. R. Bettany, and J. W. B. Stewart. 1981. Particle size fractions and their use in studies of soil organic matter: I. The nature and distribution of forms of carbon, nitrogen, and sulfur. Soil Sci. Soc. Am. J. 45:767–772.

Baldwin, I. T., R. K. Olson, and W. A. Reiners. 1983. Protein binding phenolics and the inhibition of nitrification in subalpine balsam fir soils. Soil Biol. Biochem. 15:419–423.

Bazin, M. J., A. Rutili, A. Gaines, and J. M. Lynch. 1991. Humic acid relieves pH-inhibition of nitrification in continuous-flow columns. FEMS Microbiol. Ecol. 85:9–14.

Beauchamp, E. G., W. D. Reynolds, D. Brasche-Villeneuve, and K. Kirby. 1986. Nitrogen mineralization kinetics with different soil pretreatments and cropping histories. Soil Sci. Soc. Am. J. 50:1478–1483.

Belser, L. W. 1977. Nitrate reduction to nitrite, a possible source of nitrite for growth of nitrite-oxidizing bacteria. Appl. Environ. Microbiol. 34:403–410.

Belser, L. W., and E. L. Mays. 1980. Specific inhibition of nitrite oxidation by chlorate and its use in assessing nitrification in soils and sediments. Appl. Environ. Microbiol. 39:505–510.

Belser, L. W., and E. L. Schmidt. 1978a. Diversity of ammonia oxidizing nitrifier population of a soil. Appl. Environ. Microbiol. 36:584–588.

Belser, L. W., and E. L. Schmidt. 1978b. Serological diversity within a terrestrial ammonia-oxidizing population. Appl. Environ. Microbiol. 36:589–593.

Belser, L. W., and E. L. Schmidt. 1981. Inhibitory effect of nitropyrin on three genera of ammonia-oxidizing nitrifiers. Appl. Environ. Microbiol. 41:819–821.

Bolton, H., Jr., J. L. Smith, and R. E. Wildung. 1990. Nitrogen mineralization potentials of shrub-steppe soils with different disturbance histories. Soil Sci. Soc. Am. J. 54:887–891.

Bonde, T. A., and T. Lindberg. 1988. Nitrogen mineralization kinetics in soil during long-term aerobic laboratory incubations: A case study. J. Environ. Qual. 17:414–417.

Bonde, T. A., and T. Rosswall. 1987. Seasonal variation of potentially mineralizable nitrogen in four cropping systems. Soil Sci. Soc. Am. J. 51:1508–1514.

Bondietti, E., J. P. Martin, and K. Haider. 1972. Stabilization of amino sugar units in humic-type polymers. Soil Sci. Soc. Am. Proc. 36:597–602.

Bremner, J. M., and L. A. Douglas. 1971. Inhibition of urease activity in soils. Soil Biol. Biochem. 3:297–307.

Bremner, J. M., and G. W. McCarty. 1993. Inhibition of nitrification in soil by allelochemicals derived from plants and plant residues. Soil Biochemistry 8:181–218. Dekker, NY.

Broadbent, F. E. 1986. Empirical modeling of soil nitrogen mineralization. Soil Sci. 141:208–213.

Cabrera, M. L. 1993. Modeling the flush of nitrogen mineralization caused by drying and rewetting soils. Soil Sci. Soc. Am. J. 57:63–66.

Cabrera, M. L., and D. E. Kissel. 1988a. Potentially mineralizable nitrogen in disturbed and undisturbed soil samples. Soil Sci. Soc. Am. J. 52:1010–1015.

Cabrera, M. L., and D. E. Kissel. 1988b. Evaluation of a method to predict nitrogen mineralization from soil organic matter under field conditions. Soil Sci. Soc. Am. J. 52:1027–1031.

Cameron, R. S., and A. M. Posner. 1979. Mineralisable organic nitrogen in soil fractionated according to particle size. J. Soil Sci. 30:565–577.

Cassman, K. G., and D. N. Munns. 1980. Nitrogen mineralization as affected by soil moisture, temperature, and depth. Soil Sci. Soc. Am. J. 44:1233–1237.

Clarhom, M., B. Popovic, T. Rosswall, B. Soderstrom, B. Sohlenius, H. Staaf, and A. Wiren. 1981. Biological aspects of nitrogen mineralization in humus from a pine forest podzol incubated under different moisture and temperature conditions. Oikos 37:137–145.

Crasswell, E. T., and S. A. Waring. 1972a. Effect of grinding on the decomposition of soil organic matter. I. The mineralization of organic nitrogen in relation to soil type. Soil Biol. Biochem. 4:427–433.

Crasswell, E. T., and S. A. Waring. 1972b. Effect of grinding on the decomposition of soil organic matter. II. Oxygen uptake and nitrogen mineralization in virgin and cultivated cracking clay soils. Soil Biol. Biochem. 4:435–442.

Deans, J. R., J. A. E. Molina, and C. E. Clapp. 1986. Models for predicting potentially mineralizable nitrogen and decomposition rate constants. Soil Sci. Soc. Am. J. 50:323–326.

De Beer, D., J. C. Van Den Heuvel, and S. P. P. Ottengraf. 1993. Microelectrode measurements of the activity distribution in nitrifying bacterial aggregates. Appl. Environ. Microbiol. 59:573–579.

De Boer, W., P. J. A. Klein Gunnewiek, M. Veenhuis, E. Bock, and H. J. Laanbroek. 1991. Nitrification at low pH by aggregated chemolithotrophic bacteria. Appl. Environ. Microbiol. 57:3600–3604.

DeLuca, T. H. and D. R. Keeney. 1993. Soluble organics and extractable nitrogen in paired prairie and cultivated soils in central Iowa. Soil Sci. 155:219–228.

Dommergues, Y. R., L. W. Belser, and E. L. Schmidt. 1978. Limiting factors for microbial growth and activity in soil. *In* M. Alexander (ed.), Advances in Microbial Ecology 2:49–104. Plenum Press, N.Y.

Doxtander, K. G., and M. Alexander. 1966. Nitrification by heterotrophic soil microorganisms. Soil Sci. Soc. Am. Proc. 30:351–355.

Ellert, B. H., and J. R. Bettany. 1988. Comparison of kinetic models for describing net sulfur and nitrogen mineralization. Soil Sci. Soc. Am. J. 52:1692–1702.

Eylar, O. R., and E. L. Schmidt. 1959. A survey of heterotrophic micro-organisms from soil for ability to form nitrite and nitrate. J. Gen. Microbiol. 20:473–481.

Fliermans, C. B., B. B. Bohlool, and E. L. Schmidt. 1974. Autecological study of the chemoautotroph *Nitrobacter* by immunofluorescence. Appl. Microbiol. 27:124–129.

Fox, R. H., and W. P. Piekielek. 1978a. Field testing of several nitrogen availability indexes. Soil Sci. Soc. Am. J. 42:747–750.

Fox, R. H., and W. P. Piekielek. 1978b. A rapid method for estimating the nitrogen-supplying capabilityofasoil.SoilSci. Soc. Am. J. 42:751–753.

Fox, R. H. and W. P. Piekielek. 1983. Response of corn to nitrogen fertilizer and the prediction of soil nitrogenavailability with chemical tests in Pennsylvania. Pa. Agric. Exp. Stn. Bull. 843.

Gandeza, A. T., S. Shoji, and I. Yamada. 1991. Simulation of crop response to polyolefin-coated urea: I. Field dissolution. Soil Sci. Soc. Am. J. 55:1462–1467.

Goovaerts, P., and C. N. Chiang. 1993. Temporal persistence of spatial patterns for mineralizable nitrogen and selected soil properties. Soil Sci. Soc. Am. J. 57:372–381.

Goring, C. A. I. 1962. Control of nitrification by 2-chloro-6-(trichloro-methyl) pyridine. Soil Sci. 93:211–218.

Gould, W. D., F. D. Cook, and J. A. Bulat. 1978. Inhibition of urease activity by heterocyclic sulfur compounds. Soil Sci. Soc. Am. J. 42:66–72.

Hankinson, T. R., and E. L. Schmidt. 1988. An acidophilic and a neutrophilic *Nitrobacter* strain isolated from the numerically predominant nitrite-oxidizing population of an acid forest soil. Appl. Environ. Microbiol. 54:1536–1540.

Hastings, R. C., M. T. Ceccerini, N. Miclaus, J. R. Saunders, M. Bazzicalupo, and A. J. McCarthy. 1997. Direct molecular biological analysis of ammonia oxidising bacteria populations in cultivated soil plots treated with swine manure. FEMS Microbiol. Ecol. 23:45–54.

He, X.-T., F. J. Stevenson, R. L. Mulvaney, and K. R. Kelley. 1988. Extraction of newly immobilized ^{15}N from an Illinois Mollisol using aqueous phenol. Soil Biol. Biochem. 20:857–862.

Jones, R. D., and A. P. Schwab. 1993. Nitrate leaching and nitrite occurrence in a fine-textured soil. Soil Sci. 155:272–282.

Juma, N. G., and E. A. Paul. 1984. Mineralizable soil nitrogen: Amounts and extractability ratios. Soil Sci. Soc. Am. J. 48:76–80.

Juma, N. G., E. A. Paul, and B. Mary. 1984. Kinetic analysis of net nitrogen mineralization in soil. Soil Sci. Soc. Am. J. 48:753–757.

Kelley, K. R., and F. J. Stevenson. 1985. Characterization and extractability of immobilized ^{15}N from the soil microbial biomass. Soil Biol. Biochem. 17:517–523.

Kelley, K. R., and F. J. Stevenson. 1987. Effects of carbon source on immobilization and chemical distribution of fertilizer nitrogen in soil. Soil Sci. Soc. Am. J. 51:946–951.

Kladivko, E. J., and D. R. Keeney. 1987. Soil nitrogen mineralization as affected by water and temperature interactions. Biol. Fert. Soil. 5:248–252.

Kowalchuk, G. A., J. R. Stephen, W. Deboer, J. I. Prosser, T. M. Embley, and J. W. Woldendorp. 1997. Analysis of ammonia-oxidizing bacteria of the beta subdivision of the class proteobacteria in coastal sand dunes by denaturing gradient gel electrophoresis and sequencing of pcr-amplified 16S ribosomal DNA fragments. Appl. Environ. Microbiol. 63:1489–1497.

Krogmeier, M. J., G. W. McCarty, and J. M. Bremner. 1989. Potential phytotoxicity associated with the use of soil urease inhibitors. Proc. Nat. Acad. Sci. 86:1110–1112.

Landschoot, P. J., and D. V. Waddington. 1987. Response of turfgrass to various nitrogen sources. Soil Sci. Soc. Am. J. 51:225–230.

Learch, R. N., K. A. Barbarick, L. E. Sommers, and D. G. Westfall. 1992. Sewage sludge proteins as labile carbon and nitrogen sources. Soil Sci. Soc. Am. J. 56:1470–1476.

Liang, C. N., and M. A. Tabatabai. 1978. Effects of trace elements on nitrification in soils. J. Environ. Qual. 7:291–293.

Malcolm, R. L. 1990. The uniqueness of humic substances in each of soil, stream, and marine environments. Anal. Chem. Acta 232:19–30.

Malhi, S. S., and W. B. McGill. 1982. Nitrification in three Alberta soils: Effect of temperature, moisture and substrate concentration. Soil Biol. Biochem. 14:393–399.

Marion, G. M., and C. H. Black. 1987. The effect of time and temperature on nitrogen mineralization in Arctic tundra soils. Soil Sci. Soc. Am. J. 51:1501–1508.

Matulewich, V. A., P. F. Strom, and M. S. Finstein. 1975. Length of incubation for enumerating nitrifying bacteria present in various environments. Appl. Environ. Microbiol. 29:265–268.

McCarty, G. W., and J. M. Bremner. 1986. Effect of phenolic compounds on nitrification in soil. Soil Sci. Soc. Am. J. 50:920–923.

McClung. G., and W. T. Frankenberger, Jr. 1985. Soil nitrogen transformations as affected by salinity. Soil Sci. 139:405–411.

Mengel, K., and H. Schmeer. 1985. Effect of straw, cellulose, and lignin on the turnover and availability of labelled ammonium nitrate. Biol. Fert. Soils 1:175–181.

Morrissey, R. F., E. P. Dugan, and J. S. Koths. 1974. Inhibition of nitrification by incorporation of selected heavy metals in soil. Abst. Ann. Mtg. Am. Soc. Microbiol. 1974:2.

Myers, R. J. K., C. A. Campbell, and K. L. Weier. 1982. Quantitative relationship between net nitrogen mineralization and moisture content of soils. Can. J. Soil Sci. 62:111–124.

Navarro, E., P. Simonet, P. Normand, and R. Bardin. 1992. Characterization of natural populations of *Nitrobacter* spp. using PCR/RFLP analysis of ribosomal intergenic spacer. Arch. Microbiol. 157:107–115.

Piccolo, A., L. Campanella, and B. M. Petronio. 1990. Carbon-13 nuclear magnetic resonance spectra of soil humic substances extracted by different mechanisms. Soil Sci. Soc. Am. J. 54:750–756.

Poovarodom, S., R. L. Tate III, and R. A. Bloom. 1988. Nitrogen mineralization rates of the acidic, xeric soils of the New Jersey Pinelands: Field rates. Soil Sci. 145:257–263.

Post, W. M., J. Pastor, P. J. Ainke, and A. G. Stangenberger. 1985. Global patterns of soil nitrogen storage. Nature (London) 317:613–616.

Schimel, J. P., M. K. Firestone, and K. S. Killham. 1984. Identification of heterotrophic nitrification in a Sierran forest soil. Appl. Environ. Microbiol. 48:802–806.

Schnitzer, M., and H. Kerndorff. 1980. Effects of pollution on humic substances. J. Environ. Sci. Health. 15B:431–456.

Schnitzer, M., and H. Kodama. 1992. Interactions between organic and inorganic components in particle-size fractions separated from four soils. Soil Sci. Soc. Am. J. 56:1099–1105.

Serna, M. D., and F. Pomares. 1992. Evaluation of chemical indices of soil organic nitrogen availability in calcareous soils. Soil Sci. Soc. Am. J. 56:1486–1491.

Severson, G. R., and R. L. Mahler. 1988. Influence of soil water potential and seed-banded sulfur-coated urea on Spring barley emergence. Soil Sci. Soc. Am. J. 52:529–534.

Smith, S. J., L. B. Young, and G. E. Miller. 1977. Evaluation of soil nitrogen mineralization potentials under modified field conditions. Soil Sci. Soc. Am. J. 41:74–76.

Sollins, P. G. Spycher, and C. A. Glassman. 1984. Net nitrogen mineralization from light- and heavy-fraction forest soil organic matter. Soil Biol. Biochem. 16:31–37.

Somville, M. 1984. Use of nitrifying activity measurements for describing the effect of salinity on nitrification in the Scheldt Estuary. Appl. Environ. Microbiol. 47:424–426.

Stanford, G., M. H. Frere, and R. A. V. Pol. 1975. Effect of fluctuating temperatures on soil nitrogen mineralization. Soil Sci. 119:222–226.

Stanley, P. M., and E. L. Schmidt. 1981. Serological diversity of *Nitrobacter* spp. from soil and aquatic habitats. Appl. Environ. Microbiol. 41:1069–1071.

Starr, J. L., T. B. Parkin, and J. J. Meisinger. 1992. Sample size consideration in the determination of soil nitrate. Soil Sci. Soc. Am. J. 56:1824–1830.

Stephen, J. R., A. E. McCaig, Z. Smith, J. I. Prosser, and T. M. Embley. 1996. Molecular diversity of soil and marine 16S rRNA gene sequences related to beta-subgroup ammonia-oxidizing bacteria. Appl. Environ. Microbiol. 62:4147–4154.

Stevenson, F. J. 1994. Humus Chemistry: Genesis, Composition, Reactions. John Wiley & Sons, NY. 443 pp.

Stevenson, F. J. 1986. Cycles in Soil: Carbon, Nitrogen, Phosphorus, Sulfur, Micronutrients. John Wiley & Sons, NY. 380 pp.

Talpaz, H., P. Fine and B. Bar-Yosef. 1981. On the estimation of N-mineralization parameters from incubation experiments. Soil Sci. Soc. Am. J. 45:993–996.

Tate, R. L., III. 1977. Nitrification in histosols: A potential role for the heterotrophic nitrifier. Appl. Environ. Microbiol. 33:911–914.

Tate, R. L., III. 1987. Soil Organic Matter: Biological and Ecological Effects. John Wiley & Sons, NY. 291 pp.

Verstraete, W., and M. Alexander. 1973. Heterotrophic nitrification in samples of natural ecosystems. Environ. Sci. Tech. 7:39–42.

Vitousek, P. M., and P. A. Matson. 1985. Causes of delayed nitrate production in 2 Indiana USA forests. For. Sci. 31:122–131.

Walters, D. T., M. S. Aulakh, and J. W. Doran. 1992. Effects of soil aeration, legume residue, and soil texture on transformations of macro- and micronutrients in soil. Soil Sci. 153:100–107.

Westermann, D. T., and S. E. Crothers. 1980. Measuring soil nitrogen mineralization under field conditions. Agron. J. 72:1009–1012.

Nitrogen Fixation: The Gateway to Soil Nitrogen Cycling

A reliable source of fixed nitrogen is a requirement for sustaining all biomass components, above- and belowground, in terrestrial ecosystems. In sites with significant accumulations of organic matter, the majority of this nutrient is provided by the internal cycling of nitrogen between organic and plant-available inorganic forms, as was described for climax forest ecosystems in Chapter 11. In contrast, the biological communities of more nutrient-limited soil ecosystem types may be totally reliant upon external sources for support of living community development. Such conditions are characteristic of pioneer sites, such as volcanic soils, which contain essentially no reservoirs of fixed nitrogen. Although there is a clear-cut difference between climax and pioneer systems, care must be taken in not undervaluing the importance of fixed nitrogen inputs in either system. Due to the inevitable losses of fixed nitrogen from *all* soils, some external supply of fixed nitrogen is always required in terrestrial sites to maintain long-term productivity. Otherwise, soil fixed nitrogen reserves would slowly decline until biomass productivity is reduced or even precluded. Therefore, it can be stated that a portion of the foundation for sustained productivity of soil-based systems is the insurance of renewal of the fixed nitrogen pool through existence of a functional nitrogen-fixing microbial population *in situ*.

Fortunately for the development of highly productive, sustainable terrestrial systems, fixed nitrogen inputs do balance losses. Total losses of fixed nitrogen from land ecosystems of approximately 160 to 225 Tg yr^{-1} are replaced by inputs of 214 to 262 Tg yr^{-1} (Knowles, 1981). The magnitude of various pathways of fixed nitrogen into and out of land ecosystems is demonstrated in Fig. 12.1. Primary routes for nitrogen loss from soils are volatilization, runoff, and denitrification. By far the largest nitrogen source is biological nitrogen fixation. Fertilizer application (anthropogenic input) is a distant

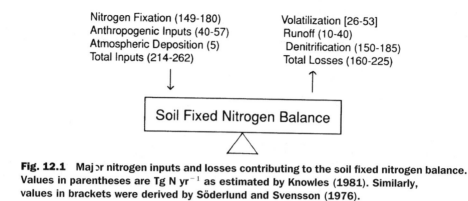

Fig. 12.1 Major nitrogen inputs and losses contributing to the soil fixed nitrogen balance. Values in parentheses are Tg N yr^{-1} as estimated by Knowles (1981). Similarly, values in brackets were derived by Söderlund and Svensson (1976).

second. This conclusion is based in part on the fact that the vast majority of the world's soils do not receive inputs of fertilizer. Understandably, most of the anthropogenically fixed nitrogen entering the soil ecosystem (estimated to be 30 Tg yr^{-1} by Burns and Hardy, 1975) is applied to agricultural and urban soils (i.e., yards and recreational areas). Interestingly, a less obvious fact is that due to the intensive management of nitrogen-fixing crops, the majority of the biologically fixed nitrogen enters the land-based nitrogen reserves in cropped lands (Burns and Hardy, 1975).

Symbiotic nitrogen fixation may be exploited to reduce use of industrially fixed nitrogen (see Chapter 13 for discussion of symbiotic nitrogen fixation). Although intensive agricultural systems are usually sustained through liberal use of industrially fixed nitrogen, economic and environmental pressures dictate reduction of fertilizer use and maximization of *in situ* biological nitrogen fixation. Most commonly, this entails cultivation of legumes, crops commonly involved in symbiotic associations with nitrogen-fixing bacteria. Similar plant-microbe partnerships involved in nitrogen fixation are actinorhizal associations. In contrast to the legume interactions, which tend to be identified with temperate agricultural systems, actinorhizal symbioses are more generally linked to native or "wild land" situations. (It must be noted that many of the vast number of underappreciated legumes are tree species, which therefore would be associated with these less-managed ecosystems.) For these nonagricultural ecosystems, considerations of balancing use of industrial and biological nitrogen sources are of less importance since the plants involved in the symbioses are rarely harvested.

With legume crops, the greatest gain in fixed nitrogen results from return of all new plant biomass to the soil. Removal (i.e., harvesting) of aboveground portions of the plant leads to little increase in soil nitrogen. Some benefit results from retention of the legume root system within soil (Alexander, 1977), but the bulk of the fixed nitrogen is contained within the harvested crop. Therefore, the best situation for cultivated legumes that can be anticipated would be retention of sufficient fixed nitrogen through incorporation of the root systems alone to maintain or minimize fixed nitrogen losses. In contrast to this situation with cropped legumes (e.g., soybeans, beans, alfalfa, and peas), a less appreciated, perhaps even more beneficial situation results from growth of legumenous trees. General soil fertility gains are realized for growth of the nitrogen-fixing trees (e.g., see, Budowski and Russo, 1997; Jonsson et al., 1996; Sanginga et al., 1996).

These introductory comments reveal the critical nature of nitrogen fixation in maintenance of native as well as managed ecosystems. Consideration of the balance between fixed nitrogen supplies, losses, and inputs in a soil ecosystem is especially important for the development of management or reclamation plans for damaged or mismanaged soil sites. Thus, this Chapter is presented with the primary objective of evaluating nitrogen fixation as a process and of elucidating the nuances of variation of ecosystem properties on nitrogen-fixation rates in managed and native soil systems.

12.1 Biochemistry of Nitrogen Fixation

An in-depth examination of the biological, biochemical, and genetic aspects of biological nitrogen fixation exceeds the purview of primary interest of most soil microbiologists. Yet a foundational understanding of some essential properties of the enzymes and the organisms involved is necessary to comprehend the properties of the soil ecosystem that control the dynamics of nitrogen fixation in native ecosystems. Primary traits of biological nitrogen fixation controlling the yields of fixed nitrogen in native soils and the kinetics of this production to be considered herein are the complexity of the enzyme (nitrogenase) catalyzing the reduction of dinitrogen (N_2) to ammonium, the quantities and sources of the energy required to drive the process, and the diversity of microbes responsible for the process. For more detailed analyses of the biochemical and genetic aspects of nitrogen fixation processes, see basic biochemistry and general microbiology textbooks and recent reviews of the topic (e.g., Brewin, 1991; Burris, 1991; Caetano-Anollés and Gresshoff, 1991).

12.1.1 The Process

Dinitrogen, because of the triple bond between the two nitrogen atoms, is a nearly inert molecule with respect to the metabolic potentialities of the biological community. The only enzymologically catalyzed reaction involving dinitrogen is accomplished by a limited number of bacterial strains. As will become evident when individual genera are discussed, these bacteria constitute a highly diverse physiological group with the capacity to reduce dinitrogen to ammonium ion as their primary common trait. It should also be stated that the ability to transform dinitrogen to ammonium is limited to bacteria; that is, no fungi, plants, or animals possess this capability.

In nitrogen fixation, the nitrogen atom is reduced from its most oxidized state (N_2) to its most reduced form (NH_4^+) by the following reaction:

$$N_2 + 8H^+ + \text{energy} \rightarrow 2NH_3 + H_2$$

Although the nitrogen atom passes through a number of oxidation states in between dinitrogen and ammonium, no free intermediates between dinitrogen and ammonia are produced. All intermediates are retained within the cell of the bacterium on the enzyme catalyzing the reaction (nitrogenase). Thus, the only direct product of nitrogen fixation encountered by soil microbes is ammonium.

Energy Consumption During Nitrogen Fixation: Considering the bonding energy associated with the triple bond of dinitrogen, it is reasonable to conclude that nitrogen fixation is an energy-intensive process. Under ideal conditions, a minimum of 16 molecules of MgATP are required for each atom of dinitrogen cleaved and reduced. Since ideal conditions are rarely achieved in native ecosystems, the energy source requirement is generally between 20 and 30 molecules of MgATP (see Burris, 1991). To put this quantity of energy into perspective, recall that the total oxidation of one molecule of glucose yields a maximum of 38 ATP molecules (see Chapter 4). Thus, a microbe-fixing nitrogen must devote nearly all of the energy provided by the oxidation of a glucose molecule to the reduction of one dinitrogen molecule to 2 ammonium ions. The ATP molecules supporting nitrogen fixation are derived primarily either directly from photosynthetic processes (photoautotrophs that fix nitrogen) or from decomposition of organic compounds (nitrogen-fixing heterotrophs supported indirectly from photosynthetic processes).

Knowledge of this energy requirement allows the soil microbiologist to deduce the sites of activity and the relative importance of the various metabolic groups of nitrogen-fixing bacteria. For example, as will be discussed below, this energy requirement limits the quantities of dinitrogen that can be reduced by microbes catalyzing soil organic matter as their energy source. For maximal conversion of dinitrogen to ammonium, a stable, abundant energy source is necessary. Such reliable supplies occur naturally in the rhizosphere (i.e., photosynthetically fixed carbon contained in root exudates or provided by symbiotic associations) or through direct conversion of light energy for the process (cyanobacteria and photosynthetic bacteria). The quantity of easily metabolized organic matter in soils in generally is rarely sufficient to support high levels of nitrogen fixation.

This energy requirement essentially dictates that the primary sites for nitrogen fixation in soil are limited to rhizosphere. Potential exceptions to this observation could be soil sites receiving large quantities of fixed carbon — that is, spill sites or systems receiving large quantities of industrial effluents. For example, Neilson and Sparell (1976) isolated nitrogen-fixing enterobacteria from paper-mill process waters. In conclusion, in native ecosystems, highest nitrogen-fixation contributions to soil ecosystems are associated with situations where energy inputs are provided by photosynthesis (nitrogen fixation by blue-green bacteria) or through transfer of plant photosynthate directly to nitrogen-fixing bacteria or actinomycetes (symbiotic nitrogen fixation).

Hydrogen Production and Hydrogenase Activity: A related concern involving the stoichiometry of reduction of dinitrogen to ammonium involves the inefficiency associated with loss of reducing power (H^+) through the formation of hydrogen gas. It appears to be logical that the most efficient process would be the incorporation of all reducing power into the nitrogenous product, but that does not occur. Formation of hydrogen gas appears to be illogical also from the fact that hydrogen is a specific competitive inhibitor of nitrogenase. Although there has been considerable research involving optimization of utilization of reducing power in dinitrogen reduction, the synthesis of hydrogen gas appears to be an obligatory part of the nitrogen-fixation reaction (Burris, 1991).

The quantities of electron flow lost to the microbe through hydrogen ion reduction can be significant. For example, nodules formed by *Rhizobium meliloti* or *Rhizobium trifolii* on their respective hosts have been shown to lose at least 17 percent of their electrons through this mechanism (Ruiz-Argüeso et al., 1979). The mean conversion of reducing power to hydrogen gas for each of these organisms was considerably greater than this minimum value. For the alfalfa nodulator, *R. meliloti*, the average proportion of electrons lost was 25 percent. The comparable value for the clover-nodulating organism (*R. trifolii*) was 35 percent.

Although there appears to be no way of preventing the formation of this hydrogen gas during nitrogen fixation, a means of improving the efficiency of the nitrogen-reducing process for the microorganism is to recover the energy contained in hydrogen gas. Hydrogenase is an enzyme that catalyzes the oxidation of hydrogen gas with water as the final product. This oxidative reaction is coupled with ATP formation or with reduction of ferredoxin or flavodoxin. These electron acceptors are of particular interest since their reduction can lead to increased ammonium ion yields. The electrons transferred to ferredoxin and flavodoxin are returned to the dinitrogen reductase portion of the nitrogenase enzyme. Examples of the occurrence of hydrogenase activity in nitrogen-fixing bacteria include *Rhodopseudomonas capsulata* (Colbeau et al., 1980), where high hydrogenase activities were also found *Bradyrhizobium japonicum* (Keyser et al. 1984; Merberg and Maier, 1983), and *R. leguminosarum* (Nelson and Salminen, 1982). The proportion of each of these species producing hydrogenase is extremely variable. For example, in the study conducted by Keyser et al. (1984), a small portion of the *B. japonicum* strains produced hydrogenase (i.e., were Hup$^+$). Of 972 *B. japonicum* isolates from 65 soybean fields located in 12 states (USA) studied, only 20 percent produced hydrogenase. Frequency of occurrence of these hydrogenase producers varied considerably between strains. None of the isolates in serogroup 135 were Hup$^+$, whereas 93 percent were positive in serogroup 122.

12.1.2 The Enzyme, Nitrogenase

Reduction of dinitrogen to ammonium in nitrogen-fixing bacteria is catalyzed by an enzyme system consisting of two distinct proteins, dinitrogenase (also known as MoFe protein or protein I) and dinitrogenase reductase (alias Fe protein or protein II). Dinitrogenase is a large protein (220 to 240 K daltons) that binds and reduces dinitrogen. Dinitrogenase reductase transfers the electrons to dinitrogenase.

Dinitrogenase and dinitrogenase reductase are highly conserved proteins between the various strains of nitrogen fixers. This fact is supported by the observation that purified dinitrogenase and dinitrogenase reductase produced from a variety of nonrelated bacterial species can be combined to produce active nitrogen fixation. Thus, a limited set of basic properties can be attributed to all nitrogenase enzymes.

All nitrogenases are extremely oxygen sensitive. Since nitrogen fixation is catalyzed by strictly anaerobic bacteria, facultative aerobes, microaerophilic organisms, and strictly aerobic bacteria, a major proportion of the organisms involved must possess a mechanism to protect the oxygen-labile protein. Two adaptive mechanisms to protect nitrogenase are available to the microbes. Either a means can be developed that results in the exclusion of molecular oxygen from sites of active nitrogenase, or nitrogenase molecules inherently resistant to oxygen can be selected. The latter option has not been found to occur.

At last seven individual means for exclusion of molecular oxygen from the environment of nitrogenase enzymes have been observed. These include:

- *Respiratory protection:* The rate of consumption of oxygen by individual cells may be enhanced so that the oxygen levels in its microenvironment are reduced to acceptable levels. For example, *Azotobacter* sp. cells have been observed to oxidize more fixed carbon in the presence of oxygen than is needed to satisfy microbial energy requirements. This protective mechanism appears to be operative in a variety of free-living organisms, including such metabolically diverse organisms as *Azotobacter vinelandii* (e.g., Shah et al., 1973) and cyanobacteria (e.g., Murry et al., 1984; Peschek et al., 1991).

- *Conformational protection:* An alternative to developing an enzyme that is inherently resistant to oxygen is to develop a nitrogenase that changes to a protective conformation in the presence of oxygen. This has been observed with *Azotobacter*. In the protected state, nitrogen fixation is totally precluded, but the activity rapidly returns when the oxygen tension is reduced to acceptable levels. This alteration of nitrogenase structure results from association of a small protein (approx. 24,000 dalton molecular weight) with the nitrogenase protein.

- *Oxygen regulation of nitrogenase synthesis:* Nitrogenase is a complex of large-protein molecules. Synthesis of these molecules thus requires expenditure of significant portions of the cells' energy resources. One means of protection of the nitrogen-fixing apparatus is to prevent its production under unfavorable conditions. Molecular oxygen represses nitrogenase synthesis in *Klebsiella pneumoniae*, *Azotobacter chroococcum*, and some strains of rhizobia.

- *Gum production:* For oxygen to interfere with dinitrogen reduction, the inhibitor and susceptible enzyme must interact. Nitrogenase can be protected by production of a physical barrier to molecular oxygen. Gums or polysaccharides accumulated external to the cell wall may provide such a protective wall. Microorganisms may synthesize extracellular polysaccharides that would reduce the diffusion rate of oxygen in the vicinity of the cell wall. This has been proposed to occur for *Azotobacter* spp. and *Derxia gummosa*.

- *Heterocyst production:* Nitrogen fixation can be an especially incompatible process in photosynthetic organisms in that molecular oxygen is a product of carbon fixation in aerobic, nitrogen-fixing cyanobacteria. With these organisms, the oxygen-labile enzymes may be isolated in structures with thick cell walls (heterocysts) to reduce contact of nitrogenase with molecular oxygen. (For further discussion of the protection of nitrogenase in cyanobacteria, see Fay, 1992, or Yoon and Golden, 1998.)

- *Leghemoglobin production:* In *Rhizobium*-legume nodules, molecular oxygen is complexed by substances with high oxygen affinities in order to reduce intracellular oxygen concentrations. A hemoglobin-like molecule (leghemoglobin) is synthesized in the cytosol around the packets of bacteria (see Chapter 13 for a description of nodule structure). This complexing of molecular oxygen results in reduction of oxygen at the bacteroid surface to about 10 nM. Further protection of nitrogenase in the bacteroids is provided by a terminal oxidase system that has an unusually high affinity for molecular oxygen.

- *Migration to suitable environment:* Aerotactic organisms may migrate to positions in their microenvironment suitable for nitrogen fixation. This has been observed with *Azospirillum* spp. (Barak et al., 1982).

- *Other mechanisms:* In aquatic ecosystems, as exemplified by cyanobacteria associated with rice cultivation, nitrogen-fixing cells may be protected by clumping of the cells, or oxygen-generating and nitrogen-fixing processes may be separated temporally. For example, cyanobacterial cells have been shown to form clusters. Molecular oxygen consumption on the surface of the cluster is sufficient to reduce oxygen concentrations inside the grouping to noninhibitory levels.

12.1.3 Measurement of Biological Nitrogen Fixation in Culture and in the Field

Accurate quantification of fixed nitrogen inputs into terrestrial systems due to biological nitrogen fixation is difficult to achieve. Even though large quantities of dinitrogen are reduced to ammonium by the soil biological community annually (e.g., Knowles, 1981), the actual change in quantities of fixed nitrogen in individual soil microsites is relatively small compared to the background concentrations of both ammonium and dinitrogen, which are common chemical components of both soil air and soil water as well as the general atmosphere. This situation necessitates sensitive and accurate measurement of small changes in reactants and products. Such procedures for use with complex soil samples are of limited availability.

A variety of techniques adaptable for quantification of dinitrogen or ammonium ions in terrestrial ecosystems are available. Unfortunately, their use to evaluate changes in fixed nitrogen concentrations in native soil samples or in growth media of axenically grown microbial cultures is problematic. To a large degree, use of direct measurements of changes in ammonium or dinitrogen concentrations for quantification of fixed nitrogen production yields equivocal data. It is not unusual for gains or loss of these nitrogenous substances in soil samples where nitrogen fixation is actively occurring to be less than or approximately equal to the intrinsic variability of the procedure used to quantify them. Thus, the sought-after value is obscured within the background "noise" associated with the data (i.e., the standard deviation or standard error of the mean of the results).

To overcome the methodological limitations imposed by the ubiquity of dinitrogen and ammonium, a variety of indirect as well as more direct procedures have historically been employed to estimate quantities of nitrogen fixed in culture and in soil samples. Methods utilized have included growth in microbiological media lacking a fixed nitrogen source to determine the capability of microbial isolates to fix nitrogen, assessment of changes in the distribution of fixed nitrogen in known soil nitrogen pools (nitrogen balance studies), tracing movement of nitrogen atoms with ^{15}N-labeled substrates, and use of acetylene reduction as an indicator of nitrogen-fixation capacity.

Nitrogen-Free Cultivation and Nitrogen Balance Procedures: Two logical procedures for assessing nitrogen-fixing potential are (1) determination of the capacity of a microorganism to grow in media lacking a source of fixed nitrogen and (2) assessment of the changes in nitrogen pools in a soil site. These methods were the most commonly used early methods for assessing nitrogen-fixation potentials. Unfortunately, both methods yield equivocal results, at best. These techniques, although applied to very different systems, are examined together because they are both supported by one faulty assumption: that all nitrogen inputs and losses in a soil site or sample or a culture medium are understood and can be accurately quantified. Thus, changes either in quantities of inorganic fixed nitrogen in the field or in the amount of the microbial biomass produced in laboratory cultures were assumed to have resulted from biological nitrogen fixation. The tenets of this assumption are rarely achieved in either the field or the laboratory.

From the view of assessment of nitrogen-fixing activity of axenically grown microbial cells, it has long been possible to prepare laboratory media with nitrogen-free chemicals and reasonably pure water. Since it is also known that a fixed nitrogen source is obligatory for cell growth and reproduction, it can logically be concluded that if a microbe grows in a growth medium composed of nitrogen-free water and chemicals, it must be fixing nitrogen. Superficially, it appears that there are no other sources of fixed nitrogen. It is not difficult to find a variety of papers published prior to the 1950s documenting nitrogen-fixing capabilities of a variety of bacterial species through use of this method. Unfortunately, conclusions reached using nitrogen-free growth media for microbial culture are equivocal at best and are more commonly erroneous.

Assuming that the fixed nitrogen has been adequately removed from substituents of the growth media or that the quantities of microbial biomass produced exceeds those that would be allowed by the trace nitrogenous contaminants of the substituents of growth medium, it would seem apparent that the only other source of fixed nitrogen for microbial replication would be biological fixation. Unfortunately, fixed nitrogen is a common component of the atmosphere of laboratories and incubators. The tendency to disregard this air pollution has resulted in a misplaced confidence in the laboratory results. Volatile nitrogenous compounds are easily transferred via the atmospheric route into the nitrogen-free media in laboratory situations. It is easy to envision conditions where a culture of a microbe producing a volatile nitrogenous compound is growing adjacent to a culture containing a nitrogen-free medium. Furthermore, biological products of neighboring cultures would not be the only source of contamination for such studies utilizing nitrogen-free growth media. Although scrupulously clean chemical and water are generally employed in preparing for culture of the test microbes, incorporation of ammonium into the microbial culture media could easily occur any time ammonia-based cleaning solvents are used in the laboratory or when a bottle of ammonium hydroxide is opened in the laboratory.

In the more complex field situations, the extent of biological nitrogen fixation has been historically estimated by quantifying changes in quantities of nitrogen contained within the various soil nitrogen pools. Nitrogen balance sheets were prepared. With this method, fixed nitrogen concentrations are measured directly in soil samples. Ammonium is usually extracted from the sample, steam distilled and titrated. Nitrate and nitrite ions are then reduced to ammonium and similarly assayed. Finally, organic nitrogen is converted to ammonium using the Kjeldahl digestion procedure. Two requirements that are basic to any nitrogen balance-sheet analysis are (1) a diligence in analyzing all nitrogen sources and (2) a capacity to assess accurately small changes in generally sizable nitrogen pools. Neither condition is met in these analyses.

Error results from difficulties in quantifying fixed nitrogen losses through denitrification and erosion as well as inputs from atmospheric sources (e.g., inputs of volatized ammonium) The problems associated with quantification of slight changes in preexisting fixed nitrogen pools were discussed above. Applicability of nitrogen balance data to total ecosystems is further reduced by the inaccuracies introduced by extrapolation of data with high standard deviations collected by analysis of a single or a limited number of field samples to account for all nitrogen inputs into the study site. Use of nitrogen balance-sheet–type studies are most commonly limited to well-defined systems experiencing major changes in the size of the nitrogen reservoirs, such as the closed system represented by a soil sample incubated in a beaker in the laboratory.

Use of ^{15}N Tracers: A nearly ideal means of compensating for the tendency of background nitrogen concentrations to obscure small changes in distribution of atoms among the various reservoirs of soil nitrogen is to trace the movement of labeled atoms within the system of interest. Sufficient labeled dinitrogen can be added to the soil sample to overcome any difficulties associated with the natural background levels of the nitrogen isotope contained therein. There are no radioactive nitrogen isotopes with a sufficiently long half-life to be practical for use in studies of nitrogen fixation in soil environments. The heavy isotope ^{15}N is useful to trace nitrogen transformations in natural soil samples. Assays for ^{15}N in soil nitrogen pools are sensitive and accurate.

The primary assumption underlying use of ^{15}N for quantification of nitrogen fixation is that the labeled atom is utilized by the microbes indiscriminately; that is, the microbes are not capable of differentiating between the ^{15}N label and the more abundant ^{14}N isotope. A further consideration is that an apparent microbial discrimination in nitrogen atom sources must not be imposed on the system by incomplete homogenization of the amended labeled nitrogenous substrate with indigenous soil nitrogen pools. This means that the ^{15}N-labeled dinitrogen molecules must be reduced to ammonium proportionally to their contribution to the total dinitrogen pool, not by their occurrence in different locations within soil microsites. Unfortunately, when using ^{15}N to study nitrogen fixation, it must be remembered that some discrimination favoring ^{15}N atoms by nitrogenase in leguminous root nodules has been shown (see Burris, 1991). Furthermore, it is not possible to distribute evenly an externally supplied ^{15}N-labeled mixture within the heterogeneous, complex soil structure. Even with these difficulties, ^{15}N has commonly been used as a tracer for nitrogen in nitrogen fixation studies.

One means of counteracting the isotope discrimination problem is to use ^{15}N-depleted dinitrogen. For this method, the reduction on the atom percent ^{15}N in the ammonium present in the sample is measured in samples incubated in an atmosphere containing dinitrogen from which the ^{15}N has been removed or highly reduced.

Other difficulties associated with large-scale utilization of ^{15}N to quantify nitrogen fixation in environmental samples are the requirement of a mass spectrometer to quantify enrichment or depletion of the ^{15}N in the ammonium pool extracted from the test system, the cost involved with purchase of the isotopically labeled substances, and the rather long incubation times in gas-tight systems required to ensure that the labeled dinitrogen sources are not diluted by atmospheric sources.

Acetylene Reduction: To overcome the methodological limitations involved with utilization of ^{15}N-labeled dinitrogen, a surrogate compound, reduced by nitrogenase at rates comparable to those associated with dinitrogen reduction, could be useful. Ideally, this substance should be quantified easily by methods readily available in most laboratories and reasonably inexpensive. Fortunately, nitrogenase is not totally specific for dinitrogen as a substrate. This enzyme can also reduce nitrous oxide, cyanide, methyl isocyanide, azide, acetylene, and cyclopropene (Fig. 12.2). The reduction of these compounds can serve as a measure of nitrogenase. The most useful alternative substrate for nitrogenase is acetylene in that it is not normally present in atmospheric samples, it is relatively inexpensive, and it can be quantified using gas chromatographs, which are more common in laboratories than are mass spectrometers.

Since most studies of nitrogen fixation have an objective of assessing or predicting the quantities of dinitrogen that could be reduced in the samples, it would be useful if the quantities of ethylene produced from acetylene could be extrapolated to nitrogen-fixation potential. The stoichiometry of the reduction of dinitrogen or acetylene by nitrogenase is as follows:

$$N_2 + 8H^+ \rightarrow 2NH_3 + H_2$$

$$3C_2H_2 + 6H^+ \rightarrow 3C_2H_4$$

Considering the hydrogen ions incorporated into the acetylene and dinitrogen by nitrogenase, reduction of 3 acetylene atoms to 3 ethylene atoms should be

$$2H^+ + 2e^- \longrightarrow H_2$$

$$N_2 + 8H^+ + 8e^- \longrightarrow 2NH_3 + H_2$$

$$N_2O + 2H^+ + 2e^- \longrightarrow N_2 + H_2O$$

$$CN^- + 7H^+ + 6e^- \longrightarrow CH_4 + NH_3$$

$$N_3^- + 3H^+ + 2e^- \longrightarrow N_2 + NH_3$$

$$C_2H_2 + 2H^+ + 2e^- \longrightarrow C_2H_4$$

Fig. 12.2 **Some alternative substrates reduced by the nitrogenase system.**

equivalent to reduction of 1 dinitrogen to 2 ammonia atoms; that is, production of 3 ethylene molecules should represent the reduction of 1 dinitrogen.

Unfortunately, extrapolation of data collected from analysis of acetylene to estimate field nitrogen fixation rates is more complicated than this stoichiometric comparison predicts. Comparison of data derived from the assessment of nitrogen fixation in parallel samples using ^{15}N-labeled dinitrogen and acetylene suggests that this ratio of 3:1 (ethylene to dinitrogen) can be attained at times in culture and soil, but more commonly the ratio ranges from 0.75 to 4.5 or greater. Therefore, caution is necessary in interpretation of the acetylene-derived data. Due to this data inconsistency, nitrogen-fixation potential measured by the acetylene procedure is generally designated *nitrogen fixation (acetylene)* so that readers may realize that the results are derived from extrapolation of results from an acetylene-based assay rather than from an assessment of dinitrogen reduction directly.

This discrepancy between predicted equivalence of acetylene and dinitrogen reduction from basic chemical principles and actual field results can be explained to a large degree by the different solubilities of dinitrogen and acetylene in water. Dinitrogen is poorly soluble in water. Thus, nitrogenase is rarely saturated by its primary substrate. In contrast, acetylene is highly water soluble. An acetylene concentration of 0.2 atmospheres is usually sufficient to saturate nitrogenase. Therefore, when acetylene is used as a substrate for nitrogenase, all nitrogenase enzyme molecules in the sample are measured. (Since the substrate is at saturating concentrations in the reaction mixture, the reaction rate is proportional to the quantities of enzyme present.) When the nonsaturating concentrations of dinitrogen are used to assess nitrogen-fixing capacity, some of the enzyme is inactive and therefore undetected. (In the latter situation, the reaction rate is substrate-, not enzyme-concentration limited.)

A further complication in using the acetylene-reduction procedure to estimate biological nitrogen-fixation rates in field samples can result from reduction of the ethylene produced by nitrogenase. To accurately quantify nitrogenase activity, all of the ethylene produced from reduction of acetylene must be detected. In some soil ecosystems, the ethylene produced by nitrogenase is oxidized by soil microorganisms. Some soil microbes are capable of using ethylene as a carbon and energy source. Thus, to assure that the ethylene detected during measurements of nitrogenase activity represents the total amount produced by nitrogenase, a control sample for disappearance of ethylene must be included in the study.

12.2 General Properties of Soil Diazotrophs

The Microbes: Nitrogen fixation is catalyzed solely by bacteria either living among the general soil microbial population or in symbiotic associations [i.e., *Rhizobium*-legume or actinorhizal symbioses (see Chapter 13) or with fungi (lichens)]. Rhizosphere-associated nitrogen-fixing bacteria (e.g., *Azos-*

pirillum sp. growing in and around the root) can also be defined as existing in a loosely symbiotic relationship with the plant. Nitrogen-fixing microorganisms, which are called **diazotrophs**, constitute a large group of marginally related bacteria. In contrast to the situation with nitrifying bacteria where the capacity to nitrify is sufficiently definitive that can be used as a taxonomic trait, the extreme physiological and structural diversity of diazotrophs preclude using a single metabolic property, synthesis of nitrogenase, as a generic defining trait. Grouping nitrogen-fixing organisms into one or a few bacterial genera based on their capacity to reduce dinitrogen to ammonium is counterproductive. These organisms are found in such diverse groups as eubacteria, the photosynthetic cyanobacteria (formerly known as the blue-green algae), and actinomyctes.

The only clear commonalities among the diazotrophs is their general classification as bacteria and their ability to fix nitrogen. Diazotrophs are aerobes, facultative anaerobes, or anaerobes. They may derive their energy from organic substances, inorganic compounds, or directly from solar energy. Diazotrophs grow as single cells or chains of individual cells (e.g., rhizobia, clostridia), as loosely associated colonies (e.g., some blue-green algae), and as myclial structures (e.g., the actinomycetes). This metabolic and physical variability enables these organisms to contribute to fixed-nitrogen pools in essentially any ecosystem where microbes are capable of growth and reproduction (e.g., see Table 12.1).

Table 12.1 **Examples of Diazotrophs Commonly Found in Soil, Rhizosphere, and Soil-Linked Aquatic Systems**

Habitat and Metabolic Grouping	Microbial Groups
Nonassociated, aerobes, heterotrophs	*Azotobacter* spp.
	Bacillus spp.
	Beijerinckia spp.
	Derxia spp.
Nonassociated, aerobes, lithotrophs	*Thiobacillus* spp.
Nonassociated, anaerobes, heterotrophs	*Clostridium* spp.
	Desulfovibrio spp.
Aquatic (swamps, bogs, rice fields), aerobes, phototrophs	Cyanobacteria (blue-green-algae)
Aquatic (swamps, bogs, rice fields), anaerobes, phototrophs	*Chaetomium* spp.
	Rhodopseudomonas spp.
	Thiospirillum spp.
Rhizosphere (loosely associated), facultative aerobes, heterotrophs	*Enterobacter* spp.
	Erwinia spp.
	Klebsiella spp.
Rhizosphere (loosely associated), aerobes, heterotrophs	*Azospirillum* spp.
	Pseudomonas spp.
Root nodules (legumes or angiosperms)	*Frankia* spp.
	Rhizobium spp.

Population Densities of Diazotrophs in Soils: Although they are highly adaptable to soil physical and chemical conditions, diazotrophic bacteria are rarely the dominant populations in terrestrial ecosystems. Exceptions are found in situations where development of a heterotrophic soil microbial community is limited by availability of fixed nitrogen. Should sufficient energy exist in the microsite for function of diazotrophs (e.g., solar or chemical energy predominantly, since significant accumulation of organic matter would encourage growth of heterotrophic populations), diazotrophs would have a competitive advantage over the general heterotrophic population. Dominance of nitrogen-fixing bacteria in the developing soil ecosystem could prevail until sufficient fixed nitrogen accumulates in the site to support an active heterotrophic microbial population. Once sufficient organic matter resources are accumulated in a soil to support an active heterotrophic community, the dominance of the diazotrophs declines. Nitrogen-fixing bacteria are generally poor competitors. Thus, high population densities of biological nitrogen fixers are generally localized and transient. It can be reasonably assumed that at least minimal populations of nitrogen-fixing procaryotes are found in all soil sites where life processes are possible. This conclusion is supported by the diversity of metabolic groupings of organisms capable of fixing nitrogen and the essential role of fixed nitrogen in biological community development.

Physical and Chemical Limitations to Diazotrophs: The type of biological nitrogen-fixing community involved and the quantities of ammonium ion produced in an ecosystem are controlled by temperature, pH, moisture saturation of the ecosystem, and the extent of the energy supply available to support diazotroph metabolism. Each of these soil properties must be considered when designing experiments to study *in situ* diazotrophs as well as when developing site-reclamation plans where biological nitrogen fixation is to be encouraged. For example, aeration is an important consideration in predicting nitrogen-fixation potential at the site in that free-living anaerobic diazotroph activity exceeds that of aerobes. Yet since plants are sensitive to oxygen deprivation of their roots, optimization of nitrogen fixation in anaerobic microsites of soil aggregates would be the method of choice for augmenting ammonium production by free-living diazotrophs.

Similarly, for heterotrophic diazotrophs living among the general soil microbial populations, high concentrations of carbonaceous substrates are critical. Therefore, in organic-matter–deficient soils, a low population of nitrogen-fixing, heterotrophic bacteria is supported. The relative contributions of each of these groups of organisms to total nitrogen fixed is proportional to the quantities of energy available to them.

An example of the control of free-living diazotroph population density with available energy source is provided by work of Azam et al. (1988), where a high correlation between amounts of glucose added to soil and nitrogen-fixation rates was detected. The relationship of nitrogen contributions by the various diazotrophs based on relationship to their energy source is dramatically demonstrated in comparing relative fixed-nitrogen contributions in agricultural soils. According to Burns and Hardy (1975), 100 to 300 kg N ha^{-1} year^{-1} is produced in these soils from the *Rhizobium*-legume association, compared to 0.4 to 0.8 kg N ha^{-1} year^{-1} from free-living soil bacteria. Furthermore, the efficiency of conversion of energy available in the ecosystem to fixed nitrogen is necessarily less than would be measured in axenic culture, since some of the energy in the growth substrates is lost to competing microbes in soil and to the mechanisms that are necessary to deal with the stresses of the soil environment.

Availability of Newly Fixed Nitrogen to Plants: A further consideration regarding the impact of nitrogen-fixation activity by free-living diazotrophs on total soil fertility relates to the availability of the limited amounts of ammonium ion produced to the higher plant community. Nitrogen fixed by these organisms is only indirectly available to the biological community. Nitrogen fixed by the diazotrophs is initially incorporated into their biomass. For this fixed nitrogen to become generally available to the soil community, the diazotroph cell must die and its biomass mineralized.

12.3 Free-Living Diazotrophs

From a consideration of the wide variety of ecosystems in which free-living bacteria diazotrophs are found, it can be concluded that a vast diversity of diazotrophs must exist. Indeed, bacteria capable of fixing nitrogen while growing independently in soil are distributed among at least 26 genera, 11 families, and 3 orders of bacteria. It must be noted that within each individual genus, the number of species capable of fixing nitrogen is small, and that not every strain of a bacterial species containing diazotrophs will necessarily be capable of fixing nitrogen.

To provide an indication of the limitations to growth and their contributions to total fixed nitrogen in soil ecosystems, representative examples of aerobic, anaerobic, and autotrophic nitrogen-fixing bacteria will be discussed. The relationship of some of these organisms to the properties of their natural habitats is analyzed in the next section. The individual species examined are not necessarily the most important in any given ecosystem or in soil ecosystems in general. In many cases, they simply represent the more commonly studied. Quite commonly, the frequency of isolation and evaluation of individual diazotrophs is determined by their ease of isolation and their propensity for reproducible growth of the organism in axenic culture rather than the proportion of the soil nitrogen-fixing population that they constitute.

Aerobic, Free-Living Diazotrophs: The three most studied free-living, strict aerobes are *Azotobacter chroococcum*, *Azotobacter beijerinckia* (*Beijerinkia* sp.), and *Derxia gumosa*. From the number of reports in the literature, the conclusion could easily be reached that *Azotobacter chroococcum* is the most widespread species in neutral or alkaline soils. This conclusion is tentative at best and may easily be in error. The frequency that this organism is the topic of study may relate to the fact that it has been commonly studied in general bacteriology classes. Thus, soil microbiologists were well versed in the methods for isolating and growing this bacterial species.

Azotobacter chroococcum is a strict aerobe with a temperature optimum of about 30°C. The numbers of these organisms in soil range from undetectable to several thousand g^{-1} soil. It is unusual to find large populations of this organism in soil, thus *Azotobacter chroococcum* is unlikely to be an important contributor to total ecosystem fixed-nitrogen inputs, except in some specialized situations (e.g., soils receiving inputs of fixed carbon substances that can provide energy for growth of the *Azotobacter* spp.). *Azotobacter beijerinckia* is dominant in acidic soils — pH values as low as 3.0. This organism is common in tropical soils, rarely being found in temperate soils. It has also been reported in to occur in high Arctic soils (Jordan and McNicol, 1978), whereas both *Azotobacter chroococcum* and *Azotobacter beijerinckia* populations have been detected in Antarctic soils (see Vishniac, 1993). Similarly, *Derxia gummosa* is common in tropical soils of South America, with pH values ranging from about 4.5 to 6.5.

A variety of nonsymbiotic diazotrophs are frequently found in rhizosphere soils. These organisms include *Azospirillum* species plus a variety of enteric bacterial species. *Azospirillum lipoferum*, which was originally described by Beijerink in 1922 and named *Spirillum lipoferum*, forms loose symbiotic associations on grasses. These spiral-shaped gram negative organisms typically grow around roots as well as penetrate the roots to grow intercellularly. These organisms use root exudates for their carbon and energy source while fixing nitrogen. The latter group of nitrogen-fixing bacteria — the enterics — includes *Klebsiella* sp. These bacteria are described below with the anaerobic and facultatively anaerobic diazotrophs and in Section 12.4.1. As will be discussed in Section 12.4.3, small increases in plant biomass productivity have resulted from inoculation of roots with various strains of *Azospirillum lipoferum*.

Anaerobes and Facultative Anaerobic Bacteria: The most commonly encountered anaerobic diazotrophs are *Clostridium* spp. These organisms are nearly universally present in poorly drained soils as well as in anaerobic microsites of arable soils. *Klebsiella* spp. are the most commonly observed facultative anaerobic nitrogen-fixing bacteria. These organisms are widely distributed in aquatic and terrestrial environments, especially in rhizosphere soils. In arable lands, clostridial populations commonly range from 10^2 to 10^5 propagules g^{-1} soil, although under optimal conditions, populations as high as 10^6 g^{-1} have been detected. In contrast, *Klebsiella* sp. have been found in rhizosphere samples at population densities of 10^6 to 10^8 g^{-1} root tissue (Evans et al., 1972). The nitrogen-fixation efficiency of these organisms is comparable to that reported for aerobic organisms (2 to 20 mg N fixed g^{-1} carbohydrate consumed). In some systems, such as forest soils, it has been suggested that nitrogen fixation is probably limited to anoxic microsites (e.g., Limmer and Drake, 1996).

Cyanobacteria: Soil microbiologists tend to disregard the input of algae to soil biogeochemical cycles, but blue-green algae can provide significant inputs of fixed nitrogen into land-based ecosystems. These photosynthetic organisms are found primarily in surface soil crusts. Significant algal populations may develop on soil surfaces, especially in the presence of a thin layer of free-standing water. Approximately one-tenth of the established genera of cyanobacteria contain diazotrophs. The absolute contribution of these organisms to soil nitrogen economy is difficult to assess since their distribution is frequently localized, but significant contributions have been measured in rice field soils (see Section 12.4).

Due to the photosynthetic nature of these organisms, the quantities of nitrogen fixed are independent of an available external chemical energy supply. Yields of fixed nitrogen due to cyanobacteria activity can be highly significant. In rice paddy soils, contributions up to 30 kg N ha^{-1} year^{-1} are commonly reported. Rychert and Skujins (1974) found that algae and lichens in soils crusts of the Great Basin Desert could produce fixed nitrogen at rates up to 84 g N ha^{-1} year^{-1} (acetylene) in the laboratory. The rates they observed would extrapolate to 10 to 100 kg N ha^{-1} year^{-1} were comparable efficiency achieved in field samples.

12.4   Function of Diazotrophs in Native Ecosystems

The relative contributions of nonsymbiotic nitrogen-fixing bacteria to overall soil ecosystem productivity can be gleaned from the foregoing discussion, whereas the importance of symbiotic associations is detailed in Chapter 13. Properties of rhizosphere-associated nitrogen fixation by free-living bacteria and interactions in wetland-type ecosystems will be emphasized here. The latter analyses includes native swamps, bogs, and marshes as well as agriculturally managed rice soils. Selection and function of diazotrophs for function in these two types of ecosystem are particularly instructional in that conditions in each of these systems types exemplify extremes in soil properties conducive to the function of nitrogenase. Both wetland soils and rhizosphere sites can contain elevated concentrations of readily metabolizable organic matter. Also, due to generally lowered oxygen tensions in both sites, there is an increased probability of development of anaerobic microsites in the wetland soils and in the rhizosphere.

12.4.1   Nitrogen-Fixing Bacteria Among Rhizosphere Populations

The rhizosphere is home for a variety of heterotrophic diazotrophs exemplified by members of the family *Enterobacteriaceae* [including *Enterobacter agglomerans*, *Klebsiella pneumoniae*, *Enterobacter cloacae*, and *Erwinia herbicola* (e.g., Haahtela et al., 1981; Lindberg and Granhall, 1984; Pedersen et al., 1978)], *Bacillus* species (e.g., Lindberg and Granhall, 1984), and *Azospirillum lipoferum* (e.g., Haahtela et al., 1981). The presence of the enterobacteria and the *Azospirillum* species conform to a priori notions regarding this ecosystem; that is, the organisms are heterotrophs and are capable of functioning under reduced oxygen tensions or in the total absence of this gas. Indeed, since both aerobic and anaerobic microsites occur in the vicinity of the growing root, even the isolation of strict aerobes, such as the example of a *Bacillus* species cited above, from this environment is not problematic.

A perhaps more significant concern for a student of the environment relates to the significance of the contribution by diazotrophs residing in the rhizosphere to the fixed-nitrogen resources of the growing plant. Although situations where a significant role of rhizosphere diazotrophs in provision of plant fixed nitrogen have been demonstrated (Dobereiner, 1997), the importance of free-living diazotrophs in the rhizosphere in plant nitrogen metabolism is more commonly described as minimal, but significant. Acetylene and ^{15}N-labeled dinitrogen are reduced in the rhizosphere albeit at minimal levels. For example, Giller et al. (1988) found little to no nitrogen fixation in the rhizosphere of sorghum (*Sorghum bicolor*) and millet (*Pennisetum americanum*). With the sorghum, measurable nitrogen fixation was detected on only one occasion, whereas with the millet, nitrogen fixed was less than 1 percent of the plant nitrogen accumulated. Berestetskii and Vasyuk (1983) found weak nitrogen fixation activity with spring wheat (*Triticum aestivum*), *Poa pratensis*, and *Phleum pratense*, but higher activity with millet. Pedersen et al. (1978) estimate that the maximum rate of nitrogen fixation (extrapolated from acetylene-reduction data) in spring wheat and sorghum was 2.5 g nitrogen per hectare per day. Interestingly, although diazotrophs functioning in rhizosphere or rhizoplane ecosystems rely on root exudates to meet fixed carbon and energy requirements, these nitrogen-fixation rates are within the range generally observed for free-living diazotrophs, in less nutritionally luxuriant soil sites. This observation suggests that although a priori it could be anticipated that the rhizosphere-residing nitrogen-fixing bacteria should not be energy limited, the actual nitrogen-fixation rates detected indicate that competition with the general rhizosphere microbial community for carbon and energy supplies limits the capacity of the diazotrophs to exploit nutrients in root exudates.

The contribution of rhizosphere nitrogen fixation to the nitrogen budget of higher plants is minimal (see Giller and Day, 1985, for a further discussion of this topic). From the view of managing soil biologically fixed nitrogen to decrease anthropogenic intervention and increase system sustainability, the question of whether the nitrogen-producing capacity of the rhizosphere ecosystem can be increased emerges. Although optimal conditions for function of nitrogenase can be anticipated to occur in the rhizosphere, either microsite conditions vary from these anticipated situations, or the efficiency of nitrogen fixation by the diazotrophs in the fields is not sufficient to lead to meaningful fixed-nitrogen contributions to the plant. (Selection and use of highly efficient diazotrophs strains will be discussed below.)

Difficulties with heterogeneity of the microenvironment of the rhizosphere encountered by nitrogen-fixing bacteria are exemplified by evaluating the impact of variation of oxygen tension of the root environment on *in situ* nitrogen fixation. Haahtela et al. (1983) found that anaerobic conditions were required for maximal expression of nitrogenase activity of *Klebsiella* sp. and *Enterobacter* sp. isolated from plants. Similarly, in roots of corn (*Zea mays*), maximal acetylene reduction was noted at partial pressures of oxygen of 1 to 2 kPa. The activity was strongly inhibited with oxygen tensions above or equal to 6 kPa (Alexander et al., 1987). Alexander and Zuberer (1989) found that with ^{15}N-labeled dinitrogen, ammonium production was 200fold greater at 2 kPa molecular oxygen than at 10 kPa. Thus, it could be predicted that a controlling factor for nitrogen fixation in the rhizosphere would be the development of anaerobic microsites. Total anaerobiosis in this habitat is clearly precluded by the requirement for oxygen for respiration of the root tissue.

Management of Rhizosphere Diazotroph Populations: A potential means of reducing fertilizer expense for cropped (nonlegume) systems and anthropogenic intervention in reclamation projects is to optimize *in situ* inputs of biologically fixed nitrogen through manipulation of rhizosphere populations. Since habitats favorable to growth and development of diazotrophs exist around the plant root, the objective of such studies can be reduced to selection of efficient nitrogen-fixing strains of rhizosphere bacteria that are capable of competing with indigenous microbes when returned to the root ecosystem. An apparently optimistic expectation for achievement of such goals is possible in that selection of natural genetic variants and production of genetic engineered strains with improved nitrogen fixation efficiency can readily be accomplished in the laboratory. Furthermore, the potential exists for the manipulation of rhizosphere populations for the successful inoculation of these laboratory-cultivated strains into field situations. Application of these procedures is exemplified by studies of *Azospirillum* spp. These organisms are normal components of the rhizosphere microbial community, fix nitrogen, and are amenable to laboratory culture.

As with any inoculation procedure, albeit soil or rhizosphere inoculation, the initial concerns with amendment of axenically cultivated strains are (1) can the desired effect be observed in the ecosystem following addition of the test strain (e.g., stimulation of plant biomass synthesis), and (2) does the alteration of the ecosystem result from the anticipated activity of the amended microbe or is an alternative mechanism operative? Once responses to these questions are acquired, then considerations regarding reproducibility, stability, and magnitude of the amendment response must be addressed.

Inoculation of a range of grass varieties with *Azospirillum* strains does increase plant biomass production (e.g., Bashan and Holguin, 1997; Desalamone et al., 1996; Gaskins et al., 1977; Smith et al., 1976; Subba-Rao et al., 1978). Several mechanisms may be proposed to explain this increase in plant growth by inoculation with cultures of *Azospirillum*. Explanations of the stimulation of biomass production include (1) augmented ammonium supplies produced by the *Azospirillum* sp. (i.e., nitrogen fixation), (2) plant hormone synthesis by the bacterium (e.g., Tien et al., 1979), (3) alteration of root structure (e.g., Fallik et al., 1988; Jain and Patriquin, 1984; Hadas and Okon, 1987), and (4) response to increased plant nutrients provided by mineralization of dead *Azospirillum* cells. (See Vandebroek and Vanderleyden, 1995, for a review of the genetics of phytohormone production, nitrogen fixation, and mechanisms of plant root attachment.)

It is probable that all of these mechanisms function in part with regard to *Azospirillum* inoculation of grasses since nitrogenase synthesis and activity appear not to provide the primary explanation for increases in plant productivity. Estimates of fixed nitrogen produced by *Azospirillum* sp. in the rhizosphere range from 2 kg N ha^{-1} (e.g., von Bülow and Döbereiner, 1975) to as much as 90 kg N ha^{-1} (Döbereiner et al., 1972; Döbereiner et al., 1973) with inoculated grasses in South America. Other studies have reported low levels of nitrogen fixation in temperate grasses (e.g., Barber et al., 1976; Harris et al., 1989; Okon et al., 1983). Alternatively, some impact of plant hormone production by the *Azospirillum* sp. used as an inoculant is logical. The bacterially produced plant hormones would stimulate root biomass production (e.g., Tien et al., 1979). Bashan and Dubrovsky (1996) have shown mixed results of inoculation with *Azospirillum* spp. on shoot-to-root ratio. In about half the cases, this ratio was increased, whereas in the remainder the ratio was decreased. Cation accumulation by the plant may (Lin et al., 1983) or may not (Bashan et al., 1990) be meaningful.

. These data suggest limitations in yields of fixed nitrogen in inoculated rhizospheres. As our understanding of the genetics mechanisms associated with nitrogen fixation in *Azospirillum* and with ability of the bacterium to compete with indigenous rhizosphere populations improves, the probability of developing an economic and effective rhizosphere-inoculation procedure will increase. Also, along with using genetic modification to improve competitive ability of *Azospirillum* cells introduced into the rhizosphere, fungicide or bacteriocides may be added with the diazotroph to increase the probability of establishment of the culture (Bashan, 1986a). With this procedure, the inhibitors are used to limit the activity of indigenous populations until the foreign bacterial strain can become established, thereby giving the amended organism a competitive advantage.

Associated with selection of appropriate strains for inoculation of root or seed tissue is the development of an effective inoculation method. Viable cells must be delivered to the appropriate habitat on the root at a time of root development, when maximal benefits to the plant are accrued and the conditions in the microsite wherein the microbes are anticipated to function are optimal for nitrogen fixation. *Azotobacter* populations can become established on grass roots through inoculation of either seeds or root tissue, although seed inoculation appears to be the more effective means of delivery (Bashan, 1986b). One advantage of seed inoculation is that the amended bacterial strain is present as the nascent root tissue is produced. Thus, instead of having to compete with established bacterial populations, as would be the situation when roots are inoculated directly, the competition is reduced to a race between seed and soil flora to colonize the newly emergent root tissue.

Should the objective be to inoculate root tissue directly, the capability of the added cells or their progeny to spread from inoculated to noninoculated roots becomes important. Roots of infected plants can serve as a vector for transmission of the exogenous bacterial strain to noninfected populations (Bashan and Levanony, 1987; Bashan and Levanony, 1989). In fact, weeds growing between rows of crop plants can serve to transport the added bacterial strain (Bashan and Levanony, 1989). Thus, although a time delay in spread of the inoculant must occur with direct inoculation of plant roots in the field, diffusion of inoculated bacterial strains throughout a field can be anticipated to result eventually. For this process, active bacterial cultures may be added to the soil surface adjacent to the plant stem and washed into the rhizosphere, plant roots could be dipped into the inoculum prior to planting, or the inoculum could be banded adjacent to the plant in a region where the roots of the growing plant would be anticipated to penetrate. As has been shown with rhizobial inoculation of legumes (see Chapter 13), a variety of carriers may be used for introduction of *Azospirillum* inoculum into soil (e.g., Bashan, 1986c). Each of these techniques would be of less efficiency than seed inoculation (i.e., from a time as well as quantity of inoculum that would have to be used), but with highly competitive and nitrogen-fixation–efficient diazotroph strains, such procedures may become more feasible.

An interesting variation on the observed effect of the inoculation of nonlegumes with *Azospirillum* spp. is the situation where legumes are inoculated with a mixture of the appropriate rhizobial strain for the legume plus *Azospirillum* sp. (e.g., Bashan and Holguin, 1997; Burdman et al., 1997; Galal, 1997). With appropriate ratios of the two microbial strains, nodulation and nitrogen fixation of the association are increased over that seen with plants inoculated with the rhizobial strain alone.

12.4.2 Nitrogen Fixation in Flooded Ecosystems

Wetland soils are sites of limited but active biological nitrogen fixation. The quantities of fixed nitrogen gained by this source are within the range commonly reported to be produced by free-living microbes in soils in general. For example, Waugmen and Bellamy (1980) estimate nitrogen-fixation inputs due to heterotrophic bacteria in German mires (using the acetylene-reduction method) to range from 0.07 to 2.1 g N m^{-1} $year^{-1}$. Slightly higher fixed-nitrogen contributions have been reported for Gulf coast salt marshes (USA) [4.5 and 15 g N m^{-2} $year^{-2}$ (Casselman et al., 1981)]. Since biomass productivity is low in swampy environments due to the prevailing flooded conditions, biologically fixed nitrogen can be a major source of this essential plant nutrient.

Free-living bacteria are responsible for nitrogen fixation in swampy soils. As might be anticipated considering the oxygen-limiting conditions of marshes, anaerobic bacteria are major participants in this process. Dicker and Smith (1980) found that *Clostridium* sp., *Azotobacter* sp., and *Desulfovibrio* sp. accounted for a major portion of the nitrogen-fixing activity in a Delaware (USA) salt marsh, although large populations of *Azotobacter* sp. were present in the samples. Cooccurrence of obligate aerobes and anaerobes in this ecosystem appears to be contradictory, but consideration of the heterogeneity of the soil profile (i.e., water-saturated soil overlain by flowing water) reveals that it is reasonable to anticipate occurrence of both aerobic and anoxic microsites. Oxygen-bearing water flowing into the system provides for the sustenance of aerobic microsites, whereas activity of the aerobic bacteria in the system creates anaerobic sites. Furthermore, sequestering of water into non-flowing situations within the soil profile also results in development of anaerobic habitats within the soil profile (stagnation).

The diversity of diazotrophs present in swampy soils indicates existence of a variety of energy sources supportive of microbial growth. Metabolic energy supply in these ecosystems is provided by (1) root exudates (e.g., from the *Spartina alterniflora* Loisel. plants in the salt marshes), (2) accumulations of partially decomposed organic matter under anoxic conditions (i.e., products of anaerobic decomposition of biomass substituents), and (3) solar energy (e.g., growth of cyanobacteria in the aquatic portion of the site).

Rice paddies provide an excellent example of the function of these various diazotroph groups in a managed, flooded ecosystem. A diverse population of autotrophic bacteria are found in the flood waters (e.g., Habte and Alexander 1980a; Kulasooriya and Silva, 1981). For example, 73 strains of cyanobacteria distributed taxonomically among 21 genera were isolated from rice soils in central Sri Lanka (Kulasooriya and De Silva, 1981). Similarly, the rhizosphere contains an active heterotrophic diazotroph population (e.g., Habte and Alexander, 1980b; Yoo et al., 1986). It should also be noted that these heterotrophic strains may also be active on aboveground plant surfaces (Ito et al., 1980). The activities of the latter populations were shown to be approximately 2.5-fold more active at fixing nitrogen than rhizosphere populations (Ito et al., 1980).

Wetland systems with their low, but significant biologically fixed nitrogen inputs provide an even more vivid picture of the complexity of diazotrophic communities within a single ecosystem than was provided by the rhizosphere examples discussed above. This diversity in physiological ability provides a genotypic foundation for long-term sustenance of the system under ever-changing conditions. Balance between fixed-nitrogen inputs from heterotrophic or autotrophic populations, aerobes or anaerobes, rhizosphere or leaf sheath populations is dependent upon such site properties as water influx rate, depth of flooding (i.e., drained to submerged), diurnal variation of temperature, and light influx.

12.5 Conclusions

The capacity to fix nitrogen is widely distributed among the bacterial genera, although only a limited number of species are actually capable of catalyzing the process. Nitrogenase, the enzyme responsible, is highly conserved among this diverse group of bacteria. Two characteristics of this reductive process dictate the yield of fixed nitrogen in various terrestrial environments. First, nitrogen fixation is an energy-intensive process. Thus, large quantities of dinitrogen are fixed only in situations where there is a high energy supply (e.g., symbiotic associations involving formation of root nodules and through action of photosynthetic bacteria). Diazotrophs growing in competition with the general heterotrophic soil microbial community contribute small quantities of fixed nitrogen to the biological community. Second, nitrogenase is an oxygen-labile enzymes. Hence, each diazotroph must expend resources to maintain anoxic conditions in the environment of active nitrogen fixation. These two overall traits of the nitrogen-fixation process and the diverse group of bacteria capable of catalyzing the reaction combined with the extremely heterogeneous soil ecosystem select for a highly diverse indigenous population of diazotrophs in any terrestrial ecosystem. For example, the rhizosphere can contain aerobic, facultative anaerobic, and strictly anaerobic diazotrophs when the oxygen sensitivity of the organisms alone is considered. This diversity of *in situ* populations and variations in phenotypic expression of nitrogen-fixing efficiency allow cautious optimism for genetic manipulation and selection of diazotrophs that can be introduced into soil systems to improve contributions to soil fixed nitrogen resources.

References

Alexander, D. B., and D. A. Zuberer. 1989. $^{15}N_2$ fixation by bacteria associated with maize roots at a low partial O_2 pressure. Appl. Environ. Microbiol. 55:1748–1753.

Alexander, M. 1977. Introduction to Soil Microbiology. John Wiley & Sons, NY.

Alexander, D. B., D. A. Zuberer, and D. M. Vietor. 1987. Nitrogen fixation (C_2H_2 reduction) associated with roots of intact *Zea mays* in fritted clay at reduced oxygen tensions. Soil Biol. Biochem. 19:1–6.

Azam, G., T. L. Mulvaney, and F. J. Stevenson. 1988. Quantification and potential availability of non-symbiotically fixed ^{15}N in soil. Biol. Fert. Soils 7:32–38.

Barak, R., I. Nur, Y. Okon, and Y. Henis. 1982. Aerotactic response of *Azospirillum brasilense*. J. Bacteriol. 152:643–649.

Barber, L. E., J. D. Tjepkema, S. A. Russell, and H. J. Evans. 1976. Acetylene reduction (nitrogen fixation) associated with corn inoculated with *Spirillum*. Appl. Environ. Microbiol. 32:108–113.

Bashan, Y. 1986a. Enhancement of wheat root colonization and plant development by *Azospirillum brasilense* following temporary depression of rhizosphere microflora. Appl. Environ. Microbiol. 51:1067–1071.

Bashan, Y. 1986b. Significance of timing and level of inoculation with rhizosphere bacteria of wheat plants. Soil Biol. Biochem. 18:297–302.

Bashan, Y. 1986c. Alginate beads as synthetic inoculant carriers for slow release of bacteria that affect plant growth. Appl. Environ. Microbiol. 51:1089–1098.

Bashan, Y., and J. G. Dubrovsky. 1996. *Azospirillum* spp. participation in dry matter partitioning in grasses at the whole plant level. Biol. Fert. Soils 23:435–440.

Bashan, Y., S. K. Harrison, and R. E. Whitmoyer. 1990. Enhanced growth of wheat and soybean plants inoculated with *Azospirillum brasilense* is not necessarily due to general enhancement of mineral uptake. Appl. Environ. Microbiol. 56:769–775.

Bashan, Y., and G. Holguin. 1997. *Azospirillum*-plant relationships — environmental and physiological advances (1990–1996). Can. J. Microbiol. 43:103–121.

Bashan, Y., and H. Levanony. 1987. Horizontal and vertical movement of *Azospirillum brasilense* CD in the soil and along the rhizosphere of wheat and weeds in controlled and field environments. J. Gen. Microbiol. 133:3473–3480.

Bashan, Y., and H. Levanony. 1989. Wheat root tips as a vector for passive vertical transfer of *Azospirillum brasilense* CD. J. Gen. Microbiol. 135:2899–2908.

Berestetskii, O. A., and L. F. Vasyuk. 1983. Nitrogen-fixing activity in rhizosphere and on roots of nonleguminous plants. Biol. Bull. 10:32–37.

Brewin, N. J. 1991. Development of the legume root nodule. Annu. Rev. Cell Biol. 7:191–226.

Budowski, G., and R. Russo. 1997. Nitrogen-fixing trees and nitrogen fixation in sustainable agriculture — research challenges. Soil Biol. Biochem. 29:767–770.

Burdman, S., J. Kigel, and Y. Okon. 1997. Effects of *Azospirillum brasilense* on nodulation of common bean (*Phaseolus vulgaris* L.). Soil Biol. Biochem. 29:923–929.

Burns, R. C., and R. W. F. Hardy. 1975. Nitrogen Fixation in Bacteria and Higher Plants. Springer-Verlag, NY.

Burris, R. H. 1991. Nitrogenases. J. Biol. Chem. 266:9339-9342.

Caetano-Anollés, G., and P. M. Gresshoff. 1991. Plant genetic control of nodulation. Annu. Rev. Microbiol. 45:345–382.

Casselman, M. E., W. H. Patrick, Jr, and R. D. DeLaune. 1981. Nitrogen fixation in a gulf coast salt marsh. Soil Sci. Soc. Am. J. 45:51–56.

Colbeau, A., B. C. Kelley, and P. M. Vignais. 1980. Hydrogenase activity in *Rhodopseudomonas capsulata*: Relationship with nitrogenase activity. J. Bacteriol. 144:141–148.

Desalamone, I. E. E., J. Döbereiner, S. Urquiaga, and R. M. Boddey. 1996. Biological nitrogen fixation in *Azospirillum* strain-maize genotype associations as evaluated by the N-15 isotope dilution technique. Biol. Fert. Soils 23:249–256.

Dicker, H. J., and D. W. Smith. 1980. Enumeration and relative importance of acetylene-reducing (nitrogen-fixing) bacteria in a Delaware salt marsh. Appl. Environ. Microbiol. 39:1019–1025.

Döbereiner, J. 1997. Biological nitrogen fixation in the tropics — social and economic contributions. Soil Biol. Biochem. 29:771–774.

Döbereiner, J., J. M. Day, and P. J. Dart. 1972. Nitrogenase activity in the rhizosphere of sugarcane and some other tropical grasses. Plant Soil 37:191–196.

Döbereiner, J., J. M. Day, and P. J. Dart. 1973. Rhizosphere associations between grasses and nitrogen fixing bacteria: Effect of O_2 on nitrogenase activity in the rhizosphere of *Paspalum notatum*. Soil Biol. Biochem. 5:157–159.

Evans, H. J., N. E. R. Campbell, and S. Hill. 1972. Asymbiotic nitrogen-fixing bacteria from the surfaces of nodules and roots of legumes. Can. J. Microbiol. 18:13–21.

Fallik, E., Y. Okon, and M. Fisher. 1988. Growth response of maize roots to *Azospirillum* inoculation: Effect of soil organic matter content, number of rhizosphere bacteria and timing of inoculation. Soil Biol. Biochem. 10:45–50.

Fay, P. 1992. Oxygen relations of nitrogen fixation in cyanobacteria. Microbiol. Rev. 56:340–373.

Galal, Y. G. M. 1997. Dual inoculation with strains of *Bradyrhizobium japonicum* and *Azospirillum brasilense* to improve growth and biological nitrogen fixation of soybean (*Glycine max* L.) Biol. Fert. Soils 24:317–322.

Gaskins, M. H., M. U. Garcia, T. M. Tien, and D. H. Hubbell. 1977. Nitrogen fixation and growth substance produced by *Spirillum lipoferum* and their effects on plant growth. Plant Physiol. 59:128.

Giller, K. E., and J. M. Day. 1985. Nitrogen fixation in the rhizosphere: Significance in natural and agricultural systems. Pp. 127–147. *In* A. E. fitter (ed.), Ecological Interactions in Soil: Plants, Microbes, and Animals. Blackwell Scientific Publications, Boston.

Giller, K. E., S. P. Wani, J. M. Day, and P. J. Dart. 1988. Short-term measurements of uptake of nitrogen fixed in the rhizospheres of sorghum (*Sorghum bicolor*) and millet (*Pennisetum americum*). Biol. Fert. Soils 7:11–15.

Haahtela, K., K. Kari, and V. Sundman. 1983. Nitrogenase activity (acetylene reduction) of root-associated, clod-climate *Azospirillum*, *Enterobacter*, *Klebsiella*, and *Pseudomonas* species during growth on various carbon sources and at various partial pressures of oxygen. Appl. Environ. Microbiol. 45:563–570.

Haahtela, K., T. Wartiovaara, V. Sundman, and J. Skujins. 1981. Root-associated N_2 fixation (acetylene reduction) by *Enterobacteriaceae* and *Azospirillum* strains in cold-climate spodosols. Appl. Environ. Microbiol. 41:203–206.

Habte, M., and M. Alexander. 1980a. Nitrogen fixation by photosynthetic bacteria in lowland rice culture. Appl. Environ. Microbiol. 39:342–347.

Habte, M., and M. Alexander. 1980b. Effect of rice plants on nitrogenase activity in flooded soils. Appl. Environ. Microbiol. 40:507–510.

Hadas, R., and Y. Okon. 1987. Effect of *Azospirillum brasilense* inoculation on root morphology and respiration in tomato seedlings. Biol. Fert. Soils 5:241–247.

Harris, J. M., J. A. Lucas, M. R. Davey, G. Lethbridge, and K. A. Powell. 1989. Establishment of *Azospirillum* inoculant in the rhizosphere of winter wheat. Soil Biol. Biochem. 21:59–64.

Ito, O., D. Cabrera, and I. Watanabe. 1980. Fixation of dinitrogen-15 associated with rice plants. Appl. Environ. Microbiol. 39:554–558.

Jain, D. K., and D. G. Patriquin. 1984. Root hair deformation, bacterial attachment, and plant growth in wheat-*Azospirillum* association. Appl. Environ. Microbiol. 48:1208–1213.

Jonsson, K., L. Stahl, and P. Hogberg. 1996. Tree fallows — a comparison between five tropical tree species. Biol. Fert. Soils 23:50–56.

Jordan, D. C., and P. J. McNicol. 1978. Identification of *Beijerinckia* in the high Arctic (Devon Island, Northwest Territories). Appl. Environ. Microbiol. 35:204–205.

Keyser, H. H., D. F. Weber, and S. L. Uratsu. 1984. *Rhizobium japonicum* serogroup and hydrogenase phenotype distribution in 12 states. Appl. Environ. Microbiol. 47:613–615.

Knowles, R. 1981. Denitrification. *In* E. A. Paul and J. N. Ladd (eds.), Soil Biochemistry 5:323–369. Dekker, NY.

Kulasooriya, S. A. and R. S. Y. De Silva. 1981. Mutivariate interpretation of the distribution of nitrogen fixing blue-green algae in rice soils in central Sri Lanka. Ann. Bot. (Lond.) 47:31–52.

Limmer, C., and H. L. Drake. 1996. Non-symbiotic N_2-fixation in acidic and pH-neutral forest soils — aerobic and anaerobic differentials. Soil Biol. Biochem. 28:177–183.

Lin, W., Y. Okon, and R. W. F. Hardy. 1983. Enhanced mineral uptake by *Zea mays* and *Sorghum bicolor* roots inoculated with *Azospirillum brasilense*. Appl. Environ. Microbiol. 45:1775–1779.

Lindberg, T., and U. Granhall. 1984. Isolation and characterization of dinitrogen-fixing bacteria from the rhizosphere of temperate cereals and forage grasses. Appl. Environ. Microbiol. 48:683–689.

Merberg, D., and R. J. Maier. 1983. Mutants of *Rhizobium japonicum* with increased hydrogenase activity. Science 220:1064–1065.

Murry, M. A., A. J. Horne, and J. R. Benemann. 1984. Phsiological studies of oxygen protection mechanism in the heterocysts of *Anabaena cylindrica*. Appl. Environ. Microbiol. 47:449–454.

Neilson, A. H., and L. Sparell. 1976. Acetylene reduction (nitrogen fixation) by *Enterobacteriaceae* isolated from paper mill process waters. Appl. Environ. Microbiol. 32:197–205.

Nelson, L. M., and S. O. Salminen. 1982. Uptake hydrogenase activity and ATP formation in *Rhizobium leguminosarum* bacteroids. J. Bacteriol. 151:989–995.

Okon, Y., P. G. Heytler, and R. W. F. Hardy. 1983. N_2 fixation by *Azospirillum* and its incorporation into host *Setaria italica*. Appl. Environ. Microbiol. 46:694–697.

Pedersen, W. L., K. Chakrabarty, R. V. Klucas, and A. K. Vidaver. 1978. Nitrogen fixation (acetylene reduction) associated with roots of winter wheat and sorghum in Nebraska. Appl. Environ. Microbiol. 35:129–135.

Peschek, G. A., K. Villgrater, and M. Wastyn. 1991. 'Respiratory protection' of nitrogenase in dinitrogen-fixing cyanobacteria. Plant Soil. 137:17–24.

Ruiz-Argüeso, T. R. J. Maier, and J. H. Evans. 1979. Hydrogen evolution from alfalfa and clover nodules and hydrogen uptake by free-living *Rhizobium meliloti*. Appl. Environ. Microbiol. 37:582–587.

Rychert, R. C., and J. Skujins. 1974. Nitrogen fixation by blue-green-algae-lichin crusts in the great basin desert. Soil Sci. Soc. Am. Proc. 38:768–771.

Sanginga, N., S. K. A. Danso, and F. Zapata. 1996. Field measurements of nitrogen fixation in leguminous trees used in agroforestry systems — influence of N-15-labeling approaches and reference trees. Biol. Fert. Soils 23:26–32.

Shah, V. K., J. L. Pate. and W. J. Brill. 1973. Protection of nitrogenase in *Azotobacter vinelandii*. J. Bacteriol. 115:15–17.

Smith, R. L., J. H. Bouton, S. C. Shank, K. H. Quesenberry, M. E. Tyler, J. R. Milam, M. H. Gaskins, and R. C. Littell. 1976. Nitrogen fixation in grasses inoculated with *Spirillum lipoferum*. Science 193:1003–1005.

Söderlund, R., and B. H. Svensson, 1976. The global nitrogen cycle. *In* B. H. Svensson and R. Söderlund (eds.), Nitrogen Phosphorus, and Sulfur — Global Cycles. SCOPE Rep. 7. Ecol. Bull. (Stockholm) 22:23–73.

Subba-Rao, N. S., K. V. B. R. Tilak, M. Lakshmi Kumari, and D. S. Singh. 1978. Response of crops to *Spirillum lipoferum* inoculation. Pp. 20. *In* W. H. Orme-Johnson and W. E. Newton (eds.), Steinbock-Kettering International Symposium on Nitrogen Fixation. University of Wisconsin, Madison.

Tien, T. M., M. H. Gaskins, and D. H. Hubbell. 1979. Plant growth substances produced by *Azospirillum brasilense* and their effect on the growth of pearl millet (*Pennisetum americanum* L.). Appl. Environ. Microbiol. 37:1016–1024.

Vanderbroek, A., and J. Vanderleyden. 1995. Genetics of the *Azospirillum*-plant root association — Review. Crit. Rev. Plant Sci. 14:445–466.

Vishniac. H. S. 1993. The microbiology of Antarctic soils. Pp. 297–241. *In* E. I. Friedmann (ed.), Antarctic Microbiology. Wiley-Liss, NY.

Von Bülow, J. F. W., and J. Döbereiner. 1975. Potential for nitrogen fixation in maize genotypes in Brazil. Proc. Natl. Acad. Sci. U.S.A. 72:2389–2393.

Waugman, G. J., and D. J. Bellamy. 1980. Nitrogen fixation and the nitrogen balance in peatland ecosystems. Ecology 61:1185–1198.

Yoo, I. D., T. Fuji, Y. Sano, K. Komagata, T. Yoneyama, S. Iyama, and Y. Hirota. 1986. Dinitrogen fixation of rice and *Klebsiella associations*. Crop Sci. 26:297–301.

Yoon, H.-S. and J. W. Golden. 1998. Heterocyst pattern formation controlled by a diffusible peptide. Science 282:935–938.

Symbiotic Nitrogen Fixation

Root-nodule–based nitrogen-fixing associations include *Rhizobium*-legume and actinorhizal symbioses. These interactions between diazotrophic bacteria and plants are major biological contributors of fixed nitrogen in soil-based ecosystems. The predominance of this fixed nitrogen source results from (1) the agricultural exploitation of *Rhizobium*-legume symbioses and (2) the advantage conferred on these symbiotic diazotrophs by the ready availability of plant-produced photosynthate. That is, nitrogen fixation is an energy-intensive process and those heterotrophic organisms living in the legume or actinorhizal nodule have more energy available to devote to nitrogen fixation than do diazotrophs living nonassociated or loosely linked with plants (i.e., growing in the rhizosphere).

Actinorhizal associations are interactions of nonleguminous angiosperms with actinomycetes of the genus *Frankia*. *Rhizobium*-legume symbioses are the most studied, largely perhaps because of their agricultural and economic importance, but from a total ecosystem viewpoint, actinorhizal symbioses are also major contributors to terrestrial nitrogen cycling. The latter symbioses are characteristic more of nonmanaged ecosystems. Commonly considered actinorhizal associations are bayberry (*Myricia pensylvanica*) and *Casuarina*, a nonleguminous tree found in tropical and semiarid countries, plus a variety of ferns. Actinorhizal associations are common to acidic or physically or chemically stressed soils. In contrast, nitrogen-fixing legumes are more common in agricultural soils and other ecosystems with soils of pH of 5 or greater. The capability of actinorhizal associations to function in acid-stressed sites suggests a potential for their use in acid-affected reclamation sites.

Both actinorhizal and *Rhizobium*-legume symbioses will be analyzed in this chapter. The primary objectives for this study are to elucidate the properties of the bacteria and the host plant that determine the efficacy and predominance of the associations in ecosystems and to use these interactions as models for evaluation of the potential for their manipulation for management of bio-geochemical cycles in soil ecosystems. Topics were selected for discussion to stress the environmental aspects of nitrogen fixation in nodulated plant root systems. Biochemical and biological aspects of the topic will be analyzed only where needed to provide a foundation for understanding their impact on the occurrence of nitrogen fixation by these symbiotic associations in native ecosystems. For a more detailed discussion of the biochemistry or biology of actinorhizal or *Rhizobium*-legume interactions, refer to essentially any basic general microbiology text or biochemistry text. Several review articles are also cited below that provide an overview of these topics in greater detail than can be provided herein.

13.1 *Rhizobium*-Legume Association

Due to the extensive cultivation of legumes, the greatest documented contribution of fixed nitrogen into land-based systems results from the infection of legume roots by species of the bacterial genera *Rhizobium* and *Bradyrhizobium*. A tremendous potential for contribution of fixed nitrogen to soil ecosystems exists among the legumes. There are approximately 700 genera and about 13,000 species of legumes, only a portion of which have been examined for the capability to fix nitrogen. Estimates are that the rhizobial symbiosis with the somewhat greater than 100 agriculturally important legumes contributes nearly half the annual quantity of biological fixed nitrogen entering soil ecosystems.

The bacterial symbionts are gram negative, nonsporeforming, aerobic rods. Some strains are nonmotile, but typically rhizobia are motile. Rhizobia metabolize several carbohydrates, possibly with the production of acid, but no gas. The symbiotic association with legumes is a distinguishing classification trait of the genus. Rhizobial species are grouped into the genera *Rhizobium, Bradyrhizobium*, and *Azorhizobium*—stem-nodulating organisms (e.g., see Adebayo et al., 1989). Speciation of these bacteria is traditionally based on host specificity as defined by cross-inoculation grouping. Strain-specific DNA probes are proving useful in studying the occurrence of these organisms in soil (e.g., Watson et al., 1995) as well as in studying their taxonomy and phylogeny (e.g., Agius et al., 1997; Anyango et al., 1995, Young and Haukka, 1996). For example, sequence of small subunit ribosomal RNA support division of rhizobia into three genera (Young and Haukka, 1996). As is detailed in the next section, serological typing, DNA analysis, and legume infection have all been used to detect, identify, and quantify rhizobia. Particularly reassuring is the fact that the groupings of rhizobia delineated by the three types of analyses generally agree.

13.1.1 Grouping of Rhizobial Strains

From the view of a microbiologist, rhizobia constitute a somewhat nondescript group of bacteria. Classification schemes based on data derived from use of traditional physiological tests are inadequate to separate the genus into meaningful groupings. An alternative trait common to rhizobia that provides a reproducible method for their classification that was testable by methods available to early microbiologists is host specificity. The resultant association of strains into groups based on host specificity is called cross-inoculation groups. **Cross-inoculation groups** are by definition a collection of leguminous species that develop nodules when exposed to bacteria obtained from the nodules of any member of that group. More than 20 cross-inoculation groups have been established. Of the 7 most studied groups, 6 are sufficiently described to designate the responsible bacteria as species. The most-studied cross-inoculation groups are:

- *The alfalfa group:* alfalfa (*Medicago* spp.) and sweet clover (*Melilotus* spp.) nodulated by *R. meliloti.*

- *The clover group:* clovers (*Trifolium* spp.) nodulated by *R. trifolii.*

- *The pea group:* pea (*Pisum* spp.) and vetch (*Vicia*) nodulated by *R. leguminosarum.*

- *The bean group:* beans (*Phaseolus* spp.) nodulated by *R. phaseoli.*

- *The soybean group:* soybeans (*Glycine* spp.) nodulated by *Bradyrhizobium japonicum. Bradyrhizobium japonicum* and *Bradyrhizobium* spp. (symbionts of the cowpea group) were previously classified as slow-growing rhizobia, *Rhizobium japonicum.*

- *The cowpea group:* This group has not achieved a species designation, but these organisms nodulate a variety of legumes including cowpeas (*Vigna* sp.), Kudzu (*Pueraria* sp.), peanuts (*Arachis* sp.), and lima beans (*Phaseolus* sp.).

The cross-inoculation system has provided a useful classification system for rhizobial strains, but some major problems have historically been associated with its application. Some of these are logistical. For example, considerable greenhouse space is necessary to grow the legume seedlings required for an accurate determination of the identity of the *Rhizobium* strain of interest. Of perhaps greater concern in using this method to group newly isolated rhizobial strains is the potential for some organisms to nodulate more than one legume species. For example, soybean and cowpea groups contain strains of rhizobia isolated on legume groups that nodulate other legume species. This variation in nodulation specificity is termed **symbiotic promiscuity**. Thus, the validity of the cross-inoculation system has been challenged in some situations. The procedure works reasonably well for the frequently studied rhizobia-nodulating common, agricultural crops of temperate regions. More difficulties may be encountered as a greater proportion of the as yet little-studied rhizobia that nodulate tropical legumes are classified. Fortunately for the classification system, many rhizobia do nodulate legumes outside of their particular class, *but* these nodules are generally nonfunctional in nitrogen fixation (i.e., they are not effective). Thus, the designation of species or group based on formation of effective nodules is sustained. It should be noted that not all of the literature reports of symbiotic promiscuity represent true diversity in legume species specificity. In some cases, apparent host diversity has been the result of contamination of the rhizobial cultural with other rhizobial strains (Leps et al., 1980).

Cross-inoculation groups have provided a reasonably stable philosophical basis for the taxonomic scheme for grouping rhizobial strains, but these groupings are gradually falling into disrepute. A variety of methods, including evaluation of DNA hybridization, numerical taxonomic procedures, DNA base ratios, serology, antibiotic resistance, genetic markers, and nutritional traits, are being applied to a taxonomy in transition. Use of strain-specific antibodies allows identification of axenically grown rhizobia. Cross-inoculation groups are serologically distinct. Serological identification of rhizobial strains is quick, inexpensive, and reliable. A variety of methods have been used to analyze the genetic structure of rhizobial strains. These techniques, as indicated above, have reinforced the concepts of species designations derived by more classical methods (e.g., Brunel et al., 1996; Madrzak et al., 1995; Tesfaye et al., 1997), reveal strain variation within individual species (e.g., Dye et al., 1995; Hernandezlucas et al., 1995; Urtz and Elkan, 1996; Vanberkum et al., 1995), as well as detect occurrence of specific strains in soils and nodules (e.g., Desa et al., 1997; Hartmann and Amarger, 1991; Sessitsch et al., 1997; Simon et al., 1996). As the nuances of the implications of the application of these procedures to the study of rhizobial taxonomy become better understood, a more inclusive grouping of temperate and tropical strains will emerge. (See Elkan and Bunn, 1992, for a recent analysis of the status of rhizobial taxonomy.)

13.1.2 Rhizobial Contributions to Nitrogen Fixation

Rhizobia are capable of fixing nitrogen in axenic culture. Pagan et al. (1975) developed a defined medium and culture procedure that allows nitrogen fixation by free-living rhizobial strains. With this method, the bacteria are grown to a high density on yeast extract and mannitol agar, suspended in sterile water, and spread on a defined agar medium. Other studies have shown that nitrogenase activity of *Rhizobium japonicum* growing in defined medium is dependent upon the carbon source metabolized (Kurz and Larue, 1975) and oxygen tension of the growth medium (Tjepkema and Evans, 1975). Micro-aerophilic conditions are required for induction of nitrogen-fixing activity by the rhizobial cells. Kiester and Evans (1976) found that the optimal oxygen level for nitrogen fixation by *Rhizobium japonicum* and a *Rhizobium* sp. was approximately 0.1 percent in the gas phase. Carbon dioxide is also required by the rhizobial cells. Aguilar and Favelukes (1982) found an obligatory need for carbon dioxide for nitrogen fixation under microaerophilic conditions. This effect of carbon dioxide is mediated through ribulose bisphosphate carboxylase activity (Manian et al., 1984). Once the environmental conditions are optimized, the rate of nitrogen fixation by free-living cells is comparable to that observed with bacteroides isolated from soybean nodules.

13.1.3 Nodulation of Legumes

Rhizobia in soil are capable of infecting roots of susceptible plants to produce root nodules (Fig. 13.1). These nodules have a structure characteristic for the various symbiotic associations. Both the plant and the infecting bacteria control the development of the nodule structure. This summary of the process of formation of legume root nodules will be divided into three topic areas: (1) properties of the plant and microbe, facilitating contact and binding of rhizobial cells to the host root, (2) recognition of susceptible host roots and attachment, and (3) root invasion processes and development of effective

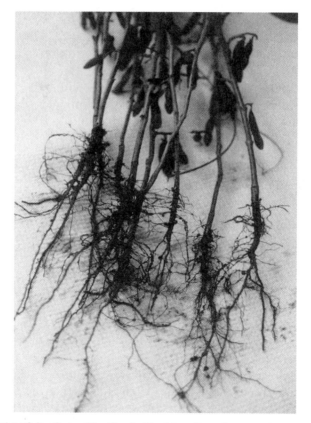

Fig. 13.1 Root nodules formed by *Bradyrhizobium japonicum* and soybeans

nodules. A variety of excellent reviews describing the biological, genetic, and biochemical aspects of nodule formation have been published (e.g., see Brewin, 1991; Caetano-Anollés and Gresshoff, 1991a). Thus, only the salient details of the process will be outlined here. Selected primary research publications are cited to exemplify studies of the interaction of rhizobial strains and their hosts.

Encounter Between Susceptible Roots and Rhizobia: Although their population densities may be low, rhizobia are found in most soils, even those sites where legumes have not grown for many decades. Hence, limitations in frequency of nodule formation in native ecosystems tend to result more from the rate of encounter of infecting bacteria and host than from presence of the symbiotic partners. Considering the heterogeneity of soil, the size of the bacterial cell, and the low proportion of the A horizon of the soil profile occupied by plant roots, random encounter between stationary bacterial cells and growing roots must be considered to be an inefficient means of inducing nodulation. Fortunately, a variety of basic properties of the soil system enhance this naturally low probability. First, in undisturbed soils, rhizobial cells are concentrated in the vicinity of the decaying root from previous growing seasons. The probability of infection of nascent root tissue is therefore enhanced by the propensity of newly developing root systems to grow into preexisting root channels.

A further increase in nodulation frequency could result from the movement or migration of infective bacteria that does occur within the soil profile. This migration of the rhizobial cells is primarily passive. That is, it usually does not involve directed motility of the cells (Issa et al., 1993a, b). The bacteria are carried by water flowing through soil macropores. **Macropores** are large passages in the soil matrix that allow the ready movement of air and water. These pores include the spaces between soil aggregates as well as the channels created by growth of roots (root channels) and movement of animals (e.g., earthworm channels). Passive transport of soil bacteria with leaching water provides for dispersion of cells over relatively large distances (in the A horizon, millimeter to centimeter distances). Note that this predominance of passive transport does not imply that directed motility of bacteria plays no role in nodulation. Compared to the situation with passive transport of cells, active motility of individual bacterial cells is of greater importance in movement of the cells in the microsite in response to exudate production by the growing root (chemotaxis; see below).

This differentiation of impact between passive and active aspects of bacterial mobility in soil does not preclude conference of a competitive advantage to rhizobial cells by the capacity for active motility. It must be noted that a number of reports exist in the literature supporting as well as opposing the conclusion that motile strains of rhizobia have an enhanced potential to occupy legume nodules compared to nonmotile bacteria. The seemingly contradictory conclusions result to a large degree from variation in the properties of the laboratory- or greenhouse-incubated soils used in the experiments as well as from the various population densities of bacteria used in the experiments. It is indisputable that motility of rhizobia is not required for nodule formation, yet a number of studies have shown both enhanced occurrence of nodulation by mobile strains compared to comparable nonmotile strains (e.g., Gulash et al., 1984; Hunter and Fahring, 1980; Malek, 1992; Napoli and Albersheim, 1980) and no benefit of motility (Ames et al., 1980; Issa et al. 1993a, b; Liu et al., 1989).

Much of the contradictory results are attributable to variation in the importance of chemotaxis of infective rhizobia under the various experimental conditions. In situations where both motile and nonmotile strains have equal access to susceptible root tissue, no benefit is gained from mobility. In contrast, if the soil moisture levels and population densities used in the experiment are such that the microorganisms are required to traverse a short distance from their habitat in the soil to the root, then an actively motile organism would gain a benefit over one that must rely on passive diffusion. Thus, it can be concluded that directed motility (i.e., chemotactic attraction) is most important to nodulation once the microbe has been moved passively by mass water flow into the vicinity of the susceptible portion of the root. It is assumed that flowing water is more important in moving the bacteria over distances up to several centimeters or more, whereas microbial motility primarily affects interactions of the bacteria with roots for cells located within a few millimeters or less from the root.

This model of the relative importance of passive and active motility of bacterial strains in nodule formation rests upon the assumption that rhizobia are capable of responding chemotactically to root exudates. Rhizobia are attracted to root exudate components (e.g., Bhagwat and Thomas, 1982). Exudate components found to attract various rhizobial strains include sugars and amino acids (e.g., Bergman et al., 1988), proteins (e.g., Currier and Strobel, 1977), dicarboxylic acids (e.g., Barbour et al., 1991), aromatic compounds (e.g., Kape et al., 1991), and flavonoids (e.g., Dharmatilake and Bauer, 1992). [It should be noted that attraction to the general components of root exudates (e.g., sugars and amino acids and their polymers) is not specific to the rhizobia, but a variety of rhizosphere bacteria are attracted to the root by these compounds. To a major degree, the response of the rhizobial population to root exudates is part of a more generalized stimulation of bacterial population densities in the rhizosphere.]

There is an impact of some root exudate components and expression of nodulating genes by the *Rhizobium* cell. Deletions in the *nif-nod* region of the symbiotic megaplasmid eliminated the chemotactic attraction of *Rhizobium meliloti* to plant roots by the flavonoid luteolin (Caetano-Anollés et al., 1988). The *nif-nod* region is constituted of genes involved in nodule formation and nitrogen fixation. For further discussion of the induction of these genes by components of root exudates and their role in nodule initiation, see Brewin (1991).

Host Recognition and Attachment: Reaction to root exudate components may allow the rhizobial strains to reach the specific portion of the root structure most susceptible to *Rhizobium* infection (that portion of the root behind the apical meristem at the site of emergence of the root hair, i.e., the zone of root elongation) more efficiently. The next step in nodule formation requires recognition and attachment to the root. Recognition of susceptible host tissue involves an interaction between proteins (lectins) on the plant root surface and rhizobial exopolysaccharides. This process could be likened to the specificity and interactions associated with an antibody-antigen reaction. Early suggestion of the role of lectins in the recognition process was presented by Bohlool and Schmidt (1974). They found that a lectin from soybean combined specifically with the soybean-nodulating bacterium (*Bradyrhizobium japonicum*) and that this lectin did not combine with other representative rhizobia. Sherwood et al. (1984) found that the location of the lectin-binding polysaccharides of *Rhizobium trifolii* on the cell surface varied with age of the culture. Random distribution of the binding agent as well as polar concentration of the substance were detected after different periods of cellular growth. The quantity and location of these receptors on the bacteria directly correlated with their attachment in short-term studies of clover root hair binding. For example, cells from 3- or 21-day-old cultures attached nearly exclusively in a polar fashion, whereas 5-day-old cells with the binding factors randomly distributed on the cell surface attached randomly to the root tissue.

Rhizobial exopolysaccharides appear to serve multiple purposes in nodulation (see Reuber et al., 1991, for a more detailed discussion of the role of exopolysaccharides in nodule formation). The varied roles include assistance in entangling the bacterial cell in the mucigel of the plant root, inhibition of host defenses through masking of the bacterial surface, and encapsulation of the bacterial cell for its protection in the infection thread. For example, after attachment and orientation of the *R. trifoli* cell with the clover root surface, an accumulation of extracellular microfibrils associated with lateral and polar surfaces of the bacterial cell occurs (Dazzo et al., 1984). The formation of these fibrils results in an attachment of the bacteria to the root hair sufficiently strong to withstand the hydrodynamic shear forces associated with high-speed vortexing of the mixture. (See Brewin, 1991 for further discussion of the role of these substances in nodule formation.)

Nodule Formation: Once the rhizobial cell has attached to the legume root, an infection thread spreading down the root hair is formed. Development of this structure is preceded by deformation of the root hair that is induced by exopolysaccharides formed by the rhizobial cell (e.g., see Battisti et al., 1992; Ervin and Hubbell, 1985; Higashi and Abe, 1980; van Brussel et al., 1992). The invagination of root hair wall forming the cellulosic infection thread serves as the means of invasion of the bacterial cells into the central portions of what becomes the nodule. The invading rhizobial cells pass through the infection thread to the root cells adjacent to the root hair. These root cells subsequently become infected with the invading rhizobial cells. Only tetraploid cells are infected and become the primordia of the nodule. It must be noted that only a small portion (usually less than 5 percent) of the invaded root hairs develop nodules. Division of the infected cells results in formation of the nodule. Bacteria multiply within the tetraploid cells, forming swollen, variable shaped and branched cells known as **bacteroids**, which are separated into packets within the nodule by a **peribacteroid membrane**. There are typically 4 to 6 bacteria per packet. Nitrogen fixation within the root nodules occurs only after formation of the bacteroids.

The nodulation cycle could be said to be completed by death and deterioration of the nodule and release of the bacteroids and bacteria contained therein into the soil. The bacteroids are apparently incapable of division when released into soil by death and decay of the nodule. Data collected by Paau et al. (1980) indicates that *Rhizobium meliloti* bacteroids in alfalfa nodules are degraded during nodule senescence. The nodule apparently always contains small numbers of dormant rod-shaped cells that are capable of existence free in the soil environment.

13.1.4 Plant Controls of Nodulation

The host plant is not a passive participant in this process involving "rearrangement of its architecture" by the *Rhizobium* strain. Expression of several plant genes is necessary for establishment of the symbiotic interaction. In their survey of the literature, Caetano-Anollé and Gresshoff (1991a) provide a listing of symbiotic legume mutants for soybean [*Glycine max* (L.) Merr.], alfalfa (*Medicago sativa* L.), red clover (*Trifolium pratense* L.), crimson clover (*Trifolium incarnatum* L.), pea (*Pisum sativum* L.), chickpea (*Cicer arietinum* L.), and peanut (*Archis hypogaea* L.). For some of the legumes, only one or two genes affecting nodulation have thus far been elucidated, but a total of 22 genes interacting with the bacterial symbiont in nodule formation were listed for pea. These observations support the conclusion that the genetic-based contribution of the host plant to root nodule formation is as complex or more so than that of the invading bacterium.

An interesting point of control by the host plant is its determination and optimization of the number of nodules formed. Most obvious among these properties is the impact of untimely fertilization of the crop on nodulation. Such inappropriate addition of fixed nitrogen to the crop suppresses nodule formation.

A less appreciated interaction is autoregulation. With autoregulation of nodulation, the plant controls the quantity and timing of nodule formation. After the first infection of roots, no further increase in number of root nodules occurs. The benefit to the host plant in preventing formation of unnecessary nodules is obvious, since nodulation requires the commitment of large quantities of plant resources. This inhibition of nodulation is reversible in that removal of nodules can induce a new round of nodule formation. This process requires expression of specific, as yet undescribed plant genes.

Furthermore, the rhizobia involved in the first round of infection must have the genetic capability to induce nodule formation and to grow within the nodules. George and Robert (1991) assessed the capacity of a number of mutants of *Rhizobium leguminosarum* bv. phaseoli to induce an autoregulatory response of *Phaseolus vulgaris* L. They found that all mutants unable to form nodules failed to induce a suppressive response of the host. The host response correlated with the ability of the bacteria to grow inside the nodules but not with the capacity to initiate nodule formation or to fix nitrogen. That is, the host response required that the infecting bacterium be infective, but not effective.

The interaction of autoregulatory responses and bacterial population density to yield optimal nodulation was examined in soybean [*Glycine max* (L.) Merr. cv. Bragg] following inoculation with *Bradyrhizobium japonicum* USDA 110 (Caetano-Anollés and Gresshoff, 1991b). The roots were initially susceptible to infection 3 to 4 days following germination. The number of nodules formed was proportional to the bacterial density up to an optimal concentration. The maximal population density resulting in an increase in the quantities of nodules formed was greater for a super-nodulating soybean variant than was observed with the wild type. Thus, the plant genome can be manipulated to increase nodulation of this agronomically important crop.

13.2 Manipulation of *Rhizobium*-Legume Symbioses for Ecosystem Management

The foregoing evaluation of the *Rhizobium*-legume symbiosis suggests the existence of a relationship that not only is of interest for basic science studies but has a variety of practical implications. The ubiquity of legumes as well as their significance as food crops suggests both economic and environmental benefits from management of this interaction. Prime questions in such considerations are as follows:

- Do *Rhizobium*-legume symbioses add significant quantities of fixed nitrogen to a soil ecosystem?

- Is there potential to improve the nitrogen inputs in a manner that will also augment ecosystem productivity?

- Are these improvements feasible with currently available methods?

Encouragement of *Rhizobium*-legume symbioses development does increase soil fixed nitrogen resources. The balance between nitrogen inputs and outputs to land systems and the identity of the sources of this fixed nitrogen (Burns and Hardy, 1975) support the conclusion that *Rhizobium*-legume symbioses are a major source of fixed nitrogen in land-based systems. About

64 percent of the 139×10^9 kg N_2 biologically fixed in terrestrial systems is produced in agricultural systems. Since the bulk of biological nitrogen fixation in agricultural systems is derived from the cultivation of legumes, these symbioses can be concluded to provide well over half the biological source of fixed nitrogen. Data collated by Alexander (1977) provide a prospective on fixed nitrogen contributions of individual crops on a more localized bases. Inputs of fixed nitrogen for alfalfa, red clover, pea, soybean, cowpea, and vetch range from 65 to 335 kg N_2 fixed ha^{-1} year^{-1}.

Once it has been concluded that these symbiotic associations are major contributors of fixed nitrogen in soil systems, the point of concern must be redirected to a consideration of the potential to increase fixed nitrogen inputs by optimization of this process. That is, the possibility of improving plant biomass yields or plant community sustainability through use of more efficient combinations of rhizobia and legumes must be examined. A diverse population of rhizobial strains already exists throughout the soil profile in agricultural and nonagricultural soils [e.g., a leguminous tree, black locust (*Robina pseudoacacia* L.) (Batzli et al., 1992); *Rhizobium leguminosarum* in agricultural soil (e.g., Bottomley and Dughri, 1989; Brockman and Bezdicek, 1989); soils under acacia trees (*Acacia albida*) (Dupuy and Dreyfus, 1992); soils from an alfalfa field (Jenkins et al., 1987); the woody legume, mesquite (*Prospopis glandulosa*) in desert soils (Jenkins et al., 1989; Waldon et al., 1989) and in tropical soils of Hawaii (Woomer et al., 1988)]. Furthermore, the indigenous rhizobial population can be quite diverse. For example, Noel and Brill (1980) used sodium dodecyl sulfate (SDS) gel electrophoresis to evaluate nodule occupants from soybeans collected from two Wisconsin (USA) locations. At one location, 19 gel types were distinguished and classified into three groups.

Nodule occupancy by these indigenous rhizobia is not the result of superior nitrogen-fixing capacity. It is derived more from a capacity of the bacterial strain to survive in soil. Additionally, both the population density and location of the bacterial habitat must allow optimal contact with susceptible roots. The proportion of the total soil rhizobial population composed of any individual strain is controlled by soil properties as well as plant growth traits. For example, the predominance of any particular rhizobial strain in the soils studied by Noel and Brill (1980) depended on the time and depth of nodule formation on the host plant. Also, Mahler and Bezdicek (1980) found that serogroup variability of *Rhizobium leguminosarum* strains in soils was affected by soil temperature, moisture, and position on the slope in the catena. Both microclimate and macroclimate variation altered strain distribution of *Rhizobium leguminosarum* strains in peas (*Pisum sativum* L.) in a Palouse (Washington, USA) site. Subsequent studies of 192 *R. leguminosarum* strains from this site showed that they grouped into 3 serogroups and 18 plasmid profile groups. Similar variety and distribution has been observed for *R. meliloti* (Jenkins and Bottomley, 1985). The latter individuals identified a variety of protein profiles for *Rhizobium* isolates from alfalfa (*Medicago sativa* L.) with SDS-gel electrophoresis.

Thus, it can be concluded that in any given soil system supporting growth of legumes, a variety of rhizobial strains are available for nodulation of root tissue. The adequacy of this indigenous population to form optimal numbers of nodules on legume populations generally depends on plant density. This principle is best exemplified by contrasting native and intensively cultivated legumes. In native ecosystems, indigenous rhizobial population densities are generally sufficient to support adequate nodulation of leguminous species to maintain ecosystem fertility and biomass productivity. With high-yield, cultivated agricultural systems or with monoculture systems, such as alfalfa fields, nitrogen fixation is commonly seen to be augmented by inoculation with exogenously grown rhizobia. Therefore, it is reasonably concluded that indigenous soil population densities were insufficient to support optimal nodulation. Hence, addition of laboratory-produced or axenically grown field *Rhizobium* sp. isolates to soils and legume seeds to stimulate crop productivity has long been a common agricultural practice. The increased nodulation of the legume roots and subsequent augmentation of biologically produced fixed nitrogen result in part from the higher probability of encounter between the host root and infective bacteria as well as the utilization of rhizobial isolates selected for an enhanced nitrogen-fixing capability.

13.3 Rhizobial Inoculation Procedures

Primary concerns in analysis of the benefits of manipulation of soil rhizobial populations have been (1) development of an effective delivery system, (2) determination of the interactions with the soil environment and soil microbial population affecting survival of the inoculant, and, more recently, (3) elucidation of properties of the inoculant strain affecting its competition with indigenous rhizobial variants for nodule occupancy. (See Brockwell and Bottomley, 1995, for a review of this topic.)

13.3.1 Inocula Delivery Systems

Axenically grown rhizobial strains capable of efficient nitrogen fixation are of little value if they do not form effective associations in the field. The first step in effective use of laboratory-derived/produced rhizobial strains is the physical conveyance of viable bacterial cells to the soil. In a scientific experiment, this step is reasonably easy since a microbiologist can quite easily prepare an active bacterial culture that can immediately be added to crop roots or seeds. Difficulties are encountered in transferring this "technology" into a "real-world" situation where the viable bacterial cells may be abused tremendously before introduction to their new environment. A commercially prepared inoculum must be capable of surviving for long periods of time under somewhat adverse conditions between growth and delivery to the agriculturist and of enduring potential maltreatment on site before the legume seeds are inoculated.

A variety of delivery systems for rhizobial inocula have been evaluated and implemented in field situations. Examples include, ground peat or an oil carrier (Kremer and Peterson, 1983) with or without fertilizer (Kremer et al., 1982), irrigation water (Ciafardini et al., 1992), water (Crist et al., 1984), coal-based carriers (Crawford and Berryhill, 1983), soils (Chao and Alexander, 1984), and fluid gels (Jawson et al., 1989). A variety of commercial inoculants are available (e.g., Hume and Blair, 1992). Variables that must be evaluated include pretreatment of the support media [e.g., sterilization of the peat (Strijdon and Van Rensburg, 1981)] and methods to prepare the inoculum for storage [e.g., drying of inocula and humidity for storage (Mary et al., 1985)].

13.3.2 Survival of Rhizobial Inocula

A related aspect to optimization of fixed nitrogen inputs from *Rhizobium*-legume symbioses is maximization of inoculum survival once they are amended to the soil system. Although the bacterial strain or variant selected for legume inoculation may have been derived from the soil-plant environment, cultivation under laboratory conditions necessarily induces changes in the phenotypic and genotypic properties of the cell. These alterations may or may not affect the capacity of the organism to adapt to the soil environment and to compete with indigenous populations. Axenically grown laboratory rhizobial strains added to legume seeds or amended directly to soil are stressed by the physical and chemical properties of the harsh soil environment as well as by interactions with the general soil microbial community. The capacity of the organism to overcome these barriers to survival determines its probability of existing in the soil site for sufficient duration to encounter a susceptible legume root and initiate infection.

Rhizobial strains amended to soils may survive for several years. For example, Crozat et al. (1982) found that their strains of *Rhizobium* (*Bradyrhizobium*) *japonicum* introduced into a variety of French soils survived at high rates ($\geqslant 10^4$ bacteria g^{-1} dry soil after 5 years). Similarly, Brunel et al. (1988), found that *Bradyrhizobium japonicum* USDA strains 125-Sp, USDA 138, and USDA 138-Sm survived for 8 to 13 years following their release into soil. Alien rhizobial strains (i.e., axenically grown and introduced or reintroduced into the soil system) persisted through passage though legume nodules produced in the soil. Survival of a portion of the introduced population may result from a capacity to adapt to the prevailing soil conditions and to compete with indigenous microbes. Viteri and Schmidt (1987), using fluorescent antibody techniques to study variation in population densities of *Bradyrhizobium japonicum* seroclusters 123 and serogroups 110 and 128, found that amendment of soil samples with metabolizable sugars resulted in not only an increase in the general populations but growth in the various rhizobial strains. This observation supports the conclusion that rhizobia are capable of competing with the general soil flora. That is, these bacteria do not require the exclusive protection of the plant nodule to maintain soil populations.

Primary soil physical and chemical properties that affect survival and nodule formation that are of concern to environmental microbiologists include soil pH, metal interactions, salinity, and moisture. Although the effect of these soils properties on nodulation has been most studied with agronomically important crops, they are also a major consideration when developing procedures to encourage fixed nitrogen inputs into reclaimed or managed ecosystems. It is generally desirable to minimize anthropogenic intervention in reclamation sites. That is, it is useful to develop site-resident biological mechanisms for control of fixed nitrogen inputs rather than to rely upon long-term fertilization plans. Thus, selection and control of an indigenous biological population capable of fixing nitrogen is essential.

Rhizobia and Acid Soils: Soil acidity affects the legume plant and the infecting rhizobia, as well as the product of their interaction, the symbiotic association. Generally host plant growth is less sensitive to extremes of acidity than is the *Rhizobium* species. Infection of host roots and survival of the bacterial symbiont are generally limited to pH greater than 5.0, whereas the host plants may grow at values as low as 4.0. This pH sensitivity is of concern to the agricultural community because of the need for sustained growth of symbiotic associations in acid soils, and to the environmental science community from the view of determining the potential for use of these highly effective fixed nitrogen sources in acid-affected sites, such as soils receiving acid mine slags or acid mine drainage. Examples of agricultural concerns include the difficulties associated with maintenance of alfalfa (Barber, 1980; Lowendorf and Alexander, 1983a), lucern (Munns and Fox, 1977; Pijnenborg et al., 1991), and clover (Richardson et al., 1988a) stands in acidic soils. Similar sensitivity to acidic soils has been observed for tropical legumes (18 legume species) as occurs with temperate legumes.

Both direct and indirect effects of acidic conditions on rhizobial activity are observed. Bacterial cells are affected by the elevated hydrogen ion concentration of acidic soils (direct effect) as well as by the alteration of metal solubility resulting from reduction in the soil pH (indirect effect) (e.g., Franco and Munns, 1982; Keyser and Munns, 1979a). The latter interaction can be due to increased metal concentrations (e.g., inhibition of growth by augmented aluminum concentrations under acidic conditions) or result from decreased soluble metal levels (e.g., reduction of the essential mineral molybdenum by increased acidity).

A differential sensitivity of the rhizobial strains to the various changes in soil cation concentration resulting from increased soil acidity is observed. Keyser and Munns (1979b) in a study of cation effect on growth of *Rhizobium* (*Bradyrhizobium*) *japonicum* found that aluminum toxicity and acidity itself had a greater impact on rhizobial growth than did manganese toxicity and calcium deficiency. It must be noted that aluminum toxicity and acid sensitivity independently affect rhizobial survival in soil (e.g., Hartel and Alexander, 1983; Hartel et al., 1983). It appears that survival of the cowpea [*Vigna unguiculata* (L.) Walp.] rhizobia is pH but not aluminum sensitive. Hartel et al. (1983) concluded that aluminum content of soil was not a major factor controlling cowpea rhizobial survival, but it did significantly alter the nodulation process.

Although it has been proposed that survival of indigenous populations of *Rhizobium trifolii* in extremely acidic soils can occur through refuge in soil microsites of less extreme acidity (Richardson and Simpson, 1988), rhizobial strains with increased tolerance of acidic conditions can be isolated (e.g., Lowendorf and Alexander, 1983b; Keyser et al., 1979; Thornton and Davey, 1983; Tiwari et al., 1992) and constructed genetically (Chen et al., 1991). Increasing acid tolerance of rhizobial strains under axenic condition does result in augmented survival in sterile acid soils (Lowendorf et al., 1981). Thus, the potential exists to select or engineer rhizobial strains in the laboratory that are both efficient nitrogen fixers and acid tolerant for use in acid-affected soil sites. Augmented survival in the acid soils could increase their probability of nodule occupancy (Dughri and Bottomley, 1983). Furthermore, interactions between the acidic conditions and cation concentrations can also affect nodule formation by rhizobia. For example, acidic conditions and levels of calcium and aluminum ions reduce induction of the nodulation genes in *Rhizobium leguminosarum* biovar *trifolii* (Richardson et al., 1988b). Thus, it can be concluded that the concentrations of fixed nitrogen provided by the *Rhizobium*-legume symbioses in acidic and/or cation-affected soils depends upon the capacity to maintain adequate population densities of infective bacteria in the soil for optimal encounter between the bacterium and susceptible plant roots.

No single cell property can be selected to optimize the acid tolerance or aluminum sensitivity of rhizobia. For example, DNA repair mechanisms (Johnson and Wood, 1990), extracellular polysaccharide production (Cunningham and Munns; 1984a), and ability to maintain intracellular pH (O'Hara et al., 1989) all appear to affect aluminum and/or acid tolerance of rhizobial strains. Variation between rhizobial strains is exemplified by studies of the role of extracellular polysaccharide production in protection of the bacterial cell. Cunningham and Munns (1984a) found that acid tolerance of rhizobial strains for six legume genera (*Cicer, Phaseolus, Leucaena, Lens, Melilotus,* and *Trifolum*) all correlated with extracellular polysaccharide production. The quantities of extracellular polysaccharide produced by 20 strains of *R. phaseoli* varied from 18 to 75 mg per 100 ml culture. The amounts of polysaccharide produced correlated with acid tolerance. Data from further studies (Cunningham and Munns, 1984b) indicated that the apparent protection did not relate to buffering capacity of the extracellular polysaccharide or its ability to chelate aluminum ions. In contrast, Kingsley and Bohlool (1982) evaluated the role of extracellular polysaccharide produced by *R. leguminosarum* bv. *phaseoli* (CIAT899) with mutant strains incapable of producing the polysaccharide. No difference in aluminum tolerance of the wild-type strain from the mutant were detected. Results from this study indicate that neither intracellular or extracellular products of the *R. leguminosarum* strain alleviated the aluminum toxicity of sensitive rhizobial strains. These studies demonstrate the potential to increase the utility of the *Rhizobium*-legume symbioses for augmenting fixed nitrogen inputs into acidic soil ecosystems as well as the complexity of the genetic control of this acid and aluminum sensitivity of the infective bacteria.

Although most interest in the effect of toxic metals on nitrogen fixation by *Rhizobium*-legume associations has involved growth of agriculturally important legumes in acidic soils, some consideration of the impact of metals is necessary in evaluating the use of these associations in metal-affected sites or in soils amended with materials that may contain environmentally significant levels of metals (e.g., Dahlin et al., 1997; McGrath et al., 1988; Smith, 1997). For example, McGrath et al. (1988) determined the potential for reduction of fixed nitrogen yields from nodulated *Trifolium repens* in soils receiving metal-contaminated sewage sludge. Plants grown on the contaminated sludge-amended soil had reduced nitrogen concentrations and reduced yields compared to those produced on uncontaminated control soils. The root nodules on the affected plants were numerous but ineffective in nitrogen fixation.

Soil Moisture Impact on Nodulation: The occurrence of rhizobial populations in desert soils and the effective nodulation of legumes growing therein (e.g., Jenkins et al., 1987; 1989; Waldon et al., 1989) attest to the fact that rhizobia are capable of existing in soils with low moisture levels. Population densities do tend to be lowest under the most desiccated situation, with increasing densities as the moisture stress is relieved. The wide range of moisture levels characteristic of ecosystems where legumes have been shown to fix nitrogen suggests that rhizobial strains with varying sensitivity to soil moisture can be selected. Laboratory studies have shown that sensitivity to moisture stress varies with bacterial strain with a variety of rhizobial strains and species [e.g., with *Rhizobium leguminosarum* bv. *trifolii* (Boonkerd and Weaver, 1982; Fuhrmann et al., 1986), with *Rhizobium meliloti* (Busse and Bottomley, 1989), with cowpea rhizobia (Osa-Afiana and Alexander, 1982a), and with *Bradyrhizobium japonicum* (Mahler and Wollum, 1980)]. Thus, it is reasonable to assume that rhizobial strains can be selected with moisture stress tolerance within the range of its legume host.

The moisture tolerance of rhizobia is also affected by other soil properties or components. For example, survival of rhizobia in drying soils can be extended by increasing contents of soil organic matter and decreased by elevated soluble aluminum (Chao and Alexander, 1982) and by the proportion of clay in the soil (Osa-Afiana and Alexander, 1982b). Furthermore, organisms produced *in situ* tend to be better adapted to moisture stress than axenically grown organisms (Pena-Cabriales and Alexander, 1979). Optimization of soil moisture for growth of the host plant, which generally is more sensitive to moisture stress than are the bacteria, results in maximization of development of *Rhizobium*-legume fixed nitrogen inputs into the soil system.

Soil Salinity: An allied concern in regards to soil moisture in reclamation of environmentally affected sites can be the effect of salinity on survival of rhizobia in soil systems. For example, amendment of soils with products of our industrial society can raise the soil salinity [e.g., sludge-borne salts (Madariaga and Angle, 1992)]. As was indicated in Chapter 5, increasing salts may have a detrimental effect on soil microbial populations as the result of direct toxicity as well as through osmotic stresses.

Rhizobia vary in efficacy of osmoregulation — response to salt stress (e.g., Busse and Bottomley, 1989; Surange et al., 1997; Zahran, 1997). This capacity of adaptation of the microbial population was exhibited in a study of the survival of *Bradyrhizobium japonicum* in sludge-amended soils. The soluble salts of the sludge (not the heavy metals) were shown to be primarily responsible for a short-term reduction in bradyrhizobial populations following sludge application to soil. Singleton et al. (1982) found many *Rhizobium* strains with enhanced salinity tolerance compared to most agriculturally important varieties. In a study of distribution of fast-growing and slow-growing mesquite nodulating rhizobia in desert ecosystems, it was found that the distribution of the rhizobial strains within the rooting system was related to total soil salts (Jenkins et al., 1989).

This analysis of some soil properties affecting survival, growth, and nodulation capacities of soil rhizobial populations demonstrates the versatility and adaptability of these essential soil populations. The genotypic and phenotypic flexibility exists within these organisms to adapt to many of the stressed soil conditions normally encountered in our soil system. This observation does not imply efficient nitrogen-fixing capabilities under these conditions. The legume may be nodulated, but fixed nitrogen production may be below the level required for optimal growth of the host plant. In these situation, symbiotic associations, such as the actinorhizal interaction, may better adapted or more adaptable for use in site management or reclamation.

13.3.3 Biological Interactions in Legume Nodulation

Rhizobia are capable of growth in native soil samples when the soils have been amended with fixed carbon sources supportive of their metabolism (e.g., Pena-Cabriales, and Alexander, 1983a; Viteri and Schmidt, 1987) or when they receive nutrients from growing roots (e.g., Pena-Cabriales and Alexander, 1983b). Success of this nutrient-based competition with indigenous populations depends on the capacity of the alien microbe to become established in an unoccupied niche. Rhizobia are best suited for competition for the newly created niche of the nitrogen-fixing nodule. Life as a member of the "free-living soil community" is more difficult. Rhizobia not only must compete for limited nutrients, but interactions with indigenous heterotrophic microbes and predators reduce their capacity to maintain population densities at sufficient levels to assure contact with susceptible legume roots (e.g., Li and Alexander, 1986; Pugashetti et al., 1982; Ramirez and Alexander, 1980) (see Chapter 7 for further discussions regarding microbial competition).

The outcome of competition between indigenous microbes can be manipulated. Aside from the brute-force method of overwhelming the system with massive population densities of the laboratory-grown rhizobial strain, use of fungicides and antibiotics can enhance the probability of rhizobial colonization (Hossain and Alexander, 1984; Li and Alexander, 1986). Hossain and Alexander (1984) augmented soybean rhizosphere colonization with *Bradyrhizobium japonicum* through use of benomyl (a fungicide) and the antibiotics streptomycin and erythromycin. *Rhizobium* strains used with these procedures must be resistant to the inhibitors.

13.4 Nodule Occupants: Indigenous Versus Foreign

Two sources of rhizobia for root infection can occur in many managed agricultural soils: (1) native populations and (2) crop inoculants. The industrially produced rhizobial strains used to inoculate crops must be capable of competing for nutrients with indigenous populations until they encounter a susceptible root. Additionally they also participate in an intense race with native strains for occupancy of nodules. A study of the history of use of *Rhizobium* inoculants for cropping of legumes leaves the impression that it has been assumed that the soil-amended strain necessarily predominated in nodule occupation. Perhaps a more pragmatic assessment of the situation would include the idea that the investigator and growers have not been as concerned with the identity of the nodule occupant as they were with augmentation of fixed nitrogen resources and the concurrent improvement in crop yields. If crop yields increased, a benefit from inoculation with highly efficient rhizobial strains was assumed to have occurred. Unfortunately, detailed examination of legume nodules reveals that it is not unusual for the laboratory-cultivated rhizobial strains to lose the competition for a place in the nodule to the less efficient but better adapted indigenous soil populations.

Nodule occupancy frequencies for nonnative rhizobial strains ranging from nearly zero to essentially 100 percent have been reported. For example, Van Rensburg and Strijdom (1985) compared *Rhizobium* nodule occupancy frequency of indigenous and axenically grown strains in *Trifolium* spp., *Medicago* spp., *Glycine max*, and *Lotus pedunculatus*. They collected 420 nodules from crops growing in fields that had been inoculated four to eight years previously with highly efficient rhizobial strains. The rate of nodule occupancy for the inoculated strain ranged from 17.7 percent in the *Glycine max* (soybean) to 100 percent with the *Lotus pedunculatus*. Klubek et al. (1988) in a study of *Bradyrhizobium japonicum* nodule occupancy in the midwestern United States detected frequency of occurrence of inoculated strains as low as 0.3 percent. Detection of highly variable capacity of laboratory-grown strains to compete with indigenous populations is commonly reported (e.g., Materon and Hagedorn, 1992; McDermott and Graham, 1989; Moawad and Bohlool, 1984).

The most obvious explanation for the variation in frequency of nodule occupancy of soil-amended rhizobial strains involves the probability of encounter of the added organism with susceptible portions of the legume root. In a study of seven legume species in Hawaii (Thies et al., 1991a), it was shown that the proportion of the nodules occupied by inoculated rhizobial strains depended on the population density of the indigenous population. Indigenous organisms would be anticipated to be better adapted to the soil system and perhaps even more appropriately distributed within the soil profile than alien organisms. The latter conclusion is based upon the observation that the physical location of indigenous organisms depends in part on the distribution of the root system in which they grew and the homogenization of the soil associated with intensive cultivation systems. Thies et al. (1991b) found that 59 percent of the observed variation in inoculation response could be attributed to the population density of the indigenous population.

Properties of rhizobia that enhance nodule occupancy rates are diverse and, to a major extent, poorly defined. A primary requirement is the ability to maintain sufficient population densities in the soil to allow a reasonable probability of encounter with host roots. Therefore, selection of inoculant strains must include consideration of the properties of the soil in which they are to be used (e.g., moisture/salinity stresses and pH), as well as the capability to compete with indigenous microbes for available nutrients and to withstand incursions of predators and parasites. Further considerations relate to the efficiency of the nodulation process. Initiation and development of effective nodules is a complex process. Many of the rhizobial cells attracted to the appropriate portion of the legume root fail to complete the infection process. To increase the probability of the inoculant with improved nitrogen-fixing efficiency contributing to augmentation of ecosystem fixed nitrogen resources, variants with increased capability of completing the nodule formation process are desirable.

13.5   Actinorhizal Associations

When the potential to form nitrogen-fixing nodules on plant roots is considered, the *Rhizobium*-legume association is frequently discussed nearly at the exclusion of considering any other interactions, yet actinorrhizal interactions (nonleguminous nitrogen-fixing symbiotic associations between actinomycete of the genus *Frankia* and a variety of woody angiosperms) are major contributors to nitrogen inputs in forests, wetlands, fields, and disturbed sites of temperate and tropical regions. Actinorhizal associations involve more than 160 species of angiosperms classified among 6 or 7 orders. As is shown in Table 13.1, a wide variety of genera of angiosperms are capable of forming actinorhizal associations. Not all genera in the various families cited are active in nitrogen fixation, and not all species of the cited genera form nitrogen-fixing associations.

Contributions of fixed nitrogen to native as well as managed ecosystems by these symbioses are comparable to those of the more studied *Rhizobium*-legume interactions. The nitrogen gains due to these interactions vary with

Table 13.1 Examples of Angiosperm Families
Involved in Nitrogen-Fixing Actinorhizal Association

Family	Genera
Betulaceae	Alnus
Casiaromaceae	Casuarina
Coriariaceae	Coriaria
Elaeagnaceae	Elaeagnus
	Hypophae
	Shepherdia
Myricaceae	Myrica
	Comptonia
Rhamnaceae	Ceanothus
	Discaria
Rosaceae	Cercocarpus
	Dryas
	Pushia

soil, climate, and plant age. Typical contributions represented by *Alnus* associations are 12 to 2000 kg N ha^{-1} and *Hippophae*, 27 to 179 kg N ha^{-1}. See Baker and Mullen (1992) for a review of the contributions of this association to global nitrogen cycling.

Furthermore, the energy requirements (i.e., photosynthetate contributions from host plants) and nitrogenase activity of actinorhizal symbioses are comparable to those of the *Rhizobium*-legume associations (Gauthier et al., 1981; Tjepkema and Winship, 1980). Note that as with rhizobia, *Frankia* strains are capable of fixing nitrogen growing in axenic culture on defined media (Gauthier et al., 1981).

It has been recognized for more than 90 years that the microbial symbiont is an actinomycete. Early identification was based on microscopic detection of actinomycete-like mycelial structures in nodule tissue (see Quispel, 1990 for a discussion of this topic). The greatest limitation to study of actinorrhizal associations has been, until recently, the inability to culture the microbial partner exclusive of the plant nodule. Until the early 1980s, the microbial partner was generally regarded to be an unculturable actinomycete. Most hypotheses regarding participation of the microbial symbiont were developed, therefore, from observation of the cells in the nodule or from biochemical studies of partially purified microbial biomass extracted from crushed nodules.

Currently, the *Frankia* sp. can be isolated and cultured (e.g., Baker et al., 1979; Benson, 1982). Procedures generally involve surface sterilization of the nodules with such substances as chlorox, hydrogen peroxide, or dilute silver nitrate. The sterilant is washed from the nodules and the nodules are crushed. The extraneous debris may be removed (Baker et al., 1979, used sucrose gradients) or laboratory culture media may be inoculated directly with the nodule extract. A variety of nitrogen-free media containing antibiotics have been used to culture the actinomycete. A key in culture of these organisms is the necessity of inhibiting the growth of fast-growing contaminants. Colonies of the slow-growing *Frankia* spp. are easily overgrown. Thus, antibiotics such as cycloheximide and nystatin are amended to the growth media (e.g., Benson, 1982). Once isolated, the proof that the axenically grown actinomycete participates in symbiotic produced fixed nitrogen results from the successful nodulation of uninfected host plants with the cultured microbial strain (e.g., Diem et al., 1982; Mansour et al., 1990; Rogers and Wollum, 1974; Van Dijk and Sliumer-Stolk, 1990).

Polymerase chain reaction amplification and related DNA analytical techniques portend to be valuable tools for elucidation of the diversity of *Frankia* strains involved in actinorhizal symbioses (e.g., Beayzova and Lechevalier, 1992; Sellstedt et al., 1992; Simonet et al., 1990; 1991) as well as the variability of these nitrogen-fixing actinomycetes in soil (e.g., Hahn et al., 1990a; Hilger and Myrold, 1991; Myrold et al., 1990) and in host nodules (Bloom et al., 1989; Hahn et al., 1990b).

Although at this time it is not recommended that the genus *Frankia* be divided into individual species (Lechevalier and Lechevalier, 1989), axenically grown *Frankia* strains are extremely diverse. For example, DNA hybridization studies (Bloom et al., 1989) revealed that highly divergent populations of *Frankia* could be isolated from bayberry (*Myrica pensylvanica*) nodules. In fact, *Frankia* strains isolated from the same nodule were shown to be not only phenotypically different by also genotypically diverse. Similarly, restriction fragment length polymorphism analysis (RFLP) has been demonstrated to divide *Frankia* strains into groups that correlated with host specificity groups separated by cross-inoculation studies (Nittayarjarn et al., 1990). (As with rhizobia, *Frankia* strains can be separated into groups based on the specificity towards their host plant.)

13.6 Conclusions

Root-nodule forming symbiotic associations are the predominant contributors of fixed nitrogen to terrestrial ecosystems. Both rhizobial and *Frankia* sp. strains form specific associations with legume and angiosperm hosts, respectively. Although adequate populations of infective diazotrophs commonly occur among soil microbes, the potential exists for manipulation of these bacterial populations to increase their contribution of fixed nitrogen to the ecosystem. For rhizobial populations, it has been shown that sufficient genotypic variability for resistance to soil pH, moisture and salinity extremes as well as for nitrogen-fixing efficiency exist to allow for development of rhizobial strains that will support increased fixed nitrogen inputs compared to those provided by the activity of indigenous populations. Selection of such strains and their utilization in the field also require consideration of those phenotypically expressed traits that improve the microbes' capacity to compete with soil heterotrophic populations as well as indigenous rhizobia. As our understanding of the ecological associations of *Frankia* spp. develops, similar manipulations to increase fixed nitrogen production by actinorhizal symbioses in managed and wild-land ecosystems will be possible.

References

Adebayo, A., I. Watanabe, and J. K. Ladha. 1989. Epiphytic occurrence of *Azorhizobium caulinodans* and other rhizobia on host and nonhost legumes. Appl. Environ. Microbiol. 55:2407–2409.

Aguilar, O. M., and G. Favelukes. 1982. Requirement for carbon dioxide for nonsymbiotic expression of *Rhizobium japonicum* nitrogenase activity. J. Bacteriol. 152:510–513.

Agius, F., C. Sanguinetti, and J. Monza. 1997. Strain-specific fingerprints of *Rhizobium loti* generated by pcr with arbitrary and repetitive sequences. FEMS Microbiol. Ecol. 24:87–92.

Alexander, M. 1977. Introduction to Soil Microbiology. John Wiley & Sons, NY.

Ames, P., S. A. Schluederberg, and K. Bergman. 1980. Behavioral mutants of *Rhizobium meliloti*. J. Bacteriol. 141:722–727.

Anyango, B., K. J. Wilson, J. L. Beynon, and K. E. Geller. 1995. Diversity of rhizobia nodulating *Phaseolus vulgalis L.* in two Kenyan soils with contrasting pHs. Appl. Environ. Microbiol. 61:4016–4021.

Baker, D. D., and B. C. Mullin. 1992. Actinorhizal symbiosis. Pp. 259–292. *In* G. Stacey, R. H. Burris, and H. J. Evans (eds.), Biological Nitrogen Fixation. Chapman and Hall, NY.

Baker, D., J. G. Torrey, and G. H. Kidd. 1979. Isolation by sucrose-density fractionation and cultivation in vitro of actinomycetes from nitrogen-fixing root nodules. Nature (London) 281:76–78.

Barber, L. E. 1980. Enumeration, effectiveness, and pH resistance of *Rhizobium meliloti* populations in Oregon Soils. Soil Sci. Soc. Am. J. 44:537–539.

Barbour, W. M., D. R. Hattermann, and G. Stacey. 1991. Chemotaxis of *Bradyrhizobium japonicum* to soybean exudates. Appl. Environ. Microbiol. 57:2635–2639.

Battisti, L., J. C. Lara, and J. A. Leigh. 1992. Specific oligosaccharide form of the *Rhizobium meliloti* exopolysaccharide promotes nodule invasion in alfalfa. Proc. Natl. Acad. Sci. USA 89:5625–5629.

Batzli, J. M., W. R. Graves, and P. van Berkum. 1992. Diversity among rhizobia effective with *Robinia pseudoacacia* L. Appl. Environ. Microbiol. 58:2137–2143.

Beayzova, M., and M. P. Lechevalier. 1992. Low-frequency restriction fragment analysis of *Frankia* strains (*Actinomycetales*). Int. J. Syst. Bacteriol. 42:422–433.

Benson, D. R. 1982. Isolation of *Frankia* strains from alder actinorhizal root nodules. Appl. Environ. Microbiol. 44:461–465.

Bergman, K., M. Gulash-Hoffee, R. E. Hovestadt, R. C. Larosiliere, P. G. Ronco II, and L. Su. 1988. Physiology of behavioral mutants of *Rhizobium meliloti:* Evidence for a dual chemotaxis pathway. J. Bacteriol. 170:3249–3254.

Bhagwat, A. A., and J. Thomas. 1982. Legume-*Rhizobium* interactions: Cowpea root exudate elicits faster nodulation response by *Rhizobium* species. Appl. Environ. Microbiol. 43:800–805.

Bloom, R. A., B. C. Mullin, and R. L. Tate III. 1989. DNA restriction patterns and DNA-DNA solution hybridization studies of *Frankia* isolates from *Myrica pensylvanica* (Bayberry). Appl. Environ. Microbiol. 55:2155–2160.

Bohlool, B. B., and E. L. Schmidt. 1974. Lectins: A possible basis for specificity in the *Rhizobium*-legume root nodule symbiosis. Science 185:269–271.

Boonkerd, N., and R. W. Weaver. 1982. Survival of cowpea rhizobia in soil as affected by soil temperature and moisture. Appl. Environ. Microbiol. 43:585–589.

Bottomley, P. J., and M. H. Dughri. 1989. Population size and distribution of *Rhizobium leguminosarum* bv. *Trifolii* in relation to total soil bacteria and soil depth. Appl. Environ. Microbiol. 55:959–964.

Brewin, N. J. 1991. Development of the legume root nodule. Annu. Rev. Cell Biol. 7:191–226.

Brockman, F. J., and D. F. Bezdicek. 1989. Diversity within serogroups of *Rhizobium leguminosarum* biovar *Viceae* in the Palouse region of Eastern Washington as indicated by plasmid profiles, intrinsic antibiotic resistance, and topography. Appl. Environ. Microbiol. 55:109–115.

Brockwell, J., and P. J. Bottomley. 1995. Recent advances in inoculant technology and prospects for the future. Soil Biol. Biochem. 27:683–697.

Brunel, B., J.-C. Cleyet-Marel, P. Normand, and R. Bardin. 1988. Stability of *Bradyrhizobium japonicum* inoculants after introduction into soil. Appl. Environ. Microbiol. 54:2636–2642.

Brunel, B., S. Rome, R. Ziani, and J. C. Cleyetmarel. 1996. Comparison of nucleotide diversity and symbiotic properties of *Rhizobium meliloti* populations from annual *Medicago* species. FEMS Microbiol. Ecol. 19:71–82.

Burns, R. C., and R. W. F. Hardy, 1975. Nitrogen fixation in bacteria and higher plants. Springer-Verlag, NY. 189 pp.

Burris, R. H. 1991. Nitrogenases. J. Biol. Chem. 266:9339–9342.

Busse, M. D., and P. J. Bottomley. 1989. Growth and nodulation responses of *Rhizobium meliloti* to water stress induced by permeating and nonpermeating solutes. Appl. Environ. Microbiol. 55:2431–2436.

Caetano-Anollés, G., D. K. Crist-Estes, and W. D. Bauer. 1988. Chemotaxis of *Rhizobium meliloti* to the plant flavone luteolin requires functional nodulation genes. J. Bacteriol. 170:3164–3169.

Caetano-Anollés, G., and P. M. Gresshoff. 1991a. Plant genetic control of nodulation. Annu. Rev. Microbiol. 45:345–382.

Caetano-Anollés, G., and P. M. Gresshoff. 1991b. Efficiency of nodule initiation and autoregulatory responses in a supernodulating soybean mutant. Appl. Environ. Microbiol. 57:2205–2210.

Chao, W.-L. and M. Alexander. 1982. Influence of soil characteristics on the survival of *Rhizobium* in soils undergoing drying. Soil Sci. Soc. Am. J. 46:949–952.

Chao, W.-L., and M. Alexander. 1984. Mineral soils as carriers for *Rhizobium* inoculants. Appl. Environ. Microbiol. 47:94–97.

Chen, H., A. E. Richardson, E. Gartner, M. A. Djordjevic, R. J. Roughley, and B. G.

Rolfe. 1991. Construction of an acid-tolerant *Rhizobium leguminosarum* biovar *trifolii* strain with enhanced capacity for nitrogen fixation. Appl. Environ. Microbiol. 57:2005–2011.

Ciafardini, G., G. Marinelli, and R. Missich. 1992. Soil biomass of *Bradyrhizobium japonicum* inoculated via irrigation water. Can. J. Microbiol. 38:584–587.

Crawford, S. L., and D. L. Berryhill. 1983. Survival of *Rhizobium phaseoli* in coal-based legume inoculants applied to seeds. Appl. Environ. Microbiol. 45:703–705.

Crist, D. K., R. E. Wyza, K. K. Mills, W. D. Bauer, and W. R. Evans. 1984. Preservation of *Rhizobium* viability and symbiotic infectivity by suspensions in water. Appl. Environ. Microbiol. 47:895–900.

Crozat, Y., J. C. Cleyet-Marel, J. J. Giraud, and M. Obaton. 1982. Survival rates of *Rhizobium japonicum* populations introduced into different soils. Soil Biol. Biochem. 14:401–405.

Cunningham, S. D., and D. N. Munns. 1984a. The correlation between extracellular polysaccharide production and acid tolerance in *Rhizobium*. Soil Sci. Soc. Am. J. 48:1273–1276.

Cunningham, S. D., and D. N. Munns, 1984b. Effects of rhizobial extracellular polysaccharide on pH and aluminum activity. Soil Sci. Soc. Am. J. 48:1276–1279.

Currier, W. W., and G. A. Strobel. 1977. Chemotaxis of *Rhizobium* spp. to a glycoprotein produced by birdsfoot trefoil roots. Science 196:434–435.

Dahlin, S., E. Witter, A. Martensson, A. Turner, and E. Bääth. 1997. Whereas the limit — changes in the microbiological properties of agricultural soils at low levels of metal contamination. Soil Biol. Biochem. 29:1405–1415.

Dazzo, F. B., G. L. Truchet, J. E. Sherwood, E. M. Hrabak, M. Abe, and S. H. Pankratz. 1984. Specific phases of root hair attachment in the *Rhizobium trifolii*-clover symbiosis. Appl. Environ. Microbiol. 48:1140–1150.

Desa, N. M. H., L. D. Kattah, L. Seldin, M. J. V. Vasconcelos, and E. Paiva. 1997. Genomic heterogeneity within bean nodulating *Rhizobium* strains isolated from cerrado soils. Soil Biol. Biochem. 29:1011–1014.

Dharmatilake, A. J., and W. D. Bauer. 1992. Chemotaxis of *Rhizobium meliloti* towards nodulation gene inducing compounds from alfalfa roots. Appl. Environ. Microbiol. 58:1153–1158.

Diem, H. G., D. Gauthier, and Y. Dommergues. 1982. Isolation and cultivation in vitro of an infective and effective strain of *Frankia* isolated from nodules of *Casuarina* sp. C. R. Seances Acad. Sci. Ser. III. Sci. Vie 295:759–763.

Dughri, M. H., and P. J. Bottomley. 1983. Effect of acidity on the composition of an indigenous soil population of *Rhizobium trifolii* found in nodules of *Trifolium subterraneum* L. Appl. Environ. Microbiol. 46:1207–1213.

Dupuy, N. C., and B. L. Dreyfus. 1992. *Bradyrhizobium* populations occur in deep soil under the leguminous tree *Acacia albida*. Appl. Environ. Microbiol. 58:2415–2419.

Dye, M., L. Skot, L. R. Mytton, S. P. Harrison, J. J. Dooley, and A. Cresswell. 1995. A study of *Rhizobium leguminosarum* biovar *trifolii* populations from soil extracts using randomly amplified polymorphic DNA profiles. Can. J. Microbiol. 41:336–344.

Elkan, G. H., and C. R. Bunn. 1992. The rhizobia. Pp. 2197–2213. *In* A. Balows, H. G. Trüper, M. Dwarkin, W. Harder, and K.-H. Schleifer (eds.), The Prokaryotes, 2nd Ed. Springer Verlag, NY.

Ervin, S. E., and D. H. Hubbell. 1985. Root hair defomations associated with fractionated extracts from *Rhizobium trifolii*. Appl. Environ. Microbiol. 49:61–68.

Franco, A. A., and D. N. Munns. 1982. Acidity and aluminum restraints on nodulation, nitrogen fixation, and growth of *Phaseolus vulgaris* in solution culture. Soil Sci. Soc. Am. J. 46:296–301.

Fuhrmann, J., C. B. Davey, and A. G. Wollum II. 1986. Desiccation tolerance of clover rhizobia in sterile soils. Soil Sci. Soc. Am. J. 50:639–644.

Gauthier, D., H. G. Diem, and Y. Dommergues. 1981. In vitro nitrogen fixation by two actinomycete strains isolated from *Causuarina* nodules. Appl. Environ. Microbiol. 41:306–308.

George, M. L. C., and F. M. Robert. 1991. Autoregulatory response of *Phaseolus vulgaris* L. to symbiotic mutants of *Rhizobium leguminosarum* bv. phaseoli. Appl. Environ. Microbiol. 57:2687–2692.

Gulash, M. P. Ames, R. C. Larosiliere, and K. Bergman. 1984. Rhizobia are attracted to localized sites on legume roots. Appl. Environ. Microbiol. 48:149–152.

Hahn, D., R. Kester, M. J. C. Starrenburg, and A. D. L. Akkermans. 1990a. Extraction of ribosomal RNA from soil for detection of *Frankia* with oligonucleotide probes. Arch. Microbiol. 154:329–335.

Hahn, D., M. J. C. Starrenburg, and A. D. L. Akkermans. 1990b. Oligonucleotide probes that hybridize with rRNA as a tool to study *Frankia* strains in root nodules. Appl. Environ. Microbiol. 56:1342–1346.

Hartel, P. G., and M. Alexander. 1983. Growth and survival of cowpea rhizobia in acid aluminum-rich soils. Soil Sci. Soc. Am. J. 47:502–506.

Hartel, P. G., A. M. Whelan, and M. Alexander. 1983. Nodulation of cowpeas and survival of cowpea rhizobia in acid, aluminum-rich soils. Soil Sci. Soc. Am. J. 47:514–517.

Hartmann, A., and N. Amarger. 1991. Genetic diversity of an indigenous *Rhizobium meliloti* field population assessed by plasmid profiles, DNA fingerprinting, and insertion sequence typing. Can. J. Microbiol. 37:600–608.

Hernandezlucas, I., L. Segovia, E. Martinezromero, and S. G. Pueppke. 1995. Phylogenetic relationships and host range of *Rhizobium* spp. that nodulate *Phaseolus vulgaris* L. Appl. Environ. Microbiol. 61:2775–2779.

Higashi, S., and M. Abe. 1980. Promotion of infection thread formation by substances from *Rhizobium*. Appl. Environ. Microbiol. 39:297–301.

Hilger, A. B., and D. D. Myrold. 1991. Method for extraction of *Frankia* DNA from soil. Agric. Ecosys. Environ. 34:107–113.

Hossain, A. K. M., and M. Alexander. 1984. Enhancing soybean rhizosphere colonization by *Rhizobium japonicum*. Appl. Environ. Microbiol. 48:468–472.

Hume, D. J., and D. H. Blair. 1992. Effects of numbers of *Bradyrhizobium japonicum* applied in commercial inoculants on soybean seed yield in Ontario. Can. J. Microbiol. 38:588–593.

Hunter, W. J., and C. J. Fahring. 1980. Movement by *Rhizobium* and nodulation of legumes. Soil Biol. Biochem. 12:537–542.

Issa, S., L. P. Simmonds, and M. Wood. 1993a. Passive movement of chickpea and bean rhizobia through soils. Soil Biol. Biochem. 25:959–965.

Issa, S., M. Wood, and L. P. Simmonds. 1993b. Active movement of chickpea and bean rhizobia in dry soil. Soil Biol. Biochem. 25:951–958.

Jawson, M. D., A. J. Franzluebbers, and R. K. Berg. 1989. *Bradyrhizobium japonicum* survival in and soybean inoculation with fluid gels. Appl. Environ. Microbiol. 55:617–622.

Jenkins, M. B., and P. J. Bottomley. 1985. Evidence for a strain of *Rhizobium meliloti* dominating the nodules of alfalfa. Soil Sci. Soc. Am. J. 49:326–328.

Jenkins, M. B., R. A. Virginia, and W. M. Jarrell. 1987. Rhizobial ecology of the woody legume mesquite (*Prospis glandulosa*) in the Sonoran desert. Appl. Environ. Microbiol. 33:36–40.

Jenkins, M. B., R. A. Virginia, and W. M. Jarrell. 1989. Ecology of fast-growing and slow-growing mesquite-nodulating rhizobia in Chihuahuan and Sonoran desert ecosystems. Soil Sci. Soc. Am. J. 53:543–549.

Johnson, A. C., and M. Wood. 1990. DNA, a possible site for action of aluminum in *Rhizobium* spp. Appl. Environ. Microbiol. 56:3629–3633.

Kape, R., M. Parniske, and D. Werner. 1991. Chemotaxis and *nod* gene activity of *Bradyrhizobium japonicum* in response to hydroxycinnamic acids and isoflavonoids. Appl. Environ. Microbiol. 57:316–319.

Keyser, H. H., and D. N. Munns. 1979a. Tolerance of rhizobia to acidity, aluminum, and phosphate. Soil Sci. Soc. Am. J. 43:519–523.

Keyser, H. H., and D. N. Munns. 1979b. Effects of calcium, manganese, and aluminum on growth of rhizobia in acid media. Soil Sci. Soc. Am. J. 43:500–503.

Keyser, H. H., D. N. Munns, and J. S. Hohenberg. 1979. Acid tolerance of rhizobia in culture and in symbiosis with cowpea. Soil Sci. Soc. Am. J. 43:719–722.

Kiester, D. L., and W. R. Evans. 1976. Oxygen requirement for acetylene reduction by pure cultures of rhizobia. J. Bacteriol. 127:149–153.

Kingsley, M. T., and B. B. Bohlool. 1992. Extracellular polysaccharide is not responsible for aluminum tolerance of *Rhizobium leguminosarum* bv. Phaseoli (CIAT899). Appl. Environ. Microbiol. 58:1095–1101.

Klubek, B. P., L. L. Hendrickson, R. M. Zablotowicz, J. E. Skwara, E. C. Varsa, S. Smith, T. G. Islieb, J. Maya, M. Valdes, F. B. Dazzo, R. L. Todd, and D. D. Walgenback. 1988. Competitiveness of selected *Bradyrhizobium japonicum* strains in Midwestern USA soils. Soil Sci. Soc. Am. J. 52:662–666.

Kremer, R. J., and H. L. Peterson. 1983. Effects of carrier and temperature on survival of *Rhizobium* spp. in legume inocula: Development of an improved type of inoculant. Appl. Environ. Microbiol. 45:1790–1794.

Kremer, R. J., J. Polo, and H. L. Peterson. 1982. Effect of suspending agent and temperature on survival of *Rhizobium* in fertilizer. Soil Sci. Soc. Am. J. 46:539–542.

Kurz, W. G. W., and T. A. Larue. 1975. Nitrogenase activity in rhizobia in absence of plant tissue. Nature (London) 256:407–409.

Lechevalier, M. P., and H. A. Lechevalier. 1989. Genus *Frankia* Brunchorst 1885, 174[AL]. Pp. 2410–2417. *In* S. T. Williams, M. E. Sharpe, and J. G. Holt (eds.),

Bergy's manual of systematic bacteriology, vol. 4. The Williams & Wilkins, Baltimore.

Leps, W. T., G. P. Roberts, and W. J. Brill. 1980. Use of two-dimensional polyacrylamide electrophoresis to demonstrate that putative *Rhizobium* cross-inoculation mutants actually are contaminants. Appl. Environ. Microbiol. 39:460–462.

Li, D.-M., and M. Alexander. 1986. Bacterial growth rates and competition affect nodulation and root colonization by *Rhizobium meliloti*. Appl. Environ. Microbiol. 52:807–811.

Liu, R., V. M. Tran, and E. L. Schmidt. 1989. Nodulating competitiveness of a nonmotile Tn7 mutant of *Bradyrhizobium japonicum* in nonsterile soil. Appl. Environ. Microbiol. 55:1895–1900.

Lowendorf, H. S., and M. Alexander. 1983a. Selecting *Rhizobium meliloti* for inoculation of alfalfa planted in acid soils. Soil Sci. Soc. Am. J. 47:935–938.

Lowendorf, H. S., and M. Alexander. 1983b. Identification of *Rhizobium phaseoli* strains that are tolerant or sensitive to soil acidity. Appl. Environ. Microbiol. 45:737–742.

Lowendorf, H. S., A. M. Baya, and M. Alexander. 1981. Survival of *Rhizobium* in acid soils. Appl. Environ. Microbiol. 42:951–957.

Madariaga, G. M., and J. S. Angle. 1992. Sludge-borne salt effect on survival of *Bradyrhizobium japonicum*. J. Environ. Qual. 21:276–280.

Madrzak, C. J., B. Golinska, J. Kroliczak, K. Pudelko, D. Lazewska, B. Lampka, and M. J. Sadowsky. 1995. Diversity among field populations of *Bradyrhizobium japonicum* in Poland. Appl. Environ. Microbiol. 61:1194–1200.

Mahler, R. L., and D. F. Bezdicek. 1980. Serogroup distribution of *Rhizobium leguminosarum* in peas in the Palouse of Eastern Washington. Soil Sci. Soc. Am. J. 44:292–295.

Mahler, R. L., and A. G. Wollum, II. 1980. Influence of water potential on the survival of rhizobia in Goldsboro loamy sand. Soil Sci. Soc. Am. J. 44:988–992.

Malek, W. 1992. The role of motility in the efficiency of nodulation by *Rhizobium meliloti*. Arch. Microbiol. 158:26–28.

Manian, S. S., R. Gumbleton, A. M. Buckley, and R. O'Gara. 1984. Nitrogen fixation and carbon dioxide assimilation in *Rhizobium japonicum*. Appl. Environ. Microbiol. 48:276–279.

Mansour, S. R., A. Dewedar, and J. G. Torrey. 1990. Isolation, culture, and behavior of Frankia strain HFPCgI4 from root nodules of *Casuarina glauca*. Bot. Gaz. 151:490–496.

Mary, P., D. Ocin, and R. Tailliez. 1985. Rates of drying and survival of *Rhizobium meliloti* strains during storage at different relative humidities. Appl. Environ. Microbiol. 50:207–211.

Materon, L. A., and C. Hagedorn. 1992. Competitiveness of *Rhizobium trifolii* strains associated with red clover (*Trifolium pratense* L.) in Mississippi soils. Appl. Environ. Microbiol. 44:1096–1101.

McDermott, T. R., and P. H. Graham. 1989. *Bradyrhizobium japonicum* inoculant mobility, nodule occupancy, and acetylene reduction in the soybean root system. Appl. Environ. Microbiol. 55:2493–2498.

McGrath, S. P., P. C. Brookes, and K. E. Giller. 1988. Effects of potentially toxic metals in soil derived from past applications of sewage sludge on nitrogen fixation by *Trifolium repens* L. Soil Biol. Biochem. 20:415–424.

Moawad, H., and B. B. Bohlool. 1984. Competition among *Rhizobium* spp. for nodulation of *Leucaena leucocephala* in two tropical soils. Appl. Environ. Microbiol. 48:5–9.

Munns, D. N., and R. L. Fox. 1977. Comparative lime requirements of tropical and temperate legumes. Plant Soil 46:533–548.

Myrold, D. D., A. B. Hilger, and S. H. Strauss. 1990. Detecting *Frankia* in soils using PCR. P. 429. *In* P. M. Gresshoff, Le. E. Roth, G. Stacey, and W. E. Newton (eds.), Nitrogen fixation: Achievements and objectives. Chapman and Hall, NY.

Napoli, C., and P. Albersheim. 1980. Infection and nodulation of clover by nonmotile *Rhizobium trifolii*. J. Bacteriol. 141:979–980.

Nittayarjarn, A., B. C. Mullin, and D. D. Baker. 1990. Screening of symbiotic frankiae for host specificity by restriction fragment length polymorphism analysis. Appl. Environ. Microbiol. 56:1172–1174.

Noel, K. D., and W. J. Brill. 1980. Diversity and dynamics of indigenous *Rhizobium japonicum* populations. Appl. Environ. Microbiol. 40:931–938.

O'Hara, G. W., T. J. Goss, M. J. Dilworth, and A. R. Glenn. 1989. Maintenance of intracellular pH and acid tolerance in *Rhizobium meliloti*. Appl. Environ. Microbiol. 55:1870–1876.

Osa-Afiana, L. O., and M. Alexander. 1982a. Differences among cowpea rhizobia in tolerance to high temperature and desiccation in soil. Appl. Environ. Microbiol. 43:435–439.

Osa-Afiana, L. O., and M. Alexander. 1982b. Clays and the survival of *Rhizobium* in soil during desiccation. Soil Sci. Soc. Am. J. 46:285–288.

Paau, A. S., C. B. Bloch, and W. J. Brill. 1980. Developmental fate of *Rhizobium meliloti* bacteroids in alfalfa nodules. J. Bacteriol. 143:1480–1490.

Pagan, J. D., J. J. Child, W. R. Scowcroft, and A. H. Gibson. 1975. Nitrogen fixation by *Rhizobium* culture on a defined medium. Nature (London) 256:406–407.

Pena-Cabriales, J. J., and M. Alexander. 1979. Survival of *Rhizobium* in soils undergoing drying. Soil Sci. Soc. Am. J. 43:962–966.

Pena-Cabriales, J. J., and M. Alexander. 1983a. Growth of *Rhizobium* in soil amended with organic matter. Soil Sci. Soc. Am. J. 47:241–245.

Pena-Cabriales, J. J., and M. Alexander. 1983b. Growth of *Rhizobium* in unamended soil. Soil Sci. Soc. Am. J. 47:81–84.

Pijnenborg, J. W. M., T. A. Lie, and A. J. B. Zehnder. 1991. Nodulation of lucerne (*Medicago sativa* L.) in an acid soil: Effects of inoculum size and lime-pelleting. Plant Soil 131:1–10.

Pugashetti, B. K., J. S. Angle, and G. H. Wagner. 1982. Soil microorganisms antagonistic towards *Rhizobium japonicum*. Soil Biol. Biochem. 14:45–49.

Quispel, A. 1990. Discoveries, discussions, and trends in research on actinorhizal root nodule symbioses before 1978. Pp. 15–33. *In* C. R. Schwintzer and J. D. Tjepkema (eds.), The Biology of *Frankia* and actinorhizal plants. Academic Press, San Diego.

Ramirez, C., and M. Alexander. 1980. Evidence suggesting protozoan predation on *Rhizobium* associated with germinating seeds and in the rhizosphere of beans (*Phaseolus vulgaris* L.). Appl. Environ. Microbiol. 40:492–499.

Reuber, T. L., A. Urzainqui, J. Glazebrook, J. W. Reed, and G. C. Walker. 1991. *Rhizobium meliloti* exopolysaccharides. Structures, genetic analyses, and symbiotic roles. Ann. NY Acad. Sci. 646:61–68.

Richardson, A. E., A. P. Anderson, G. S. James, and R. J. Simpson. 1988a. Consequences of soil acidity and the effect of lime on the nodulation of *Trifolium subterraneum* L. growing in an acid soil. Soil Biol. Biochem. 20:439–445.

Richardson, A. E., and R. J. Simpson. 1988. Enumeration and distribution of *Rhizobium trifolii* under subterranean clover-based pasture growing in an acid soil. Soil Biol. Biochem. 20:431–438.

Richardson, A. E., R. J. Simpson, M. A. Djordjevic, and B. G. Rolfe. 1988. Expression of nodulation genes in *Rhizobium leguminosarum* biovar *trifoli* is affected by low pH and by Ca and Al ions. Appl. Environ. Microbiol. 54:2541–2548.

Rogers, R. O., and A. G. Wollum II. 1974. Virulens if the *Alnus* endophyte after *in vitro* cultivation. Soil Sci. Soc. Am. J. 38:756–759.

Sellstedt, A., B. Wullings, U. Nyström, and P. Gustafsson. 1992. Identification of *Casuarina-Frankia* strains by use of polymerase chain reaction (PCR) with arbitrary primers. FEMS Microbiol. Lett. 93:1–6.

Sessitsch, A., G. Hardarson, A. D. L. Akkermans, and W. M. Devos. 1997. Characterization of *Rhizobium etli* and other *Rhizobium* spp. that nodulate *Phaseolus vulgaris* L. in an Austrian soil. Molecular Ecol. 6:601–608.

Sherwood, J. E., J. M. Vasse, F. B. Dazzo, and G. L. Truchet. 1984. Development and trifoliin A-binding ability of the capsule of *Rhizobium trifolii*. J. Bacteriol. 159:145–152.

Simon, T., S. Klaova, and K. Petrzik. 1996. Identification of *Rhizobium* strains and evaluation of their competitiveness. Folia Microbiol. 41:65–72.

Simonet, P., P. Normand, A. Moiroud, and R. Bardin. 1990. Identification of *Frankia* strains in nodules by hybridization of polymerase chain reaction products with strain-specific oligonucleotide probes. Arch. Microbiol. 153:235–240.

Simonet, P., M.-C. Grosjean, A. K. Misra, S. Nazaret, B. Cournoyer, and P. Normand. 1991. *Frankia* genus-specific characterization by polymerase chain reaction. Appl. Environ. Microbiol. 57:3278–3286.

Singleton, P. W., S. A. El Swaify, and B. B. Bohlool. 1982. Effect of salinity on *Rhizobium* growth and survival. Appl. Environ. Microbiol. 44:884–890.

Smith, S. R. 1997. *Rhizobium* in soils contaminated with copper and zinc following long-term application of sewage sludge and other organic wastes. Soil Biol. Biochem. 29:1475–1489.

Strijdon, B. W., and H. J. Van Rensburg. 1981. Effect of steam sterilization and gamma irradiation of peat on quality of *Rhizobium* inoculants. Appl. Environ. Microbiol. 41:1344–1347.

Surange, S., A. G. Wollum, N. Kumar, and C. S. Nautiyal. 1997. Characterization of *Rhizobium* from root nodules of leguminous trees growing in alkaline soils. Can. J. Microbiol. 43:891–894.

Tesfaye, M., D. J. Petersen, and F. B. Holl. 1997. Comparison of partial 23s rDNA sequences from *Rhizobium* species. Can. J. Microbiol. 43:526–533.

Thies, J. E., P. W. Singleton, and B. B. Bohlool. 1991a. Influence of the size of indigenous rhizobial populations on establishment and symbiotic performance of introduced rhizobia on field-grown legumes. Appl. Environ. Microbiol. 57:19–28.

Thies, J. E., P. W. Singleton, and B. B. Bohlool. 1991b. Modeling symbiotic performance of introduced rhizobia in the field by use of indices of indigenous population size and nitrogen status of the soil. Appl. Environ. Microbiol. 57:29–37.

Thornton, F. C., and C. B. Davey. 1983. Acid tolerance of *Rhizobium trifolii* in culture media. Soil Sci. Soc. Am. J. 47:496–501.

Tiwari, R. P., W. G. Reeve, and A. R. Glenn. 1992. Mutations conferring acid sensitivity in the acid-tolerant strains of *Rhizobium meliloti* WSM419 and *Rhizobium leguminoarum* biovar *vicae* WMS710. FEMS Microbiol. Lett. 100:107–112.

Tjepkema, J., and H. J. Evans. 1975. Nitrogen fixation by free-living *Rhizobium* in a defined liquid medium. Biochem. Biophys. Res. Commun. 65:625–628.

Tjepkema, J. D., and L. J. Winship. 1980. Energy requirement for nitrogen fixation in actinorhizal and legume root nodules. Science 209:279–280.

Urtz, B. E., and G. H. Elkan. 1996. Genetic diversity among *Bradyrhizobium* isolates that effectively nodulate peanut (*Arachis hypogaea*). Can. J. Microbiol. 42:1121–1130.

Vanberkum, P., D. Beyene, F. T. Vera, and H. H. Keyser. 1995. Variability among *Rhizobium* strains originating from nodules of *Vicia faba*. Appl. Environ. Microbiol. 61:249–253.

van Brussel, A. A. N., R. Bakhuizen, P. C. van Spronsen, H. P. Spaink, T. Tak, B. J. J. Lugtenberg, and J. W. Kijne. 1992. Induction of pre-infection thread structures in the leguminous host plant by mitogenic lipo-oligosaccharides of *Rhizobium*. Science 257:70–72.

van Dijk, C., and A. Sliumer-Stolk. 1990. An infective strain type of *Frankia* in the soil of natural stands of *Alnus glutinosa* (L.) Gaertner. Plant Soil 127:107–121.

van Rensburg, H. J., and B. W. Strijdom. 1985. Effectiveness of *Rhizobium* strains used in inoculants after their introduction into soil. Appl. Environ. Microbiol. 49:127–131.

Viteri, S. E., and E. L. Schmidt. 1987. Ecology of indigenous soil rhizobia: Response of *Bradyrhizobium japonicum* to readily available substrates. Appl. Environ. Microbiol. 53:1872–1875.

Waldon, H. B., M. B. Jenkins, R. A. Virginia, and E. E. Harding. 1989. Characteristics of woodland rhizobial populations from surface and deep-soil environments of the Sonoran Desert. Appl. Environ. Microbiol. 55:3058–3064.

Watson, R. J., C. H. Crockett., T. Martin, and R. Heys. 1995. Detection of *Rhizobium meliloti* cells in field soil and nodules by polymerase chain reaction. Can. J. Microbiol. 41:816–825.

Woomer, P., P. W. Singleton, and B. B. Bohlool. 1988. Ecological indicators of native rhizobia in tropical soils. Appl. Environ. Microbiol. 54:1112–1116.

Young, J. P. W., and K. E. Haukka. 1996. Diversity and phylogeny of rhizobia. New Phytol. 133:87–94.

Zahran, H. H. 1997. Diversity, adaptation and activity of the bacterial flora in saline environments. Biol. Fertil. Soils 25:211–223.

Chapter **14**

Denitrification

Some would use a term such as pollutant when considering the occurrence of nitrate in soil, whereas others would prefer to describe soil nitrate as an essential nutrient for ecosystem sustenance. One group would seek to minimize levels of the alleged environmentally offensive substance, whereas those interested in plant biomass production would seek to optimize nitrate production. Both groups would underscore the necessity of understanding processes leading to losses of soil nitrate to achieve their objectives.

To a great degree, the existence of such divergent opinions regarding the value of nitrate in soil results from the propensity of this compound to be in the wrong place at the wrong time at undesirable concentrations. Nitrate ions in the root zone of a food crop during maximal biomass production are inarguably beneficial, whereas migration of that same nitrate below the root zone to underlying aquifers transforms a beneficial soil chemical component into an undesirable water contaminant.

Societal implications associated with the timing and location of accumulation of significant soil nitrate loadings reach far beyond this conflict of development of a soil nitrate balance sheet and agreement on the best nutrient management plans for maintenance of an ecosystem. For example, traditional considerations have included concerns that nitrate contents of soil leachates or runoff waters may alter regional water quality (e.g., augmented lake eutrophication rates) and that there are potential public health difficulties (methemoglobenemia) associated with elevated nitrate loadings in drinking waters. Other concerns include management of fixed nitrogen amendment to soil systems to minimize societal costs and unwanted environmental impacts. Excessive losses of fertilizer nitrogen, whether added to agricultural, urban, or reclaimed soils, alter the economics of biomass production as well as increase the energy expenditures associated with industrial nitrogen fixation.

Furthermore, increased utilization of industrially fixed nitrogen may have an environmental impact far beyond the localized site. Fertilizer production requires inputs of large quantities of fossil fuels. Utilization of these resources to produce plant nutrients not only decreases the reserves of a nonrenewable commodity but also increases greenhouse gas production through its conversion to carbon dioxide and energy

Denitrification Defined: If nitrogen fixation is used as a reference for the initiation point of the nitrogen cycle, denitrification is its termination. **Denitrification** is the biological reduction of nitrogen oxides to nitrous oxide and/or dinitrogen. The process is catalyzed exclusive by a diverse group of bacteria that use nitrate or nitrite as terminal electron acceptors in their respiratory processes. Since the electrons are more readily transferred to molecular oxygen by the microbes involved, denitrification can occur only under anoxic conditions. From this basic definition of denitrification, the complexity of controlling its occurrence in heterogeneous soils becomes evident. For denitrification to occur, the microorganisms must have an energy supply (usually an organic carbon source) and reside in an anoxic soil microsite, yet most soil systems are predominantly aerobic and the primary processes producing nitrate also have an obligatory requirement for molecular oxygen. That is, within the soil system itself, the primary source of the initial substance in the denitrification pathway, nitrate, is autotrophic nitrification, an obligatorily aerobic process.

These limited examples of the role of denitrification in controlling levels of nitrate occurring in soil and the availability of this primary plant nutrient within soil ecosystems underscore the necessity of developing an understanding of the basic properties of the process and the factors in the soil environment that control the rate of reduction of nitrate to dinitrogen. Thus, this chapter will be developed with the overall objective of elucidating the basic biological and biochemical properties of denitrification and the processes determining the extent of its occurrence in terrestrial ecosystems.

14.1 Pathways for Biological Reduction of Soil Nitrate

Soil fixed nitrogen resources may be conserved through both assimilatory and dissimilatory nitrate-reductive processes or they may be lost from the soil ecosystem by dissimilatory nitrate reduction. In this context, "conserved" refers to retention within the soil ecosystem (generally as biomass, organic matter, or ammonium). Assimilatory and dissimilatory nitrate reduction both involve the transfer of electrons to nitrogen oxides, but they differ in the ultimate fate of the reduced nitrogen atom. With assimilatory reduction, nitrate is reduced to ammonium, which is subsequently incorporated into cellular biomass. Thus, with this process, the quantity of nitrogen reduced is proportional to cellular requirements for biomass production. In contrast, for dissimilatory nitrate reduction, the nitrogenous compounds are accepting electrons in support of cellular respiration. The final products (dinitrogen, nitrous oxide, or ammonium) are released from the cell and accumulate in the environment in concentrations far beyond those necessary for biomass synthesis. Note from the view of retention of fixed nitrogen within the soil ecosystem, dissimilatory processes can lead to losses (production of dinitrogen and nitrous oxide) or conservation (production of ammonium). These two fates of the nitrate-reductive processes (dissimilatory vs. assimilatory reduction) are characterized by commonalties of some of the intermediates and products (e.g., ammonium and nitrate). They are also distinguished by differences in the specific enzymes catalyzing the reductive reactions.

Since assimilatory reduction of nitrate is closely linked to biomass production, it is catalyzed by all soil inhabitants capable of using nitrate as a nitrogen source. Ammonium, the terminal product of the process, is incorporated into plant and microbial biomass. Also, although a portion of the cellular reducing power is transferred to the nitrogen, the cells rely upon other pathways as terminal electron acceptors (e.g., molecular oxygen and fermentative products).

Three commonly evaluated microbial processes are classed under the title of dissimilatory nitrate reduction. These are distinguished by their respective products: (1) nitrite, (2) ammonium, and (3) nitrous oxide and dinitrogen (denitrification).

The first of these dissimilatory reductions of nitrate involves the simple reduction of nitrate to nitrite. The general ability of many bacterial species to catalyze the process has led to the use of nitrate reduction to nitrite as a reliable taxonomic test for grouping bacterial isolates into genera and species. This reductive process is also significant in some *in situ* soil systems. In a soil site overlain by oxygen-bearing water, a nitrite and nitrate cycle may develop that provides energy for nitrite oxidizers (see Fig. 11.3). Under anoxic conditions, nitrate is reduced to nitrite by portions of the general soil bacterial population. This nitrite then diffuses from its source in the anoxic microsites to aerobic microsites where the autotrophic nitrite oxidizers utilize it in their energy generating processes. Evidence for the occurrence of such interactions in native sites is the occurrence of an imbalance between ammonium and nitrite oxidizer populations in some soil ecosystems. The nitrite oxidizer populations are greater than would be anticipated to occur where ammonium oxidizers constitute the sole provider of their energy resource, nitrite.

The second and third members of our list of dissimilatory nitrate-reduction processes, dissimilatory ammonium production and denitrification, have several common characteristics. Both processes are coupled to organic matter decomposition (i.e., oxidation of fixed carbon compounds provides cellular energy and reducing power), the nitrogen oxides that are reduced serve as terminal electron acceptors, and growth yields of the bacteria are increased as a result of the energy provided by passage the electrons produced by oxidation of the energy source through an electron transport pathway that includes cytochromes (see Section 14.2 for discussion of the role of cytochromes in denitrification).

Although not previously considered to be a major soil process, ammonium production from nitrate reduction has been shown to occur in soils and sediments (e.g., Buresh and Patrick, 1978; Fazzolari et al., 1990; Koike and Hattori, 1978; Smith and Zimmerman, 1981). Dissimilatory ammonium production is catalyzed by a variety of enteric bacteria, *Bacillus* species, and *Clostridium* species in soil (e.g., Caskey and Tiedje, 1980; Prakash and Sadana, 1973; Rehr and Klemme, 1989; Smith and Zimmerman, 1981). As with denitrification, nitrous oxide has been found to be a product of metabolism of dissimilatory ammonium producers, but the nitrous oxide appears to be produced as a side reaction in these organisms (e.g., Smith, 1982).

14.2 Biochemical Properties of Denitrification

During biological denitrification, nitrate is transformed to dinitrogen by a series of reductive reactions, as follows:

$$2HNO_3 \rightarrow 2HNO_2 \rightarrow [2NO] \rightarrow N_2O \rightarrow N_2$$

Nitric oxide is generally enclosed in brackets in depiction of this reaction sequence since it is not usually detected as a free intermediate. There is also some controversy whether nitric oxide is an true intermediate in the process or whether its presence represents a side reaction (e.g., Betlach and Tiedje, 1981; Hochstein and Tomlinson, 1988; Hollocher et al., 1980; Garber and Hollocher, 1981; 1982). In either case, the significant aspect of the reaction sequence from the soil microbiologist's viewpoint is that nitrate is reduced to nitrous oxide and dinitrogen with the possible transient accumulation of nitrite plus the occasionally detected transient accumulation of nitric oxide (e.g., Cady and Bartholomew, 1960; Cooper and Smith, 1963). As will be indicated below, a variable ratio—depending upon the soil chemical properties—of nitrous oxide to dinitrogen is produced.

Carbon and Energy Sources for Denitrifiers: The electrons for nitrogen oxide reduction are provided by oxidation of a variety of fixed carbon compounds. Most commonly, glucose is added to soil samples as a carbon source by investigators in laboratory investigations of denitrification processes, but it must be remembered that a large number of organic compounds are catalyzed during this process. This array of oxidizable substrates includes many complex organic compounds, some of which are of concern from the view of soil contamination [e.g., acetone (Platen and Schink, 1989), anthranilic acid (Braun and Gibson, 1984), cresol (Bossert and Young, 1986), and toluene (Evans et al., 1992). For a review of this topic consult Evans and Fuchs (1988)]. Since the electrons for these oxidations are transferred through a chain of acceptors similar to the pathway leading ultimately to the reduction of molecular oxygen by aerobic heterotrophs, the growth yields of microbes oxiding organic carbon substances while using nitrogen oxides as terminal electron acceptors are comparable with the two classes or acceptors (molecular oxygen and nitrogen oxides).

Using the traditional definition of anaerobic processes (i.e., energy-yielding catabolic reactions obligatorily occurring in the absence of molecular oxygen), dentrifiers could be considered to be facultative anaerobes. Yet many soil microbiologists do not consider denitrification to be a true anaerobic process in that the primary overall difference between nitrate respiration and oxygen-based respiration is that in the former electron transport process, nitrate is merely replacing oxygen as the final electron acceptor. Denitrifiers are thus classed as aerobes. They use molecular oxygen as a terminal electron acceptor when oxygen is available and nitrogen oxides when it is not. Both processes are accomplished through use of cytochromes in an electron transport chain (e.g., Ballard and Ferguson, 1988; Bolgiano et al., 1989) (Fig. 14.1). If oxygen is present, different cytochromes are used for its reduction to water, and dissimilatory nitrate reduction is inhibited (e.g., Hernandez and Rowe, 1987).

Induction of Synthesis of Nitrogen Oxide Reductases: A further modification of cellular metabolism of the denitrifier growing in aerobic situations is that the requisite enzymes for denitrification are not synthesized or are produced in limited quantities. Nitrogen oxide reductase synthesis is repressed by molecular oxygen. Once the oxygen is depleted in the microsite where the putative denitrifier resides, synthesis of these denitrifier enzymes is induced. Thus, development of a maximally functional denitrification system once anoxic conditions are imposed requires about 40 minutes to three hours. For some denitrifiers, the time lapse necessary for synthesis of nitrogen oxide reductases is reduced by an induction of synthesis of the first enzyme in the pathway, nitrate reductase, as the available molecular oxygen supply is being depleted. Full induction of the pathway occurs with complete exhaustion of molecular oxygen resources. Thus, the microbe is able to respond to the absence of molecular oxygen by reducing nitrate to nitrite, but it avoids synthesis of the full array of denitrification enzymes should anaerobiosis not be achieved. This differential synthesis of denitrification enzymes can result in a transient accumulation of nitrite.

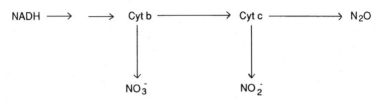

Fig. 14.1 Summary of the electron transport pathway leading from oxidation of fixed carbon compounds to nitrate, nitrite, or nitrous oxide as terminal electron acceptors.

14.3 Environmental Implications of Nitrous Oxide Formation

At one time, the transient accumulation of nitrous oxide in reaction vessels wherein biological denitrification was occurring was ignored or considered to be a product of the closed reaction system. It is now known that in native terrestrial environments, nitrous oxide is a common product of denitrification. For example, fluxes of nitrous oxide from soils have been shown to range from 7 to 165 kg N ha^{-1} $year^{-1}$ in drained histosols in South Florida (Terry et al., 1981a). Nitrous oxide evolution from mineral soil-based systems is usually less than reported from these drained organic soils. Annual losses from irrigated soils in California (USA) range from 19.6 to 41.8 kg N ha^{-1} (Ryden and Lund, 1980). Lesser yields have been detected from cropped soils (Benckiser et al., 1996; Goodroad et al., 1984; Mosier and Hutchinson, 1981) and prairie soils (Mosier et al., 1981). Intensity and nature of fertilization used in cropping systems also may affect the quantity of nitrous oxide produced (e.g., Loro et al., 1997; Mulvaney et al., 1997; McTaggart et al., 1997). Noncropped, native ecosystems also produce more limited quantities of nitrous oxide during plant growth seasons (Goodroad and Keeney, 1984).

These data show that nitrous oxide production by soil processes is highly variable. Soil properties controlling the quantity of this product yielded by soil microbes include soil redox potential (Letey et al., 1981; Smith et al., 1983), moisture tension (Davidson and Swank, 1986; Freney et al., 1979; Regina et al., 1996; Terry et al., 1981b), and nitrate and oxygen concentrations (Blackmer and Bremner, 1978; 1979, Firestone et al., 1979; 1980; Terry and Tate, 1980) as well as the time of day that the measurements are taken (Blackmer et al., 1982). Generally, the proportion of gaseous nitrogen products composed of nitrous oxide increases with increasing soil acidity, reduced soil temperature, and augmented soil nitrate levels, but the exact ratio of nitrous oxide to dinitrogen for a specific soil system varies with the combination of the soil chemical and physical properties existent therein and is therefore not reliably predicted.

Understanding the potential for production of nitrous oxide by soil ecosystems has greater importance than simply allowing a better comprehension of soil nitrogen transformations. Nitrous oxide is an ozone-depleting gas. The reactions involved in this process are presented in Fig. 14.2. It must be noted that not all of the nitrous oxide evolving from the soil surface results from activity of denitrifiers. A variety of soil heterotrophs as well as autotrophic nitrifiers produce this gas as a metabolic product or by-product (e.g., Blackmer et al., 1980; Bleakley and Tiedje, 1982; Davidson et al., 1986; Smith and Zimmerman, 1981). Therefore, the soil properties affecting nitrous oxide fluxes do not correlate only with those favoring biological denitrification. Complexity in interpreting the data is added by the fact that nitrous oxide is also produced by chemodenitrification as well as the various biological sources (see review by Bremner, 1997, and examples of quantification of the processes in the field by Heinrich and Haselwandter, 1997, and Nielsen et al., 1996).

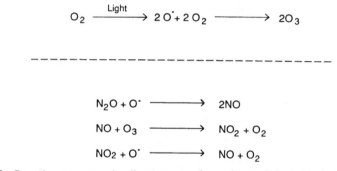

$$O_2 \xrightarrow{\text{Light}} 2O^{\cdot} + 2O_2 \longrightarrow 2O_3$$

- -

$$N_2O + O^{\cdot} \longrightarrow 2NO$$

$$NO + O_3 \longrightarrow NO_2 + O_2$$

$$NO_2 + O^{\cdot} \longrightarrow NO + O_2$$

Fig. 14.2 Reaction sequence leading to ozone formation and the role of nitrous oxide in its destruction.

14.4 Microbiology of Denitrification

Assessment of Soil Denitrifier Populations: The population density of denitrifiers in soil is commonly estimated through use of most-probable-number (MPN) procedures (e.g., Focht and Joseph, 1973; Volz, 1977). These methods are based on detection of the conversion of nitrate or nitrite to gaseous end products in liquid culture media (e.g., Fig. 14.3). Most-probable-number methods are inherently imprecise (see chapter 2) for a discussion of the limitations of MPN procedures). More accurate and sensitive methods for enumeration of these organisms will become available as the characteristics of the genome encoding denitrification and the similarities in these gene sequences between the various denitrifiers are elucidated, thereby enabling the utilization of pcr procedures for the quantification of denitrifiers in soil (e.g., Linne von Berg and Bothe, 1992; Ye et al., 1993).

Denitrifier population density data do provide an indication of the number of organisms present in a soil ecosystem that are capable of catalyzing the process. Unfortunately, in most soils, population densities of denitrifiers do not

Quantification of Denitrifiers

1. Prepare a serial dilution of a fresh soil sample [eg., 10^{-1}, 10^{-2} . . . 10^{-7}]
2. Inoculate 5 tubes containing 9 ml of nitrate broth with 1.0ml of dilution.
3. Incubate the inoculated media at 30°C for two weeks.
4. Quantify residual nitrate in tubes. Presence of any nitrate = −.
5. Calculate MPN using standard tables.

Fig. 14.3 Outline of a most-probable-number procedure for estimating denitrifier populations in soil (see Focht and Joseph, 1973 for example of method details).

correlate with denitrification rates. Soil microbial communities commonly contain several million denitrifiers per gram dry soil. Because of the fact that the respiration of these organisms is not limited solely to nitrate reduction (recall that they can also reduce molecular oxygen), their presence in the system indicates a potential for denitrification only should conditions favor its occurrence. The denitrifier populations occurring in any particular soil with an oxygen-containing atmosphere may have developed through oxygen-based respiration.

General Traits of Denitrifiers: Denitrifiers are a biochemically and taxonomically diverse group of bacteria. Although some denitrifiers are chemoautotrophs (e.g., using hydrogen or reduced sulfur compounds as energy sources) and others are photoautotrophs, most of these organisms generally derive their energy from oxidation of fixed carbon substrates, including single carbon compounds. The primary substrate and end product of the process are nitrate and dinitrogen, but some denitrifiers can reduce only nitrite (nitrite dependent denitrifiers), whereas others lack nitrous oxide reductase, thereby producing nitrous oxide as the terminal product. The variety of metabolic types of bacteria capable of denitrifying defies any effort to group these organisms into one or a few bacterial genera.

Further difficulty with evaluation of the occurrence of the "denitrifier trait" in classical literature results from the fact that many organisms classified historically as denitrifiers are not really denitrifiers. Care must be exercised in classing organisms as denitrifiers based strictly on the conversion of nitrate or nitrite to nitrous oxide or dinitrogen independent of the extent of the reaction. Many bacterial strains are capable of reducing nitrate and of producing *limited* quantities of dinitrogen or nitrous oxide. Although some of these organisms have been termed denitrifiers in the past, these marginal nitrate reducers are *not* true denitrifiers. To be a true denitrifier, a microbe must meet the following criteria:

- At least 80 percent of the nitrate or nitrite reduced by the bacterium must be converted to dinitrogen and nitrous oxide.

- There must be an increased growth yield due to the reduction of nitrate, nitrite, or nitrous oxide. This trait is the primary requirement for classing an organism as a denitrifier in that it shows that nitrogen oxide reduction is a dissimilatory process.

- The conversion of nitrate to nitrous oxide and dinitrogen must occur at a high rate. That is, the process must be central to cellular intermediary metabolism, not just a side reaction providing a minor pathway for electron transport.

- The presence of cytochrome cd or dissimilatory nitrite reductase should be demonstrable in the microbial cells.

Generic Identity of Denitrifiers: Most of the denitrifiers isolated from soils and waters are members of the genera *Pseudomonas* or *Alcaligenes*, but a large number of other bacterial genera contain strains of denitrifiers (e.g., Gamble et al., 1977). As with diazotrophs, denitrification is not a sufficiently definitive trait to allow separation of all denitrifiers into one or a limited number of genera. Not only are the individual bacterial strains of denitrifiers metabolically distinct, but the properties of the enzymes associated with the denitrification process itself are highly variable. For example, in a comparison of denitrification by *Pseudomonas stutzeri, Pseudomonas aeruginosa*, and *Paracoccus denitrificans* it was found that the rates of anoxic growth of the organisms varied 1.5-fold, gas production varied over 8-fold, and cell yield differed by 3-fold (Carlson and Ingraham, 1983). The metabolic and taxonomic diversity of these organisms is exemplified by the following partial listing of some of the bacterial genera containing denitrifiers (for a more complete listing of denitrifiers, see Knowles, 1981, 1982).

- The genus *Achromobacter* is cited as containing a variety of denitrifiers, including methane oxidizers and nitrite-dependent denitrifiers. (These organisms have now been grouped among the species of the genus *Alcaligenes*.)

- *Alcaligenes eutrophus*, an organism originally classified as *Hydrogenomonas eutrophus*, is capable of autotrophic growth using hydrogen as an energy source, carbon dioxide for carbon, and nitrate as a terminal electron acceptor (e.g., Pfitzner and Schlegel, 1973). This bacterial species is a facultative autotroph since it also uses organic carbon as an energy source. Other common denitrifiers in the genus *Alcaligenes* include *A. denitrificans* and *A. odorans*. Another facultative autotroph that oxidizes hydrogen for energy is *Paracoccus denitrificans*.

- Several nitrogen-fixing bacteria are also capable of denitrifying. These diazotrophs include some strains of *Azospirillum brasilense* (e.g., Chauret et al., 1992; Neyra et al., 1977) and some *Rhizobium* species (e.g., Chan et al., 1989; Coyne and Focht, 1987; Daniel et al., 1980; van Berkum and Keyser, 1985; Zoblotowicz et al., 1978)

- Some denitrifiers are also thermophilic (e.g., Hollocher and Kristjánsson, 1992).

- Some strains of *Chromobacter* denitrify using nitrate or nitrite as terminal electron acceptors, reducing them to nitrous oxide and dinitrogen, whereas *Chromobacter violacium* can only reduce nitrogen oxides to nitrous oxide (Bazylinski et al., 1986).

- *Halobacterium marismortui* is a halophilic denitrifier that was isolated from the Dead Sea.

- *Hyphomicrobium* strains denitrify with methanol as the primary energy source. The practical significance of understanding the environmental limitations to activity of these strains relates to the fact that methanol has been used as an additive to sewage effluent to provide the energy source necessary for maximization of denitrification rates for removal of nitrate from the effluent.

- The photosynthetic bacterium *Rhodopseudomonas sphaeroides* forma sp. *denitrificans* is both a denitrifier and a diazotroph (Dunstan et al., 1982).

- *Thiobacillus denitrificans* is an example of a sulfur-oxidizing chemoautotroph that grows in the absence of oxygen with nitrate as a terminal electron acceptor. The energy sources for this organism include sulfide, elemental sulfur, and thiosulfate. *T. denitrificans* can also use molecular oxygen as a terminal electron acceptor.

14.5 Quantification of Nitrogen Losses from an Ecosystem via Denitrification

Historically, the most commonly used techniques for assessing denitrification as with nitrogen fixation have involved (1) determination of nitrogen balances in the ecosystem and (2) quantification of distribution and movement of ^{15}N-labeled fixed nitrogen through soil reservoirs. Currently, more commonly used laboratory-based assays of denitrification are (1) assessment of disappearance of nitrogen oxides from soil slurries and (2) use of acetylene inhibition of nitrous oxide reductase. The principles, advantages, and deficiencies of each of these techniques will be assessed herein.

14.5.1 Nitrogen Balance Studies

As was discussed previously in relationship to quantification of nitrogen fixation in field sites, assessment of changes in distribution of nitrogen among the various soil reservoirs as a means of quantifying process rates appears to be conceptually simple and logically attractive. For application of nitrogen balance methods to quantification of denitrification, the amount of nitrogen denitrified is concluded to be the unaccounted for or missing nitrogen. The quantity of nitrogen denitrified is that preexisting in the system plus any inputs (e.g. nitrogen fixation and fertilizer inputs) minus known losses (e.g., nitrogen in crops, nitrogen leached, ammonium volatilized). The procedure is also called the *difference method* because all nitrogen occurring in the system is subtracted from the sum of that previously occurring therein plus any amendments.

Use of the nitrogen balance method for estimation of denitrification processes in soils is predicated upon three assumptions:

- All of the organic and fixed nitrogen compounds contained in soil can be accurately quantified.

- The assay procedures available for quantifying soil nitrogen compounds are sufficiently sensitive to allow for detection of small changes in the various soil nitrogen compounds. Furthermore, it is assumed that the soil, water, and air samples analyzed are representative of the ecosystem and that no changes in the nitrogen levels existing in these samples at the site occurred between collection and analysis.

- The sole means of loss of gaseous nitrogen from soil is denitrification.

Unfortunately, none of these assumptions is valid. Reasonably sensitive assay procedures are available to assay soil organic and inorganic nitrogen pools. Their application to soil nitrogen cycle analysis is limited by the fact that changes in soil nitrogen reserves are commonly less than the precision of the assay method. Further problems with the direct quantification of nitrogen species in a particular soil site result from the fact that soil is an open system. This means that water-soluble substances such as nitrate, nitrite, and simple organic nitrogenous compounds move freely into and out of the portion of the ecosystem of interest. Inability to accurately account for all fixed nitrogen inputs and losses from the soil site results in appreciable statistical variability in nitrogen balance data. This variation also results in invalidation of the second assumption.

Probably the biggest difficulty with application of nitrogen balance studies to assessment of denitrification is the third assumption (all nitrogen volatilization results from denitrification). There are at least three major routes of nitrogen volatilization from soil. Nitrogen may be lost through:

- Nonbiological losses of ammonia (ammonia volatilization)

- Chemical decomposition of nitrite

- Denitrification as dinitrogen and nitrous oxide

Ammonia volatilization can be a major route for loss of fixed nitrogen from soil. Up to 25 percent of the ammonia applied to soil as fertilizers or formed microbiologically may be lost as ammonia. Fortunately, for most soil systems, this process of nitrogen volatilization is a minor problem since ammonia losses are insignificant below pH 7.0. (Recall that at acidic pHs, the equilibrium between ammonium and ammonia is shifted so that ammonium predominates.) Ammonia volatilization is particularly apparent in the use of farmyard manure and at times in association with fertilization with urea. Major environmental concern with this process is linked more to the migration of soil fixed nitrogen to nearby bodies of water than with its interference with quantification of denitrification.

Nitrite decomposition (chemodenitrification) is rarely a major route of exodus of fixed nitrogen from soils. The most commonly documented situation for accumulation of meaningful nitrite concentrations is linked to alkaline soil conditions where nitrite accumulates due to ammonia toxicity to *Nitrobacter* spp. This phenomenon occurs in soils fertilized with anhydrous ammonia. In this process, nitrite reacts chemically with soil organic matter to produce dinitrogen and nitrous oxide plus some nitric oxide. In the absence of molecular oxygen, the nitric oxide may be evolved, but it is generally oxidized to nitrogen dioxide as soon as it is exposed to molecular oxygen. Although this reaction can occur in soil, nitrite will more commonly be biologically denitrified in soil. Thus, the relative importance of the three pathways of nitrogen volatilization from soils may be ranked, with denitrification being the major means of nitrogen loss, ammonia volatilization being ranked, a distant second, and nitrite decomposition occurring only as a minor contributor.

The uncertainties associated with the nitrogen balance procedure in estimating soil nitrogen losses due to denitrification were exemplified by data from a study in which this procedure was compared to results from a parallel quantification of nitrogen dynamics using ^{15}N to trace changes in soil nitrogen reservoirs (Rolston et al., 1979). In the more moist soils studied, actual denitrification losses were generally 0 to 30 kg N ha^{-1} less than amounts calculated from the nitrogen balance procedures. These differences ranged from 12 to 65 kg N ha^{-1} for drier soils. The authors concluded that with their soils, the uncertainties with the nitrogen balance procedure resulted primarily from large sampling variability in nitrogen leaching results (although variability of soil residual nitrate and organic nitrogen were also large). These results further emphasize the conclusion that even in soils where nitrogen volatilization due to chemical processes is minimal, the accuracy of the nitrogen balance procedure is limited due to difficulties in precisely accounting for all soil nitrogen forms present and leaving the soil system.

14.5.2 Use of Nitrogen Isotopes to Trace Soil Nitrogen Movement

Although the short-lived radioactive nitrogen isotope (^{13}N) has been used to quantify denitrification processes (Firestone et al., 1979; Hollocher et al., 1980; Speir et al., 1995a; 1995b; Tiedje et al., 1979), due to the short half-life of the radioactive isotope, the heavy isotope of nitrogen (^{15}N) is more amenable to long-term evaluation of denitrification reactions in soils. For these studies, ^{15}N-labeled fixed nitrogen sources (organic or inorganic) are amended to field soils or incubated soil samples (greenhouse or laboratory) and the ^{15}N-labeled dinitrogen or nitrous oxide quantified.

As with the evaluation of biological nitrogen fixation, implementation of this procedure is predicated on a series of assumptions, the most significant being (1) the denitrifier populations do not discriminate between nitrogen isotopes and (2) the soil-amended ^{15}N-labeled fixed nitrogen is homogeneously mixed with native soil nitrogen resources and is thus of equivalent availability to the denitrifier population. As was noted with biological nitrogen fixation, satisfaction of these assumptions is also questionable for assessing denitrification.

Discrimination of nitrogen isotopes by denitrifiers has been demonstrated (e.g., Blackmer and Bremner, 1977). Clear microbial capacity to discriminate between ^{14}N and ^{15}N-labeled nitrate was observed by these workers. Significant nitrogen isotope effects were detected in both reduction of nitrate to nitrite, and nitrite to gaseous end products.

Furthermore, because of the heterogeneity of the soil system, uniform mixing of an external nitrogen supply with indigenously formed fixed nitrogen is difficult at best. Vanden Heuvel et al. (1988) found that the magnitude of this error increased with the range of ^{15}N enrichment between native soil N isotope distributions and amended nitrogen (for a constant mean ^{15}N enrichment), and that it varied depending on the number of nitrate pools existent in the system. The error was not affected by overall nitrogen gas evolution rate or the ^{15}N enrichment of the amended fertilizer. The theoretical basis for this potential for overestimation and underestimation of nitrous oxide and dinitrogen evolution was documented through mathematical modeling by Boast et al. (1988).

Even though these cited studies demonstrate significant difficulties with utilization of ^{15}N to quantify losses of fixed nitrogen from soil ecosystems due to denitrification, its application to field, laboratory, and greenhouse studies is appropriate when the limitations are considered. The use of ^{15}N is appropriate for field experiments [especially for long-term studies as exemplified by Focht and Stolzy (1978)] when the limitations are considered during data interpretations. ^{15}N labeling of nitrogen species in greenhouse and laboratory incubation studies has also been particularly useful. In the latter situations, extensive disruption of soil structure during sample preparation is common and the conditions for denitrification are more controlled.

14.5.3 Disappearance of Endogenous or Added Nitrogen Oxides

Since nitrate, nitrite, and nitrous oxide are readily quantified from soil, it is reasonable to consider that the rate of change in concentrations (native or amended) of these chemicals can serve as an indication of denitrification rates in laboratory-incubated soil samples (e.g., Bremner and Shaw, 1958a). This conclusion is based on the assumption that all of the nitrate or nitrite lost from the soil sample is denitrified. Except in systems where extensive dissimilatory ammonium production is occurring, alternative fates of the nitrogen oxide (generally nitrate) tend not to interfere with this analysis procedure. Clearly, some nitrate could be incorporated into microbial biomass if the assay is conducted over a time frame sufficiently long for microbial growth to occur. Thus, in this situation, denitrification rates would be underestimated by the amount of nitrate assimilated by the microbial biomass. Since the assays are conducted under laboratory conditions, plant uptake and leaching are not important because plants are excluded from the system and the reaction vessel is a closed system.

Greater problems with the use of changes in nitrogen oxide levels as an assessment of denitrification kinetics in soil are derived from the fact that in most soil systems, nitrate concentrations are a few $\mu g\,g^{-1}$ soil at best.

Generally, nitrite is not detected in native soil samples. Thus, soils are generally amended with exogenous nitrate or nitrite for a meaningful estimate of the capacity of the soil microbial community to denitrify nitrogen oxides to be attained. To assure that the totality of the denitrification enzymes is measured (that is, enzyme and not substrate is the limiting factor), enzyme-saturating concentrations of nitrate or nitrous oxide are generally added to soil with this technique. Therefore, since these nitrogen oxide concentrations are artificially elevated above those normally occurring in the field, denitrification rates assessed using this procedure must be understood to be potential rates rather than actual field values. The potential values would be equivalent to field denitrification rates only in the unusual situations when field nitrate concentrations are sufficient to saturate denitrification enzymes present.

14.5.4 Acetylene Block Method for Assessing Denitrification Processes in Soil

Incubation of soil samples or amendment of the soil atmosphere *in situ* with 10^{-2} atmospheres of acetylene inhibits the reduction of nitrous oxide to dinitrogen (Yoshinari and Knowles, 1976; Yoshinari et al., 1977). Since commonly available gas chromatography procedures may be used to quantify nitrous oxide, the acetylene block method provides a sensitive method for measurement of denitrification enzymes. See Fig. 14.4 for an outline of the procedure. The validity of this technique for determination of denitrification kinetics in soil slurries (e.g., Smith et al., 1978) and in field situations [e.g., soil cores (Parkin et al., 1985a) and direct field measurements (e.g., Aulakh et al., 1991; Mosier et al., 1986)] has been confirmed through the comparison of data derived from use of acetylene block methods with those from ^{15}N and ^{13}N enrichments. Acetylene inhibition of nitrous oxide reduction is a quick and reliable procedure for estimation of *in situ* levels of denitrification enzymes in soil slurries, soil cores, and in the field. The initially constant nitrous oxide

Acetylene Block Method for Estimating Soil Denitrification Potential

1. Place 10g soil sample into 60 ml serum bottle
2. Add water or KNO_3 solution to soil
3. Purge and evacuate bottles with Argon or Argon + 10% acetylene
4. Incubate for appropriate time interval with mixing
5. Assay headspace N_2O with gas chromatographic procedure

Fig. 14.4 Outline of a general procedure using the acetylene block method for quantifying denitrification potential in soil samples incubated in the laboratory. Samples receiving nitrate amendment provide estimates of maximum denitrification potential of the sample. Incubation times may vary from one hour to several days, depending upon the experimental objectives.

evolution rate in soil slurries corresponds to preexisting denitrifier enzyme in the soil sample (Smith and Tiedje, 1979a).

A problem of providing an inhibitory level of acetylene throughout the soil sample is most frequently encountered in field sites and, to some degree, soil column experiments. In the field situation and with soil columns, care must be taken to assure complete admixture of the acetylene gas with the soil atmosphere. This limitation to the assay procedure is usually not important when using soil slurries since the suspensions are agitated at sufficient rates to guarantee distribution of the inhibitory gas throughout the mixture.

Acetylene is toxic to ammonium oxidizers (e.g., Berg et al., 1982). Thus, in long-term field or laboratory studies where indigenous nitrification is relied upon for the nitrate source for denitrifiers, denitrification would be reduced or precluded. Implementation of this procedure in the field or with long-term incubated soil columns requires either duplication of sampling sites so that assays would not be repeated at the same specific site receiving acetylene or sufficient time between assays to allow for recovery of the nitrifier population to occur.

Additional difficulties may be encountered in the field should the acetylene itself stimulate carbon mineralization, thereby augmenting nitrate reduction (Haider et al., 1983). Nitrate amendment to soil samples may stimulate both aerobic and anaerobic acetylene oxidation. Based on this observation, Haider et al. (1983) conclude that repeated application of acetylene to the same field plot should be avoided.

It must be noted that these complications are not universally associated with field implementation of the acetylene inhibition procedure. Ryden and Dawson (1982) found no difficult with its repeated use on grassland soils. In conclusion, these data further underscore the necessity of conducting appropriate controls to assure that the impact of acetylene on the soil system is direct alteration of nitrous oxide evolution through inhibition of nitrous oxide reductase, and not indirectly through alteration of the activity of microbial populations dissociated from nitrate reduction.

Two additional factors to be considered in regards to the use of acetylene to estimate denitrification potential in soil are the capacity to estimate nitrous oxide production from both autotrophic nitrifiers and denitrifiers in the same soil sample, and the existence of potential difficulties due to enhanced nitric oxide decomposition in soil receiving acetylene. Nitrifiers and denitrifiers are inhibited by different concentrations of acetylene. Therefore variation in both concentration of acetylene and duration of exposure (Kester et al., 1996) or concentration (Inbushi et al., 1996; Webster and Hopkins, 1996) have proven valuable in estimating the quantity of nitrous oxide produced by each source in soils and sediments. (Alternatively, Stevens et al., 1997, used differential ^{15}N-labeling of nitrate and ammonium pools to distinguish the two nitrous oxide sources.)

An additional concern in situations where nitric oxide is a by-product of denitrification is the impact of acetylene on the decomposition of the nitric oxide. Increasing acetylene levels in soil have been shown to increase the decomposition of nitric oxide by a factor of as much as 5- to 557-fold (Bollmann and Conrad, 1997a). The consequence of the augmented nitric oxide decomposition is an underestimation of the associated denitrification rate (Bollmann and Conrad, 1997b)

14.6 Environmental Factors Controlling Denitrification Rates

The primary factors that have been shown to control denitrification processes in the field are (1) the nature and amount of organic matter available as energy sources to the denitrifiers, (2) the soil nitrate concentration, (3) the aeration/ moisture status of the soil, (4) soil pH, and (5) soil temperature. Each soil property and examples of their effect in native soil systems will be evaluated herein.

14.6.1 Nature and Amount of Organic Matter

Most denitrifiers in soil are heterotrophs. Thus, the primary energy source for the process is plant biomass in most soil ecosystems. Although colloidal organic matter is a primary soil component, amendment of soil with metabolizable carbon generally does stimulate denitrification (e.g., de Catanzaro and Beauchamp, 1985; Stanford et al., 1975b). For such experiments, saturating concentrations of nitrate are usually amended to soil slurries with the carbon and energy source (frequently glucose) varied. The increase in denitrification rate due to the augmented energy source generally resembles Michaelis kinetics (e.g., Bowman and Focht, 1974; Reddy et al., 1982). Bowman and Focht (1974) found a Michaelis constant of 500 μg glucose ml^{-1} when nitrate reduction was assessed in a Coachella fine sand. Further evidence for carbon limitations of denitrification rates is derived from the observation that denitrification rates generally correlate with total organic carbon and available carbon levels in soil. Available carbon has commonly been estimated by measurement of biological oxygen demand of soil water extracts (e.g., Beauchamp et al., 1980; Burford and Bremner, 1975; Katz et al., 1985; Stanford et al., 1975c). There are rare situations in field sites where carbon does not limit denitrification. This exception is usually associated with mineral soils already receiving high organic matter inputs (e.g., sludge-amended soils).

The generally observed correlation of denitrification rates with available organic matter leads to the prediction that treatment of soil to increase availability of native soil organic matter should result in an increase in the denitrification rate. Therefore, soil manipulations that alter the accessibility of indigenous organic matter to microbes can also change the indigenous denitrification rate measured. Freezing (e.g., McGarity, 1962) and air drying (e.g., Patten et al., 1980) of soil samples can augment the denitrification rate detected with the soil samples. Thus, even with the most careful handling of field soil samples, the possibility that the laboratory-derived results are an overestimation of field denitrification rates must be considered.

An interesting possibility based on the observed proportionality of denitrification rates to available fixed carbon substrates is that the denitrification rate may be elevated in the rhizosphere compared to nonrhizosphere soil. Root exudates can be predicted to provide the energy source for the denitrifiers, and anoxic microsites would be anticipated to occur in the vicinity of the roots where denitrifiers can function. Addition of macerated roots to soils samples as well as invasion of soil by root tissue has been shown to stimulate denitrification (e.g., Garcia, 1975; Prade and Trolldenier, 1988). Others, (Haider et al., 1985; 1987; Smith and Tiedje, 1979b) did not observe a rhizosphere effect on denitrification. Since the denitrifiers and the plants could be competing for the same nitrate pool, it could be hypothesized that the reason for the disparate observations is that different nitrate concentrations were used in the studies. But, Smith and Tiedje (1979b) found that variation of soil nitrate concentrations did not alter the denitrification rate in their rhizosphere samples. No rhizosphere effect on denitrification rate was found with high nitrate concentrations. A reduction of denitrification occurred in the presence of low nitrate. Since the reduced nitrate concentrations used in their studies better reflected the generally encountered field situations, it could be predicted that denitrification rates in the rhizosphere should be reduced over those of nonplant-affected soils. These data exemplify the complexity of the interactions of physical, chemical, and biological factors in controlling denitrification in the rhizosphere. Although there is a clear impact of fixed carbon concentration on activity of denitrifiers, their function is also delimited by nitrate concentrations (for which they would be competing with plants as well as other bacterial populations), and anoxic microsites.

14.6.2 Nitrate Concentration

Since denitrification is an enzymatically catalyzed process, the reaction rate of the process is anticipated to follow Michaelis-Menten kinetics. That is, the rate of nitrate reduction to dinitrogen and nitrous oxide should increase until a saturating concentration is reached. Once enzyme-saturating concentrations of the substrate have been reached, two alternatives are generally observed. There can be no effect of the augmented substrate levels, or the substances may become inhibitory. The latter situation is generally detected in study of denitrification. That is, high nitrate concentrations tend to be inhibitory to nitrogen oxide reductases.

Nitrate reduction in soil is generally found to be either a zero or first order process. For example, McGarity (1961) found denitrification rate to be independent of nitrate concentration (60 to 472 μg g^{-1}) in some South Australian soils. Bowman and Focht (1974) observed first order kinetics with a Coachella fine sand. The Michaelis constant for the latter study was 170 μg nitrate ml^{-1}. Murray et al. (1989) measured Michaelis constants for mixed denitrifier populations from agricultural soils of 1.8 to 13.7 μM nitrate. Combinations of kinetic relationships may also be observed. Reddy et al. (1978) measured both first and zero order reaction kinetics with flooded soils. In their study, nitrate disappearance from the waters above the soil was zero order, whereas within the soil matrix, nitrate consumption followed first order kinetics. In this situation, a nonbiological parameter entered into the reaction kinetics — diffusion of nitrate from the surface waters into the sediment soils. Diffusion limited the quantities of nitrate available to the soil microbes. Therefore, the available nitrate was reduced by the microbial community as quickly as it entered into their microsite. Myrold and Tiedje (1985) suggest process rate limitations due to diffusion of nitrate to denitrifiers in large soil aggregates.

At high concentrations, nitrate inhibits nitrous oxide reductase. This inhibition varies with soil pH and history of the soil. Greater effect of high nitrate is detected at low pH, whereas extended flooding of the soil relieves the inhibitory effect of nitrate on the reductase.

14.6.3 Aeration/Moisture

As was noted above when the biochemistry of denitrification was analyzed, oxygen represses the synthesis of the reductases associated with denitrification. As soil oxygen tensions are reduced, reduction of nitrate to nitrous oxide and dinitrogen increases (e.g., Allison et al., 1960; Cady and Bartholomew, 1961). Because of the slow rate of diffusion of molecular oxygen in water compared to air, this effect of aeration on denitrification rates in soils is highly related to soil moisture. In general, denitrification is not detected below 60 percent water-holding capacity. Above this value, denitrification rate generally correlates with soil moisture.

The field implication of this observation is that it can be anticipated that the majority of nitrate is denitrified in soil under flooded conditions. Generally low denitrification rates are observed continuously in a drained soil system, but maximal activity is associated with increases in moisture due to rainfall (Sexstone et al., 1985a). In the field, as soil moisture increases, losses of nitrate through denitrifier activity increases (e.g., Kroeckel and Stolp, 1988; Pilot and Patrick, 1972). This sensitivity of denitrifiers to increasing soil moisture explains a significant amount of the augmented loss of nitrate from no-till soils due to denitrifier activity compared to that occurring in conventionally tilled agricultural systems (Aulakh et al., 1984a; 1984b; Rice and Smith, 1982). This increase in denitrification results from augmentation of the number of anaerobic microsites in the soil. That is, the low amount of denitrification occurring in drained soils is occurring in anaerobic microsites and is *not* the result of activity of molecular oxygen resistant denitrifiers. No-till agricultural practices result in formation of more anaerobic centers within large soil aggregates where denitrification can occur (Hojberg et al., 1994; Sexstone et al., 1985b). Anaerobiosis is not the sole factor controlling the denitrification rate in the aggregate in that this rate does not correlate with the size of the anaerobic zone. Other contributing parameters are nitrate diffusion rate into the anaerobic microsites and the rate of molecular oxygen consumption by the aerobic organisms growing on the surface of the aggregate.

14.6.4 pH

Although denitrification occurs within the pH range of approximately 3.9 to 9.0, maximum nitrogen oxide reduction occurs from pH 7.0 to 8.0. The rate decreases as the pH is lowered (e.g., Bremner and Shaw, 1958b; Focht, 1974; Koshkinen and Keeney, 1982; Waring and Gilliam, 1983). Acidophilic or acid-tolerant denitrifier populations are apparently selected in soils with histories of low pH. Parkin et al. (1985b) noted two distinctly different pH optima with soils of pH 3.9 and 6.3, which approximated the native pH values of the soils. It should be noted that the acidic soil used in the latter study had a 20-year history of low pH, which would have provided more than adequate time for development of an acid-resistant soil microbial population.

Because of the occurrence of chemical denitrification in extremely acidic soils (e.g., Bollag et al., 1973), assessment of denitrification in such soils is difficult. Fortunately, the major products of the chemical and biological processes differ. Nitric oxide and nitrogen dioxide predominate from chemical reduction of nitrite, whereas nitrous oxide and dinitrogen are biologically produced (Bollag et al., 1973). Unfortunately, dinitrogen and nitrous oxide are also reported among the products of chemodenitrification (e.g., Bulla et al., 1970; Reuss and Smith, 1965). When the two processes are separated, significant biological reduction of nitrous oxides can be shown to occur. Muller et al. (1980) found that biological denitrification rates varied from 0.12 to 53.8 μg day^{-1} in a variety of low pH (pH $\geqslant$ 3.6) spodosols and peats collected from southern Finland. Complete inhibition of the reduction of nitrous oxide to dinitrogen was found in 99.3 percent of these soil samples. Similarly, Gilliam and Gambrell (1978) found that acidity (pH values as low as 4.5) was not a serious limitation to denitrification in Atlantic coastal plain soils (USA). As with the previous study, nitrous oxide was a major product of denitrification in these soils (Weier and Gilliam, 1986). Similarly, in a study of a silt loam soil with pH varying from 4.6 to 6.9, nitrous oxide composed 83 percent of the gaseous nitrogen products of denitrification at pH 4.6 and 5.4. At pH 6.9, dinitrogen was the predominant product.

Frequently, several soil physical or chemical properties can interact to retard denitrification under acidic conditions. For example, Dubey and Fox (1974) found that low pH combined with low soil organic matter precluded denitrification in humid tropical soils of Puerto Rico. Similarly, George and Antoine (1982) noted that the pH optimum varied with temperature in their studies of a salt marsh soil. In their study, nitrous oxide was produced only at low pH and nitrate concentrations.

14.6.5 Temperature

The minimum temperature for denitrification is generally associated with the occurrence of free water. Smid and Beauchamp (1976) extrapolated their data from study of denitrification in the Ap horizon of a Huron clay loam to suggest denitrification at or near 0°C. Incubation of the same soil under anaerobic conditions resulted in a complete inhibition of denitrification at 5°C with a gradual increase in the rate to 30°C (Bailey and Beauchamp, 1973). Jacobson and Alexander (1980) in a study of two soils maintained in an anaerobic atmosphere found no nitrate-reducing activity at 1°C. Nitrate was slowly reduced at 7°C. As with the previous studies, the rate increased with increasing soil temperature. These data suggest a minimum temperature for denitrification approaching the freezing point for water with an apparent impact of the presence of molecular oxygen on the limiting temperature. The maximum temperature is approximately that value limiting biological activity in soil in general, 75°C. The optimum temperature (60 to 70°C) is well above the normally occurring soil temperature range. In the range of increasing temperature, the nitrate-reduction rate increases at a rate approximating that of general soil biological processes. That is, a Q_{10} of approximately 2 describes the relationship between increasing temperature and reaction rate (e.g., Stanford et al., 1975a).

These data indicate that significant denitrification can be anticipated to occur in any soil ecosystem at temperatures where general biological activity can be anticipated to occur. The Q_{10} value of about 2 indicates that the process is reasonably sensitive to slight fluctuations in soil temperature. Thus, not only should a seasonal variation in this process be anticipated, but a daily impact should be anticipated. That is, losses of fixed nitrogen as nitrous oxide and dinitrogen will necessarily vary with daily heating and cooling of the upper portions of the soil profile, so time of day is a consideration when developing protocols for field studies.

14.6.6 Interaction of Limitations to Denitrification in Soil Systems

A strict consideration of each of the soil properties listed above and their impact on denitrification processes provides a limited view of the variation of nitrate reduction in field sites. In reality, each soil property is sufficiently variable across the soil landscape to provide significant landscape-scale variation in denitrification rates. Landscape analysis should be incorporated into models of soil fixed nitrogen losses through denitrification since topography has a major impact on *in situ* denitrification rates (Ball et al., 1997; Pennock et al., 1992). In the Pennock et al. (1992) study, when the site was considered as a whole, variables that were most influential on nitrogen losses were volumetric water content and soil redox potential. In the level portions of the site, volumetric water content was most highly correlated with denitrification, whereas on the shoulder and foot slopes, respiration and bulk density were the most influential. Parkin (1987) found that "hot spots" of denitrification activity were associated with pockets of particulate organic matter. Parsons et al. (1991) found coefficients of variation between 74 and 268 percent for spatial variability of nitrogen gas loss from agricultural soils of central Kentucky (USA). Increases in nitrate-reduction activity related to increases in soil moisture and soil respiration. These data support the conclusion that for a complete understanding of denitrification kinetics in a soil system to be developed, a study of the heterogeneity of the system must be developed, particularly with regard to soil moisture, organic matter, and most likely to some degree nitrate distribution.

14.7 Concluding Comments

An understanding of the nuances of the impact of soil properties on biological denitrification and management of this process to regulate soil nitrate loadings is essential for optimization of agricultural production as well as for minimization of the potential for soil and water pollution. This terminal stage in the nitrogen cycle (return of fixed nitrogen to atmospheric reservoirs) is catalyzed by a diverse group of soil bacteria that are capable of using light, mineral (inorganic), or organic compounds as energy sources. The only readily discernible commonalties among denitrifiers are (1) they are solely bacteria, (2) they are capable of reducing nitrogen oxides in the absence of free oxygen, and (3) these nitrogen oxides serve as terminal electron acceptors for bacterial respiration. Thus, the primary determinants of denitrification rates in the field are the availability of energy sources, nitrate, and the absence of molecular oxygen from the microenvironment of the bacteria. The diversity of the organisms involved leads to the prediction that if life is possible in the system, a denitrifier should exist that is capable of functioning therein. This adaptability and ubiquity of denitrifiers supports the conclusion that denitrification is a process amenable to management for control of soil nitrate loadings.

References

Allison, F. E., J. N. Carter, and L. D. Sterling. 1960. The effect of partial pressure of oxygen on denitrification in soil. Soil Sci. Soc. Am. Proc. 24:283–285.

Aulakh, M. S., D. A. Rennie, and E. A. Paul. 1984a. Gaseous nitrogen losses from soils under zero-till as compared with conventional-till management systems. J. Environ. Qual. 13:130–136.

Aulakh, M. S., D. A. Rennie, and E. A. Paul. 1984b. The influence of plant residues on denitrification rates in conventional and zero tilled soils. Soil Sci. Soc. Am. J. 48:790–794.

Aulakh, M. S., J. W. Doran, and A. R. Mosier. 1991. Field evaluation of four methods for measuring denitrification. Soil Sci. Soc. Am. J. 55:1332–1338.

Bailey, L. D., and E. G. Beauchamp. 1973. Effects of temperature on NO_3^- and NO_2^-

reduction, nitrogenous gas production, and redox potential in a saturated soil. Can. J. Soil Sci. 53:213–218.

Ball, B. C., G. W. Horgan, H. Clayton, and J. P. Parker. 1997. Spatial variability of nitrous oxide fluxes and controlling soil and topographic properties. J. Environ. Qual. 26:1399–1409.

Ballard, A. L., and S. J. Ferguson. 1988. Respiratory nitrate reductase from *Paracoccus denitrificans*. Evidence for two *b*-type haems in the γ subunit and properties of a water-soluble active enzyme containing α and β subunits. Eur. J. Biochem. 174:207–212.

Bazylinski, D. A., E. Palome, N. A. Blakemore, and R. P. Blakemore. 1986. Denitrification by *Chromobacterium violacium*. Appl. Environ. Microbiol. 52:696–699.

Beauchamp, E. G., C. Gale, and J. C. Yoemans. 1980. Organic matter availability for denitrification in soils of different textures and drainage classes. Commun. Soil Sci. Plant Anal. 11:1221–1233.

Benckiser, G., R. Eilts, A. Linn, H. J. Lorch, E. Sumer, A. Weiske, and F. Wenzhofer. 1996. N_2O emissions from different cropping systems and from aerated, nitrifying and denitrifying tanks of a municipal waste water treatment plant. Biol. Fertil. Soils 23:257–265.

Berg, P., L. Klemedtsson, and T. Rosswall. 1982. Inhibitory effect of low partial pressures of acetylene on nitrification. Soil Biol. Biochem. 14:301–303.

Betlach, M. R., and J. M. Tiedje. 1981. Kinetic explanation for accumulation of nitrite, nitric oxide, and nitrous oxide during bacterial denitrification. Appl. Environ. Microbiol. 42:1074–1084.

Blackmer, A. M., and J. M. Bremner. 1977. Nitrogen isotope discrimination in denitrification of nitrate in soils. Soil Biol. Biochem. 9:73–77.

Blackmer, A. M., and J. M. Bremner. 1978. Inhibitory effect of nitrate on reduction of N_2O to N_2 by soil microorganisms. Soil Biol. Biochem. 10:187–191.

Blackmer, A. M., and J. M. Bremner. 1979. Stimulatory effect of nitrate on reduction of N_2O to N_2 by soil microorganisms. Soil Biol. Biochem. 11:313–315.

Blackmer, A. M., J. M. Bremner, and E. L. Schmidt. 1980. Production of nitrous oxide by ammonia-oxidizing chemoautotrophic microorganisms in soil. Appl. Environ. Microbiol. 40:1060–1066.

Blackmer, A. M., S. G. Robbins, and J. M. Bremner. 1982. Diurnal variability in rate of emission of nitrous oxide from soils. Soil Sci. Soc. Am. J. 46:937–942.

Bleakley, B. H., and J. M. Tiedje. 1982. Nitrous oxide production by organisms other than nitrifiers or denitrifiers. Appl. Environ. Microbiol. 44:1342–1348.

Boast, C. W., R. L. Mulvaney, and P. Baveye. 1988. Evaluation of nitrogen-is tracer techniques for direct measurement of denitrification in soil: I, theory, Soil Sci. Soc. Am. J. 52:1317–1322.

Bolgiano, B., L. Smith, and H. C. Davies. 1989. Electron transport reactions in a cytochrome c-deficient mutant of *Paracoccus denitrificans*. Biochim. Biophys. Acta 973:227–234.

Bollag, J.-M., S. Drzymala, and L. T. Kardos. 1973. Biological versus chemical nitrite decomposition in soil. Soil Sci. 116:44–50.

Bollag, J.-M., and G. Tung. 1972. Nitrous oxide release by soil fungi. Soil Biol. Biochem. 4:271–276.

Bollmann, A., and R. Conrad. 1997a. Enhancement by acetylene of the decomposition of nitric oxide in soil. Soil Biol. Biochem. 29:1057–1066.

Bollmann, A., and R. Conrad. 1997b. Acetylene blockage technique leads to underestimation of denitrification rates in oxic soils due to scavenging of intermediate nitric oxide. Soil Biol. Biochem. 29:1067–1077.

Bossert, I. D., and L. Y. Young. 1986. Anaerobic oxidation of p-cresol by a denitrifying bacterium. Appl. Environ. Microbiol. 52:1117–1122.

Bowman, R. A., and D. D. Focht. 1974. The influence of glucose and nitrate concentrations upon denitrification rates in sandy soils. Soil Biol. Biochem. 6:297–301.

Braun, K., and D. T. Gibson. 1984. Anaerobic degradation of 2–aminobenzoate (anthranilic acid) by denitrifying bacteria. Appl. Environ. Microbiol. 48:102–107.

Bremner, J. M. 1997. Sources of nitrous oxide in soils. Nutr. Cycl Agroecosyst. 49:7–16.

Bremner, J. M., and K. Shaw. 1958a. Denitrification in soil. I. Methods of investigation. J. Agr. Sci. 51:22–39.

Bremner, J. M., and K. Shaw. 1958b. Denitrification in soil. II. Factors affecting denitrification. J. Agr. Sci. 51:40–52.

Bulla, L. A., Jr., C. M. Gilmour, and W. B. Bollen. 1970. Non-biological reduction of nitrite in soil. Nature (London) 225:664.

Buresh, R. J., and W. H. Patrick, Jr. 1978. Nitrate reduction to ammonium in anaerobic soil. Soil Sci. Soc. Am. J. 42:913–918.

Burford, J. R., and J. M. Bremner. 1975. Relationships between the denitrification capacities of soil and total, water-soluble and readily decomposible soil organic matter. Soil Biol. Biochem. 7:389–394.

Cady, F. B., and W. V. Bartholomew. 1960. Sequential products of anaerobic denitrification in Norfolk soil material. Soil Sci. Soc. Am. Proc. 24:477–482.

Cady, F. B., and W. V. Bartholomew. 1961. Influences of low pO_2 on denitrification processes and products. Soil Sci. Soc. Am. Proc. 25:362–365.

Carlson, C. A., and J. L. Ingraham. 1983. Comparison of denitrification by *Pseudomonas stutzeri*, *Pseudomonas aeruginosa*, and *Paracoccus denitrificans*. Appl. Environ. Microbiol. 45:1247–1253.

Caskey, W. H., and J. M. Tiedje. 1980. The reduction of nitrate to ammonium by a *Clostridium* sp. isolated from soil. J. Gen. Microbiol. 119:217–223.

Chan, Y.-K., L. R. Barran, and E. S. P. Bromfield. 1989. Denitrification activity of phage types representative of two populations of indigenous *Rhizobium meliloti*. Can. J. Microbiol. 35:737–740.

Chauret, C., W. L. Barraquio, and R. Knowles. 1992. Molybdenum incorporation in denitrifying *Azospirillum brasilense* Sp7. Can. J. Microbiol. 38:1042–1047.

Cooper, G. S., and R. L. Smith. 1963. Sequence of products formed during denitrification in some diverse western soils. Soil Sci. Soc. Am. Proc. 27:659–662.

Coyne, M. S., and D. D. Focht. 1987. Nitrous oxide reduction in nodules: Denitrification or N_2 fixation? Appl. Environ. Microbiol. 53:1168–1170.

Daniel, R. M., I. M. Smith, J. A. D. Phillip, H. D. Ratcliffe, J. W. Drozd., and T. A. Bull. 1980. Anaerobic growth and denitrification by *Rhizobium japonicum* and other rhizobia. J. Gen. Microbiol. 120:517–521.

Davidson, E. A., and W. T. Swank. 1986. Environmental parameters regulating gaseous nitrogen losses from two forested ecosystems via nitrification and denitrification. Appl. Environ. Microbiol. 52:1287–1292.

Davidson, E. A., W. T. Swank, and T. O. Perry. 1986. Distinguishing between nitrification and denitrification as sources of gaseous nitrogen production in soil. Appl. Environ. Microbiol. 52:1280-1286.

de Catanzaro, J. B., and E. G. Beauchamp. 1985. The effect of some carbon substrates on denitrification rates and carbon utilization in soil. Biol. Fert. Soil 1:183–187.

Dubey, H. D., and R. H. Fox. 1974. Denitrification losses from humid tropical soils of Puerto Rico. Soil Sci. Soc. Am. Proc. 38:917–920.

Dunstan, R. H., B. C. Kelley, and D. J. D. Nicholas. 1982. Fixation of dinitrogen derived from denitrification of nitrate in a photosynthetic bacterium, *Rhodopseudomonas sphaeroides* forma sp. *denitrificans*. J. Bacteriol. 150:100–104.

Evans, P. J., W. Ling, N. J. Palleroni, and L. Y. Young. 1992. Quantification of identification by strain T1 during anaerobic degradation of toluene. Appl. Microbiol. Biotechnol. 37:136–140.

Evans, W. C., and G. Fuchs. 1988. Anaerobic degradation of aromatic compounds. Ann. Rev. Microbiol. 42:289–317.

Fazzalari, E., A. Mariotti, and J. C. Germon. 1990. Dissimilatory ammonia reduction vs. denitrification in vitro and in inoculated agricultural soil samples. Can. J. Microbiol. 36:786–793.

Firestone, M. K., R. B. Firestone, and J. M. Tiedje. 1980. Nitrous oxide from soil denitrification: Factors controlling its biological production. Science 208:749–751.

Firestone, M. K., M. S. Smith, R. B. Firestone, and J. M. Tiedje. 1979. The influence of nitrate, nitrite, and oxygen on the composition of the gaseous products of denitrification in soil. Soil Sci. Soc. Am. J. 43:1140–1144.

Focht, D. D. 1974. The effect of temperature, pH, and aeration on the production of nitrous oxide and gaseous nitrogen — a zero order kinetic model. Soil Sci. 118:173–179.

Focht, D. D., and H. Joseph. 1973. An improved method for the enumeration of denitrifying bacteria. Soil Sci. Soc. Am. Proc. 37:698–699.

Focht, D. D., and L. H. Stolzy. 1978. Long-term denitrification studies in soils fertilized with $(^{15}NH_4)_2SO_4$. Soil Sci. Soc. Am. J. 42:894–898.

Freney, J. R., O. T. Denmead, and J. R. Simpson. 1979. Nitrous oxide emission from soils at low moisture contents. Soil Biol. Biochem. 11:167–173.

Gamble, T. M., M. R. Betlach, and J. M. Tiedje. 1977. Numerically dominant denitrifying bacteria from world soils. Appl. Environ. Microbiol. 33:926–939.

Garber, E. A. E., and T. C. Hollocher. 1981. ^{15}N tracer studies on the role of NO in denitrification. J. Biol. Chem. 256:5459–5465.

Garber, E. A. E., and T. C. Hollocher. 1982. ^{15}N, ^{18}O tracer studies on the activation of nitrite by denitrifying bacteria. J. Biol. Chem. 257:8091–8097.

Garcia, J.-L. 1975. Effet rhizosphere du riz sur la denitrification. Soil Biol. Biochem. 7:139–141.

George, U. S., and A. D. Antoine. 1982. Denitrification potential of a salt marsh soil: Effect of temperature, pH and substrate concentration. Soil Biol. Biochem. 14:117–125.

Gilliam, J. W., and R. P. Gambrell. 1978. Temperature and pH as limiting factors in loss of nitrate from saturated Atlantic coastal plain soils. J. Environ. Qual. 7:526–532.

Goodroad, L. L., and D. R. Keeney. 1984. Nitrous oxide emission from forest, marsh, and prairie ecosystems. J. Environ. Qual. 13:448–452.

Goodroad, L. L., D. R. Keeney, and L. A. Peterson. 1984. Nitrous oxide emissions from agricultural soils in Wisconsin. J. Environ. Qual. 13:557–561.

Haider, K., A. R. Mosier, and O. Heinemeyer. 1983. Side effects of acetylene on the conversion of nitrate in soil. Z. Pflanzenernaehr. Bodenk. 146:623–633.

Haider, K., A. Mosier, and O. Heinemeyer. 1985. Phytotron experiments to evaluate the effect of growing plants on denitrification. Soil Sci. Soc. Am. J. 49:636–641.

Haider, K., A. Mosier, and O. Heinemeyer. 1987. The effect of growing plants on denitrification at high soil nitrate concentrations. Soil Sci. Soc. Am. J. 51:97–102.

Heinrich, M., and K. Haselwandter. 1997. Denitrification and gaseous nitrogen losses from an acid spruce forest soil. Soil Biol. Biochem. 29:1529–1537.

Hernandez, D., and J. J. Rowe. 1987. Oxygen regulation of nitrate uptake in denitrifying *Pseudomonas aeruginosa*. Appl. Environ. Microbiol. 53:745–750.

Hochstein, L. I., and G. A. Tomlinson. 1988. The enzymes associated with denitrification. Ann. Rev. Microbiol. 42:231–261.

Hojberg, O., N. P. Revsbech, and J. M. Tiedje. 1994. Denitrification in soil aggregates analyzed with microsensors for nitrous oxide and oxygen. Soil Sci. Soc. Am. J. 58:1691–1698.

Hollocher, T. C., E. Garber, A. J. L. Cooper, and R. E. Reiman. 1980. [13]N, ^{15}N isotope and kinetic evidence against hyponitrite as an intermediate in denitrification. J. Biol. Chem. 255:5027–5030.

Hollocher, T. C., and J. K. Kristjánsson, 1992. Thermophilic denitrifying bacteria: A survey of hot springs of Southwestern Iceland. FEMS Microbiol. Ecol. 101:113–119.

Inubushi, K., H. Naganuma, and S. Kitahara. 1996. Contribution of denitrification and autotrophic and heterotrophic nitrification to nitrous oxide production in andosols. Biol. Fertil. Soils 23:292–298.

Jacobson, S. N., and M. Alexander. 1980. Nitrate loss from soil in relation to temperature, carbon source and denitrifier populations. Soil Biol. Biochem. 12:501–505.

Katz, R., J. Hagin, and L. T. Kurtz. 1985. Participation of soluble and oxidizable soil organic compounds in denitrification. Biol. Fert. Soils 1:209–213.

Kester, R. A., W. Deboer, and H. J. Laanbroek. 1996. Short exposure to acetylene to distinguish between nitrifier and denitrifier nitrous oxide production in soil and sediment samples. FEMS Microbiol. Ecol. 20:111–120.

Knowles, R. 1981. Denitrification. *In* E. A. Paul and J. N. Ladd (eds.) Soil Biochemistry 5:323–369. Dekker, NY.

Knowles, R. 1982. Denitrification. Microbiol. Rev. 46:43–70.

Koike, I., and A. Hattori. 1978. Denitrification and ammonia formation in anaerobic coastal sediments. Appl. Environ. Microbiol 35:278–282.

Koskinen, W. C., and D. R. Keeney. 1982. Effect of pH on the rate of gaseous products of denitrification in a silt loam soil. Soil Sci. Soc. Am. J. 46:1165–1167.

Kroeckel, L., and H. Stolp. 1988. Influence of oxygen on denitrification and aerobic respiration in soil. Biol. Fert. Soils 1:189–193.

Letey, J., N. Valoras, D. D. Focht, and J. C. Ryden. 1981. Nitrous oxide production and reduction during denitrification as affected by redox potential. Soil Sci. Soc. Am. J. 45:727–730.

Linne von Berg, K.-H., and H. Bothe. 1992. The distribution of denitrifying bacteria in soils monitored by DNA-probing. FEMS Microbiol. Ecol. 86:331–340.

Loro, P. J., D. W. Bergstrom, and E. G. Beauchamp. 1997. Intensity and duration of denitrification following application of manure and fertilizer to soil. J. Environ. Qual. 26:706–713.

McGarity, J. W. 1961. Denitrification studies on some South Australian soils. Plant Soil 14:1–21.

McGarity, J. W. 1962. Effect of freezing of soil on denitrification. Nature (London) 196:1342–1343.

McTaggart, I. P., H. Clayton, J. Parker, L. Swan, and K. A. Smith. 1997. Nitrous oxide emissions from grassland and spring barley, following N fertiliser application with and without nitrification inhibitors. Biol. Fertil. Soils 25:261–268.

Mosier, A. R., and G. L. Hutchinson. 1981. Nitrous oxide emissions from cropped fields. J. Environ. Qual. 10:169–173.

Mosier, A. R., W. D. Guenzi, and E. Z. Schweizer. 1986. Field denitrification estimates by nitrogen-15 and acetylene inhibition techniques. Soil Sci. Soc. Am. J. 50:831–833.

Mosier, A. R., M. Stillwell, W. J. Parton, and R. G. Woodmansee. 1981. Nitrous oxide emissions from a native shortgrass prairie. Soil Sci. Soc. Am. J. 45:617–619.

Muller, M. M., V. Sundman, and J. Skujins. 1980. Denitrification in low pH spodosols and peats determined with the acetylene inhibition method. Appl. Environ. Microbiol. 40:235–239.

Mulvaney, R. L., S. A. Khan, and C. S. Mulvaney. 1997. Nitrogen fertilizers promote denitrification. Biol. Fertil. Soils 24:211–220.

Murray, R. E., L. L. Parsons, and M. S. Smith. 1989. Kinetics of nitrate utilization by mixed populations of denitrifying bacteria. Appl. Environ. Microbiol. 55:717–721.

Myrold, D. D., and J. M. Tiedje. 1985. Diffusional constraints on denitrification in soil. Soil Sci. Soc. Am. J. 49:651–657.

Neyra, C. A., J. Döbereiner, R. Lalande, and R. Knowles. 1977. Denitrification by N_2-fixing *Spirillum lipoferum*. Can. J. Microbiol. 23:300.

Nielsen, T.H., L. P. Nielsen, and N. P. Revsbech. 1996. Nitrification and coupled nitrification-denitrification associated with a soil-manure interface. Soil Sci. Soc. Am. J. 60:1829–1840.

Parkin, T. B. 1987. Soil microsites as a source of denitrification variability. Soil Sci. Soc. Am. J. 51:1194–1199.

Parkin, T. B., A. J. Sextone, and J. M. Tiedje. 1985a. Comparison of field denitrification rates determined by acetylene-based soil core and nitrogen-15 methods. Soil Sci. Soc. Am. J. 49:94–99.

Parkin, T. B., A. J. Sexstone, and J. M. Tiedje. 1985b. Adaptation of denitrifying populations to low soil pH. Appl. Environ. Microbiol. 49:1053–1056.

Parsons, L. L., R. E. Murrray, and M. S. Smith. 1991. Soil denitrification dynamics: Spatial and temporal variations of enzyme activity, populations, and nitrogen gas loss. Soil Sci. Soc. Am. J. 55:90–95.

Patten, D. K., J. M. Bremner, and A. M. Blackmer. 1980. Effects of drying and air-dry storage of soils on their capacity for denitrification of nitrate. Soil Sci. Soc. Am. J. 44:67–70.

Pennock, D. J., C. van Kessell, R. E. Farrell, and R. A. Sutherland. 1992. Landscape-scale variations in denitrification. Soil Sci. Soc. Am. J. 56:770–776.

Pfitzner, J., and H. G. Schlegel. 1973. Denitrification in *Hydrogenomonas eutropha* strain H16. Arch. Mikrobiol. 90:199–211.

Pilot, L., and W. H. Patrick Jr. 1972. Nitrate reduction in soils: Effect of soil moisture tension. Soil Sci. 114:312–316.

Platen, H., and B. Schink. 1989. Anaerobic degradation of acetone and higher ketones via carboxylation by newly isolated denitrifying bacteria. J. Gen. Microbiol. 135:883–891.

Prade, K., and G. Trolldenier. 1988. Effect of wheat roots on denitrification at varying soil air-filled porosity and organic-carbon content. Biol. Fert. Soil 7:1–6.

Prakash, O., and J. C. Sadana. 1973. Metabolism of nitrate in *Achromobacter fischeri*. Can. J. Microbiol. 19:15–25.

Reddy, K. R., W. H. Patrick Jr., and R. E. Phillips. 1978. The role of nitrate diffusion in determining the order and rate of denitrification in flooded soil: I. Experimental results. Soil Sci. Soc. Am. J. 42:268–272.

Reddy, K. R., P. S. C. Rao, and R. E. Jessup. 1982. The effect of carbon mineralization on denitrification kinetics in mineral and organic soils. Soil Sci. Soc. Am. J. 46:62–68.

Regina, K., H. Nykanen, J. Silvola, and P. J. Martikainen. 1996. Fluxes of nitrous oxide from boreal peatlands as affected by peatland type, water table level and nitrification capacity. Biogeochemistry 35:401–418.

Rehr, B., and J.-H. Klemme. 1989. Formate dependent nitrate and nitrite reduction to ammonia by *Citrobacter freundii* and competition with denitrifying bacteria. Antonie Van Leeuwenhoek 56:311–321.

Reuss, J. O., and R. L. Smith. 1965. Chemical reactions of nitrites in acid soils. Soil Sci. Soc. Am. Proc. 29:267–270.

Rice, C. W., and M. S. Smith. 1982. Denitrification in no-till and plowed soils. Soil Sci. Soc. Am. J. 46:1168–1173.

Rolston, D. E., F. E. Broadbent, and D. A. Goldhamer. 1979. Field measurement of denitrification: II. Mass balance and sampling uncertainty. Soil Sci. Soc. Am. J. 43:703–708.

Ryden, J. C., and K. P. Dawson. 1982. Evaluation of the acetylene-inhibition technique for the measurement of denitrification in grassland soils. J. Sci. Food Agric. 33:1197–1296.

Ryden, J. C., and L. J. Lund. 1980. Nitrous oxide evolution from irrigated land. J. Environ. Qual. 9:387–393.

Smid, A. E., and E. G. Beauchamp. 1976. Effects of temperature and organic matter on denitrification in soil. Can. J. Soil Sci. 56:385–391.

Sexstone, A. J., T. B. Parkin, and J. M. Tiedje. 1985a. Temporal response of soil denitrification rates to rainfall and irrigation. Soil Sci. Soc. Am. J. 49:99–103.

Sexstone, A. J., N. P. Revsbech, T. B. Parkin, and J. M. Tiedje. 1985b. Direct measurement of oxygen profiles and denitrification rates in soil aggregates. Soil Sci. Soc. Am. J. 49:645–651.

Smith, C. J., M. F. Wright, and W. H. Patrick. Jr. 1983. The effect of soil redox potential and pH on the reduction and production of nitrous oxide. J. Environ. Qual. 12:186–188.

Smith, M. S. 1982. Dissimilatory reduction of NO_2^- to NH_4^+ and N_2O by a soil *Citrobacter* sp. Appl. Environ. Microbiol. 43:854–860.

Smith, M. S., M. K. Firestone, and J. M. Tiedje. 1978. The acetylene inhibition method for short-term measurement of soil denitrification and its evaluation using nitrogen-13. Soil Sci. Soc. Am. J. 42:611–615.

Smith, M. S., and J. M. Tiedje. 1979a. Phases of denitrification following oxygen depletion in soil. Soil Biol. Biochem. 11:261–267.

Smith, M. S., and J. M. Tiedje. 1979b. The effect of roots on soil denitrification. Soil Sci. Soc. Am. J. 43:951–955.

Smith, M. S., and K. Zimmerman. 1981. Nitrous oxide production by nondenitrifying soil nitrate reducers. Soil Sci. Soc. Am. J. 45:865–871.

Spier, T. W., H. A. Kettles, and R. D. More. 1995a. Aerobic emissions of N_2O and N_2 from soil cores—measurement procedures using N-13-labelled NO_3^- and NH_4^+. Soil Biol. Biochem. 27:1289–1298.

Spier, T. W., H. A. Kettles, and R. D. More. 1995b. Aerobic emissions of N_2O from soil cores—factors influencing production from N-13-labelled NO_3^- and NH_4^+. Soil Biol. Biochem. 27:1299–1306.

Stanford, G., S. Dzienia, and R. A. Vander Pol. 1975a. Effect of temperature on denitrification rate in soils. Soil Sci. Soc. Am. Proc. 39:867–870.

Stanford, G., J. O. Legg, S. Dzienia, and E. C. Simpson. Jr., 1975b. Denitrification and associated nitrogen transformations in soils. Soil Sci. 120:147–152.

Stanford, G., R. A. Vander Pol, and D. Dzienia. 1975c. Denitrification rates in relation to total and extractable soil carbon. Soil Sci. Soc. Am. Proc. 39:284–289.

Stevens, R. J., R. J. Laughlin, L. C. Burns, J. R. M. Arah, and R. C. Hood. 1997. Measuring the contributions of nitrification and denitrification to the flux of nitrous oxide form soil. Soil Biol. Biochem. 29:139–151.

Terry, R. E., and R. L. Tate III. 1980. The effect of nitrate on nitrous oxide reduction in organic soils and sediments. Soil Sci. Soc. Am. J. 44:744–746.

Terry, R. E., R. L. Tate III, and J. M. Duxbury. 1981a. Nitrous oxide emissions from drained, cultivated organic soils of South Florida. Air Pollut. Contr. Assoc. J. 31:1173–1176.

Terry, R. E., R. L. Tate III, and J. M. Duxbury. 1981b. The effect of flooding on nitrous oxide emissions from an organic soil. Soil Sci. 132:228–232.

Tiedje, J. M., R. B. Firestone, M. K. Firestone, M. R. Betlach, M. S. Smith, and W. H. Caskey. 1979. Methods for the production and use of nitrogen-13 in studies of denitrification. Soil Sci. Soc. Am. J. 43:709–715.

van Berkum, P., and H. H. Keyser. 1985. Anaerobic growth and denitrification among different serogroups of soybean rhizobia. Appl. Environ. Microbiol. 49:772–777.

Vanden Heuvel, R. M., R. L. Mulvaney, and R. G. Hoeft. 1988. Evaluation of nitrogen-15 tracer techniques for direct measurement of denitrification in soil: II. Simulation studies. Soil Sci. Soc. Am. J. 52:1322–1326.

Volz, M. G. 1977. Denitrifying bacteria can be enumerated in nitrite both. Soil Sci. Soc. Am. J. 41:549–551.

Waring, S. A., and J. W. Gilliam. 1983. The effect of acidity on nitrate reduction and denitrification in lower coastal plain soils. Soil Sci. Soc. Am. J. 47:246–251.

Webster, E. A., and D. W. Hopkins. 1996. Contributions from different microbial processes to N_2O emission from soil under different moisture regimes. Biol. Fertil. Soils 22:331–335.

Weier, K. L., and J. W. Gilliam. 1986. Effect of acidity on denitrification and nitrous oxide evolution from Atlantic coastal plain soils. Soil Sci. Soc. Am. J. 50:1202–1205.

Ye, R. W., M. R. Fries, S. G. Bezborodnikov, B. A. Averill, and J. M. Tiedje. 1993. Characterization of the structural gene encoding a copper-containing nitrite reductase and homology of this gene to DNA of other denitrifiers. Appl. Environ. Microbiol. 59:250–254.

Yoshinari, T., R. Hynes, and R. Knowles. 1977. Acetylene inhibition of nitrous oxide reduction and measurement of denitrification and nitrogen fixation in soil. Soil Biol. Biochem. 9:177–183.

Yoshinari, T., and R. Knowles. 1976. Acetylene inhibition of nitrous oxide reduction by denitrifying bacteria. Biochem. Biophys. Res. Commun. 69:705–710.

Zoblotowicz, R. M., D. L. Eskew, and D. D. Focht. 1978. Denitrification in *Rhizobium*. Can. J. Microbiol. 24:757–760.

C h a p t e r 15

Sulfur, Phosphorus,
and
Mineral Cycles

Biogeochemical cycles can easily be grouped into two categories based on the significance of soil minerals to nutrient availability. With the carbon and nitrogen cycles, the soil-based portions of the cycles are strongly biologically based. Once carbon dioxide and dinitrogen are incorporated into biomass, the bulk of the cycling processes involve movement of the carbon and nitrogen atoms between organic matter and water-soluble organic and inorganic nutrient pools. Nearly all of the soil carbon and nitrogen are contained within the organic components in the ecosystem. In contrast, with sulfur, phosphorus, and metals, major reservoirs in soil necessarily include the soil mineral fractions—for example, pyrite and phosphatic rocks. Frequently, availability of these latter plant nutrients for biomass synthesis is determined by their rates of dissolution and formation of the mineral components.

Thus, the primary objective of this chapter is to elucidate the major biological processes associated with these nutrient cycles functioning at the intersection of soil minerals and the living community. Emphasis will be directed toward the biological aspects of the processes, but those reactions involved in conversion of various minerals to organic forms will be elucidated.

15.1 Sulfur in the Soil Ecosystem

Sulfur atoms occur in soil in a wide variety of organic and inorganic compounds. Sulfur is found in humus, biomass, and minerals. Volatile sulfurous substances are among the gases of the soil atmosphere, and ionic forms contribute to the salt content of interstitial water. The primary water-soluble ionic sulfur form in soils is sulfate. As a gas, sulfur atoms are detected in hydrogen sulfide, which may occur at significant concentrations in the atmosphere of waterlogged soils. Sulfur contributes to the soil particulate fraction in both soil organic matter and soil minerals. Furthermore, sulfur-bearing minerals include the original source of sulfur, pyrite (FeS_2), occurring in igneous rocks, and gypsum ($CaSO_4.2H_2O$) and epsomite ($MgSO_4.7H_2O$).

Quantities of organic and mineral sulfur in soil are extremely variable, ranging from a value as low as 0.002 percent in highly leached, weathered soils of humid regions to concentrations as high as 5 percent in calcareous and saline soils of arid and semiarid regions (Stevenson, 1986). Much of the soil sulfur is derived from internal sources, but external inputs such as fertilizers, air pollutants (e.g., sulfur dioxide), acid mine drainage (sulfuric acid), and a variety of xenobiotic compounds (e.g., pesticides) may also be significant.

The importance of transformations of the sulfur atom to life-supporting processes is documented by the fact that not only are sulfur-containing organic compounds [e.g., biotin, coenzyme A, thiamine, glutathione, cysteine, cystine, methionine (Fig. 15.1)] essential cellular components, but inorganic sulfur substances serve as energy sources [e.g., sulfide (S^{2-}), elemental sulfur (S^0)], and electron acceptors [e.g., sulfate (SO_4^{2-})] for soil microbes.

A primary property of the sulfur atom contributing to its utility and versatility in biological systems is its wide range of commonly occurring oxidation states. Common valence states of sulfur relating to biological active molecules are -2 (e. g., in organic compounds as sulfhydrals and in mineral sulfides), 0 (elemental sulfur), and $+6$ (e.g., organosulfates and mineral sulfates). See Table 15.1 for further documentation of the variation in valence of sulfur atoms in commonly detected soil sulfur compounds. Microbes are the primary mediators for cycling the sulfur atom among these various oxidation states, although spontaneous chemical processes contribute to these transformations to a limited degree.

15.2 Biogeochemical Cycling of Sulfur in Soil

A generalized schematic of the sulfur cycle is depicted in Fig. 15.2. As with those biogeochemical cycles previously evaluated, sulfur atoms can be seen to be transferred between a variety of organic and inorganic reservoirs. Primary soil-resident sulfur pools are organic residues (biomass plus its decomposition

Table 15.1 Oxidation State of Major Sulfur Compounds in Soil

Compound	Oxidation State
Organic S (R-SH)	-2
Sulfide (S^{2-})	-2
Elemental S (S^0)	0
Thiosulfate ($S_2O_3^{2-}$)	$+2$ (Average/S)
Tetrathionate ($S_4O_6^{2-}$)	$+2$ (Average/S)
Sulfur Dioxide (SO_2)	$+4$
Sulfite (SO_3^{2-})	$+4$
Sulfate (SO_4^{2-})	$+6$

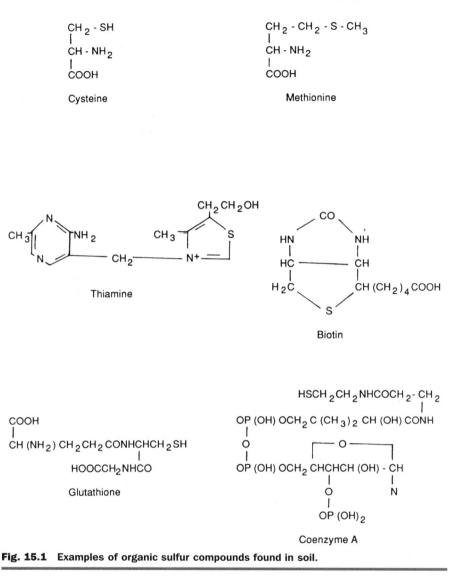

Fig. 15.1 Examples of organic sulfur compounds found in soil.

products and humic substances derived from them) as well as soil minerals. In contrast to the nitrogen cycle, except in limited situations (e.g., anthropogenically generated sulfur dioxide produced primarily from the burning of fossil fuels), atmospheric contributions of sulfur compounds to this biogeochemical cycle are limited. This diversity in sulfur reservoirs and the intermixing of biological and mineral sulfur pools in soil suggests a complexity in sulfur cycling in soil not previously encountered in our study of carbon and nitrogen cycles. This situation is exemplified by the observation that three mini-cycles can be highlighted in this conceptual model of the soil sulfur cycle:

Atmosphere

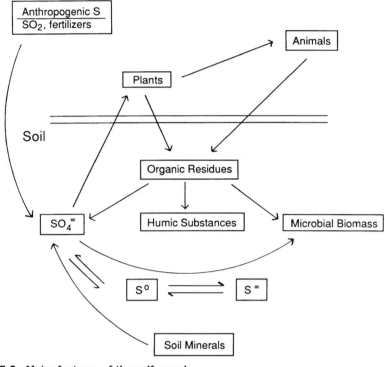

Fig. 15.2 Major features of the sulfur cycle.

- *Cycling of sulfur between sulfide (and elemental sulfur) reservoirs and sulfate:* This process involves biological reduction of sulfate under anoxic conditions as well as oxidation of sulfide and elemental sulfur. The latter energy-generating process is catalyzed by soil microbes that require molecular oxygen or nitrate as their terminal electron acceptor. Sulfate in the absence of molecular oxygen is reduced to sulfide. These processes are analogous to nitrification and denitrification in the nitrogen cycle.

- *Cycling of sulfur between soil biomass reservoirs and soluble sulfate (assimilation and mineralization):* Sulfate incorporated into biomass may be derived from mineralization of native soil organic matter and plant and animal residues. Organic sulfur of biomass may be mineralized prior to its incorporation into newly synthesized cells. (The term *may* is used in this situation since microbial cells may incorporate plant biomass substituents directly into their cellular biomass). This situation can be exemplified by the mineralization of organic wastes (e.g., farmyard manure) in soil. The organic sulfur contained therein is mineralized to sulfate and then reincorporated (immobilized, assimilated) into microbial or plant biomass.

- *Complexing of sulfide and sulfate into soil minerals and their solubilization to biological available substances:* Some mineral sulfur may be lost from the above mineralization-assimilation cycle through transfer to water-insoluble reservoirs. Sulfide accumulation in soil minerals has been defined as **sulfidization** whereas the exposure of sulfide-bearing minerals to oxidizing conditions resulting in production of sulfate is **sulfurization**. These processes may involve dissolution or precipitation of the minerals through chemical processes (e.g., precipitation of sulfide by iron or solubilization of minerals under acidic conditions) as well as biological oxidation or reduction of the primary soil mineral components. This portion of the overall sulfur cycle is highlighted because it represents the formation of an inorganic sulfur pool with long-term implications on the geological portion of the biogeochemical cycle.

This conceptual model of the sulfur cycle is developed with an emphasis on the primary sulfur resources in soil ecosystems. Examples of other sulfur-containing reservoirs, which may be of significant environmental importance, at least on a localized basis, include (1) various waste products of society that may be incorporated into soil (e.g., sludges, composts, garbage, and fly ash), (2) agricultural chemicals (including pesticides), and (3) air pollutants (e.g., products of fossil fuel combustion in electrical power generation and in automobile exhausts). These substances enter the soil sulfur cycle through reactions similar to those already depicted. For example, sulfur dioxide from fossil fuel consumption enters the interstitial water, where it is oxidized to sulfate. All biodegradable organic sulfur compounds are mineralized, as is shown for native biomass of the ecosystem.

The most obvious atmospheric contribution to the sulfur cycle is the previously mentioned sulfur dioxide, but it must be understood that particulate (dry deposition) sulfur inputs may also be significant on a localized basis. As with the sulfur dioxide, particulate sulfur inputs (primarily as elemental sulfur, but also as various sulfates) contribute to soil acidity by increasing sulfate concentrations. An extreme example of the impact of deposition of elemental sulfur on an ecosystem is provided by studies of forest ecosystems in Alberta, Canada, receiving large amounts of industrially produced sulfur. Kennedy et al. (1988) and Maynard et al. (1986) both report dramatic declines in above-ground vegetation quantity and diversity due to increasing soil sulfur loadings. Sulfur inputs reported by Maynard et al. (1986) ranging from 4100 to 51,400 μg g^{-1} soil resulted in decreasing soil pH (4.4 to 2.4) and augmentation of the *Thiobacillus thiooxidans* populations. Visser and Parkinson (1989), in their study of forest soils receiving comparable elemental sulfur loadings, found a decline in microbial biomass that correlated with soil pH and a parallel reduction in the ability of the soil community to mineralize glucose. The environmental impact of the perturbation of the sulfur cycle resulting from deposition of unusually large quantities of elemental sulfur is a logical outgrowth of the capacity of the native soil microbial population to use this sulfur substance as an energy source.

15.3   Biological Sulfur Oxidation

Chemoorganotrophic and chemolithotrophic bacteria derive their energy from oxidation of organic and mineral substrates. The large range of oxidation states of the sulfur atom (-2 to $+6$) make it a nearly ideal candidate as a primary energy source for resourceful soil microbes. In soil ecosystems, a variety of soil microbes have the capacity to derive their energy from oxidation of inorganic sulfur compounds. This exploitation of sulfur oxidation is a minor biogeochemical process in that terrestrial ecosystems are predominantly driven by heterotrophic processes. Sulfur oxidation can be an important energy source for some portions of the terrestrial ecosystem, such as exotic deep sea hydrothermal vents. Both sulfide and elemental sulfur are substrates in this energy-yielding process.

The diversity of the oxidation status of the sulfur atom allows for the concurrent oxidation and reduction of the same compound by the same microbe. In this disproportionation process, the sulfur atom is serving both as an electron donor and electron acceptor for the microorganism. Examples of the process are demonstrated by the disproportionation of thiosulfate and sulfite by *Desulfovibrio sulfodismutans*. Sulfate and sulfide are formed from the sulfur compounds by the following reactions:

$$S_2O_3^{2-} + H_2O \rightarrow SO_4^{2-} + H_2S$$
$$4SO_3^{2-} + 2H^+ \rightarrow 2SO_4^{2-} + H_2S$$

15.3.1   Microbiology of Sulfur Oxidation

Oxidation of sulfur-containing compounds, especially elemental sulfur, has a major impact on overall ecosystem function, primarily due to the fact that the terminal product of the reaction, sulfate, results in acidification of the soil. Elemental sulfur has long been used as a soil amendment when reduction of soil pH is desired. Environmental degradation has been linked with excessive oxidation of sulfur, as can result from mining of sulfide-bearing ores or atmospheric deposition of elemental sulfur due to anthropogenic activities — as noted above. Hence, an understanding of the biological interactions involved with the oxidation of sulfur compounds and the nuances of environmental management of the process is essential for proper stewardship of soil resources, as well as for reclamation of mismanaged systems.

The Thiobacilli: Although a diverse group of soil microbes are capable of oxidizing sulfur and sulfide for energy, members of the bacterial genus *Thiobacillus* are the primary mediators of this process in soil ecosystems. As is indicated below, some studies suggest that heterotrophic bacteria and fungi may be of some importance in this biogeochemical process, but the significance of their contribution has yet to be elucidated.

The thiobacilli are chemoautotrophic or facultatively chemoautotrophic bacteria that are widely distributed in soil ecosystems. These generally motile, gram negative bacteria are distinguished from other members of the soil community by their capacity to catalyze at least some portion of the oxidative continuum of sulfide, elemental sulfur, thiosulfate, and/or tetrathionate to sulfate. Among the energy substrates of the five most studied members of this genus (*T. thiooxidans, T. ferrooxidans, T. thioparus, T. novalis,* and *T. denitrificans*), elemental sulfur, thiosulfate, tetrathionate, and sulfide are most commonly listed. It must be noted that each species of the genus *Thiobacillus* is capable of oxidizing a specific portion of the spectrum of oxidizable sulfur compounds occurring in soil. For example, not all species or strains within a given species are capable of oxidizing elemental sulfur. Furthermore, not all *Thiobacillus* species are limited to deriving their energy from oxidation of sulfur compounds; fixed carbon and ferrous iron may also be oxidized. The 18 described species can be divided into three groups based on their sensitivity to acidity and capacity to oxidize fixed carbon, neutrophilic chemoautotrophs, neutrophilic facultative autotrophs, and acidophiles (Kuenen et al., 1992). Of the most studied members of this genus, *T. novalis* is a facultative chemoautotroph and *T. ferrooxidans* recovers some energy from ferrous iron oxidation (Table 15.2).

These sulfur-oxidizing bacteria area further distinguished by the nature of their terminal electron acceptor and their resistance to acidic conditions. The latter trait is particularly valuable to the survival of members of the genus in that extremely acidic soils (pH 1 to 3 are commonly reported) may result from their formation of sulfuric acid. Most members of the genus are obligate aerobes, although *Thiobacillus denitrificans* uses oxygen or nitrate as its terminal electron acceptor and *T. ferrooxidans* transfers electrons to ferric ion

Table 15.2 **Metabolic Properties of Commonly Studied *Thiobacillus Species***

Thiobacillus Species	Optimum pH	Energy Substrates	Electron Acceptors
T. thiooxidans	2.0–3.5	S^0, $S_2O_3^{2-}$, $S_4O_6^{2-}$	O_2
T. ferrooxidans	2.0–3.5	S^0, $S_2O_3^{2-}$, Fe^{2+}	O_2
T. thioparus	Approx. 7.0	S^0, S^{2-}, $S_2O_3^{2-}$, $S_4O_6^{2-}$	O_2
T. novalis	Approx. 7.0	$S_2O_3^{2-}$, organic compounds	O_2
T. denitrificans	Approx. 7.0	S^0, S^{2-}, $S_2O_3^{2-}$, $S_4O_6^{2-}$	O_2, NO_3^-

under anoxic conditions. A wide range of tolerance to acidity is characteristic of the genus. For example, *T. thiooxidans* and *T. ferrooxidans* have pH optima between 2.0 and 3.5. In contrast, *T. thioparus, T. denitrificans,* and *T. novellus* grow optimally at near-neutral pH values (Table 15.2). (For a more general discussion of the properties and function of *Thiobacillus* species, see Alexander, 1977; Harrison, 1984; Kuenen et al., 1992; Stevenson, 1986; or Visniac and Santer, 1957.)

The kinetics of formation of sulfate by these microbes can be exemplified by a listing of the reactions involved in elemental sulfur, thiosulfate, and tetrathionate oxidation by *T. thiooxidans:*

$$2S + 3O_2 + 2H_2 \rightarrow 2H_2SO_4$$

$$Na_2S_2O_3 + 2O_2 + H_2O \rightarrow H_2SO_4 + Na_2SO_4$$

$$2Na_2S_4O_6 + 7O_2 + 6H_2O \rightarrow 6H_2SO_4 + 2Na_2SO_4$$

T. ferrooxidans oxidizes elemental sulfur and thiosulfate by reactions similar to those listed above for *T. thiooxidans,* but this oxidation is combined with the oxidation of ferrous to ferric iron as follows:

$$12FeSO_4 + 3O_2 + 6H_2O \rightarrow 4Fe_2(SO_4)_3 + 4Fe(OH)_3$$

This species also contains enzymes capable of reducing ferric iron (Fe^{3+}) (Sugio et al, 1987), tetravalent manganese (Mn^{4+}), (Sugio et al., 1988a), cupric (Cu^{2+}) (Sugio et al. 1990), and Molybdic (Mo^{6+}) (Sugio et al., 1988b) ions. The environmental significance of the latter processes has not been elucidated.

The oxidative process for *T. denitrificans* is coupled to the reduction of nitrate as follows:

$$5S + 6KNO_3 + 2H_2O \rightarrow K_2SO_4 + 4KHSO_4 + 3N_2$$

The environmental significance of the latter process is that it provides a mechanism for sulfur oxidation in anoxic environments. Alternatively, ferric ion may also serve as a terminal electron acceptor for sulfur-oxidizing *T. ferrooxidans* growing in the absence of molecular oxygen. Pronk et al. (1992) have shown a linear relationship between ferrous iron accumulation and cell density in *T. ferrooxidans* reducing ferric iron and oxidizing elemental sulfur. Suzuki et al. (1990) have also shown that several strains of *T. ferrooxidans* are capable of oxidizing elemental sulfur under anoxic conditions when using ferric ion as a terminal electron acceptor. Some of the strains oxidized the sulfur with ferric ion as the electron acceptor at rates comparable to those of the aerobic process, but with others, sulfur oxidation occurred at a rate 50 percent or more less than that occurring in the presence of oxygen.

A variety of other metal-mobilizing, sulfur-oxidizing *Thiobacillus* species have been evaluated, [e.g., *T. acidophilus* (Meulenbert et al., 1992), *T. prosperus* (Huber and Stetter, 1989), *T. versutus* (Beffa et al., 1991), and *T.*

cuprinus (Huber and Stetter, 1990)]. Other autotrophic bacteria have been described which catalyze sulfur oxidation under aquatic or more specialized conditions. These include phototrophic bacteria (Brune, 1989) and *Sulfolobus* species which function in hot springs (Alexander, 1977).

Heterotrophic Sulfur Oxidizers: Several heterotrophic microbes also catalyze sulfur oxidation. *Aspergillus niger, Mucor flavus,* and *Trichoderma harzianum,* for example, oxidize elemental sulfur (Grayston et al., 1986). These fungi produced substantial amounts of sulfate from sulfur oxidation with thiosulfate (*Aspergillus niger*) or thiosulfate and tetrathionate (*Mucor flavus*) accumulating as intermediates. *Aureobasidium pullulans* (de Bary) Arnaud, isolated from sycamore phylloplane that had been exposed to atmospheric pollution, oxidized elemental sulfur to dithionite, tetrathionate, and sulfate (Killham et al. 1981). Similar products are produced by yeast isolates of the genus *Rhodotorula* (Kurek, 1979, 1985) and bacteria (Pepper and Miller, 1978) and actinomycete (Yagi et al., 1971) isolates. These heterotrophic microorganisms utilize carbonaceous substrates for their carbon and energy source and generally accumulate oxidized sulfur compounds after the period of active growth. Thus, their role in sulfur oxidation in native soil ecosystems is questioned.

Indirect evidence suggests that there is at least some role for the heterotrophs in elemental sulfur oxidation in soil. Lawrence et al. (1988) evaluated the effect of elemental sulfur fertilization on a variety of measures of sulfur oxidation in two Grey Luvisolic soils from Canada. In one soil, increases in *Thiobacillus* populations were induced by elemental sulfur amendment, whereas in the other soil no change in these populations was detected. Heterotrophic sulfur oxidizers were the most abundant sulfur oxidizers in both soils. In a related study (Lawrence and Germida, 1988), elemental sulfur oxidation in 28 Canadian soils correlated linearly with soil microbial biomass C (r = 0.68, p < 0.01) and soil respiration (r = 0.88, p < 0.01). Further evidence implicating a heterotrophic contribution to the sulfur oxidation included (1) augmentation of heterotrophic biomass through glucose amendment increased sulfur oxidation rates and (2) inhibition of general heterotrophic microbial activity reduced sulfur oxidation rates.

In contrast, Lee et al. (1987) in their study of 48 New Zealand soils, found that elemental sulfur amendment resulted in increases of thiobacilli populations with no change or a decline in heterotrophic sulfur-oxidizing populations. These reports suggest that environmental selection is likely of great significance in determining the dominant populations involved in sulfur oxidation in soil ecosystems. Of primary importance in considering the impact of autotrophic vs. heterotrophic populations in modified sulfur cement oxidation is the requirement of oxidizable carbon substrates for function of the heterotrophs. These substrates are concentrated in surface horizons of the soil profile and are limiting in subsoils (Tate, 1987).

15.3.2 Environmental Conditions Affecting Sulfur Oxidation

Although sulfur-oxidizing microbes are considered to be ubiquitous, populations of autotrophic thiobacilli can be essentially undetectable in some soils. These populations are inducible by sulfur amendment. For example, thiobacilli population augmentation by sulfur amendment has been demonstrated with several soils from Kansas (Attoe and Olson, 1966), Canada (Janzen and Bettany, 1987a), and New Zealand (Lee et al. 1988a). The increased sulfur oxidizer population resulting from prior sulfur amendments can apparently cause increased sulfur-oxidation rates in subsequent seasons (Lee et al., 1988b). These observations relate specifically to *Thiobacillus* populations. Since, as indicated above, a variety of heterotrophic microbes are also capable of oxidizing sulfur in soil, it is reasonable to conclude that sulfur oxidation is not precluded in soils due to nonexistence of requisite microbes. Significant autotrophic or heterotrophic populations can be anticipated to occur in any biologically active soil. The rate may be limited due to initially low population densities but the number of autotrophic microbes should increase following amendment with sulfur substrates. Therefore, it can be concluded that since the requisite microbial strains that oxidize sulfur can be assumed to be present in any given soil ecosystem, oxidation of sulfur is dependent primarily upon occurrence of physical and chemical conditions conducive to the function of these microbes.

Physiologically, the microbes require an energy source (an oxidizable substrate) and an electron acceptor. For most living organisms, these are a fixed carbon supply, such as glucose, and oxygen. For elemental sulfur oxidation, the sulfur itself serves as the electron source — that is, the oxidizable substrate, and oxygen is the electron acceptor. One species of elemental sulfur oxidizer, *Thiobacillus denitrificans*, has been isolated that is capable of oxidizing sulfur under anoxic conditions. Cytochromes are used in the electron transport system, and the energy yield approaches that gained when oxygen serves as the electron acceptor. Although some stains of *Thiobacillus denitrificans* have been demonstrated to be unable to oxidize extracellularly offered elemental sulfur anaerobically (Schedel and Trueper, 1980), others function quite efficiently with nitrate as an electron acceptor and elemental sulfur as the energy source. For example, a *Thiobacillus denitrificans*-based system is sufficiently efficient that it has been proposed as a means of removing excess nitrate from waste water. Elemental sulfur serves as a cheap source of energy for the denitrification process (Batchelor and Lawrence, 1978). High percentages of denitrification (over 95 percent) have been demonstrated using this process (Hashimoto et al., 1987).

The use of elemental sulfur as an energy source in soil is limited by the capacity of the microorganism to interact with the substrate. Physical attachment of thiobacilli to sulfur particles is required for enzymatic oxidation of sulfur. This interaction of the microbes with sulfur particles is the rate-controlling factor in sulfur oxidation. This was shown clearly more than 50 years ago in a study of *Thiobacillus oxidans* by Vogler and Umbreit (1941). Espejo and Romero (1987) demonstrated that for *Thiobacillus ferrooxidans*, only bacteria attached to the sulfur particle were capable of growth. No replication occurred once the bacteria were detached from the sulfur particle. Growth of the microbes was not uniform on the sulfur particle surface, but rather it proceeded at distortions present on the surface. Staining of sulfur prills with crystal violet reveals that even after more than 4 months' incubation, intact regions remained on the substrate surface, whereas destruction of the surface had occurred quite deeply at other locations on the prill.

Thiobacillus ferrooxidans has a more versatile energy-recovery capability. When attached to sulfur particles, sulfur oxidation is the primary energy-yielding mechanism. Microbes that become detached from the sulfur particles can survive through the oxidation of ferrous ion to ferric ion. Sulfur-oxidation rate was not affected by the presence of ferrous ion in the growth medium (Espejo et al, 1988). For *Thiobacillus albertis*, specific cellular adhesions, bacterial glycocalyx, are produced to facilitate the interaction of the thiobacilli cells with the sulfur particle (Bryant et al., 1984). Elemental sulfur-oxidation by the latter *Thiobacillus* species is proportional to particle size, surface area per unit weight and the crystal microstructure of the elemental sulfur (Laishley et al., 1986).

Since the elemental sulfur particles do not dissolve in water, and direct contact of the microbial cell with the sulfur particle is necessary for oxidation to proceed, it can be concluded that surface area of the sulfur particles controls their oxidation rate in soils. This has been demonstrated repeatedly in a wide variety of soils over the last several decades (e.g., see Bertramson et al., 1950; Frederick and Starkey, 1948; Janzen and Bettany, 1987b; Laishley et al., 1986; Lee et al, 1988a, 1988b; McCaskill and Blair, 1987; Nor and Tabatabai, 1977; Watkinson, 1989). For each of these studies, a size distribution of sulfur particles was added to soil and sulfate concentration, or pH changes due to sulfur oxidation were quantified as a function of particle size. In each case, rapid oxidation was detected with small particles and limited reaction of the larger particles occurred. For example, Lee et al. (1988a) found that after 340 days, 80 to 90 percent of elemental sulfur particles less than 0.15 mm diameter were oxidized as opposed to only 24 to 55 percent of the particles size ranges greater than 0.15 mm diameter. McCaskill and Blair (1987) found that with mixtures of particle sizes, a biphasic oxidation curve could be detected. A reduction in conversion rate occurred once the 0.1- and 0.2-mm diameter particles were exhausted. Frederick and Starkey (1948) evaluated particles of sulfur in pipe sealing mixtures. They found that over a three-week incubation period, sulfur oxidation was essentially nondetectable with particle sizes greater than about 3 mm mean diameter.

Along with availability of sulfur-oxidizing microbes in soil and their growth substrates, sulfur oxidation is controlled by soil moisture, pH, and temperature. Because of the physical stability of elemental sulfur particles in soil, the impact of these factors on sulfur oxidation relates to their effect on the microbes catalyzing the reactions.

In soil, the rates of oxidation of sulfur-containing compounds and elemental sulfur tend to increase with increasing soil moisture, reach an optimum level, and then rapidly decline (e.g., Attoe and Olson, 1966; Janzen and Bettany, 1987b; Lettl et al. 1981, Moser and Olson, 1953). With slight variation dependent upon soil texture, optimum soil moisture levels are near field capacity.

Complex interactions between soil moisture, soil texture, and microbial populations alter elemental sulfur oxidation rates (for example, see Janzen and Bettany, 1987a, 1987b; Watkinson, 1989). In one study of soils with limiting oxidizer population densities (as indicated by responses to inoculation), elemental S conversion rate was not affected by clay contents ranging from 9 to 52 percent (McCaskill and Blair, 1987). This contrasts with data from Janzen and Bettany (1987b), who found that with 40 diverse Canadian soils, elemental sulfur oxidation rate was negatively correlated with soil clay contents. Watkinson (1989) found that the improved soil porosity resulting from amendment of phosphate rock to mineral soils augmented elemental sulfur oxidation. It can be concluded that the prevailing controller of elemental sulfur oxidation in soils of varying moisture and texture is the availability of the obligatory electron acceptor, molecular oxygen.

Although elemental sulfur is used to reduce soil acidity, its oxidation rate actually generally increases with soil pH. Both Attoe and Olson (1966) and Janzen and Bettany (1987b) found a positive correlation of elemental sulfur-oxidation rate and soil pH in a wide variety of soils. Further Lettl et al. (1981) noted that oxidation of elemental sulfur was stimulated by amendment of soil with calcium carbonate.

Elemental sulfur oxidation occurs between 4 and 55°C (Stevenson, 1986), a range well within that normally observed in soil systems. The sulfur oxidation rate is directly proportional to the temperature in soil systems (Ahonen and Tuovinen, 1990; Attoe and Olson, 1966; Chapman 1989; Janzen and Bettany, 1987c; Nor and Tabatabai, 1977). A wide range of Q_{10} values was reported in these studies: 1.9–3.1 (Chapman, 1989), 2.1 (Ahonen and Touvinen, 1990), and 3.2–4.3 (Janzen and Bettany, 1987c).

15.4 Biological Sulfur Reduction

As is the situation with reduction of nitrate in the nitrogen cycle, sulfate can be reduced by both assimilatory and dissimilatory paths. Assimilatory sulfate reduction is the source of sulfur for biomass synthesis. Therefore, this sulfate-reduction process is catalyzed by a variety of organisms, including higher plants, algae, fungi, and bacteria. In contrast, dissimilatory sulfate reduction, sulfidogenesis, is catalyzed by a specialized group of obligately anaerobic bacteria. In soil, the organisms involved with the latter reaction are generally found in the genera *Desulfovibrio* and *Desulfotomaculum*. As with nitrate in denitrification, sulfate is the terminal electron acceptor for dissimilatory sulfate reducers. In both dissimilatory processes, the electrons are transferred through a cytochrome-based electron-transport process to the terminal electron acceptor.

With dissimilatory sulfate reduction, sulfate is reduced through the transfer of eight electrons produced by the oxidation of a fixed carbon substrate as described by the following reaction:

$$SO_4^{2-} + 8e^- + 8H^+ \rightarrow H_2S + 2H_2O + 2OH^-$$

The sulfate atom is activated in this process through the consumption of adenosine triphosphate (ATP). A variety of fixed carbon compounds may be oxidized to provide the electrons for this process. The most commonly studied energy sources are organic acids (e.g., lactate, acetate, and fatty acids), but an increasing number of more complex substances, such as various aromatic compounds, are being added to the list of electron donors. Coupling of this process with mineralization of organic soil pollutants is being explored as a means of bioremediation of soils (see chapter 16).

Dissimilatory sulfate reduction in soil systems is distinguished by its occurrence only in specialized ecosystems. Aside from the requirement for an oxidizable substrate and sulfate, the primary environmental property delimiting function of these anaerobic bacteria is anoxic conditions. Sulfidogenesis is detected only in soil ecosystems exhibiting extremely reducing conditions. A reducing potential of about -0.25 volts is required before dissimilatory sulfate reduction occurs. Thus, this sulfate-reduction process is limited to sites similar to those where carbon dioxide reduction (methanogenesis) occurs — that is, swampy or chronically flooded soils.

Although sulfate is a common component of soil interstitial water, its presence in concentrations supportive of extensive dissimilatory sulfate reduction is limited. A typical example of a land-based ecosystem that would allow extensive sulfide generation is a marsh site inundated with sea or brackish water. Anoxic conditions prevail in the marsh soils, and sulfate is a major anion in seawater. The best indication of current or previously existent sulfidogenic conditions is detection of metal sulfide deposits in the soil profile or detection of hydrogen sulfide fluxes from the site.

It should be noted that although the primary substrate considered for sulfidogenesis is sulfate, elemental sulfur can also be reduced by a variety of microorganisms. For example, *Desulfurolobus ambivalens* has the capacity to oxidize or to reduce elemental sulfur (Zillig et al., 1986). *Desulfotomaculum termoacetoxidans* (Min and Zinder, 1990), *Desulfuromonas acetoxidans* (Gebhardt et al., 1985), and *Spirillum* 5175 N (Zoephel et al., 1988), all reduce elemental sulfur to hydrogen sulfide.

Sulfidogenesis may be manipulated to ameliorate the impact of a variety of environmental pollutants, both organic and inorganic. For example, this process is being explored as a means of reducing organic contamination of chronically flooded soils, reduce drainage water acidity, and ameliorate the difficulties associated with heavy metal contamination of soils and waters.

Anaerobic Biodegradation: A variety of xenobiotic compounds containing halogenated aromatic rings may enter soil ecosystems in undesirable concentrations. Under aerobic conditions, the halogenated substances are reasonably stable, but under anoxic conditions, bacteria have been shown to be capable of removing the halogen atom. Since at least some of these substances are mineralized by the sulfidogenic bacteria, the combination of dehalogenating and sulfate-reducing populations may be used to remove this troublesome class of compounds from a site that can be easily managed under anoxic conditions. These techniques may be particularly operative in sediments polluted with xenobiotic compounds. (This topic is addressed in greater detail in Chapter 16.)

Reduce Acidity of Acid Mine Drainage: Acid mine drainage results from exposure of previously buried sulfide-bearing soil minerals to oxidative conditions in the presence of an active sulfur-oxidizing population. The environmental impact of the highly acidified waters is further accentuated by their high metal loadings, resulting from the dissolution of the soil and mine slag minerals. Thus, soils and waters receiving these drainage waters are adversely affected not only by the acidic conditions but also by toxic levels of heavy metals. A means of reducing the environmental impact in these situations is to pass the drainage waters through a swampy area where dissimilatory sulfate reducers can convert the sulfate to sulfide. The metal concentrations of the waters are reduced by their precipitation as metal sulfides. (See Mills, 1985, for a more detailed discussion of this process.)

Reduction of Complications of Metal Contamination in Soil: Sulfidogenesis can also be exploited in situations where soils have been contaminated by heavy metals. As with the acid mine drainage problem outlined above, metal loadings of the leachate waters from the contaminated site may be removed by precipitation as metal sulfides in a sulfate-reducing system. Mobility of the metals may also be reduced in the site through creation of appropriate conditions for sulfide genesis within the soil profile. The primary difficulty with the utilization of sulfidogenesis both to remove metal-contaminants from acid mine drainage and to reclaim metal-contaminated sites is that the anoxic conditions of the purification site must be maintained. Should the soil become aerobic, sulfur-oxidizing bacteria will utilize the sulfide for an energy source, thereby mobilizing the previously stabilized heavy metals.

15.5 Mineralization and Assimilation of Sulfurous Substances

Mineralization of organic sulfur compounds is a microbially catalyzed process that is analogous to carbon and nitrogen mineralization. Although the relative rates of nitrogen, carbon, and sulfur mineralization in a specific soil system may differ (e.g., see Bettany et al., 1980, for an example of a field study, and

Tabatabai and Al-Khafaji, 1980, for a laboratory-based study demonstrating this difference), each is dependent upon activity of a variety of heterotrophic bacteria and fungi. This commonalty of mediators results in a similarity in impact of variation of environmental conditions on process rates. That is, sulfur mineralization is anticipated to occur in any ecosystem wherein heterotrophic microbes can function. Under aerobic conditions, complete mineralization of an organic substrate yields carbon dioxide, ammonium, and sulfate. Under anoxic conditions, carbon dioxide, methane, sulfide, and ammonium are primary products. Optimal conditions for mineralization of sulfur-containing organic compounds include a near-neutral pH, mesophilic temperatures, and approximately field-capacity soil moisture. It is easily concluded a priori that since sulfur mineralization occurs over essentially the totality of the physical and chemical conditions supportive of microbial life, the process is catalyzed by an extremely diverse group of common soil bacteria and fungal species.

The mineral sulfur produced through the catalytic activity of the soil decomposer community may be assimilated by the plant community or immobilized (assimilated) by the soil microbes. Since ecosystem conditions generally favor assimilation of sulfate into plant or microbial biomass, the bulk of the sulfur contained in soil systems is in the organic fractions (e.g., see David et al., 1982; Swank et al., 1984). Since these assimilatory processes are catalyzed by the plant community and the heterotrophic soil bacteria and fungi, the physical and chemical limitations to the activities would be similar to those previously described for overall soil microbial activity.

As has been observed with carbon and nitrogen mineralization, the rate of mineralization of sulfur compounds is proportional to the biodegradation susceptibility of the organic compound itself, as well as its physical availability to the degrader population. That is, the decomposition rate is directly dependent upon the capability of the microbial population to synthesize the requisite enzymes necessary for the catalysis and the proximity of the organic substance to the active microbial population. Because both biodegradation-sensitive (common cellular substituents) and more biodegradation-resistant sulfurous compounds (humified substances) are found in soil (e.g., see Anderson et al., 1981, or Bettany et al., 1979), the conversion of organic sulfur to sulfate in an aerobic soil system may or may not be proportional to the total soil organic matter content. This relationship is totally dependent upon the relative contribution of biodegradable and accessible organic sulfur to biologically stabilized (e.g., humic acid) sulfur.

15.6 The Phosphorus Cycle

Processes associated with transformations of phosphorus occurring in soil are delineated by the same principles as the other soil-based biogeochemical cycles. Mineralization and assimilation are catalyzed by the general soil biological community. The phosphorus cycle, like the sulfur cycle, involves the interconversion of mineral and organic nutrient forms, with the quantity of water-soluble phosphate being controlled not only by biological mineralization processes but also by dissolution rates from soil mineral fractions. Unique aspects of the phosphorus cycle include the facts that phosphorus does not undergo any valence changes in the cycle, and that there is no gaseous component to the cycle. Based on the similarities of the biological and chemical properties of the phosphorus cycle to those of other nutrient cycles, only the salient features of the phosphorus cycle will be described herein. For a more detailed analysis of the nuances of the phosphorus cycle, see Stevenson (1986).

In a depiction of the phosphorus cycle with biomass accentuated, the pivotal process is mineralization of organic matter into plant and microbially available phosphate (Fig. 15.3). As with the mineralization processes central to the carbon, nitrogen, and sulfur cycles, phosphorus mineralization is catalyzed by the general soil heterotrophic microbial population (e. g., Molla et al., 1984). Hence, as in the previously described process, mineralization rate is controlled by the physical and chemical factors delimiting overall microbial heterotrophic activity. Phosphate mineralization and its assimilation into biomass is generally optimal under those soil conditions suitable for aboveground plant community development. As with the previous cycles, mineralization occurs under both aerobic and anaerobic conditions, but maximal activity is detected in the presence of molecular oxygen.

A variety of relatively easily decomposed organic phosphorus substances are found in biomass. The most abundant phosphatic substances are inositol

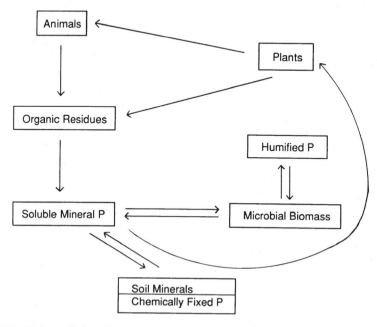

Fig. 15.3 Major soil phosphorus cycle reservoirs.

phosphates, which constitute 10 to 50 percent of the organic phosphate fraction. Other organic phosphates are phospholipids (1 to 5 percent) and nucleic acids (0.2 to 2.5 percent) (Stevenson, 1986). Lesser concentrations of phosphopyridines (e.g., NAD and NADH) and nucleotides (e.g., ATP) are detectable in soil organic phosphorus fractions.

Native phosphate in soils is derived from rock phosphate, apatite, which has the empirical formula $3(Ca_3(PO_4)_2) \cdot CaX_2$ where X equals Cl^-, F^-, OH^-, or CO_3^{2-}. Secondary minerals formed from phosphate in soil are wavellite $[Al_3(PO_4)_2(OH)_3 \cdot 5H_2O]$, variscite $[Al(PO_4) \cdot 2H_2O]$, dufrenite $[FePO_4 \cdot Fe(OH)_3]$, strengite $[Fe(PO_4) \cdot H_2O]$, and vivianite $[Fe_3(PO_4)_2 \cdot 8H_2O]$. Along with the effect of phosphate mineralization rate, the quantities of plant-available phosphate present are determined by the solubilization rate of phophatic minerals. Organophosphate levels in soil range from 15 to 80 percent (Stevenson, 1986). In low organic matter ecosystems, the solubilization rate of mineral phosphate is a major delimiter of aboveground biomass productivity.

Soil microbes are not only mediators of phosphate mineralization and immobilization processes, but they may also be instrumental in solubilization of rock phosphates. Directly synthesized products of microbial activity that solubolize rock phosphate are organic acids and carbon dioxide. The organic acids liberate phosphate through chelation of the associated metal ions. Carbon dioxide contributes to acidity of soil water by formation of carbonic acid.

As the pH of an acidic soil increases, organic acid production becomes increasingly important in determining dissolution rate of rock phosphates (e.g., Traina et al., 1986).

15.7 Microbially Catalyzed Soil Metal Cycling

Soil-resident reservoirs of biomass-nutrient cations (as well as toxic cations) are easily grouped into three categories: contained in biomass, dissolved in interstitial water, and retained in soil minerals (i.e., water-insoluble forms) (Fig. 15.4). Biomass productivity may be determined by existence of excessive quantities (toxic metals) as well occurrence of minimal levels (required nutrients). The soil biological community is instrumental in catalyzing a variety of processes that directly or indirectly result in the transfer of cations between these three soil reservoirs.

15.7.1 Interactions of Metals with Living Systems

Benefits from maintenance of appropriate levels of cations in interstitial waters result from their essential roles in cellular biochemistry. Plant-essential trace metals include iron, zinc, manganese, calcium, boron, molybdenum, and, most likely, nickel. Animals also require cobalt, chromium, and tin. In contrast, cations of major interest from the view of negative interactions with biomass production include cadmium, mercury, and lead.

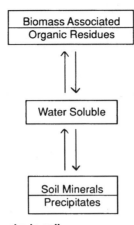

Fig. 15.4 **Primary metal reservoirs in soil.**

In most ecosystems, the balance between aboveground productivity and trace metal availability is maintained by the movement of soil-resident cations between the various reservoirs (Fig. 15.4). Limitations of these plant nutrients to overall biomass production are generally restricted to ecosystems where augmentation of biomass production above that essentially dictated by native soil mineral cycling rates is necessary (i.e., intensive agricultural production) or in a case where metals are removed from the system through harvesting of aboveground biomass. In these situations, aboveground biomass needs are met through external amendments and/or modification of soil pH to optimize availability of soil resources.

In contrast to the situation exemplified by intensive agricultural systems, metal limitations for microbial productivity in soil are generally a rarity. Soil microbes possess highly efficient mechanisms for transport of cations into the cell. Furthermore, as is discussed below, many members of the soil microbial community can produce chelators that increase water-soluble cation resources.

An increasing array of environmental problems are arising where management of the soil for stimulation of ecosystem sustenance and/or development results in alteration of metal availability to the microbial community. Historically, the most commonly cited toxicity problem involving soil cations is associated with aluminum concentrations in acidic soils. Aluminum is toxic to biological processes. As the soil pH declines, the solubility of aluminum increases. Thus, best management practices of acidic soils for crop production have included addition of lime to soil to raise the pH as well as amendment with organic matter to complex the water-soluble ions (See Stevenson, 1982, for a more complete analysis of aluminum solubility in soil.) More recent considerations involving toxic metals in soil have been associated with environmental contamination (e.g., smelter and foundry operations) and from disposal of societal wastes (e.g., land disposal of metal contaminated sludges and composts).

The implications of increased soil loadings of toxic cations on ecosystem function are best exemplified by the succession of aboveground and below-ground communities associated with metal contamination of soils by smelter operations. The availability of the toxic cation may become the controlling factor in community development, as was observed by Nordgren et al. (1983) in soils near a brass mill in South Sweden. In these soils, sampled along a copper and zinc gradient (with maximum levels of 20,000 μg copper and zinc g^{-1} soil), the influence of soil organic matter content and moisture on fungal community composition was minimal compared to the effect of heavy metal pollution on these populations.

A more general impact on ecosystem communities has been noted for heavy-metal–contaminated soils near a zinc smelter at Palmerton, Pennsylvania. Soil heavy metal loadings (litter horizon within 1 km of the smelter) in the 1970s included about 26,000 μg zinc g^{-1} soil, 10,000 μg iron g^{-1} soil, 2300 μg cadmium g^{-1} soil, and 340 μg copper g^{-1} soil (Strojan, 1978), with as much as 90 percent of the metals being retained within the top 15 cm of the soil profile (Baucher, 1973). These heavy metal loadings declined to approximately one-tenth these values at a control area about 40 km east of the smelter. In the 1978 study, consideration of the average amount of organic matter in the more highly polluted site and comparison of decomposition rates of leaf litter in litter bags led to the conclusion that organic matter decomposition rates were reduced by the heavy metal contamination (Strojan, 1978). In a 1975 report (Jordan, 1975), reduced germination of tree seeds due to high zinc concentrations was proposed as a mechanism preventing the establishment of invader tree species in regions where indigenous trees, shrubs, and herb populations were in decline from the metal contamination. Additionally, in this site, the prevalent metal-contaminant concentrations exceed those necessary to result in major reductions of total bacterial, actinomycete, and fungal populations (Jordan and Lechevalier, 1975). Major alterations of the soil microbial as well as aboveground populations resulted from the metal contamination, as demonstrated by changes in the soil metabolic diversity (Kelly and Tate, 1998).

Currently, the aboveground status of the site at Palmerton, PA can be defined as an "establishing" grass community due to overlaying of the soil with a sludge-fly ash-limestone-grass seed mixture or an essentially barren system in untreated sites (Fig. 15.5). The once-flourishing forest ecosystem has been reduced to a bare mountainside. Reclamation of the site requires redevelopment of a functional soil microbial community accompanied with establishment of aboveground plant community consisting of a grass community.

With both of these examples, extremely high metal loadings were detected in the environmental samples. It must be remembered that the toxicity of the metals is proportional to their availability to the microbial community. Thus, inhibitory concentrations of metals derived from pure culture, study where all of the toxicant is water soluble, frequently do not correspond to field

Fig. 15.5 Metal contaminate site in Palmerton, PA (USA) showing effect of organic matter, lime, and fly ash amendments on grass growth.

observations, where total soil metal concentration can be distributed between a variety of chemical forms that vary in availability to the soil microbial community (e.g., Lighthart et al. 1983). Therefore, when evaluating metal loading toxicity in native ecosystems, soil properties that alter metal distribution between water-soluble and water-insoluble species must be considered. For example, in a study of planktonic, sediment, and epilithic bacterial community adjustment to heavy metal concentrations (Dean-Ross and Mills, 1989), changes in microbial populations did not reflect the known toxicity of the contaminating metals. It was concluded that the occurrence of a high pH in the samples that reduced the solubility of the toxic metals apparently resulted in a lack of correlation between metal concentrations and response of the bacterial communities to the toxicants.

15.7.2 Microbial Response to Elevated Metal Loadings

The most obvious effect of high metal loadings on the soil microbial community is toxicity. Death of sensitive microbes and plants may result directly from dissolved cations as well as indirectly from the release of cations following catabolism of the organic matter to which they are bound. Also, alteration of soil microbial community dynamics and functions can be anticipated to result from soil contamination with heavy metals. A primary impact of metal contamination is the inhibition of catabolism of organic substances by the toxic metals. This was reflected in the studies of metal-contaminated forest sites cited above by the augmented accumulation of plant debris in regions of highest metal concentrations.

Microbial populations may be inhibited by release of organic-matter–associated cations through microbial catabolism of the organic matter. The microbial recovery of energy from the fixed carbon substances results in accumulation of toxic concentrations of the metal. This phenomenon is exemplified by studies of the decomposition of iron, aluminum, zinc, and copper salts complexed to polysaccharides (Martin et al., 1966), metal-nitrilotriacetate (calcium, manganese, magnesium, copper, zinc, cadmium, iron, and sodium chelates were examined) complexes (Firestone and Tiedje, 1975), as well as copper, mercury, zinc and cadmium salts of 2,4-dichloro-phenoxyacetic acid methyl ester (Said and Lewis, 1991). As the concentration of the metal liberated from the organic complex increased, microbial activity decreased. The degree of inhibition varied with toxicity of the metal to the decomposer populations.

Overall reduction of microbial respiration by increasing concentrations of a variety of heavy metals is exemplified by the study of Lighthart et al. (1983). Additionally, examples of the capacity of heavy metal contamination to reduce or inhibit soil biological processes include a variety of enzymatic activities (e.g., Cole, 1977; Doelman and Haanstra, 1979; 1989; Stott et al., 1985) as well as various soil nitrogen transformations [immobilization, mineralization and nitrification (Chang and Broadbent 1982), anaerobic nitrogen transformations (Blais et al., 1988), and nitrogen fixation (Wickliff et al., 1980)].

A variety of organisms must exist in soil with augmented resistance to heavy metal pollutants. For example, soil bacteria have been isolated with augmented resistance to metal contaminants, including cadmium (Bopp et al., 1983), copper (Dressler et al., 1991; Yang et al., 1993), mercury (Kelley and Reanney, 1984), nickel (Schmidt et al., 1991), and silver (Haefeli et al., 1984). Selective enrichment of metal-tolerant bacterial populations in polluted soils has been demonstrated with classical microbiological procedures (e.g., Duxbury and Bicknell, 1983) as well as DNA probe-mediated methods (e.g., Diels and Mergeay, 1990).

15.7.3 Microbial Modification of Metal Mobility in Soils

Soil properties controlling the availability of both essential nutrient cations and toxicants are primarily soil pH, redox potential, and organic matter concentrations. Soil-buffering capacity, chlorinity, inorganic cations and anions, clay mineralogy, and moisture (e.g., Babich and Stotzky, 1983) also contribute to metal solubility. The microbial community is instrumental in modifying each of these soil factors in a manner that can enhance or reduce soil interstitial water metal loadings. Furthermore, microbial interactions affecting metal mobility in soil can be grouped as indirect or direct. Indirect interactions include pH alteration, siderophore production, and modification of the physical environment (e.g., changing the redox potential). Direct interactions include valance change, substitution, methylation, transalkylation, incorporation into cell substituents, and release from organic association through mineralization.

Acidification and Chelation: Soil microbes produce a variety of inorganic and organic acids that contribute to solubilization of metals. The inorganic acids could be considered to be by-products of energy metabolism. Two of the more commonly encountered products are carbonic and sulfuric acid. Aerobic microbes yield carbon dioxide as their terminal oxidation product. This gaseous substance exists in equilibrium in soil between carbonic acid, bicarbonate, and carbonate. Carbonic acid is a weak acid capable of dissolving soil minerals.

A more dramatic example of mineral acid production is the sulfuric acid produced by oxidation of sulfides and elemental sulfur by soil thiobacilli. This process has been exploited by recover of mineral from ores (e.g., Carlson et al., 1992) and has been proposed as a means of removing mineral sulfides from municipal sludges (e.g., Jain and Tyagi, 1992). Ecosystem destruction resulting from this process when oxidation of mineral sulfides and associated dissolution of metals (typically iron) results in the production of the orange colored leach rate known as acid mine drainage (e.g., Mills, 1985).

Organic acids that may solubolize soil minerals either directly through acidification of the microsite or through chelation reactions are produced by aerobic and anaerobic soil microbial processes. Under anaerobic conditions, a variety of organic acids accumulate due to incomplete oxidation of fixed carbon substrates. Commonly produced substance under aerobic conditions are siderophores. These chelators are important in iron nutrition of bacteria and higher plants. Bacterial-synthesized siderophores may be instrumental in solubolizing significant quantities of iron for higher plant biomass synthesis. Such substances produced by rhizosphere bacteria have been shown to stimulate the associated plant productivity (e.g., Kloepper et al., 1980). Similarly, rhizobial siderophores have been demonstrated to stimulate clover productivity (Derylo and Skorupska, 1992). A large number of members of the general soil community also produce these metal chelators, for example, *Pseudomonas* species (e.g., Ankenbauer et al., 1988; Bar-Ness et al., 1991; Cody and Gross, 1987; Hoefte et al., 1991). The fact that these substances are synthesized within the soil ecosystem is supported by the capacity to isolate a variety of such substances from soil samples (e.g., Akers, 1983a; 1983b; Powell et al., 1982).

Direct and Indirect Oxidation and Reduction of Soil Minerals: Oxidation of metals by soil bacteria is exemplified by the oxidation of ferrous ion by *Thiobacillus ferrooxidans* and mercury oxidation by a variety of bacteria (e.g., Holm and Cox, 1975), as well as by the oxidation of metal oxides [e.g., conversion of arsenite to arsenate by *Alcaligenes faecalis* (Phillips and Taylor, 1976)]. Similarly, a number of bacteria are capable of reducing a number of metallic cations and metallic oxides, including iron (e.g., Lovley and Phillips, 1986), manganese and iron (e.g., Nealson and Myers, 1992), uranium (Lovley et al., 1991; Lovley and Phillips, 1992), and chromate (Eary and Rui, 1991).

Indirect modification of metal mobility by changing its valence state can result from microbial modification of the redox potential of the soil microsite. This process is envisioned in soil ecosystems where microbial energy metabolism results in exhaustion of molecular oxygen, nitrate, and other electron acceptors, creating the highly reducing condition necessary for spontaneous chemical reduction of metals (such as ferric ion reduction to ferrous iron). The result of these reactions frequently results in changes in soil physical structure. For example, iron reduction in waterlogged soils can result in gley formation (see Tate, 1987). In this situation, the precipitation of the ferrous ion by the concurrently formed sulfide cause a gray-to-black coloration of the soil. This alteration of the soil color within the profile can typically be used to determine transient occurrence of a high-water table and the resulting imposition of anoxic conditions.

An interesting indirect modification of oxidation state of metal compounds may occur in the rhizosphere of plants that actively oxygenate their root zone. These plants, such as rice, transport molecular oxygen from the aboveground tissue into the root zone. Oxidation of technetium in the rice (*Oryza sativa* L.) was shown to result from this mechanism (Sheppard and Evenden, 1991).

Bioaccumulation and Solubilization of Metal Ions Through Mineralization: As is true of all living cells, soil microbial biomass incorporates essential metals into cell substituents. Also, accumulation of metallic cations on or within living cells can greatly exceed these metabolic requirements. This bioaccumulation is delineated by the achievement of cellular loadings of metals far beyond any levels that might be associated with cellular metabolism. This process is exemplified among the bacterial populations by strontium accumulation by *Micrococcus luteus* (Faison et al., 1990) and precipitation of cadmium by *Clostridium thermoaceticum*. In the latter example, cadmium is accumulated at the cell surface as well as in the surrounding medium (Cunningham and Lundie, 1993). Similarly, cesium (Tomioka et al., 1992) and zinc (Sakurai et al., 1990) can be accumulated by bacteria. It must be noted that the cells need not be living for augmentation of cell-associated metals to occur. Kurek et al. (1982) found that dead microorganisms were capable of absorbing more cadmium than did the live cells used in their studies.

The return of biomass-associated metals to soil water-soluble or mineral forms occurs through mineralization of the microbial biomass following death of the cell. As opposed to the cycling of carbon, nitrogen, and sulfur discussed above, the return of the metal to a mineral state likely provides little benefit to the decomposer population and is generally not directly enzymatically mediated. This metal cycling could be classified as a gratuitous product of microbial cycling of carbon.

Other Soil Metal Transformations: As was indicated above, soil minerals may also be methylated and transalkylated by soil microbial populations. The attachment of alkyl groups to the metal may alter the metal mobility within the ecosystem through enhanced volatility and water solubility. For example, monomethyl and dimethylmercury are volatile. Therefore, their formation in soils can result in reduction in the level of mercury contamination at the specific soil site. (See Tate, 1987, for further discussion of the behavior of metals in soil ecosystems.)

14.7.4 Managing Soils Contaminated with Metals

Metal-contaminated soils present a unique challenge to the soil microbiologist charged with the responsibility for contributing to the development of renovation plans. Conceptually, most microbial reclamation procedures involve management of availability of the contaminant to the soil community of degraders and encouragement of mineralization of the offending substance. In contrast, reduction of elevated metal loadings involves either removal of the contaminant from the site (leaching or excavation of the soil itself) or modification of the properties of the polluted soil in a manner that encourages sequestering of the toxic metal into biologically unavailable forms. Since the three major soil properties controlling metal mobility are pH, redox potential, and organic matter content, traditional soil reclamation involving metal management has included manipulation of these parameters.

Some shifting of soil metals from water-soluble to the particulate fractions can be achieved through soil amendment with lime and/or organic matter. Increase in soil pH to near neutrality through the lime amendment encourages a change in the equilibrium between soluble and insoluble metal forms. With the addition of organic matter, the soluble metal loadings are reduced through complexing of the metal with the organic material itself. (See Tate, 1987, for a more detailed discussion of control of soil metals through these mechanisms). Either of these nonbiological management procedures provide only a temporary solution to the metal toxicity. For example, once the organic matter retaining the metal is decomposed, any associated cations are again released into the interstitial soil water. Similarly, from the view of management of soil pH, the natural tendency of soil systems is to become increasingly acidic. Thus, continued monitoring and adjustment of soil acidity may be necessary in particularly sensitive ecosystems if on-site retention of the metal contaminant is essential.

Neither of these procedures, pH adjustment and organic matter amendment, presents any difficulty to the development or continued function of the soil microbial community. Organic matter encourages the stability of soil decomposer communities through the provision of carbon and energy. Furthermore, since the soil pH is raised to values approximating neutrality (i.e., within the range of optimal function of the majority of the members of the soil biological community), negative effects on biological parameters are not anticipated.

A more intensive stewardship of metal-contaminated sites may actually involve alteration of the biological properties of the system. Activity of both the aerobic and the anaerobic microbial populations can be manipulated to encourage sequestering of metal contaminants. A logical extension of understanding the dynamics of sulfidogenesis and chemical interactions of the sulfide produced with soil metals is management of the cycle to purify soil interstitial waters. Provision of reducing conditions in the presence of sulfate ions may allow production of sufficient sulfide to reduce soluble metal loadings to acceptable levels. Metal management through control of the sulfur cycle is commonly accomplished by creation of artificial wetlands. The difficulty with this procedure is that the wetland created must be maintained in an anoxic condition. Once oxygen is introduced, sulfur oxidizers utilize the sulfide as an energy source, thereby liberating the associated metal ion.

15.8 Conclusions

The foregoing analysis of the sulfur, phosphorus, and mineral cycles in soils underscores not only their essentiality to total ecosystem biomass productivity but also a primary role in reversal of adverse situations resulting from anthropogenic activities. That is, excessively acidic soils may result from elevated influxes of reduced sulfur compounds, heavy metal contamination may preclude sustenance of soil microbial communities, and mobility of soil phosphate may accelerate eutrophication of regional surface waters. Ameliorating these unfavorable soil or water conditions, while optimizing biomass productivity in soil ecosystems containing restrictive concentrations of the essential components of these cycles, requires a firm appreciation for not only the biological aspects of the biogeochemical cycles but also the chemistry of formation and dissolution of soil minerals. These biogeochemical cycles are highlighted by the major role of soil primary and secondary minerals in maintaining concentrations of bioavailable water-soluble nutrients. This intersection of the mineral with the biological world and the implications of societal mismanagement of soil ecosystems involving soil phosphate, sulfur, and metal loadings can be managed to encourage development of sustainable ecosystems.

References

Ahonen, L., and O. H. Tuovinen. 1990. Kinetics of sulfur oxidation at suboptimal temperatures. Appl. Microbiol. 56:560–562.

Akers, H. A. 1983a. Isolation of the siderophore schizokinen from soil of rice fields. Appl. Environ. Microbiol. 45:1704–1706.

Akers, H. A. 1983b. Multiple hydroxamic acid microbial iron chelators (siderophores) in soils. Soil Sci. 135:156–159.

Alexander, M. 1977. Soil Microbiology. John Wiley & Sons, N.Y. 467 pp.

Anderson, D. W., S. Saggar, J. R. Bettany, and J. W. B. Stewart. 1981. Particle size fractions and their use in studies of soil organic matter: I. The nature and distribution of forms of carbon, nitrogen, and sulfur. Soil Sci. Soc. Am. J. 45:767–772.

Ankenbauer, R. G., T. Toyokuni, A. Staley, K. L. Rinehart Jr., and C. D. Cox. 1988. Synthesis and biological activity of pyrochelin, a siderophore of *Pseudomonas aeruginosa*. J. Bacteriol. 170:5344–5351.

Attoe, O. J., and R. A. Olson. 1966. Factors affecting rate of oxidation in soils of elemental sulfur and that added in rock phosphate-sulfur fusions. Soil Sci. 101:317–325.

Babich, H., and G. Stotzky. 1983. Physicochemical factors of natural reservoirs affect the transformation and exchange of heavy metals toxic to microbes. *In* R. Hallberg (ed.), Environmental Biogeochemistry. Ecol. Bull. (Stockholm) 35:315–323.

Bar-Ness, E., Y. Chen, Y. Hadar, H. Marschner, and V. Roemheld. 1991. Siderophores of *Pseudomonas putida* as an iron source for dicot and monocot plants. Plant Soil 130:231–241.

Batchelor, B., and A. W. Lawrence. 1978. A kinetic model for autotrophic denitrification using elemental sulfur. Water Res. 12:1075–1084.

Baucher, M. J. 1973. Contamination of soil and vegetation near a zinc smelter by zinc, cadmium, copper, and lead. Environ. Sci. Technol. 7:131–135.

Beffa, T., M. Berczy, and M. Aragno. 1991. Chemolithoautotrophic growth on elemental sulfur (S) and respiratory oxidation of S by *Thiobacillus versutus* and another sulfur-oxidizing bacterium. FEMS Microbiol. Lett. 84:285–290.

Bertramson, B. R., M. Fried, and S. L. Tisdale 1950. Sulfur studies of Indiana soils and crops. Soil Sci. 70:27–41.

Bettany, J. R., S. Saggar, and J. W. B. Stewart. 1980. Comparison of the amounts and forms of sulfur in soil organic matter fractions after 65 years of cultivation. Soil Sci. Soc. Am. J. 44:70–74.

Bettany, J. R., J. W. B. Stewart, and S. Saggar. 1979. The nature and forms of sulfur in organic matter fractions of soils selected along an environmental gradient. Soil Sci. Soc. Am. J. 43:981–985.

Blais, J. S., W. D. Marshall, and R. Knowles. 1988. Toxicity of alkyl lead salts to anaerobic nitrogen transformations in sediment. J. Environ. Qual. 17:457–462.

Bopp, L. H., A. M. Chakrabarty, and H. L. Ehrlich. 1983. Chromate resistance plasmid in *Pseudomonas fluorescens*. J. Bacteriol. 155:1105–1109.

Brune, D. C. 1989. Sulfur oxidation by phototrophic bacteria. Biochim. Biophys. Acta 975:189–221.

Bryant, R. D., J. W. Costerton, and E. J. Laishley. 1984. The role of *Thiobacillus albertis* glycocalyx in the adhesion of cells to elemental sulfur. Can. J. Microbiol. 30:81–90.

Carlson, L., E. B. Lindström, K. B. Halalberg, and O. H. Tuovinen. 1992. Solid-phase products of bacterial oxidation of arsenical pyrite. Appl. Environ. Microbiol. 58:1046–1049.

Chang, F.-H., and F. E. Broadbent. 1982. Influence of trace metals on some soil nitrogen transformations. J. Environ. Qual. 11:1–4.

Chapman, S. J. 1989. Oxidation of micronized elemental sulfur in soil. Plant Soil 116:69–76.

Cody, Y. A., and D. C. Gross. 1987. Characterization of pyoverdin$_{pss}$, the fluorescent siderophore produced by *Pseudomonas syringae* pv. *syringae*. Appl. Environ. Microbiol. 53:928–934.

Cole, M. A. 1977. Lead inhibition of enzyme synthesis in soil. Appl. Environ. Microbiol. 33:262–268.

Cunningham, D. P., and L. L. Lundie, Jr. 1993. Precipitation of cadmium by *Clostridium thermoaceticum.* Appl. Environ. Microbiol. 59:7–14.

David, M. B., J. M. Mitchell, and J. P. Nakas. 1982. Organic and inorganic sulfur constituents of a forest soil and their relationship to microbial activity. Soil Sci. Soc. Am. J. 46:847–852.

Dean-Ross, D., and A. L. Mills. 1989. Bacterial community structure and function along a heavy metal gradient. Appl. Environ. Microbiol. 55:2002–2009.

Derylo, M., and A. Skorupska. 1992. Rhizobial siderophore as an iron source for clover. Physiol. Plant 85:549–553.

Diels, L., and M. Mergeay. 1990. DNA probe-mediated detection of resistant bacteria from soils highly polluted by heavy metals. Appl. Environ. Microbiol. 56:1485–1491.

Doelman, P., and L. Haanstra. 1979. Effect of lead on soil respiration and dehydrogenase activity. Soil Biol. Biochem. 11:475–479.

Doelman, P., and L. Haanstra. 1989. Short- and long-term effects of heavy metals on phosphatase activity in soils: An ecological dose-response model approach. Biol. Fert. Soils 8:235–241.

Dressler, C., U. Kües, D. H. Nies, and B. Friedrich. 1991. Determinants encoding resistance to several heavy metals in newly isolated copper-resistant bacteria. Appl. Environ. Microbiol. 57:3079–3085.

Duxbury, T., and B. Bicknell. 1983. Metal-tolerant bacterial populations from natural and metal-polluted soils. Soil Biol. Biochem. 15:243–250.

Eary, L. E., and D. Rai. 1991. Chromate reduction by subsurface soils under acidic conditions. Soil Sci. Soc. Am. J. 55:676–683.

Espejo, R. T., B. Escobar, E. Jedlicki, P. Uribe, and R. Badilla-Ohlbaum. 1988. Oxidation of ferrous iron and elemental sulfur by *Thiobacillus ferrooxidans.* Appl. Environ. Microbiol. 54:1694–1699.

Espejo, R. T., and P. Romero. 1987. Growth of *Thiobacillus ferrooxidans* on elemental sulfur. Appl. Environ. Microbiol. 53:1907–1912.

Faison, B. D., C. A. Cancel, S. N. Lewis, and H. I. Adler. 1990. Binding of dissolved strontium by *Micrococcus luteus.* Appl. Environ. Microbiol. 56:3649–3656.

Firestone, M. K., and J. M. Tiedje. 1975. Biodegradation of metal-nitrilotriacetate complexes by a *Pseudomonas* species: Mechanism of reaction. Appl. Microbiol. 29:758–764.

Frederick, L. R., and R. L. Starkey. 1948. Bacterial oxidation of sulfur in pipe sealing mixtures. J. Am. Water Works Assoc. 40:729–736.

Gebhardt, N. A., R. K. Thauer, D. Linder, P.-M. Kaulfers, and N. Pfennig. 1985. Mechanism of acetate oxidation to carbon dioxide with elemental sulfur in *Desulfuromonas acetoxidans.* Arch. Microbiol. 141:392–398.

Grayston, S. J., W. Nevell, and M. Wainwright. 1986. Sulfur oxidation by fungi. Trans. Br. Mycol. Soc. 87:193–198.

Haefeli, C., C. Franklin, and K. Hardy. 1984. Plasmid-determined resistance in *Pseudomonas stutzeri* isolated from a silver mine. J. Bacteriol. 158:389–392.

Hashimoto, S., K. Furukawa, and M. Shioyama. 1987. Autotrophic denitrification using elemental sulfur. J. Ferment. Technol. 65:683–692.

Harrison, A. P., Jr. 1984. The acidophilic thiobacilli and other acidophilic bacteria that share their habitat. Annu. Rev. Microbiol. 38:265–292.

Hoefte, M., K. Y. Seong, E. Jurkevitch, and W. Verstraete. 1991. Pyroverdin production by the plant growth beneficial *Pseudomonas* strain 7NSK2: Ecological significance in soil. Plant Soil 130:249–257.

Holm, H. W., and M. F. Cox. 1975. Transformation of elemental mercury by bacteria. Appl. Microbiol. 29:491–494.

Huber, H., and K. O. Stetter. 1989. *Thiobacillus prosperus* sp. nov., represents a new group of halotolerant metal-mobilizing bacteria isolated from a marine geothermal field. Arch. Microbiol. 151:479–485.

Huber, H., and K. O. Stetter. 1990. *Thiobacillus cuprinus* new species, A novel facultatively organotrophic metal-mobilizing bacterium. Appl. Environ. Microbiol. 56:315–322.

Jain, D. K., and R. D. Tyagi. 1992. Leaching of heavy metals from anaerobic sewage sludge by sulfur-oxidizing bacteria. Enzyme Microb. Technol. 14:376–383.

Janzen, H. H., and J. R. Bettany. 1987a. Oxidation of elemental sulfur under field conditions in central Saskatchewan Canada. Can. J. Soil Sci. 67:609–618.

Janzen, H. H., and J. R. Bettany. 1987b. Measurement of sulfur oxidation in soils. Soil Sci. 143:444–452.

Janzen, H. H., and J. R. Bettany. 1987c. The effect of temperature and water potential on sulfur oxidation in soils. Soil Sci. 144:81–89.

Jordan, M. J. 1975. Effects of zinc smelter emissions and fire on a chestnut-oak woodland. Ecology. 56:78–91.

Jordan, J. J., and M. P. Lechevalier. 1975. Effects of zinc-smelter emissions on forest soil microflora. Can. J. Microbiol. 21:1855–1865.

Kelly, J. J. and Robert L. Tate III. 1998. Effects of heavy metal contamination and remediation on soil microbial communities in the vicinity of a zinc smelter. J. Environ. Qual. 27:609–617.

Kelley, W. J., and D. C. Reanney. 1984. Mercury resistance among soil bacteria: Ecology and transferability of genes encoding resistance. Soil Biol. Biochem. 16:1–8.

Kennedy, K. A., P. A. Addison, and D. G. Maynard. 1988. Effect of elemental sulfur on the vegetation of a lodgepole pine stand. Environ. Pollut. 51:121–130.

Killham, K., N. D. Lindley, and M. Wainwright. 1981. Inorganic sulfur oxidation by *Aureobasidium pullans*. Appl. Environ. Microbiol. 42:629–631.

Kloepper, J. W., J. Leong, M. Teintze, and M. N. Schroth. 1980. Enhanced plant growth by siderophores produced by plant growth-promoting rhizobacteria. Nature (London) 286:885–886.

Kuenen, J. G., L. A. Robertson, and O. H. Tuovinen. 1992. The genera *Thiobacillus, thiomicrospira,* and *Thiosphaera*. Pp. 2197–2213. *In* A. Balows, H. G. Trüper, M. Dwarkin, W. Harder, and K.-H. Schleifer (eds.), The Prokaryotes. Springer-Verlag, NY.

Kurek, E. 1979. Oxidation of inorganic sulfur compounds by yeast. Acta Microbiol. Pol. 28:169–172.

Kurek, E. 1985. Elemental sulfur and thiosulfate oxidation by *Rhodotorula* sp. Zentrabl. Microbiol. 140:497–500.

Kurek, E., J. Czaban, and J.-M. Bollag. 1982. Sorption of cadmium by microorganisms in competition with other soil constituents. Appl. Environ. Microbiol. 43:1011–1015.

Laishley, E. J., R. D. Bryant, B. W. Kobryn, and J. B. Hyne. 1986. Microcrystalline structure and surface area of elemental sulfur as factors influencing its oxidation by *Thiobacillus albertis.* Can. J. Microbiol. 32:237–242.

Lawrence, J. R., and J. J. Germida. 1988. Relationship between microbial biomass and elemental sulfur oxidation in agricultural soils. Soil Sci. Soc. Am. J. 52:672–677.

Lawrence, J. R., V. V. S. R. Gupta, and J. J. Germida. 1988. Impact of elemental sulfur fertilization on agricultural soils. II. Effects on sulfur-oxidizing populations and oxidation rates. Can. J. Soil Sci. 68:475–484.

Lee, A., C. C. Boswell, and J. H. Watkinson. 1988a. Effect of particle size on the oxidation of elemental sulfur: Thiobacilli numbers, Soil Sulfate, and its availability to pasture. N. Z. J. Agric. Res. 30:373–386.

Lee, A., J. H. Watkinson, and D. R. Lauren. 1988b. Factors affecting oxidation rates of elemental sulfur in a soil under a ryegrass dominant sward. Soil Biol. Biochem. 20:809–816.

Lee, A., J. H. Watkinson, G. Orbell, J. J. Bagyara, and D. R. Lauren. 1987. Factors influencing dissolution of phosphate rock and oxidation of elemental sulfur in some New Zealand soils. N. Z. J. Agric. Res. 30:373–386.

Lettl, A., O. Langkramer, and V. Lochman. 1981. Some factors influencing production of sulfate by oxidation of elemental sulfur and thiosulfate in upper horizons of spruce forest soils. Folia Microbiol. 26:158–163.

Lighthart, B., J. Baham, and V. V. Volk. 1983. Microbial respiration and chemical speciation in metal-amended soils. J. Environ. Qual. 12:543–548.

Lovley, D. R., and E. J. P. Phillips. 1986. Organic matter mineralization with reduction of ferric iron in anaerobic sediments. Appl. Environ. Microbiol. 51:683–689.

Lovley, D. R., and E. J. P. Phillips. 1992. Reduction of uranium by *Desulfovibrio desulfuricans.* Appl. Environ. Microbiol. 58:850–856.

Lovley, D. R., E. J. P. Phillips, Y. A. Gorby, and E. R. Landa. 1991. Microbial reduction of uranium. Nature (London) 350:413–416.

Martin, J. P., J. O. Ervin, and R. A. Shepherd. 1966. Decomposition of the iron, aluminum, zinc, and copper salts or complexes of some microbial and plant polysaccharides in soil. Soil Sci. Soc. Am. Proc. 30:196–200.

Maynard, D. G., J. J. Germida, and P. A. Addison. 1986. The effect of elemental sulfur on certain chemical and biological properties of surface organic horizons of a forest soil. Can. J. For. Res. 16:1050–1054.

McCaskill, M. R., and G. J. Blair. 1987. Particle size and soil texture effects of elemental sulfur oxidation. Agron. J. 79:1079–1083.

Meulenbert, R., J. T. Pronk, W. Hazeu, P. Bos, and J. G. Kuenen. 1992. Oxidation of reduced sulphur compounds by intact cells of *Thiobacillus acidophilus.* Arch. Microbiol. 157:161–168.

Mills, A. L. 1985. Acid mine waste drainage: Microbial impact on the recovery of soil and water ecosystems. Pp. 35–81. *In* R. L. Tate III and D. A. Klein (eds.), Soil Reclamation Processes. Microbiological Analyses and Applications. Dekker, NY.

Min, H., and S. H. Zinder. 1990. Isolation and characterization of a thermophilic sulfate-reducing bacterium *Desulfotomaculum thermoacetoxidans* new species. Arch. Microbiol. 153:399–404.

Molla, M. A. Z., A. A. Chowdhury, A. Islam, and S. Hoque. 1984. Microbial mineralization of organic phosphate in soil. Plant Soil 78:393–394.

Moser, U. S., and R. V. Olson. 1953. Sulfur oxidation in four soils as influenced by soil moisture tension and sulfur bacteria. Soil Sci. 76:251–257.

Nealson, K. H., and C. R. Myers. 1992. Microbial reduction of manganese and iron: New approaches to carbon cycling. Appl. Environ. Microbiol. 58:439–443.

Nor, Y. M., and M. A. Tabatabai. 1977. Oxidation of elemental sulfur in soils. Soil Sci. Soc. Am. J. 41:736–741.

Nordgren, A., E. Bååth, and B. Söderström. 1983. Microfungi and microbial activity along a heavy metal gradient. Appl. Environ. Microbiol. 45:1829–1837.

Pepper, I. L., and R. H. Miller. 1978. Comparison of the oxidation of thiosulfate and elemental sulfur by two heterotrophic bacteria and *Thiobacillus thiooxidans*. Soil Sci. 126:9–14.

Phillips, S. E., and M. L. Taylor. 1976. Oxidation of arsenite to arsenate by *Alcaligenes faecalis*. Appl. Environ. Microbiol. 32:392–399.

Powell, P. E., G. R. Cline, C. P. P. Reid. and P. J. Szaniszlo. 1982. Occurrence of hydroxamate siderophore iron chelators in soils. Nature (London) 287:833–834.

Pronk, J. T., J. C. De Bruyn, P. Bos, and J. G. Kuenen. 1992. Anaerobic growth of *Thiobacillus ferrooxidans*. Appl. Environ. Microbiol. 58:2227–2230.

Said, W. A., and D. L. Lewis. 1991. Quantitative assessment of the effects of metals on microbial degradation of organic chemicals. Appl. Environ. Microbiol. 57:1498–1503.

Sakurai, I., Y. Kawamura, H. Koike, Y. Inoue, Y. Kosako, T. Nakase, Y. Knodou, and S. Sakurai. 1990. Bacterial accumulation of metallic compounds. Appl. Environ. Microbiol. 56:2580–2583.

Schedel, M., and H. G. Trueper. 1980. Anaerobic oxidation of thiosulfate and elemental sulfur in *Thiobacillus denitrificans*. Arch. Microbiol. 124:205–210.

Schmidt, T., R.-D. Stoppel, and H. G. Schlegel. 1991. High-level nickel resistance in *Alcaligenes xylosoxidans* 31A and *Alcaligenes eutrophus* KT02. Appl. Environ. Microbiol. 57:3301–3309.

Sheppard, S. C., and W. G. Evenden. 1991. Can aquatic macrophytes mobilize technetium by oxidizing their rhizosphere? J. Environ. Qual. 20:738–744.

Stevenson, F. J. 1982. Humus Chemistry. Genesis, Composition, Reactions. John Wiley & Sons, NY. 443 pp.

Stevenson, F. J. 1986. Cycles of soil: Carbon, Nitrogen, Phosphorus, Sulfur, Micronutrients. John Wiley & Sons, NY 380 pp.

Stott, D. E., W. A. Dick, and M. A. Tabatabai. 1985. Inhibition of pyrophosphatase activity in soils by trace elements. Soil Sci. 139:112–117.

Strojan, C. J. 1978. Forest leaf litter decomposition in the vicinity of a zinc smelter. Oecologia (Berlin) 32:203–212.

Sugio, T., Y. Tsjita, K. Hirayama, K. Inagaka, and T. Tano. 1988a. Mechanism of tetravalent manganese reduction with elemental sulfur by *Thiobacillus ferrooxidans*. Agric. Biol. Chem. 52:185–190.

Sugio, T., Y. Tsujita, K. Inagaki, and T. Tano. 1990. Reduction of cupric ions with elemental sulfur by *Thiobacillus ferrooxidans*. Appl. Environ. Microbiol. 56:693–696.

Sugio, T., Y. Tsujita, T. Katagiri, K. Inagaki, and T. Tano. 1988b. Reduction of molybdic ions with elemental sulfur by *Thiobacillus ferrooxidans*. J. Bacteriol. 170:5956–5959.

Sugio, T., W. Mizunashi, K. Inagaki, and T. Tano. 1987. Purification and some properties of sulfur ferric ion oxidoreductase from *Thiobacillus ferrooxidans*. J. Bacteriol. 169:4916–4922.

Suzuki, I., T. L. Takeuchi, T. D. Yuthasastrakosol, and J. K. Oh. 1990. Ferrous iron and sulfur oxidation and ferric iron reduction activities of *Thiobacillus ferrooxidans* are affected by growth on ferrous iron and sulfur or a sulfide ore. Appl. Environ. Microbiol. 56:1620–1626.

Swank., W. T., J. W. Fiktzgerald, and J. T. Ash. 1984. Microbial transformation of sulfate in forest soils. Science 223:182–184.

Tabatabai, M. A., and A. A. Al-Khafaji. 1980. Comparison of nitrogen and sulfur mineralization in soils. Soil Sci. Soc. Am. J. 44:1000–1006.

Tate, R. L., III. 1987. Soil Organic Matter: Biological and Ecological Effects. John Wiley & Sons, NY. 291 pp.

Tomioka, N., H. Uchiyama, and O Yagi. 1992. Isolation and characterization of cesium-accumulating bacteria. Appl. Environ. Microbiol. 58:1019–1023.

Traina, S. J., G. Sposito, D. Hesterberg, and U. Kafkafi. 1986. Effect of pH and organic acids on orthophosphate solubility in an acidic, montmorillonitic soil. Soil Sci. Soc. Am. J. 50:45–52.

Visniac, W. V., and M. Santer. 1957. The thiobacilli. Bacteriol. Rev. 21:195–213.

Visser. S. and D. Parkinson. 1989. Microbial respiration and biomass in soil of a lodgepole pine stand acidified with elemental sulfur. Can. J. For. Res. 19:955–961.

Volger, K. G., and W. W. Umbreit. 1941. The necessity for direct contact in sulfur oxidation by *Thiobacillus thiooxidans*. Soil Sci. 51:331–337.

Watkinson, J. H. 1989. Measurement of the oxidation rate of elemental sulfur in soil. Aust. J. Soil res. 27:365–376.

Wickliff, C., H. J. Evans, K. R. Carter, and S. A. Russell. 1980. Cadmium effects on the nitrogen fixation system of red alder. J. Environ. Qual. 9:180–184.

Yagi, S., S. Katai, and T. Kimura. 1971. Oxidation of elemental sulfur to thiosulfate by *Streptomyces*. Appl. Microbiol. 22:157–159.

Yang, C.-H., J. A. Menge., and D. A. Cooksey. 1993. Role of copper resistance in competitive survival of *Pseudomonas fluroescens* in soil. Appl. Environ. Microbiol. 59:580–584.

Zillig, W., S. Yeats, I. Holz, A. Boeck, M. Rettenberger, F. Gropp, and G. Simon. 1986. *Desulfurolobus ambivalens* new genus, new species, An autotrophic archaebacterium facultatively oxidizing or reducing sulfur. Syst. Appl. Microbiol. 8:197–203.

Zoephel, A., M. C. Kennedy, H. Beinert, and P. M. H. Kroneck. 1988. Investigations on microbial sulfur respiration. 1. Activation and reduction of elemental sulfur in several strains of eubacteria. Arch. Microbiol. 150:72–77.

C h a p t e r **16**

Principles of
Bioremediation

A soil ecosystem requiring bioremediation can be pictured to be one in which the biological community has failed to function as might have been anticipated. In an undisturbed system, biologically decomposable substances entering the soil are mineralized. The concentration of the residual organic material is the difference between the amendment and decomposition rates. In a soil site receiving continued regular amendments, it is reasonable to assume that some equilibrium level of residual organic matter will be achieved. If the input ceases or the material is added at long time intervals, it is equally reasonable to anticipate that the added substance could be totally eliminated. In contrast, in our hypothetically polluted soil, a toxic organic substance, albeit biodecomposable, has accumulated to concentrations that inhibit and/or interfere with the normal function of the ecosystem; that is, the system failed. Note that this statement does not exempt anthropogenic management failure from consideration. People are part of and major influences on the ecosystem. Thus, the failure of the microbial community to rid its environment of a toxic organic material could be a result of its being overwhelmed by mismanagement of the soil.

Considering that the pollutant accumulated to toxic levels, we can conclude that normal soil processes are inadequate to ameliorate the situation. The inability of the soil microbes to prevent the accumulation of the organic pollutant could be due to a number of factors:

- No individual or groups of microbial species exist in the soil capable of mineralizing or detoxifying the contaminant.

- The rate of input of the pollutant may be greater than its maximum allowable decomposition rate, resulting in ecosystem degradation.

- Chemical, physical, or biological limitations of the decomposers may exist or may be created by the contamination of the soil system, thereby preventing microbial removal of the pollutant (e.g., anoxic conditions may develop preventing occurrence of obligatory aerobic processes).

- The pollutant(s) were amended to the soil at levels that are toxic to the decomposers themselves.

- Conditions may be optimal for decomposition to occur, but the decomposer population could be physically separated from the decomposable pollutant. That is, the organic compound may be physically or chemically sequestered.

- Decomposition of the pollutant may create conditions in the soil wherein further mineralization is precluded (e.g., acidification of the system and depletion of molecular oxygen).

The objective of bioremediation is to manage the soil physical, chemical, and biological properties in a manner that overcomes these limitations to biological activity and optimizes the capability of the microbial community to detoxify the soil.

The scientific basis for the management of the soil microbial community for remediation is built on the same biological and chemical foundation as occur in nonpolluted soil systems. This commonality between functional native systems and those with excessive accumulations of toxic organic chemicals is exemplified by considering the processes involved with organic matter accumulation in soil. Amendment of any soil with plant biomass fuels the activity of the biological community and results in the production of microbial biomass and humus. With a normally functioning soil system, this accumulation of organic matter has a positive influence on ecosystem stability. Incorporation of plant biomass components into soil humus results, in part, from covalent linkages with soil humic substances. The organic components may also be occluded from decomposers by diffusing into soil micropores, by noncovalent stabilization within the three-dimensional structure of humic acids, or by sorption onto soil minerals. Since these are time-dependent processes, a slowly mineralized substance has greater chance of becoming sequestered or sorbed onto or within clay or humic substances than do more biodegradation-susceptible compounds. Thus, an easily metabolized organic compound, such as glucose or a protein, is normally fully mineralized by the soil microbial community in a short time, whereas cellulose and lignin of plant biomass are commonly stabilized within the soil structure. Also, small pieces of intact plant tissue can be protected from microbial attack by becoming encased in soil aggregates. In this situation, the sequestration results from physical adhesion of the soil particulate to small pieces of plant material. From the view of the normally functioning soil and nontoxic organic substances, these organic matter stabilizing processes result in enhancement of the soil system or an improved soil quality. In contrast, similar protection of toxicants can result in ecosystem degradation and a reduction of ecosystem or environmental health. The processes are the same, but divergent consequences to the ecosystem function occur.

In the negative situation, disruption of soil aggregate structure may actually become a positive management solution. For pollutant management, there is a need to encourage rapid decomposition of toxicants to reduce the impact of sequestration on accumulation of the pollutant. With long-term polluted sites, the system may have to be managed to reduce the sequestration that has already occurred and to increase the bioavailability of the toxicant. In the latter situation, reduction in or total destruction of the soil aggregate structure could be considered in the management plan initially to liberate the hidden pollutant, followed by management of the soil to improve soil structure. The latter objective could be accelerated by introduction of plant communities.

Contamination of a soil ecosystem with biodegradable organic substances creates a unique situation for the microbes involved with their decomposition, as well as for the site manager charged with evaluating the potential for bioremediation and the associated ecosystem risk. Thus, this chapter is presented with the overall objective of providing a foundation for considering the biological, chemical, and physical principles underlying or controlling the potential for bioremediation of soil ecosystem and the ultimate limitations or risks associated with such remediated systems. This analysis will necessarily commence with a consideration of the microbes involved, but the primary emphasis will involve consideration of the physical and chemical limitations of the processes as they relate to bioavailability of the contaminant.

16.1 Foundational Concepts of Bioremediation

16.1.1 Bioremediation Defined

Soil bioremediation is defined as the management or utilization of the soil biological community to detoxify, immobilize, or mineralize organic contaminants in the soil ecosystem. Specifically, this process involves conversion of the pollutant totally to mineral forms (e.g., carbon dioxide, methane, ammonium, hydrogen sulfide, etc.), its conversion to forms with no significant negative health traits (generally viewed from a human health viewpoint, but this discussion includes the concept of "health" of the total biological community), and/or immobilization via covalent linkage to soil humic acids (humification). The biological community generally managed for bioremediation is the native soil microbial community, but higher plants (phytoremediation) may also be manipulated to enhance toxicant amelioration, especially for remediation of metal-contaminated soils. The primary emphasis of this chapter is management of soil microbes (including those of the rhizosphere) to optimize reduction of the risk associated with an organic toxicant in soil. These microbes may be indigenous to the contaminated soil system or amended to the soil during the remediation process. That is, the metabolic capabilities of the indigenous microbial community may be enhanced with genetically engineered microbial strains (gems) or other laboratory-derived or -cultured microbial strains. Site management includes manipulation of soil chemical and physical properties to optimize both bioavailability of the pollutant and the microbial capacity to transform it. Frequently bioavailability is discussed in a limited sense — that is, in relationship to human health or risk. For this discussion, bioavailability will be considered to be the availability of the organic pollutant to the totality of the biological community, especially the soil microbes involved in its detoxification and/or mineralization.

16.1.2 Conceptual Unity of Bioremediation Science

The foundation for our understanding of the processes involved in bioremediation is derived from a unity of biochemical-microbiological-ecological principles. These supporting principles for bioremediation are as follows:

- *All relationships of comparative biochemistry apply equally in the axenic culture of microorganisms and in the field soil.* A compound mineralized by one or a combination of pathways or processes in culture can be anticipated to be similarly transformed in a field soil. A caveat to this rule is that the rate and extent of the transformation is controlled by the environmental situation of the microbes involved, including stresses imposed on the microbes through their interactions with other members of the biological community. Thus, the reactions may proceed at faster or slower rates, and different intermediates may accumulate in the field due to limiting conditions (nutrient, physical, chemical, or other). Only a small proportion of the soil microbes have been cultured in the laboratory and their biochemical capacities characterized. Variations may exist in many biochemical processes that have yet to be described.

- *All microbial growth requirements are the same whether in the laboratory culture or in the field.* If a growth factor is required in culture for a specific enzyme to function, it must also be supplied in the soil system. Similarly, the microbes and enzymes are controlled by the same physical limitations (e.g., pH, temperature) in the field and culture. The microbes may be functioning at the extremes of their growth margin — hence a slower process rate is seen — but the organisms will not function beyond their normal range without some environmental compensation (a phenomenon also observed in axenic culture; see Chapter 5). For example, provision by root exudates or by other members of the microbial community of a cofactor whose de novo synthesis is precluded by high temperature may allow the microbe to function at a higher temperature.

- *All limitations resulting from ecological interactions apply, including the need to accommodate the presence of other microbes as well as adaptation to the physical and chemical properties of the microsite wherein the microbe lives.* This principles has most likely been the most overlooked when developing genetically engineered strains or when selecting variant strains of microbes for soil inoculation. A microbe amended to a soil ecosystem with the intent of stimulation of bioremediation processes must not only possess the requisite genetic information and be capable of expressing that capability *in situ*, but it must also have the capacity to become, at least temporarily, a part of the overall soil microbial community. That is, the alien microbe must be able to express the requisite biochemical activity in the community, become established in that community, and accommodate or adjust to all means of interactions (amensalism, competition, etc.) with its neighbors — at least until its presence is no long required for bioremediation.

These underlying principles of bioremediation processes underscore the fact that successful management of the soil microbial community to ameliorate contamination with organic pollutants is truly an interdisciplinary endeavor. All involved must have an appreciation of soil science in general as well as the basic principles of soil microbiology, microbial physiology, and ecology in general. This interdisciplinary aspect of bioremediation technology is further accentuated when the impact of environmental engineers in applying the principles in the field is considered.

16.1.3 Complexity of Remediation Questions

Bioremediation involves mitigation of chemical contamination of soils and thereby a reduction of the impact of the presence of the pollutants on ecosystem function. By the nature of the vast array of organic substances and of the diversity in the extent of their encounter with soil systems, the soil system managed for bioremediation can range from a few square meters to many hectares in area. The degree of intervention may range from the simple working of a compacted soil surface and amendment with fertilizer to the implementation of complex, engineered systems for pollutant retention and soil management developed to meet the needs for remediation of highly contaminated industrial sites. Definition of the breadth of chemicals contaminating soils extends far beyond the purview of this chapter, but the range in structural complexity can be gained from such reviews as that of Swobada-Colberg (1995) on xenobiotic chemicals.

Implications of the importance of bioremediation not only extend across those aspects associated with quality of the particular soil of interest, but also encompass more general concerns of environmental or ecosystem health. First concerns with bioremediation must involve mineralization of the organic substances of interest or immobilization of the inorganic substance, as is involved with phytoremediation of metal-contaminated soils, but these initial endeavors must be executed with considerations of the effect of system management on the general health of the total environment in mind. That is, management of the decomposer population has implications far beyond inducing a simple reduction in soil toxicant loading. For example, utilization of mycorrhizal fungi as mineralizers of organic contaminants also alters nutrient and water dynamics of the associated plant community. This enhancement of the mycorrhizal association will result in a stimulation of plant biomass productivity, both aboveground and belowground. This increased plant growth will result in augmented activity of the rhizosphere community and an improved soil aggregate structure (see Chapter 1), primarily due to the stimulation of biological activity via exudate production.

A perhaps expected effect of the increased plant growth relates to soil water dynamics. In a soil system without a plant community, the direction of movement of the soil water, and any solutes (contaminants) contained therein, would be predominantly downward. With the increased water use of the plant in the system, a significant portion of the soil water is drawn to the root, thereby decreasing the quantity lost through percolation beyond the root zone. Thus, less pollutant is lost from the system by leaching to groundwater and greater quantities are retained within the biologically activity zone of the soil [see Clothier and Green (1997) for a review of the impact of roots on soil water dynamics].

16.2 The Microbiology of Bioremediation

Consideration of the principles of bioremediation can be divided into two primary areas: (1) Properties of the microbes themselves affecting their interaction with and decomposition of the pollutant, and (2) properties of the ecosystem controlling the bioavailability of the toxicant. Successful biological remediation of contaminated soils relies not only on the metabolic capability of the microbes but also on the probability of a favorable encounter between the toxicant and its decomposer community. That is, success depends on the bioavailability of the pollutant. Biodegradable organic substances entering a soil ecosystem can be mineralized at a reasonable rate (assuming no other limitation to microbial activity beyond those fulfilled by the organic amendment) only as long as the pollutant is present within the soil interstitial water and existent in appropriate concentrations. Failure to fulfill the latter requirement can relate to concentrations below that which allows efficient interactions with degradative enzymes, as well as higher concentrations, which may be toxic to the microbe. Organic substances tend to have maximal concentrations in the available soil water immediately after entering the soil environment. During this time period, for metabolizable substances, the most probable fate of the organic substance is mineralization and incorporation of the carbon or metabolites into microbial biomass. A portion of the organic substance may be humified, sorbed onto soil minerals, leached through the profile, or volatilized. For bioremediation situations, this seemingly simple situation is complicated by the fact that the distribution of the organic substrate becomes more complex with time — that is, the proportion that becomes biologically unavailable increases with time in the soil system. Thus, for soil microbes, native or otherwise, to decompose soil contaminants, they must possess the genetic capability necessary to synthesize the requisite enzymes, be in an environment conducive to expression of their genetic potential, and have the capability to overcome the physical limitations of accessibility of the contaminant. The physical and chemical requirements for successful bioremediation of a soil site are the topic of the next section. This discussion will be limited to the properties of the microbial community that are requisite for successful biologically based remediation of chemically contaminated soil ecosystems.

16.2.1 Microbes as Soil Remediators

Soil microbes responsible for the decomposition of organic pollutants exist in three states:

- The decomposer populations are sufficiently high that immediate decomposition of the pollutant can occur. If the quantity of the pollutant soluble in the soil water is nontoxic and available at concentrations sufficient for effective catalysis, then degradation would be anticipated to occur — assuming that no other limitations (nutrients, oxygen, etc.) exist.

- Microbes capable of decomposing the contaminant are present in the soil system in low numbers. The pollutant may induce development of sufficient populations to allow the ultimate degradation of the pollutant. Major factors of consideration in this situation include the time necessary for the population to develop (see Chapter 4 on microbial growth kinetics in soil) and the possibility that other soil factors may limit the development of adequate microbial populations for biological remediation to occur.

- No indigenous organisms capable of decomposing the contaminant are present in the microbial community. In this situation, the potential to amend the soil with decomposer populations capable of surviving and functioning therein must be considered. This process may involve use of genetically engineered or modified organisms or simply adding soil or other amendment that contains the requisite populations. This process imposes an added limitation discussed above — the necessity of the alien microbes to overcome competitive interactions with the indigenous microbes. The existence of an "underutilized" carbon and energy source, the pollutant, in the system could provide an advantage to the amended or invading microbes.

If the appropriate organisms exist within the soil system for biological decomposition of the substance to occur, then there are several alternative processes that should be evaluated when assessing the potential for bioremediation. The microbes may metabolize the organic pollutant directly, with the product ultimately being mineralized through common intermediary metabolism pathways. The pollutant may be metabolized indirectly. In this situation, there may be no organisms present in the system capable of mineralizing the substance, but the contaminant may be transformed cometabolically and ultimately decomposed through the interaction of several members of the microbial community (a consortium). Further variation in the metabolic processes associated with transformation of the pollutant results from the microbial capability to utilize a variety of terminal electron acceptors. Thus, although aerobic decomposition of organic substances is generally the initial pathway considered, in contaminated ecosystems, which are anoxic, the capacity of bacteria to use alternative terminal electron acceptors, such as nitrate and sulfate, can be exploited to purify the soil. Each of these alternatives will be discussed in greater detail when the pathways for bioremediation are considered below.

16.2.2 Substrate-Decomposer Interactions

Microbial Mediation of Bioremediation: Description of specific pathways for biodegradation is beyond the scope of this chapter. (See Chapter 4 for a general discussion of microbial metabolism in soil.) Examples of the complexity or variety of alternative pathways through which soil organic contaminants can be decomposed are provided by Young and Cerniglia (1995). Perhaps, the biggest mental obstacle to maintaining a clean soil environment in the early industrial era was that generally held, rather optimistic concept of microbial infallibility (see Chapter 4), combined with the confidence derived from the great knowledge that had been accumulated regarding the complexity of intermediary metabolism of soil bacteria and fungi. A portion of that trust in the capabilities of the soil microbes and the breadth of our understanding of their metabolic prowess was not misplaced. A vast array of biologically synthesized organic chemicals and industrially produced organic chemicals (xenobiotic chemicals) with structures that mimic the biologically synthesized substances are metabolized in soil. Of course, the full function of the microbial community is predicated on the precepts that the proper environmental conditions must be maintained for optimal function of the soil community, and that the pollutants do not enter the system faster than the microbial community can decompose them. More problematic are those xenobiotic chemicals that do not have structures resembling biologically synthesized molecules. These substances may be totally recalcitrant, metabolized in part (DDT conversion to DDE), humified (if sufficiently reactive chemically), or lost from the system through volatilization or leaching. Also, some of these chemcials may be slowly transformed cometabolically. For example, trichloroethylene (TCE) is decomposed cometabolically by microbes using methane or other simple carbon compounds as their carbon and energy source (see Wackett, 1995, for a review of this topic). Similarly, halogenated organic compounds can be dehalogenated cometabolically and the resultant product mineralized. A common example of this process is microbial reductive dechlorination of polychlorinated biphenyls (PCBs) (e.g., Bedard and Quensen, 1995). Approximately 1.4 billion pounds of PCBs have been released and have accumulated in the environment. In anaerobic environments, the chlorine molecules can be removed from some of these substances, thereby reducing the toxic risk of the contamination.

For many structurally complex organic chemicals, degradation frequently relies on the existence of consortia or microbial assemblages to catalyze the processes. Microbial consortia are groups of microbial species that acting together are capable of mineralizing a substance that no single member can achieve alone. Perhaps the most commonly cited natural example of a group of members of the soil microbial community interacting to achieve complete mineralization of a complex substrate is the conversion of cellulose to methane. No single microbe is capable of mineralizing cellulose to methane, yet in anoxic systems, methane production from cellulose is common. A variety of anaerobic bacteria (e.g., *Clostridium* spp.) convert the cellulose to the simple organic acids that are then utilized by the methanogens in their production of methane. Similar interactions are easily envisioned to occur in the decomposition of xenobiotic substances. The dehalogenation of PCBs described above can be seen as an initial step in a complex decomposition pathway of a seemingly environmentally stable substance. A scenario can be described wherein the PCBs are dehalogenated under anaerobic conditions. The products then diffuse or are transported to an aerobic site where the biphenyl can be mineralized.

The best examples of exploitation of dehalogenation reactions for remediation of organic wastes are derived from studies of contaminated sediments, but there is no reason to anticipate that under comparable conditions such reactions would not occur in a soil ecosystem. A particularly interesting study is one in which terminal electron acceptors other than molecular oxygen were used for the degradation of monochlorobenzoate isomers by Nile River sediments (Kazumi et al., 1995). In this study, the monochlorobenzoates were degraded under denitrifying, methanogenic, iron-reducing, and sulfidogenic conditions. Conversely, a study by Fliermans et al. (1988) demonstrated the capacity of aerobic microbial consortia to convert trichloroethylene (TCE) to hydrochloric acid and carbon dioxide. In the latter example, the consortia were isolated from subsurface sediments and grown in enrichment cultures. Other examples of microbial consortia-based mineralization include decomposition of dicyclopentadiene (Stehmeier et al., 1996) and a complex mixture of aromatic compounds contained in olive oil mill effluent (Zouari and Ellouz, 1996). Similarly, in a complex biofilm, Field et al. (1995) demonstrated decomposition of aromatic pollutants by consortia of aerobic and anaerobic bacteria. Also, the anaerobic degradation of pentachlorophenol can be catalyzed by a methanogenic consortium (Juteau et al., 1995).

These examples demonstrate the variety of organic chemicals decomposed by microbial consortia and underscore the importance of the entire soil microbial community in bioremediation. The involvement in bioremediation of this community or mixed culture per se is particularly significant when it is considered that in most situations wherein the indigenous soil microbial community is utilized to purify the system, a mixture — generally complex — of organic compounds is encountered. Soils are rarely contaminated with a pure xenobiotic chemical. Even pesticides, which are commonly conceptually spoken of as if a single chemical substance were reaching soil, are commonly added to the soil-plant system in petroleum or other appropriate carrier mixture to aid dispersion and to facilitate dilution of the active ingredient to the appropriate dosage.

Role of Biofilms in Soil Remediation: This admixture of a variety of potential carbon and energy sources encourages development of interactive soil microbial communities. This observation leads to the question of the form or nature of the physical association of these microbes within the soil environment. As was indicated in Chapter 2, bacteria exist in soil as discrete microcolonies whereas filamentous organisms may extend through several microhabitats. Much of the surface area of soil particles is not inhabited since the microorganisms tend to be concentrated in organic-matter–rich areas. An efficient microbial community assemblage that is commonly exploited for wastewater purification (trickling filters) and fluidized beds is microbial biofilms. Biofilms are communities of microorganisms that develop at the interface of soil particles and water. They are primarily water (70 to 95 percent wet weight) (Flemming, 1993), with the cells attached to the particle surface through extracellular polymers (70 to 95 percent of the dry weight of the biofilm) (Flemming, 1993). Thus, the microbial cells are immobilized in a tightly organized structure, forming highly organized consortia. This physical arrangement is favorable for decomposition of complex mixtures of organic substances in systems with adequate or high nutrient levels and for optimal interaction of the community members (Costerton et al., 1995).

A question remains regarding the potential for a parallel conceptualization of the relationship of microorganisms in soils and the biofilms of trickling filters per se. That is, does a parallel community structure to the biofilm exemplified by trickling filters exist in soil, and is it important for soil biological remediation? It is easy to envision the physical association of sessile microbes in soil within an assemblage akin to biofilms. The primary difficulty arises in considering the extent of the surface area of the soil particles occupied by such films. Soils tend to be carbon-limited ecosystems that generally experience extremes in moisture, ranging from water-saturated to highly xeric conditions. With a waste purification system, such as a trickling filter or a fluidized bed, optimization requires continuously flowing water (aerobic systems) or saturated water for those engineered processes in which anaerobic metabolism may be encouraged. These conditions (i.e., flowing oxygenated or deoxygenated water) generally do not exist in soil.

In spite of this observation, the biofilm concept can be invoked in part to explain the interaction of consortia in soil as well as the efficiency of water purification in systems such as wetlands (e.g., Alvord and Kadlec, 1996). Occurrence of highly developed biofilms is limited in most soil systems, at least for extended periods, due to the impact of biofilm development of saturated hydraulic conductivity. As the input of nutrients stimulates the development of the biofilm, the soil pore diameter is diminished sufficiently that saturated hydraulic conductivity can be rapidly reduced by as much as four orders of magnitude (e.g., Vandevivere and Baveye, 1992), eventually reaching the point where flow of the nutrients into soil and maintenance of the biofilm community is precluded (Jennings et al., 1995). The rapidity of this biofilm development results from the tendency for colonization of the surface (inlet) of the soil column receiving the nutrient stream (see Vandevivere and Baveye, 1992). In highly structured soils, especially silt loams and clay soils, development of these pores results in the plugging of micropores. Thus, in high-nutrient and water-flow systems, a sandy matrix is more desirable to maintain the hydraulic conductivity (see Vandevivere et al., 1995).

Therefore, it can be concluded that the biofilm concept is useful in describing the microassemblages of communities of microorganisms on surfaces of soil particles as the result of locally elevated levels of nutrients. That is, a complex, interactive microbial community certainly exists in soil microsites. A more extensive macrofilm, per se, could be anticipated to occur only under rare circumstances. The development of more traditionally described biofilm is limited to high-nutrient situations, such as sites receiving chronic inputs of waste-laden waters, a situation less likely to exist with most bioremediation situations. For bioremediation, the microbes are encountering a varied distribution of the pollutant, some free in interstitial water (as would parallel the waste stream situation to some degree) and some bound or sequestered within soil micropores. Thus, both the quantity and availability of the carbon and energy source would limit the physical size of the developing biofilm.

Importance of Substrate Concentration to Remediation Success: To this point, the key elements for microbial decomposition of organic pollutants considered include the following:

- The necessity of the microbe to possess the requisite genetic potential
- The ability to express that metabolic capability in a soil ecosystem
- The requirement for the microbe to possess the appropriate traits for successful interaction with other members of the soil biological community — including the ability to become at least temporarily established therein should the use of nonnative organisms be required for successful site bioremediation
- The existence of consortia of microbial strains and species in assemblages or associations (microbiofilms) that optimize the capacity to decompose the complex mixtures of organic substances encountered in polluted soils whose pollutants cannot be removed by any single species or strain

A further biological trait that must be addressed relates to the impact of the quantity of organic pollutant present on microbial and enzymatic activities. There is an optimal range of pollutant concentrations that facilitates biological remediation. Thus, there is a need to assess our capability of managing the concentrations of these substances so that the quantities of pollutant are above the minimum required for induction and function of the enzymes required for their metabolism, and below any threshold for toxicity to the decomposers.

Perhaps the soil with toxic loading of pollutant is most easily managed, at least conceptually. Two difficulties can be conceived to exist with excessive soil loading of toxic organic substances: (1) The quantities of a carbon-rich substance are sufficiently high that a nitrogen or other nutrient limitation is imposed on the microbial community and/or (2) the pollutant is present at levels that are toxic to the decomposer community. In either situation, dilution or reduction of the toxicant concentration is the most easily conceived, but perhaps from an engineering viewpoint, perhaps less easily achieved. An example of toxicant loading reduction followed by biodegradation is exemplified by a feasibility study of remediation of hydrocarbon-contaminated soil from beneath a paint factory (Origgi et al, 1997). In that situation, bioventing and biofiltration of the outlet gas stream resulted in significant remediation of the limited area contaminated with naptha, toluene, and xylenes. The residual hydrocarbons were then reduced significantly through biodegradation.

A more challenging situation relates to systems containing extremely low levels of toxicants. An environmental or ecosystem risk may exist, but the concentration of the substance is below that generally required for active microbial mineralization to occur. In this situation, microbes capable of decomposing low concentrations of the pollutant must be enriched or the quantity of the substance encountered by the microbes raised to allow for biodegradation. The latter option emerges when high levels of pollutant exist in the system but the substance has low water solubility. Thus, from the microbes' view, only low levels of the essentially water-insoluble pollutant are available for metabolism.

Managing Pollutant Water Solubility: With substances of limited water solubility, surface area is a primary determinant in controlling microbial transformations. A prime example of this situation is exemplified by the oxidation rate of elemental sulfur particles (see Chapter 15). Physical contact of the microbes with the sulfur is possible because the sulfur is not toxic. Thus, the microbes literally adhere to the sulfur particle surface. This situation is more complex when considering the decomposition of hydrophobic organic compounds. Growth on the surface of a crystalline organic compound may be precluded by its toxicity. Microbes existing at the surface of the crystal may actually be inhibited or killed by the capability of the hydrophobic organic contaminant to become associated with the hydrophobic portions of the cell membrane. One means of overcoming this solubility difficulty associated with hydrophobic organic substances is through the natural production of surfactants by the microbial community, or use of soil amendment with industrially produced surfactants as an aspect of bioremediation plans.

Pertinent questions regarding the applicability of surfactants for the enhancement of bioavailability of soil organic contaminants of low water solubility include:

- Does the amendment of a contaminated soil with surfactants enhance the biodegradation rate of the organic pollutant?

- How does the degradation rate vary with surfactant type and concentration? That is, is more better? Are there advantages or disadvantages of anionic or nonionic surfactants?

- Are additional environmental problems created by the use of surfactants?

- If surfactant producing microbes are amended to soil, can they synthesize the surfactant and thereby enhance biodegradation of the pollutant of concern?

A variety of laboratory studies have been conducted with objectives of answering one or more of these questions. At this time, the general practicality of such research is still in question. A selection of representative studies will be examined to provide examples of the state of the art of such research.

Evaluation of the efficacy of surfactants in the environment has been conducted using a variety of low water solubility substances, such as polycyclic aromatic hydrocarbons (PAHs), naphthalene, and xylenes, added to sediment and soil samples (e.g., Aronstein and Alexander, 1993; Grimberg et al., 1996; Liu et al., 1995; Zhang et al., 1997). In each situation, it was generally concluded that the amendment of the hydrocarbon-contaminated solid matrix with surfactants stimulated the biodegradation rate of the water-insoluble contaminant. Some effect of surfactant usage on both organic and inorganic soil contaminants has been noted to occur. Huang et al. (1997) found that addition of an anionic surfactant to a soil containing both lead and naphthalene resulted in enhanced solubilization of the naphthalene and desorption of the lead ion, suggesting that the surfactant could be used to increase mineralization of the hydrocarbon and to enhance the capacity to leach the lead from the soil profile.

Surfactant concentration is a significant concern for reclamation enhancement in that as its concentration increases, so does the tendency to form micelles. The impact of micelle formation is that the hydrophobic pollutant may become stabilized in the micelle and thus remain unavailable to the microbial community. This was shown by Volkering et al. (1995) in their evaluation of the utility of several nonionic surfactants for enhancing the solubility of phenanthrene in batch growth experiments. In contrast, Liu et al. (1995) found that naphthalene solubilized by micelles of two nonionic surfactants was bioavailable to a mixed culture of bacteria. Zhang et al. (1997) found that phenanthrene within micells of the nonionic surfactant monorhamnolipid was less bioavailable than that contained within dirhamnolipid micelles, suggesting variation in the bioavailability due to the nature of the surfactant forming the micelles. Vanhoof and Jafvert (1996) noted that dechlorination of hexachlorobenzene varied with relationship of the surfactant level to the critical micelle concentration. For example, tween 80 decreased dechlorination at concentrations significantly above the critical micelle concentration, whereas stimulation was noted at or below the critical micelle concentration.

Further concerns with using high concentrations of surfactant relate to the potential for toxicity of the surfactant and impact on the biological system due to the degradation of the surfactant itself. Toxicity of the surfactant is readily determined through evaluation of the effect of amendment of soil with varying concentrations of surfactant on key indicators of microbial activity, such as respiration or even mineralization of the toxicant of interest. The effect of biodegradability of the surfactant is perhaps more interesting. A study demonstrating both toxicity of surfactants and the effect of biodegradation was presented by Tiehm (1994), where it was noted that toxicity of the surfactants examined decreased with increasing hydrophobicity. The nontoxic surfactants were effective in enhancing the degradation of a variety of hydrocarbons, fluorene, phenanthrene, anthracene fluoranthene, and pyrene. Interestingly, sodium dodecyl sulfate inhibited the degradation of the polycyclic aromatic hydrocarbons because it was a preferred substrate. In a related study (Tiehm et al., 1997), biodegradable surfactants were found to interfere with surfactant enhancement of biodegradation of polycyclic aromatic hydrocarbons in manufactured gas plant soil due to depletion of molecular oxygen during mineralization of the surfactant.

Surfactants can be synthesized by soil microbes. Thus, it could be proposed that it could be advantageous to amend soil not with the surfactant itself but with a microbe that could survive, grow, and produce the surfactant *in situ*. An example of this type of system was described by Barriault and Sylvestre (1993). They amended soil contaminated with Aroclor 1242 (PCBs) with a PCB-degrading bacterial strain, biphenyl as a cosubstrate to stimulate PCB degradation, and a second bacterial strain, *Alcaligenes faecalis* strain B-556, which produced a surfactant. Decomposition of the Aroclor was stimulated over that not receiving the surfactant producer.

Use of soil inoculation to stimulate bioavailability of water-soluble toxicants is more complex (beyond the regulatory difficulties) than simple amendment with the surfactant since (1) the microbial strain producing the surfactant must be capable of survival and function within the soil system, and (2) the surfactant must be produced in effective concentrations at the site where it is needed. That is, the surfactant synthesizer, the target pollutant, the organisms responsible for its decomposition, and the surfactant must all coexist in the same microsite at appropriate levels for optimal activity. If the decomposer population is not physically associated with the pollutant and the surfactant, then the solubolized pollutant must diffuse or be leached to the location within the soil matrix where the biodegradation can occur.

Biological Remediation of Soils with Low Loading of Pollutant: The above analysis of the complexities involved with the use of surfactants is predicated on the assumption that the soil site has been contaminated with reasonably high levels of the pollutants, but that the bioavailability to the decomposer community is low due to their low water solubility. Situations may also exist where the concentration of the organic chemical in interstitial water is limited simply because the pollutant is present at low levels. For this consideration, a low level of a pollutant is considered to be one for which there is environmental concern but the organic compound is present at levels that are insufficient to serve as a carbon and energy for soil microbes. In this case, the primary limitation to mineralization results from the probability of encounter of the decomposable substance and members of the decomposer population or their extracellular enzymes, and the capability of the requisite enzymes to catalyze the process. The latter situation results from the fact that enzyme efficiency is optimal at substrate concentrations approximating the Michaelis constant (K_m) of the enzyme. These levels are generally several orders of magnitude greater than the quantities of substrate considered in this discussion.

If the metabolizable substance is present in concentrations below that which can effectively react with the enzyme, then its stability in soil will mimic that of a recalcitrant substance unless mechanisms exist that will result in elevated concentrations in microsites. One such commonly encountered mechanism in soil results from sorption on clay surfaces. Two processes are involved in this situation. The biodecomposable pollutant is concentrated on the clay surfaces so that an elevated level of the substance exists in relationship to the soil surrounding the microsite. But this material, due to the fact that it is sorbed on the clay surface, is of limited availability to the microbial community. Since sorptive processes are reversible, this sorption of the pollutant on the clay surface also results in augmented levels in the water surrounding the clay particle. Thus, the microbe occupying the microsite, should it be capable of decomposing the pollutant, will mineralize the pollutant, resulting in a reduction of the soil loading of the pollutant below levels that would be generally predicted from assessment of the pollutant concentration.

A further mechanism that can result in augmented mineralization of low-level organic pollutants in soil involves stimulation of growth or activity of the decomposer population through amendment of the soil with alternative growth substrates. As was indicated above, a low-level contamination of soil is defined as one in which the substance is present in concentrations less than those necessary to sustain microbial growth—thus, the situation where the both the population of decomposers and the substance to be decomposed are sufficiently low that their encounter is improbable. Augmentation of the microbial population due to the presence of the pollutant cannot be anticipated due to the fact that it is present in concentrations below those needed to serve as carbon and energy source. An alternative in this situation is to amend the soil with a metabolizable substance that increases the general soil microbial population (e.g., Pahm and Alexander, 1993; Schowanek et al., 1997). Schowanek et al. (1997) amended soil with tetradecenyl succinic acid (TSA), a major component of a detergent builder, with and without sewage sludge. Decomposition of the low level of TSA in the absence of sludge was preceded by a 13-day lag period, whereas no lag was observed when it was accompanied by sewage sludge. Similarly, Pahm and Alexander (1993) were able to stimulate the mineralization of low concentrations of p-nitrophenol (PNP) in soil amended with PNP-degrading bacteria by the addition of glucose. Presumably, in each situation, the enhancement of the energy supply to the microbial community by the glucose and sewage sludge resulted in a sufficiently augmented energy supply for the microbial community that the pollutant concentrations could be gratuitously reduced by microbial mineralization.

16.2.3 Microbial Inoculation for Bioremediation

Given the conclusion that a soil microbial community continually adapts to its changing, evolving environment, it is logical to conclude that new members of the community are continually being recruited. These alien populations may arise from propagules that enter the soil ecosystem on soil particles carried by wind or in water or on the surfaces of invading roots. Thus, we can logically conclude that adaptation of a microbial community to chemical pollution through invasion of alien organisms should, or even must, occur, especially if soil adjacent to the contaminated site contains the appropriate decomposer populations. Therefore, the question regarding management of bioremediation systems is not whether naturally occurring modifications of the soil microbial community composition to pollution occur. The composition of the soil microbial community must be resilient to ensure survival of the ecosystem as environmental conditions evolve. The pertinent question is, can the community development be manipulated so that highly efficient decomposer populations are selected? A logical extension of this question is, can laboratory-developed (including genetically engineered microbes) microbes, amended to a soil system, become part of the functioning community, and effectively and efficiently renovate the site?

The literature is replete with reports supporting the potential use of a variety of bacteria and fungal species for bioremediation of soil (e.g., see Madsen and Kristensen, 1997; Miethling and Karlson, 1996; Sack and Fritsche, 1997; Smith et al., 1997 as recent examples of such studies). This amendment of soil with a decomposer population may also be supplemented with a selective substrate to facilitate establishment of the inoculated population (Lajoie et al., 1993; Nishiyama et al., 1993) in situations where the host receives no or minimal advantage from the mineralization process. Furthermore, Madsen and Kristensen (1997) provide an example where a surfactant is introduced with the exogenous bacterial population to stimulate decomposition of an organic contaminant. In this latter example of introduction of microbial strains or species into a soil, polycyclic aromatic hydrocarbon decomposition was stimulated by enhanced availability of the hydrocarbon due to interaction with the surfactant produced by the laboratory-cultured microbial strain. Another interesting example of such research involving managing the soil biological community for optimization of soil remediation was conducted by Siciliano and Germida (1997). They suggest that both the plant and the microbial community can be managed to enhance bioremediation.

Rhizosphere Management for Biological Remediation: When considering the interaction between plants and soil microbes from the view of enhancing bioremediation potential, the unique properties of the rhizosphere must be emphasized. Recall that the rhizosphere soil is that portion of the ecosystem where biological activity is maximized (see Chapter 8). Thus, it can be generally concluded that mineralization reactions are maximized therein. Linked with the above-cited studies demonstrating enhanced decomposition of organic pollutants by addition of metabolizable organic matter, the conclusion that bioremediation processes should be enhanced in the rhizosphere is logical. The question thus becomes, are there data suggesting benefits of rhizosphere interactions in bioremediation?

A variety of studies have suggested augmented decomposition of agricultural chemicals and nonagricultural chemicals in rhizosphere soils (see Anderson et al., 1993, for a review of this topic). For example, Reilley et al. (1996) found that plants can enhance the decomposition of polycyclic aromatic hydrocarbons in the immediate environment of the root. Their study suggested that once the toxicity of petroleum components to plants has been sufficiently reduced, incorporation of plants into a landfarming scheme can augment the bioremediation effectiveness. This stimulated activity can result simply from the general elevation of the microbial populations in the rhizosphere due to the energy and nutrient supply of the root exudates (e.g., Bollag et al., 1994). Gilbert and Crowley (1997) also note that some plant-derived terpenoids may also promote polychlorinated biphenyl degradation by an *Arthrobacter* strain in the rhizosphere.

Considering that there are numerous studies demonstrating that ectomycorrhizal fungi are capable of mineralizing complex organic compounds, it is also logical to conclude that mycorrhizal associations may also be beneficial for bioremediation (e.g., Gaskin and Fletcher, 1997). For this benefit to be maximized in a managed ecosystem, propagules of the mycorrhizal fungus must be present (either through plant inoculation or as indigenous soil populations), and the plant community must develop sufficiently for these symbiotic associations to form.

A complication that must be considered when managing both microbial and plant communities for bioremediation is the potential for the formation of plant-toxic intermediates during decomposition of the pollutant (e.g., Hoagland et al., 1994; Pfender, 1996). Bioremediation potential of the microbe-plant system may be lost if toxic intermediates that are not further metabolized accumulate in the rhizosphere.

Distribution of Alien Organisms in Soil: In the studies cited above, the primary concern of the research was whether the organisms were capable of decomposing the target organic compound within a soil matrix. That potential activity has been adequately demonstrated in test tube studies. The next question relates to the capability of applying laboratory-selected microbes to soil in a manner that allows them to pass through soil pores to the site where their action is required.

In a fluidized bed reactor, it is reasonably easy to introduce an alien microbial strain and to reach the conclusion that through the mixing process of the reactor the organism will encounter the target organic compound and mineralize or detoxify it. This capability of transporting the microbe to the site of requisite action in soil is more problematic. Generally, it is desirable to maintain as much as possible of the native soil structure during reclamation of a soil ecosystem. Also, due to the large area requiring reclamation, mixing of microbial populations throughout the contaminated profile is generally impractical. Thus, the dynamics of microbial movement within the highly structured soil environment becomes a concern for evaluating the utility of soil inoculation for bioremediation. Considerable research has been conducted regarding the transport of various microbial strains, primarily within sand columns, and a variety of mathematical relationships have been developed describing the various transport phenomena in a variety of soil systems (e.g., see Devare and Alexander, 1995; Murphy and Tate, 1996; Natsch et al., 1996). The primary conclusion from these studies is that bacteria can be transported through soil macropores preferentially during saturated flow, such as after a heavy rainfall. Limitations relate to plugging of soil pores when high populations of microbes are amended to the soil and the amended microbes are unable to enter micropores due to size constrictions. Thus, although it can be demonstrated that with proper consideration of inoculum density and flow dynamics bacteria may be added to soil and distributed therein, the bioremediation accomplished by such processes is limited by the capability of the organisms to penetrate all microsites of the soil profile containing the pollutant. The microbes can easily be envisioned to pass through and colonize macropores and some micropores, but pollutant contained within the smaller micropores would be inaccessible to the added microbial cells.

Survival of Alien Microbes in Soil: A further concern associated with utilization of laboratory-selected microbial strains for bioremediation relates to their survival beyond the time necessary for renovation of the site. It is generally considered desirable for the alien population to be eliminated once the pollutant has been exhausted or diluted to an acceptable level. Generally, to enhance biological containment of genetically modified organisms within soil systems, it is desirable to limit the survival of the organisms without limiting their beneficial effects. This process is generally accomplished through the use of controlled suicide systems (e.g., Jensen et al., 1993; Knudsen and Karlstroem, 1991; Molin and Kjellberg, 1993;). For example, Jensen et al.

(1993) describe the utilization of a two-element suicide system for the containment of *Pseudomonas putida* KT2440. Each genetic element of the suicide system operated independently to enhance the population decline once function of the microbe was no longer required. Ahrenholtz et al. (1994) describe a system where the cell population is regulated through the destruction of intracellular DNA available for horizontal gene transfer processes.

Allied with this use of genetically engineered microbes with optimized decomposition capabilities and diminished extended survival potential (after depletion of the organic pollutant) is a consideration of our capacity for detection of the microbes. Highly sensitive techniques are required to assure detection of low populations of the amended organisms. A useful but less sensitive procedure involves utilization of the BIOLOG method for assessing metabolic diversity (see Chapter 3). Vahjen et al. (1995) evaluated the use of this procedure for detection of *Corynebacterium glutamicum* inoculated with 10^6 colony-forming units g^{-1} soil. They were able to detect the presence of the inoculated organisms as long as the cell density was greater than 10^5 colony-forming units g^{-1} soil. Matheson et al. (1997) have provided an example of using DNA probes to detect the persistence of a particular strain of *Berkholderia cepacia* when as few as 10 colony-forming units occurred in a population of 10^5 nontarget colony-forming units. Generally, DNA-based probes are preferable for assessing survival of exogenous microbial populations in soil ecosystems.

An interesting application of genetically modified microbes in soil remediation is incorporation of fluorescence genes into the bacterial genome to provide an indication of the bioavailability of the organic pollutant within the soil system and of the degradation of the pollutant (Burlage et al., 1994; Heitzer et al., 1994; Rice et al., 1995; Webb et al., 1997). For these studies, reporter strains (see Chapter 3) of bacteria are developed through incorporation of the bioluminescent lux genes from *Vibrio fischeri* into the genome of the genetically engineered microbe. The genes are transcriptionally fused to the catabolic gene sequence so that they are expressed whenever the catabolic genes are expressed. That is, exposure to the contaminant results in inducible bioluminescence. The magnitude of the response and the response times were shown to be concentration dependent in a system developed with *Pseudomonas fluorescens* for the decomposition of naphthalene and salicylate (Heitzer et al., 1994). Thus, these organism serve as *in situ* indicators of the bioavailability of the pollutant (if it is present but not bioavailable, the bioluminescence genes are not induced), decomposition of the pollutant (the catabolic genes are synthesized concurrent with the bioluminescence genes), and the termination of decomposition (the genes are no longer expressed).

16.3 Soil Properties Controlling Bioremediation

The impacts of the soil physical environment on bioremediation outcome can be grouped into two categories: factors that delimit biological activity and properties that limit accessibility of the decomposer populations to the pollutant (i.e., sequestration of chemical pollutants). General analysis of the physical and chemical requirements of the soil microbial community was presented in Chapter 5. Thus, the discussion here will be limited to specific examples of ecosystem management to optimize a particular biological transformation or process.

16.3.1 Physical and Chemical Delimiters of Biological Activities

The primary soil properties managed for bioremediation considerations are soil pH (generally for the control of metal mobility), nutrient supplies (macro- and micronutrients as well as carbon and energy sources), and soil moisture/aeration. These delimiters of expression of biological activity provide key examples of how the rate, nature of the process, and ultimate fate of the pollutant can be selected by managing the soil physical environment. Due to the resilience of soil microbes, fully functional microbial communities are found in what would superficially be considered impossible conditions. Therefore, the initial planning question for biological remediation becomes, "Are the soil conditions conducive for development or existence of a fully functional microbial community?" If this question is answered in the affirmative, then concerns relate to specific effects of the physical and chemical environment on the decomposition of the substance of concern (especially the biochemical processes involved) and the bioavailability of the pollutant. If not, the prevailing conditions that preclude life must be modulated prior to implementation of a management plan for control of the microbial community to optimize bioremediation.

pH: In the previous consideration of the effect of soil pH on microbial activity (Chapter 5), it was noted that microbial growth and activity occur in both acidic and alkaline soil ecosystem. Therefore, there is no inherently bad implication of the occurrence of soil systems at the ranges of pH values commonly beyond the ranges normally encountered for agricultural production in temperate regions (slightly acidic to neutral pH). In fact, numerous tropical cropping systems (pineapple, cassava, etc.) are productive on soils with pHs of less than 4. The major requirement for reclamation management is that the full implications of pH on the processes occurring in the chemically polluted soil be considered. Examples of soil processes where pH effects need to be evaluated are nutrient cycling and effects of pH on the physical and chemical states of the organic chemical of question.

Nutrient-cycling processes most commonly managed for bioremediation relate to nitrogen availability. As indicated in Chapter 11, both nitrification and symbiotic nitrogen fixation can be affected by acidic conditions. Autotrophic nitrifiers function above pH 5.0 to 5.5, whereas the *Rhizobium*-legume symbiotic associations for nitrogen fixation require a soil pH greater than about 5.5. Thus, in acid soils where fixed nitrogen is limiting, either nitrogen fertilization is necessary, the soils will need to be limed to a level conducive for nitrogen fixation to occur with legumes, or actinorhizal associations, such as bayberry, have to be cultivated.

The greater concern involving soil pH is likely the impact of pH on the ionic state of the pollutant and the resultant effect of this charge variation on its interaction with the enzymes involved in its mineralization or detoxification, as well as its sorption to soil organic matter and clays. At pH values above the pKa, acidic groups are ionized, facilitating ionic interactions with enzymes as well as the physical environment, whereas below the pKa, hydrophobic interactions may be favored. Similarly, from an environmental health view, hydrophobic interactions tend to favor associations with membrane lipids and bioaccumulation of the toxicants. An example of the effect of pH variation on the state of an organic chemical in soil is provided by looking at the sorptive properties of atrazine and fomesafen, two pesticides, in soil (Weber, 1993). Decreasing pH resulted in higher sorption of both herbicides. Atrazine sorption resulted from ionic bonding at low pH and physical bonding at neutral pH, whereas fomesafen sorption was the product of physical forces at neutral pH and hydrophobic bonding and/or precipitation at low pH. Thus, the bioavailability of the pesticides could be anticipated to be lower at the acidic pH's than at high or neutral pH.

Nutrient Amendment and Bioremediation: Examples of requirements for nutrient amendments for bioremediation of soils are generally associated with hydrocarbon-rich contaminants, such as petroleum. It has long been known that nitrogen fertilization is necessary to stimulate degradative activities with land farming of petroleum. Even in situations with other variables such as temperature in alpine soils (Margesin and Schinner, 1997) or salinity and temperature with beaches of Prince William Sound (Pritchard, 1991) potentially limiting biological activity, use of nitrogenous fertilizers enhanced biodegradation. Other nutrient amendments may also involve provision of a carbon and energy source in situations where the pollutant can be decomposed only cometabolically [e.g., carbon tetrachloride and a variety of easily metabolized carbon sources (Lewis and Crawford, 1993)] or simply from the general stimulation of the microbial population (as indicated above). In the latter situation, composted waste materials provide an energy source for the general soil microbial populations (thereby stimulating mineralization of the contaminant), stimulate soil aggregate-forming processes resulting in an improved environment for function of the microbial community, improve the soil moisture dynamics of the system, and also result in increased humification of the organic toxicant, thereby providing an alternative path for removal of the toxicant from the system (e.g., Kastner et al., 1995; Laine and Jorgensen, 1996). Thus, recycling of composted societal by-products can assist in optimization of soil bioremediation processes.

Soil Moisture/Aeration: As was indicated in Chapter 5, the world of soil microbes can be conceptualized as an aquatic system, albeit a thin film of water covering a largely particulate matrix. The organisms are bathed in this film of water, nutrients are transported into and away from the cell by diffusion or mass transport processes, and the water itself is a requisite participant in many biochemical transformations. These observations are true whether we are dealing with a native soil ecosystem or a remediation site. Of particulate concern in managing polluted soil is the interrelationship between soil moisture and soil oxygen status. That is, decomposition rates, detoxification processes, the extent of mineralization, and the nature of the metabolic intermediates produced vary with the nature of the terminal oxygen acceptor. Much of the current bioremediation philosophy is predicated on the objective of optimizing free oxygen levels. This management strategy is based on the assumption that oxygen-based metabolism is more rapid and generally leads to accumulation of fewer metabolic intermediates than would occur under anoxic conditions. Considering that the terrestrial system is bathed in an oxygen-rich atmosphere, this philosophy is generally reasonable, although as will be indicated below, consideration of other terminal electron acceptors — for instance, nitrate, sulfate, and carbon dioxide — is useful in naturally anoxic systems where manipulation of aerobic processes is impractical.

The examination of native soil ecosystems, such as swamps and related water-saturated systems, shows that organic matter decomposition is reduced and the accumulation of partially decomposed organic intermediates is accentuated in the total absence of free oxygen. But soil microbes have a tremendous versatility in dealing with variation in oxygen tension. Reduction of oxygen tensions a percent or two from the normal atmospheric level of approximately 21.0 percent is limiting for higher animals, but in drained histosols, microbial mineralization of the accumulated organic matter continues to free oxygen tensions of less than 1 percent (see Broadbent, 1960). Similar relationships of biodegradation processes to available oxygen levels would be anticipated to occur in soils contaminated with organic chemicals, although the rate of decomposition will necessarily vary directly with oxygen tension as the oxygen levels become limiting. Examples of limitations to bioremediation by oxygen levels are common in studies of soil- and sediment-remediation processes (e.g., Hurst et al., 1997; Madsen et al., 1996). Hurst et al. (1997) noted that pentachlorophenol degradation in contaminated soil from a superfund site was precluded in the absence of oxygen, but was enhanced with soil oxygen concentrations between 2 and 21 percent. They concluded that a soil oxygen concentration of 2 to 5 percent was necessary for pentachlorophenol bioremediation. When depletion of soil oxygen levels by biological activity exceeds the diffusion rate for oxygen replenishment, the oxygen tension can be elevated through bioventing (e.g., Hurst et al., 1997) or by amendment of the soil with peroxides (e.g., Fiorenza and Ward, 1997).

In some situations, soil oxygen can be depleted to optimize processes favored by anoxic conditions (e.g., dehalogenation or reduction of nitro groups). Kaake et al. (1992) describe a soil treatment method for removal of the pesticide dinoseb (2-sec-butyl-4,6-dinitrophenol), where soils were pretreated by amendment with a starchy potato-processing by-product to reduce the redox potential and create anaerobic conditions. An anaerobic microbial consortium developed that was capable of mineralizing dinoseb without formation of polymerization products commonly seen under aerobic or microaerobic conditions. Alternatively, alternation of aerobic and anoxic conditions can be used to facilitate bioremediation. For example, a two-stage process starting with an anaerobic conditions facilitates the bioremediation of munitions compounds including 2,4,5-trinitrotoluene (TNT) (e.g., Funk et al. 1993; Breitung et al. 1996). During the first phase, ammunition residues and aminoaromatic compounds were depleted, and the subsequent reduced products were mineralized during the aerobic phase of the process.

It is given that, by far, the bulk of organic matter decomposition in soil results from oxygen-based processes. Thus, it is logical that initial considerations in developing bioremediation plans must involve evaluation of the potential for optimization of aerobic soil metabolism, but there are many contaminated soil ecosystems where utilization of alternative electron acceptors is more practical and economical due to the fact that the site is naturally anoxic. A further encouragement in this consideration is the fact that in many instances, the soil contaminants may include nitrate or sulfate, two logical terminal electron acceptor alternatives. Furthermore, dissimilatory nitrate and sulfate reduction yield energy recoveries to the microbes approximating those gained from oxygen metabolism (see Chapters 14 and 15). Thus, it is reasonable to conclude that these processes should be as efficient and rapid as the more commonly studied and applied aerobic degradation processes.

Denitrification is the most studied anoxic process for bioremediation. Denitrification, as a means of reducing soil nitrate loading, by definition is a bioremediation process (e.g., see Smith et al., 1994; Weier et al., 1994). Since a carbonaceous compound is generally mineralized as the source of electrons for denitrification, the potential exists for selection of denitrifiers that mineralize toxic organic compounds for bioremediation of soils contaminated with both nitrate and decomposable organic chemicals (see Casella and Payne, 1996, for a review of this topic). Indeed, as indicated above, a variety of potential soil toxicants are mineralized by denitrifiers (see also Crawford et al., 1998; Haner et al., 1995). A major question remaining relates to the capability of managing bioremediation in field situations under denitrifying conditions. An example of the potential for manipulation of anoxic processes is provided by bioremediation of BTEX (benzene, toluene, ethylbenzene, and xylene) in petroleum-contaminated soils. BTEX degradation has been noted with ferric iron, sulfate, or nitrate as terminal electron acceptors as well as under methanogenic conditions (Lovley, 1997). Denitrification alone is apparently not sufficient to bioremediate gasoline (BTEX)-contaminated soils due to the inability of the denitrifiers to decompose benzene (Kao and Borden, 1997).

16.3.2 Sequestration and Sorption Limitations to Bioavailability

A real limitation to the successful conclusion of biological remediation efforts relates to the quantity of residual pollutant remaining in the system. The longer the time interval since initial contamination of the soil, the greater the probability that significant quantities of the organic pollutant will not be accessible to the biological community. A key to understanding this limitation to the ultimate cleanup of the site resides in developing comprehension of the impact of soil heterogeneity on the biological processes occurring therein. As was emphasized in Chapter 1, the variety of minerals, complex organic substances, and biological entities existent in any soil results in a myriad of different microenvironments within which soil microbes function. This heterogeneity affects both the immediate and potential bioavailability of decomposable organic substances and toxicants. For an organic compound to be decomposed or for this substance to inhibit microbial activity, it must have contact with the microbe or enzymes produced by the microbe. (Recall that extracellular enzymes may transform organic matter into water-soluble products, which through mass transport or diffusion come into direct contact with the microbial cell.) Two primary properties of the soil system affect bioavailability: physical location within the soil pore structure and sorption onto soil particulate surfaces.

Due to the dynamic nature of the distribution of materials within the soil matrix, the quantity of substance sequestered within soil pores, sorbed to soil minerals, or dissolved in soil water changes with time. It is commonly observed that the bioavailability, chemical extractability, and decomposition rate of soil organic contaminants decreases with time since initial exposure (e.g., Alexander, 1995; Grant et al., 1995; Radosevich et al., 1997). This time-dependent variation in bioavailability can be demonstrated through assessment of percent uptake by earthworms, bacterial degradation, or solvent extractability (e.g, Kelsey et al., 1997). For example, Kelsey and Alexander (1997) found that less rigorous solvent extraction methods correlated better with the availability of atrazine, phenanthrene, and napthalene to earthworms in sterile soil than did the more vigorous extraction procedures commonly used to assess soil contamination. The degree of sequestration varies with time as well as with soil properties, such as clay and organic matter contents (e.g., White et al., 1997). Furthermore, the source of the microbial inoculum may also affect the extent of biodegradation that occurs in the soil system. Sandoli et al. (1996) found that microbes isolated from their sand sediments that were contaminated with phenthrene decomposed the contaminant with a shorter lag period than did those from other sources.

These observations support the hypothesis that the bioavailability of organic substances in soil to decomposition by soil microbes relates to the pore structure of the soil as well as the capacity of the substances to interact with clay minerals and colloidal organic matter. This hypothesis has been supported by use of model systems composed of glass, polystyrene, diatomite, and silica beads with varying pore structure and hydrophobicity (Nam and Alexander, 1998). This study demonstrated that both hydrophobicity and existence of nanopores in the solid matrix affected the bioavailability of phenanthrene.

It is reasonably easy to envision the nature of the effect of micropore structure on chemical-microbe interactions. Substances trapped within soil pores of mean diameters less than those of microbes and containing no microbes would not be detoxified or mineralized. The quantity of pollutant retained in micropores is controlled by the physical forces dictating soil water dynamics. Entrance into the pores is diffusion limited, as opposed to the soil macropores, where saturated flow allows for mass transport of solutes in gravitational water. Once solutes enter the micropores their exit from the system is controlled by (1) diffusion, (2) their retention within the more rigid crystalline water layer on the pore surface, and (3) their tendency to sorb onto the surfaces of clay or organic matter surfaces in the pores. The role of sorptive phenomena is documented by the observation that biodegradation in soil containing both pollutant and its decomposers is frequently proportional to the desorption rate of the pollutant (e.g., Carmichael et al., 1997). The fact that the nature of the mineral component of the soil (especially clay type) affects decomposition kinetics underscores the need to evaluate the soil mineralogy (e.g., Apitz and Meyersschulte, 1996; Haderlein et al., 1996). Noncovalent interactions of organic substances with soil humic substances can result in associations with stability resembling that of covalently bound substances (Dec and Bollag, 1997). Chemical models have shown that simple organic compounds can form associations with clay surfaces and within pores of soil humic acids that are sufficiently stable to essentially remove the toxicant or biodegradable substance for the biologically available nutrient pool (Schulten, 1995; Schulten and Schnitzer, 1997). Thus, bioavailability of organic chemicals in soil and their potential for biological remediation are controlled by the tendency of the substance to diffuse into and out of soil micropores as well as by the forces preventing diffusion of the chemical back out of the pore (sorption, noncovalent stabilization in humic acids and retention in crystalline water).

16.4 Concluding Observations

Bioremediation of polluted soils is a possibility only because of the extensive diversity of the soil microbial community and the redundancy of capabilities that exist therein. Indeed, even though there is a potential to develop laboratory-derived microbial species and strains capable of decomposing soil pollutants *in situ*, the bulk of soil biological remediation is facilitated or accomplished by the indigenous soil microbial community. Thus, success of any remediation project relies totally on a clear understanding of the biological capability inherent in the system and the physical, chemical, and biological limitations of the expression of that activity. A variety of soil properties can be varied to optimize the rate and extent of biological remediation, but the ultimate controller of the quantity of toxicant remaining in the system is its bioavailability. Retention of chemicals within soil micropores and association with soil minerals and organic matter preclude total elimination of the soil contaminant. The ultimate environmental risk of this residual toxicant is determined by the potentiality for the bioavailability to change with time. The primary unanswered question in relationship to the overall risk of retaining sequestered toxicants in a remediated soil relates to the still to be determined probability that the toxicant will be released in meaningful concentrations at a future date. Based on the above analysis of the limitations to bioremediation, a meaningful concentration would be one that poses a toxic threat to the function of the indigenous biological community or to higher plants or animals in the ecosystem.

References

Ahrenholtz, I., M. G. Lorenz, and W. Wackernage. 1994. A conditional suicide system in *Escherichia coli* based on the intracellular degradation of DNA. Appl. Environ. Microbiol. 60:3746–3751.

Alexander, M. 1995. How toxic are toxic chemicals in soil. Environ. Sci. Technol. 29:2713–2717.

Alvord, H. H., and R. H. Kadlec. 1996. Atrazine fate and transport in the Des Plaines wetlands. Ecol. Modelling 90:97–107.

Anderson, T. A., E. A. Guthrie, and B. T. Walton. 1993. Bioremediation in the rhizosphere. Environ. Sci. Technol. 27:2630–2636.

Apitz, S. E., and K. J. Meyersschulte. 1996. Effects of substrate mineralogy on the biodegradability of fuel components. Environ. Toxicol. Chem. 15:1883–1893.

Aronstein, B. N, and M. Alexander. 1993. Effect of a non-ionic surfactant added to the soil surface on the biodegradation of aromatic hydrocarbons within the soil. Appl. Environ. Microbiol. 39:386–390.

Barriault, D., and M. Sylvestre. 1993. Factors affecting PCB degradation by an implanted bacterial strain in soil microcosms. Can. J. Microbiol. 39:594–602.

Bedard. D. L., and J. F. Quensen III. 1995. Microbial reductive dechlorination of polychlorinated biphenyls. Pp. 127–216. *In* L. Y. Young and C. E. Cerniglia (eds.), Microbial transformation and Degradation of Toxic Organic Chemicals. Wiley-Liss, NY.

Bollag, J.-M., T. Mertz, and L. Otjen. 1994. Role of microorganisms in soil bioremediation. Bioremediation through rhizosphere Technology, American Chemical Society Symposium Series 563:2–10.

Broadbent, F. E. 1960. Factors influencing the decomposition of organic soils of the California Delta. Hilgardia 29:587–612.

Breitung, J., D. Bruns-Nagel, K. Steinbach, L. Kaminski, D. Gemsa, and E. von Low. 1996. Bioremediation of 2,4,6-trinitrotoluene-contaminated soils by two different aerated compost systems. Appl. Microbiol. Biotechnol. 44:795–800.

Burlage, R. S., A. V. Palumbo, A. Heitzer, and G. Sayler. 1994. Bioluminescent reporter bacteria detect contaminants in soil samples. Appl. Biochem. Biotechnol. 45:713–740.

Casella, S., and W. J. Payne. 1996. Potential of denitrifiers for soil environment protection. FEMS Microbiol. Lett. 140:1–8.

Carmichael, L. M., R. F. Christman, and F. K. Pfaender. 1997. Desorption and mineralization kinetics of phenanthrene and chrysene in contaminated soils. Environ. Sci. Technol. 31:126–132.

Clothier, B. E., and S. R. Green. 1997. Roots: The big movers of water and chemical in soil. Soil Science 162:534–543.

Costerton, J. W., Z. Lewandowski, D. E. Caldwell, D. R. Korer, and H. M. Lappin-Scott. 1995. Microbial biofilm. Annu. Rev. Microbiol. 49:711–745.

Crawford, J. J., G. K. Sims, R. L. Mulvaney, and M. Radosevich. 1998. Biodegradation of atrazine under denitrifying conditions. Appl. Microbiol. Biotechnol. 49:618–623.

Devare, M., and M. Alexander. 1995. Bacterial transport and phenanthrene biodegradation in soil and aquifer sand. Soil Sci. Soc. Am. J. 59:1316–1320.

Dec, J., and J.-M. Bollag. 1997. Determination of colvalent and noncovalent binding interactions between xenobiotic chemicals and soil. Soil Sci. 162:858–874.

Field, J. A., A. J. M. Stams, M. Kato, and G. Shraa. 1995. Enhanced biodegradation of aromatic pollutants in cocultures of anaerobic and aerobic bacterial consortia. Antonie van Leeuwenhoek. 67:47–77.

Fiorenza. S., and C. H. Ward. 1997. Microbial adaptation to hydrogen peroxide and biodegradation of hydrocarbons. J. Indust. Microbiol. Biotechnol. 18:140–151.

Flemming, H. C. 1993. Biofilms and environmental protection. Water Sci. Technol. 27:1–10.

Fliermans, C. B., T. J. Phelps, D. Ringelberg, A. T. Mikell, and D. C. White. 1988. Mineralization of trichloroethylene by heterotrophic enrichment cultures. Appl. Environ. Microbiol. 54:1709–1714.

Funk, S. B., D. J. Roberts, D. L. Crawford, and R. L. Crawford. 1993. Initial-phase optimization for bioremediation of munition compound-contaminated soils. Appl. Environ. Microbiol. 59:2171–2177.

Gaskin, J. L., and J. Fletcher. 1997. The metabolism of exogenously provided atrazine by the ectomycorrhizal fungus *Hebeloma crustuliniforme* and the host plant *Pinus ponderosa*. Phytoremediation of Soil and Water Contaminants, American Chemical Society Symposium Series. 664:152–160.

Gilbert, E. S., and D. E. Crowley. 1997. Plant compounds that induce polychlorinated biphenyl biodegradation by *Arthrobacter* sp. Strain B1B. Appl. Environ. Microbiol. 63:1933–1938.

Grant, C. L., T. F. Jemloms. K. F. Meyers, and E. F. McCormick. 1995. Holding-time estimates for soils containing explosives residues — comparisons of fortification vs field contamination. Environ. Toxicol. Chem. 14:1865–1874.

Grimberg, S. J., W. T. Stringfellow, and M. D. Aitken. 1996. Quantifying the biodegradation of phenanthrene by *Pseudomonas stutzeri* P16 in the presence of a nonionic surfactant. Appl. Environ. Microbiol. 62:2387–2392.

Haderlein, S. B., K. W. Weissmahr, and R. P. Schwarzenbach. 1996. Specific adsorption of nitroaromatic — explosives and pesticides to clay minerals. Environ. Sci. Technol. 30:612–622.

Haner, A., P. Hohener, and J. Zeyer. 1995. Degradation of p-xylene by a denitrifying enrichment culture. Appl. Environ. Microbiol. 61:3185–3188.

Heitzer, A., K. Malachowsky, J. E. Thonnard, P. R. Bienkowski, D. C. White, and G. S. Sayler. 1994. Optical biosensor for environmental on-line monitoring of naphthalene and salicylate bioavailabiltiy with an immobilized bioluminescent catabolic reporter bacterium. Appl. Environ. Microbiol. 60:1487–1494.

Hoagland, R. E., R. M. Zablotowice, and M. A. Locke. 1994. Propanil metabolism by rhizosphere microflora. Pp. 160–183. *In* Bioremediation through rhizosphere technology. ACS Symposium Series, American Chemical Society, Washington, DC.

Huang, C., J. E. Vanbenschoten, T. C. Healy, and M. E. Ryan. 1997. Feasibility study of surfactant use for remediation of organic and metal contaminated soils. J. Soil contamin. 6:537–556.

Huesemann, M. H. 1997. Incomplete hydrocarbon biodegradation in contaminated soils: Limitations in bioavailability of inherent recalcitrance. Bioremed. J. 1:27–39.

Hurst, C. J. R. C. Sims, J. L. Sims, D. L. Sorensen, J. E. Mclean, S. Huling. 1997. Soil gas oxygen tension and pentachlorophenol biodegradation. J. Environ. Engin. 123:364–370.

Jennings, D. A., J. N. Petersen, R. S. Skeen, B. S. Hooker, B. M. Peyton, D. L. Johnstone, and D. R.Yonge. 1995. Effects of variations in nutrient loadings on pore plugging in soil columns. Appl. Biochem. Biotech. 51/52:727–734.

Jensen, L. B., J. L. Ramos, Z. Kaneva, and S. Molin. 1993. A substrate-dependent biological containment system for *Pseudomonas putida* based on the *Escherichia coli* gef gene. Appl. Environ. Microbiol. 59:3713–3717.

Juteau, P., R. Beaudet, G. McSween, F. Lepine, S. Milot, and J.-G. Bisaillon. 1995. Anaerobic biodegradation of pentachlorophenol by a methanogenic consortium. Appl. Microbiol. Biotech. 44:218–224.

Kaake, R. H., D. G. Roberts, T. O. Stevens, T. O. Crawford, and D. L. Crawford. 1992. Bioremediation of soils contaminated with the herbicide 2-sec-butyl-4,6-dintirophenol (dinoseb). Appl. Environ. Microbiol. 58:1683–1689.

Kao, C. M. and R. C. Borden. 1997. Site-specific variability in BTEX biodegradation under denitrifying conditions. Groundwater 35:305–311.

Kastner, M., S. Lotter, J. Heerenklage, M. Breuer-Jammali, R. Stegmann, and B. Mahro. 1995. Fate of 14C-labeled anthracene and hexadecane in compost manured soil. Appl. Microbiol. Biotechnol. 43:1128–1135.

Kazumi, J., M. M. Häggblom, and L. Y. Young. 1995. Diversity of anaerobic microbial processes in chlorobenzoate degradation: Nitrate, iron, sulfate and carbonate as electron acceptors. Appl. Microbiol. Biotech. 43:929–936.

Kelsey, J. W., and M. Alexander. 1997. Declining bioavailability and inappropriate estimation of risk of persistent compounds. Environ. Toxicol. Chem. 16:582–585.

Kelsey, J. W., B. D. Kottler, and M. Alexander. 1997. Selective chemical extractants to predict bioavailability of soil-aged organic chemicals. Environ. Sci. Technol. 31:214–217.

Knudsen, S. M, and O. H. Karlstroem. 1991. Development of efficient suicide mechanisms for biological containment of bacteria. Appl. Environ. Microbiol. 57:85–92.

Laine, M. M., and K. S. Jorgensen. 1996. Straw compost and bioremediated soil as inocula for the bioremediation of chlorophenol-contaminated soil. Appl. Environ. Microbiol. 62:1507–1513.

Lajoie, C. A., G. J. Zylstra, M. F. DeFlaun, and P. F. Strom. 1993. Development of field application vectors for bioremediation of soils contaminated with polychlorinated biphenyls. Appl. Environ. Microbiol. 59:1735–1741.

Lewis, T. A., and R. L. Crawford. 1993. Physiological factors affecting carbon tetrachloride dehalogenation by the denitrifying bacterium *Pseudomonas* sp. Strain KC. Appl. Environ. Microbiol. 59:1635–1641.

Liu, Z. B., A. M. Jacobson, and R. G. Luthy. 1995. Biodegradation of naphthalene in aqueous nonionic surfactant systems. Appl. Environ. Microbiol. 61:145–151.

Lovley, D. R. 1997. Potential for anaerobic bioremediation of BTEX in petroleum-contaminated aquifers. J. Indust. Microbiol. Biotechnol. 18:75–81.

Madsen, T., and P. Kristensen. 1997. Effects of bacterial inoculation and nonionic surfactants on degradation of polycyclic aromatic hydrocarbons in soil. Environ. Toxicol. Chem. 16:631–637.

Madsen, E. L., C. L. Mann, and S. E. Bilotta. 1996. Oxygen limitations and aging as explanations for the field persistence of naphthalene in coal tar-contaminated surface sediments. Environ. Toxicol. Chem. 15:1876–1882.

Margesin, R., and F. Schinner. 1997. Efficiency of indigenous and inoculated cold-adapted soil microorganisms for biodegradation of diesel oil in alpine soils. Appl. Environ. Microbiol. 63:2660–2664.

Matheson, V. G., J. Manakata-Marr, G. D. Hopkins, P. L. McCarthy, J. M. Tiedje, and L. J. Forney. 1997. A novel means to develop strains-specific DNA probes for detecting bacteria in the environment. Appl. Environ. Microbiol. 63:2863–2869.

Miethling, R., and U. Karlson. 1996. Accelerated mineralization of pentachlorophenol in soil upon inoculation with *Mycobacterium chlorophenolicum* PCP1 and *Sphingomonas chlorophenolica* RA2. Appl. Environ. Microbiol. 62:4361–4366.

Molin, S., and S. Kjellberg. 1993. Release of engineered microorganisms: Biological containment and improved predictability for risk assessment. Ambio 22:242–245.

Murphy, S. L., and R. L. Tate III. 1996. Bacterial movement through soil. *In* G. Stotzky and J. M. Bollag (eds.), Soil Biochem. 9:253–286. Dekker, NY.

Nam, K., and M. Alexander. 1998. Role of nanoporosity and hydrophobicity in sequestration and bioavailability: Tests with model solids. Environ. Sci. Technol. 32:71–74.

Natsch, A., C. Keel, J. Troxler, M. Zala, N. Von Albertini, an G. Defago. 1996. Importance of preferential flow and soil management in vertical transport of a biocontrol strain of *Pseudomonas fluorescens* in structured field soil. Appl. Environ. Microbiol. 62:33–40.

Nishiyama, M., K. Senoo, and S. Matsumoto. 1993. Establishment of γ-1,2,3,4,5,6-hexachlorocyclohexane-assimilating bacterium, *Sphingomonas paucimobilis* strain SS86 in soil. Soil Biol. Biochem. 25:769–774.

Origgi, G., M. Colombo, F. Depalma, M. Rivolta, P. Rossi and V. Andreoni. 1997.

Bioventing of hydrocarbon-contaminated soil and biofiltration of the off-gas—Results of a field scale investigation. J. Environ. Sci. & Hlth. 32:2289–2310.

Pahm, M. A., and M. Alexander. 1993. Selecting inocula for the biodegradation of organic compounds at low concentrations. Microbial Ecol. 25:275–286.

Pfender, W. F. 1996. Bioremediation bacteria to protect plants in pentachlorophenol-contaminated soil. J. Environ. Qual. 25:1256–1260.

Pritchard, P.H. 1991. Bioremediation as a technology: Experiences with the Exxon Valdez oil spill. Ecological Research Series. U.S. Environmental Protection Agency. Washington, DC.

Radosevich, M., S. J. Traina, and O. H. Tuovinen. 1997. Atrazine mineralization in laboratory-aged soil microcosms inoculated with S-triazine-degrading bacteria. J. Environ. Qual. 26:206–214.

Reilley, K. A., M. K. Banks, and A. P. Schwab. 1996. Dissipation of polycyclic aromatic hydrocarbons in the rhizosphere. J. Environ. Qual. 25:212–219.

Rice, J. F., R. F. Fowler, A. A. Arrage, D. C. White, and G. S. Sayler. 1995. Effects of external stimuli on environmental bacterial strains harboring an algD-lux bioluminsecent reporter plasmid for the study of corrosive biofilms. J. Industr. Microbiol. 15:318–328.

Sack, U., and W. Fritsche. 1997. Enhancement of pyrene mineralization in soil by wood-decaying fungi. FEMS Microbiol. Ecol. 22:77–83.

Sandoli, R. L., W. C. Ghiorse, and E. L. Madsen. 1996. Regulation of microbial phenanthrene mineralization in sediment samples by sorbent-sorbate contact time, inocula and gamma irradiation-induced sterilization artifacts. Environ. Toxicol. Chem. 15:1901–1907.

Schowanek, D. R., T. C. J. Feijtel, and T. W. Federle. 1997. Effect of concentration and environmental form of tetradecenyl succinic acid on its mineralization in soil. Biodegradation 7:377–382.

Schulten, H.-R. 1995. The three-dimensional structure of soil organo-mineral complexes studyied by analytical pyrolysis. J. Anal. Appl. Pyrol 32:111–126.

Schulten, H.-R., and M. Schnitzer. 1997. Chemical model structures for soil organic matter and soils. Soil Sci. 162:115–130.

Siciliano, S. D., and J. J. Germida. 1997. Bacterial inoculants of forage grasses that enhance degradation of 2-chlorobenzoic acid in soil Environ. Toxicol. Chem. 16:1098–1104.

Smith, M. J., G. Lethbridge, and R. G. Burns. 1997. Bioavailability and biodegradation of polycyclic aromatic hydrocarbons in soils. FEMS Microbiol. Letters 152:141–147.

Smith, R. L., M. L. Ceazan, and M. H. Brooks. 1994. Autotrophic hydrogen-oxidizing, denitrifying bacteria in groundwater, potential agents for bioremediation of nitrate contamination. Appl. Environ. Microbiol. 60:1949–1955.

Stehmeier, L. G., T. R. Jack, and G. Voordouw. 1996. In vitro degradation of dicyclopentadiene by microbial consortia isolated from hydrocarbon contaminated soil. Can. J. Microbiol. 42:1051–1060.

Swobada-Colberg, N. G. 1995. Chemical contamination of the environment: Sources, types, and fate of synthetic chemicals. Pp. 27–74. *In* L. Y. Young and C. E.

Cerniglia (eds.), Microbial Transformation and Degradation of Toxic Organic Chemicals. Wiley-Liss, NY.

Tiehm, A. 1994. Degradation of polycyclic aromatic hydrocarbons in the presence of synthetic surfactants. Appl. Environ. Microbiol. 60:258–263.

Tiehm, A., M. Stieber, P. Werner, and F. H. Frimmel. 1997. Surfactant-enhanced mobilization and biodegradation o polycyclic aromatic hydrocarbons in manufactured gas plant soil. Environ. Sci. Techn. 31:2570–2576.

Vahjen, W., J.-C. Munch, and C. C. Tebbe. 1995. Carbon source utilization of soil extracted microorganisms as a tool to detect the effects of soil supplemented with genetically engineered and non-engineered *Corynebacterium glutamicum* and a recombinant peptide at the community level. FEMS Microbiol. Ecol. 18:317–328.

Vandevivere, P., and P. Baveye. 1992. Saturated hydraulic conductivity reduction caused by aerobic bacteria in sand columns. Soil Sci. Am. J. 56:1–13.

Vandevivere, P., P. Baveye, D. Sanchez de Lozada, and P. DeLeo. 1995. Microbial clogging of saturated soils and aquifer materials: Evaluation of mathematical models. Water Resources Res. 31:2173–2180.

Vanhoof, P. L., and C. T. Jafvert. 1996. Reductive dechlorination of chlorobenzenes in surfactant-amended sediment slurries. Environ. Toxicol. Chem. 15:1814–1824.

Volkering, F., A. M. Breure, J. G. Vanandel, and W. H. Rulkens. 1995. Influence of nonionic surfactants on bioavailability and biodegradation of polycyclic aromatic hydrocarbons. Appl. Environ. Microbiol. 61:1699–1705.

Wackett, L. P. 1995. Bacterial co-metabolism of halogenated organic compounds. Pp. 217–241. *In* L. Y. Young, and C. E. Cerniglia (eds.), Microbial Transformation and Degradation of Toxic Organic Chemicals. Wiley-Liss, NY.

Webb, O. F., P. R. Bienkowsi, U. Matrubutham, F. A. Evans, A. Heitzer, and G. S. Sayler. 1997. Kinetics and response of a *Pseudomonas fluorescens* HK66 biosensor. Biotechnol. Bioeng. 54:491–502.

Weber, J. B. 1993. Ionization and sorption of fomesafen and atrazine by soils and soil constituents. Pesticide Sci. 39:31–38.

Weier, K. L., J. W. Doran, A. R. Mosier, J. F. Power, and T. A. Peterson. 1994. Potential for bioremediation of high nitrate irrigation water via denitrification. J. Environ. Qual. 23:105–110.

White, J. C., J. W. Kelsey, P. B. Hatzinger, and M. Alexander. 1997. Factors affecting sequestration and bioavailability of phenanthrene in soils. Environ. Toxicol. Chem. 16:2040–2045.

Young, L.Y., and C. E. Cerniglia (eds.). 1995. Microbial Transformations and Degradation of Toxic Organic Chemicals. John Wiley & Sons, NY. 654 pp.

Zhang, Y. M., W. J. Maier., and R. M. Miller. 1997. Effect of rhamnolipids on the dissolution, bioavailabiltiy and biodegradation of phenanthrene. Environ. Sci. Technol. 31:2211–2217.

Zouari, N., and R. Ellouz. 1996. Microbial consortia for the aerobic degradation of aromatic compounds in olive oil mill effluent. J. Indust. Microbiol. 16:155–162.

Concluding Challenge

In the preceding pages, the multifaceted world of soil microbes has been revealed. It is not difficult at this point of our study to reach the conclusion that each of the individual habitats of soil microbes is nearly unique. Microsites are composed of a diversity of physical components assembled in varying degrees of structure ranging from an apparently structureless array of sand grains to the complex, ever-changing combinations of mineral matter termed soil aggregates. Furthermore, microbes encounter an infinite combination of soil organic compounds ranging from simple organic acids, amino acids, and monosacharides to the highly complex humic acids. This assemblage of mineral and physical components of the microbial world is further varied by the presence of a variety of inorganic substances many of which provide the energetic basis for growth and development of a portion of the soil microbial community known as the chemolithotrophs. An additional layer of complexity of the soil system is provided by the myriad of potential members of the microbial community and the variations in their interactions.

The complexity of the soil biological community is only partially revealed by the abbreviated illumination of the microbial world provided by this text. Growth and development of individual microbial cells and the associated alteration of soil components through cellular energy metabolism and the by-products of this activity appear to be nearly chaotic. In contrast, microbial growth in test tubes is highly ordered and easily described by a variety of somewhat simple mathematical relationships. In soil, a highly productive cell may be allowed only one to a few cellular divisions before stresses of space, nutrients, or biological interactions reduce the proliferation rate. In reality, it is difficult to envision a situation in soil where the maximal division rate under the optimal conditions of a culture tube could ever occur. Further, quiescent organisms may suddenly be spurred into a few rounds of seemingly unimpeded division by the influx of an energy source or temporary relief from growth-inhibiting conditions, but this time of luxury can just as quickly be disrupted by exhaustion of the newly found growth sources or imposition of new impediments to proliferation.

The preceding chapters present the basic principles of the discipline of soil microbiology, providing some order for the description of this highly disordered world in soil and for prediction of the behavior of the soil microbial community as the properties of its habitats are altered. With our continually growing appreciation of the complexity of the soil microbial community and the interactions between its members, it becomes readily apparent that the information contained in these pages only serves as a guidepost along the path to a more complete understanding of the home of the soil microbes and the genetic potential contained in that population. This pathway could be likened to a divided highway leading to the ultimate in understanding and wisdom in this discipline. The seemingly contradictory endeavors of delving into the minutia of the system must be combined with just as diligent a drive to conceptualize the role of the soil microbes in function of the total ecosystem. This latter system must extend far beyond the simple soil of a meadow or mineland reclamation site to the totality of terrestrial systems and perhaps beyond.

As our capacity to understand and perhaps model larger landscape units increases, our intellectual concept of an ecosystem has enlarged. Instead of an ecosystem being a limited area, reasonably homogenous from the perspective of aboveground aspects, we are becoming more aware of the fact that a true ecosystem is a mixture of contiguous but clearly distinct systems (e.g., a swamp bordering a forest system), as well as a continuum of slowly changing systems. Nowhere is this concept of an ecosystem being a complex mixture of interacting system types more clearly demonstrated than in soil, where we find subsoil processes affecting surface soil systems, which are further regulated by external inputs such as organic matter and moisture from the aboveground portions of the grander system. On a microscale, we can also develop a realistic conceptual scenario of complex interactions of a variety of microsites being summed to produce the product that we may describe as a particular soil system, such as a grassland.

With this more inclusive and less reductive view of ecosystem description comes an enhanced appreciation of the impact of soil microbiological processes on total terrestrial processes. Necessarily, the basic principles described in the preceding chapters were elucidated with examples of limited ecosystem types, sometimes simply the system that develops from a soil sample contained in a test tube. Thus, a somewhat disparate view of the impact of such processes as denitrification (and associated nitrous oxide production) on general terrestrial processes can be gained. It is easy conceptually to separate the internal portions of the soil aggregate wherein denitrification occurs from the external oxygenated regions of the same aggregate. More dramatically, conceptually, the denitrification occurring in a swamp, or even a constructed wetland, can readily be isolated from its impacts on the function of the surrounding systems, be they forest, agricultural lands, or even urban environments.

Taken in isolation, the concepts of soil microbiology could almost be considered to be evolving at a "snail's pace," yet once we move our mental contemplation from the limited world of the specific microbe or process, a multitude of major challenges to the environmental microbiologist arise. Concerns such as prediction of impacts of global warming and the role of soil microbes as they interact with the organic resources on the production of greenhouse gasses, maintenance of soil quality, not even to mention the problems of reclaiming chemically polluted soils or soils damaged from mismanagement demand an expansion of soil microbial concepts further than ever before. Solutions to these environmental problems and support for such efforts as global climate modeling can be derived to some degree from extrapolation of data and concepts derived from studies of more limited ecosystems. But the greater challenge to soil microbiologists resides in developing reliable models and data sets supportive of more general models of total ecosystem (the entire terrestrial system), predictive of the impact of anthropogenic activities and of our reclamation efforts.

This treatise was introduced with the objectives of presenting the basic principles of soil microbiology and of illuminating them with examples from the primary literature accumulated over the short history of this discipline. Any soil microbiology textbook, no matter how complete, can provide only a snapshot of the current status of our studies. A viable scientific endeavor must be continually growing. Great gains in describing the soil environment have been made, but the activities of society have created environmental challenges that may baffle even the most learned soil microbiologist. The challenges of reclaiming severely damaged soil sites, the goals of properly managing our terrestrial home, and the necessity of manipulating the soil system to provide the needs of an ever-growing, ever-changing society demand refinement of our knowledge of the capacity of the soil microbes to adapt to their continually evolving physical home and our capability to manage this minute soil community to meet both their and our needs for survival. Thus, those who have mastered the introduction to soil microbiology provided within these pages are now challenged to use the information as a form of intellectual energy to advance our knowledge base into the realms necessary to meet the demands of the twenty-first century. As society entered the twentieth century, soil microbiology, if it could really be said to exist then, would have been viewed as a diversion into elucidation of the intriguing "wee beasties" of the soil upon which we walk. In the approaching new millenium, soil microbiology must play an integral role in the development of ecosystem-sustaining stewardship practices.

Index